Pertussis

ECOLOGY AND EVOLUTION OF INFECTIOUS DISEASES SERIES

Series Editor: John M. Drake

UNIVERSITY OF GEORGIA
Center for the Ecology
of Infectious Diseases

Infectious diseases are embedded in a complex of social, environmental, and biological forces. This series provides an up-to-date synthesis of contemporary issues in the ecology and evolution of infectious diseases. Topics covered include aspects of population ecology, vector biology, animal behavior and movement, methods for inference and disease forecasting, ecoimmunology, evolution, biogeography, and others. Each book makes topical connections to pathogens of current interest. The series is intended for students and researchers working in ecology, evolution, and epidemiology with applications to the disciplines of one health (animal health, public health, and environmental health), environmental management, health policy, therapeutics, and wildlife conservation.

Pertussis: Epidemiology, Immunology, and Evolution
Pejman Rohani and Samuel V. Scarpino

Pertussis

Epidemiology, Immunology, and Evolution

EDITED BY

Pejman Rohani and Samuel V. Scarpino

OXFORD
UNIVERSITY PRESS

OXFORD

UNIVERSITY PRESS

Great Clarendon Street, Oxford, OX2 6DP,
United Kingdom

Oxford University Press is a department of the University of Oxford.
It furthers the University's objective of excellence in research, scholarship,
and education by publishing worldwide. Oxford is a registered trade mark of
Oxford University Press in the UK and in certain other countries

First Edition published in 2019
Impression: 1

Published in the United States of America by Oxford University Press
198 Madison Avenue, New York, NY 10016, United States of America

British Library Cataloguing in Publication Data
Data available

Library of Congress Control Number: 2018957725

ISBN 978–0–19–881187–9

DOI: 10.1093/oso/9780198811879.001.0001

Printed and bound by
CPI Group (UK) Ltd, Croydon, CR0 4YY

I dedicate this book to my parents in gratitude for their love, their courage, and their sacrifice.

Pej Rohani

I dedicate this book to my parents, Betty and Phil, and my wife Laura for their support, encouragement, and love.

Sam Scarpino

Foreword

The Expanded Programme on Immunization (EPI), launched by the World Health Organization (WHO) and the United Nations Children's Fund (UNICEF) in 1974, was a triumph for public health in the twentieth century. Before EPI, fewer than 5 per cent of children received six essential vaccines. Within three decades, these vaccines were being given to the majority of children born each year. Immunization was instrumental in halving the death rate of children under five between 1990 and 2015, in the drive to reach the Millennium Development Goals. Now, immunization is integral to achieving universal health coverage and to meeting the Sustainable Development Goals by 2030.

Immunization against pertussis (whooping cough), combined with diphtheria and tetanus, was at the forefront of EPI's success. Twenty per cent of all infants born in 1980 received three doses of the triple vaccine (DTP3). Coverage exceeded 80 per cent by 2006 and reached 86 per cent in 2016, more than 100 million of the children born that year. As the proportion of children vaccinated has increased, so the number of pertussis cases has plummeted, from around 2 million reported in 1980 to fewer than 200,000 annually from 1995 onwards.

No one should forget the huge toll of suffering that has been banished by immunization. Thanks to the vaccines (DTP), and to the means of their distribution (EPI), the worst outcome of *Bordetella pertussis* infection has become a largely avoidable tragedy: a child's illness that begins with a runny nose, catarrh, and cough, deteriorates to a whooping cough with paroxysms and vomiting, and then progresses to hypoxia, brain damage, and death.

Despite the enormous progress made since the 1970s, the management of pertussis in the home, in the clinic, and in whole populations still faces big obstacles today. The first and most obvious task is to vaccinate more children each year—at the very least to meet the target of 90 per cent DTP3 coverage in every country by 2020, the deadline that will mark the end of the 'decade of vaccines'. In 2016, 130 of 194 WHO member states met that national target. Among the populous low- and middle-income countries that did not do so, and where expanding coverage continues to be a challenge, are India, Indonesia, Nigeria, Pakistan, the Philippines, and South Africa.

Higher immunization rates will certainly reduce the numbers of cases and deaths from pertussis but the downward trajectory is uncertain, in part because we do not know the true burden of infection and disease. Pertussis is under-reported because the symptoms are often mild or non-specific, particularly in older children and adults, and because the most sensitive and specific diagnostic tests, based on identifying the bacterium (nucleic acid test) rather than the immunological response to infection (serology), are not yet widely available. Best estimates of the burden of disease put the annual number of cases of pertussis in children under five at more than 20 million, with over 100,000 deaths, but these figures will undoubtedly be refined as better diagnostic and surveillance methods are more widely adopted, especially in high-burden countries.

Although the numbers of pertussis cases and deaths are probably falling worldwide, case numbers have rebounded in some high income countries where high rates of vaccination had previously forced pertussis to low levels. To explain this resurgence is a live issue for research, with implications for the containment of the disease worldwide. The possibilities include interruptions in vaccination;

waning natural immunity from historic outbreaks, which were periodic before widespread immunization; waning immunity among adolescents and adults who were vaccinated during childhood; reduced vaccine efficacy associated, for example, with the switch from whole-cell (wP) to acellular pertussis (aP) vaccines; selection of bacterial clones resistant to natural or vaccine-induced immunity; and increased reporting of pertussis because of greater awareness among physicians, enhanced surveillance, and improved diagnostic tests.

Each of these factors may play a role, but their relative importance in different settings has not been thoroughly evaluated. If and when vaccine coverage falls, anywhere in the world, the persistently wide geographical distribution of *B. pertussis* means that further outbreaks are almost inevitable.

Waning immunity is more typical of bacterial infections such as *B. pertussis* than childhood viral infections (cf. measles), and surveillance data, coupled with mathematical modelling, suggest that this is a plausible explanation, though perhaps not the complete explanation, for pertussis resurgence in some parts of the United States.

Waning immunity post vaccination is a general phenomenon, but its effects have apparently been aggravated by the switch from more immunogenic and reactogenic wP vaccines to less immunogenic and safer aP vaccines. In England and Wales, for example, the switch in 2004 was followed by an outbreak in 2012, which included cases in older children and adults who probably became sources of infection for others. But the analysis of surveillance data for England and Wales is not straightforward because a rise in cases reported among individuals aged 15 years and over after 2004 was likely due to better ascertainment after the introduction of serological testing in 2001.

As to the question of whether the bacterium itself is evolving, there are certainly changes in genotype frequency over time. But the evidence that vaccine efficacy has diminished because of selection for resistant genes or antigens remains ambiguous. Nevertheless, the possibility of evolved vaccine resistance cannot yet be discounted.

To add to the complications, the resurgence of serious childhood infections such as pertussis carries the risk of amplifying vaccine scepticism and hesitancy among the general public.

These are legitimate concerns, and sound science is an indispensable part of the response.

All these questions set a big agenda for research. In this context, Pej Rohani and Sam Scarpino have assembled an authoritative collection of articles that address the challenges of pertussis in the world today—the uncertain disease burden and trends, the growing evidence for resurgence, the multiple options for vaccination and case management, and changing public perceptions. Their book covers immunology and vaccination, diagnostics and surveillance, experimental animal models, genetics and evolution, epidemiology and public health, and more besides. It is an essential and timely contribution to the science of controlling, and ultimately eliminating, a persistent and dangerous infectious disease.

Christopher Dye
Visiting Professor of Zoology, University of Oxford
Formerly Director of Strategy, World Health
Organization

Acknowledgements

This book follows from a meeting that was held at the Santa Fe Institute (SFI), New Mexico, United States, in March 2016, titled 'Re-emerging Infectious Diseases: The Challenge of Pertussis'. The meeting was organized by Aaron King, Eric Harvill, Sam Scarpino, and Pej Rohani. Participants included public health professionals, epidemiologists, microbiologists, immunologists, evolutionary biologists, and modellers. Over the course of 2 days, we heard about the state of the art in pertussis research with a concerted attempt to map out key research questions. This volume is our effort at a synthesis of that event and pertussis biology more generally. We are grateful to the participants of the SFI workshop and the authors of the chapters of this book for their integral contributions. We appreciate the financial support of the SFI, the Arizona State University-SFI Center for Biosocial Complex Systems, and the Research Coordination Network on IDEAS (Infectious Disease Evolution Across Scales), funded by a National Science Foundation award to Andrea Graham and colleagues. We thank a number of researchers who have made important contributions to understanding pertussis but whose participation in this project was not possible. They include Jim Cherry, Nicole Guiso, Fritz Mooi, Mirjam Krestzchmar, Tod Merkel, and Daniela Hozbor.

Pej Rohani would also like to thank his long-term collaborators on pertussis: Bryan Grenfell, David Earn, Helen Wearing, Aaron King, Maria Riolo, Felicia Magpantay, Matthieu Domenech de Cellès, and Ana Bento.

Sam Scarpino thanks the SFI for providing him the intellectual freedom to enter new research areas, especially pertussis, and his collaborators on whooping cough: Ben Althouse, Glen Otero, Haedi DeAngelis, Meagan Fitzpatrick, and Alison Galvani.

Contents

List of Contributors

Aideen C. Allen School of Biochemistry and Immunology, Trinity College Dublin, Dublin 2, Ireland
Email: aideenallen@gmail.com

Benjamin M. Althouse Institute for Disease Modeling, Seattle, WA 98005, USA
Email: balthouse@idmod.org

Gayatri Amirthalingam Immunisation, Hepatitis and Blood Safety Department, Public Health England, London, NW9 5EQ, UK
Email: Gayatri.Amirthalingam@phe.gov.uk

Shelly Bolotin Applied Immunization Research, Public Health Ontario, Toronto, ON M5G 1V2, Canada
Email: Shelly.Bolotin@oahpp.ca

Lisa Borkner School of Biochemistry and Immunology, Trinity College Dublin, Dublin 2, Ireland
Email: borknerl@tcd.ie

Colin S. Brown Immunisation, Hepatitis and Blood Safety Department, Public Health England, London, NW9 5EQ, UK
Email: Colin.Brown@phe.gov.uk

Natasha Crowcroft Applied Immunization Research and Evaluation, Public Health Ontario, Toronto, ON M5G 1V2, Canada
Email: natasha.crowcroft@oahpp.ca

Hester de Melker Epidemiology and Surveillance Unit, National Institute for Public Health and the Environment, 3720 BA Bilthoven, Utrecht, Netherlands
Email: hester.de.melker@rivm.nl

Violette Dirix Laboratory of Vaccinology and Mucosal Immunity, Universite Libre de Bruxelles, 1070 Brussels, Belgium
Email: vdirix@ulb.ac.be

Matthieu Domenech de Cellès Pharmacoepidemiology and Infectious Diseases, Institut Pasteur, 75015 Paris, France
Email: matthieu.domenech-de-celles@pasteur.fr

Sylvain Gandon Ecology and Evolutionary Epidemiology, CNRS, 34293 Montpellier, France
Email: sylvain.gandon@cefe.cnrs.fr

Eric T. Harvill Department of Infectious Diseases, University of Georgia, Athens, GA 30602, USA
Email: harvill@uga.edu

Aaron A. King Department of Ecology and Evolutionary Biology, University of Michigan, Ann Arbor, MI 48104, USA
Email: kingaa@umich.edu

Ruiting Lan School of Biotechnology and Biomolecular Sciences, University of New South Wales, Sydney, NSW 2052, Australia
Email: r.lan@unsw.edu.au

Bodo Linz Department of Infectious Diseases, University of Georgia, Athens, GA 30602, USA
Email: bodo.linz@uga.edu

Camille Locht Centre for Infection and Immunity of Lille, University of Lille, 59019 Lille, France
Email: camille.locht@pasteur-lille.fr

Iain MacArthur Department of Biology and Biochemistry, University of Bath, Bath, BA2 7AY, UK
Email: I.MacArthur@bath.ac.uk

Felicia M.G. Magpantay Department of Mathematics and Statistics, Queen's University, Kingston, K7L 3N6, Canada
Email: felicia.magpantay@gmail.com

Françoise Mascart Laboratory of Vaccinology and Mucosal Immunity, Universite Libre de Bruxelles, 1070 Brussels, Belgium
Email: fmascart@ulb.ac.be

Jennifer Maynard Department of Chemical Engineering, The University of Texas at Austin, Austin, TX 78712, USA
Email: maynard@che.utexas.edu

Peter McIntyre School of Public Health, University of Sydney, Westmead, NSW 2006, Australia
Email: peter.mcintyre@health.nsw.gov.au

Jodie McVernon The Peter Doherty Institute for Infection and Immunity, University of Melbourne, Melbourne, VIC 3000, Australia
Email: j.mcvernon@unimelb.edu.au

Elizabeth Miller Immunisation, Hepatitis and Blood Safety Department, Public Health England, London, NW9 5EQ, UK
Email: liz.miller@phe.gov.uk

Kingston H.G. Mills School of Biochemistry and Immunology, Trinity College Dublin, Dublin 2, Ireland
Email: kingston.mills@tcd.ie

Alicja Misiak School of Biochemistry and Immunology, Trinity College Dublin, Dublin 2, Ireland
Email: misiaka@tcd.ie

Tracy Nicholson Virus and Prion Research, United States Department of Agriculture, Ames, IA 50010, USA
Email: tracy.nicholson@ars.usda.gov

Sophie Octavia School of Biotechnology and Biomolecular Sciences, University of New South Wales, Sydney, NSW 2052, Australia
Email: s.octavia@unsw.edu.au

Glen Otero Translational Genomics Research Institute, Phoenix, AZ 85004, USA
Email: gotero@linuxprophet.com

Andrew Preston Department of Biology and Biochemistry, University of Bath, Bath, BA2 7AY, UK
Email: A.Preston@bath.ac.uk

Helen Quinn National Centre for Immunisation Research and Surveillance of Vaccine Preventable Diseases, Children's Hospital at Westmead, Westmead, NSW 2145, Australia
Email: helen.quinn@health.nsw.gov.au

Pejman Rohani Odum School of Ecology and Department of Infectious Diseases, University of Georgia, Athens, GA 30606, USA
Email: rohani@uga.edu

Samuel V. Scarpino Network Science Institute, Northeastern University, Boston, MA 02115, USA
Email: s.scarpino@northeastern.edu

Amanda L. Skarlupka Department of Infectious Diseases, University of Georgia, Athens, GA 30602, USA
Email: skarlupka@uga.edu

Tami H. Skoff Centers for Disease Control and Prevention, Atlanta, GA 30333, USA
Email: tlh9@cdc.gov

Michael R. Weigand Centers for Disease Control and Prevention, Atlanta, GA 30333, USA
Email: yrh8@cdc.gov

Mieszko M. Wilk School of Biochemistry and Immunology, Trinity College Dublin, Dublin 2, Ireland
Email: WILKM@tcd.ie

Margaret M. Williams Centers for Disease Control and Prevention, Atlanta, GA 30333, USA
Email: cux6@cdc.gov

Prologue: the pertussis problem

Pejman Rohani and Samuel V. Scarpino

History: origins, vaccines, and health burden

> And children became unconscious, and many people, old and young, died in the first epidemic because of this fainting and the intensity of cough[1,2].

As you will see in the coming chapters, most issues relating to the ecology, evolution, and immunobiology of whooping cough remain contentious. Even the origins of the bacterium that causes this disease, *Bordetella pertussis*, and the record of its first outbreaks are hotly debated in both the historical and contemporary literature. Consider that, to date, most authors refer to the 1578 account of a whooping cough outbreak in Paris by de Baillou as the earliest description of the disease[3]. However, this assertion contradicts de Baillou's own text in which he describes an epidemic of whooping cough by Horman 1519 nearly 60 years earlier in England[4,5]: 'I am foule rayed with a chyne cowgh', which Still 1931 suggests that 'indeed, more than fifty years before Ballonius [Latinized form of de Baillou] wrote, the disease was sufficiently well known in England to have a popular name, the chin-cough'[5]. Adding weight to Horman's 1519 account of whooping cough[4], *chyne cough* was included in the entry of 'Chincough', which also lists 'hooping cough' as an alternative, in James Murray's first edition of what would become the *Oxford English Dictionary*[6].

Recent genomic evidence supports the hypothesis that *B. pertussis*—along with its close relatives in the Bordetellae—have been circulating in humans for millennia[7,8]. However, early medical reports (such as the one mentioned earlier, from 1484 in Herat, Persia[1]) suggest that *B. pertussis* did not begin to cause virulent outbreaks of whooping cough until the late fifteenth or early sixteenth century[1,7,9] (see Figure 1). Indeed, prior to a 1484 outbreak in Herat, there are no reliable case reports of whooping cough—although see Magner's account of a disease similar to whooping cough in 1433 in Korea[10]—and, from an epidemiological perspective, the early outbreaks in both Herat and Europe resulted in a high prevalence of infection across a broad range of ages, a pattern consistent with newly emerging or re-emerging pathogens[1].

Although *B. pertussis* takes a backseat to smallpox, anthrax, and polio in the history of vaccinology and germ theory, early medical researchers were grappling with whooping cough in the eighteenth century. For example, Still[5] describes, 'How near von Rosenstein [a colleague of Linnaeus who many call the founder of modern paediatrics] came to the conception of bacterial infection is seen in his remark on whooping-cough: "The true cause of this disease must be some heterogeneous matter or seed which has a multiplicative power as is the case with smallpox...Whether this multiplicative miasma be a kind of insects I cannot affirm with certainty; however, we find that it is communicated by infection and that a part of it is attracted by the breath down into the lungs."' Unfortunately, and of course only adding to the controversies surrounding whooping cough, Still[5] does not provide a direct citation for the von Rosenstein quote.

Rohani, P. and Scarpino, S. V., *Prologue: the pertussis problem. In: Pertussis: epidemiology, immunology, and evolution*. Edited by Pejman Rohani and Samuel V. Scarpino: Oxford University Press (2019). © Oxford University Press. DOI: 10.1093/oso/9780198811879.003.0017

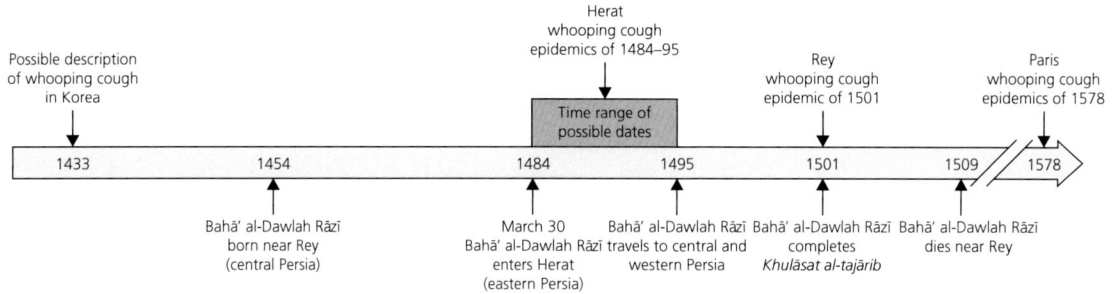

Figure 1 Timeline of historical pertussis outbreaks. Figure reproduced from[1].

Despite persistent uncertainty regarding when exactly *B. pertussis* started causing noticeable disease, by the late nineteenth century whooping cough was recognized as one of the deadliest illnesses afflicting humanity. Indeed, prior to the introduction of widespread vaccination in the 1950s, whooping cough was one of the leading causes of infant mortality and perhaps the leading cause of infant death due to infectious disease[11,12]: 'In emphasizing the importance of whooping cough as a cause of death, Gordon (1951) pointed out that in the United States during the period 1940–1948, whooping cough killed almost three times as many infants less than one year of age as measles, mumps, chicken pox, rubella, scarlet fever, diphtheria, poliomyelitis, and meningitis all together'[13]. Despite remaining a leading cause of infant mortality[12], whooping cough mortality began a steady decline in many countries as early as the 1920s (decades before the widespread adoption of vaccination). Plausible hypotheses for the decline in whooping cough mortality include improved treatment and the use of *B. pertussis* vaccine progenitors after symptom onset[11,12,14].

1950s onwards: declining incidence, candidate for eradication

Although whole-cell vaccines against *B. pertussis*—so called because they contain entire chemically inactivated bacteria—have been available since the 1920s[15,16], widespread vaccination campaigns did not begin until the 1950s[11,16,17]. Prior to the initial vaccination efforts of the 1950s, estimates from the World Health Organization (WHO) suggest that nearly all children contracted pertussis, half of whom would go on to develop symptomatic whooping cough[17]. *Bordetella pertussis* vaccines were included in the WHO Expanded Programme on Immunization, which began in 1974, and led to the lowest reported number of whooping cough cases in the United States. After the widespread adoption of vaccination against *B. pertussis*, cases and mortality plummeted—going locally extinct in many places around the globe[18–20]—leading to calls for whooping cough to be considered as a candidate for eradication[13]. However, beginning in the 1980s, the disease began a resurgence in some countries which had persistently high vaccination coverage[19]. Today, whooping cough caused by *B. pertussis* has again become a leading cause of childhood mortality[17].

Resurgence despite high vaccine uptake: highest burden of any vaccine-preventable disease

Although the number of cases in the United States, Australia, the United Kingdom, several other countries in Europe, and Israel began to increase in the 1980s and 1990s[21], global whooping cough incidence has steadily declined[17]. In addition, vaccine scares in the United Kingdom, Sweden, Germany, and a number of other countries caused dramatic increases in whooping cough prevalence and ultimately led many countries to switch from the whole-cell vaccine to acellular vaccines, which only contain purified *B. pertussis* antigens[22,23]. And, as you will see in the coming chapters, the global diversity in vaccine types (e.g. some countries still use the whole-cell vaccine and there are many acellular vaccines with different sets of antigens) and

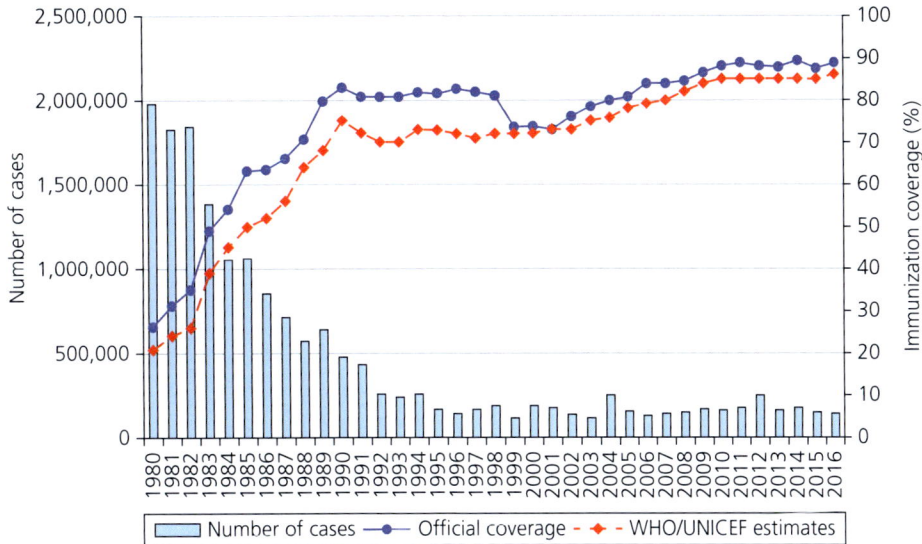

Figure 2   Global pertussis vaccination coverage and disease incidence. The prevalence (blue bars) and vaccination coverage as estimated by the WHO (blue) and UNICEF (red) from 1980 through 2016. Figure reproduced from[24].

vaccine schedules substantially complicate our ability to uncover mechanisms behind changes in whooping cough prevalence. Nevertheless, the declining global incidence of *B. pertussis* is directly tied to increasing vaccination coverage and has directly saved millions of lives[17] (see Figure 2). Despite what is largely seen as a public health success, whooping cough still causes close to 100,000 childhood deaths each year and remains a leading cause of infant mortality[17].

As of 2014, *B. pertussis*, and the disease it causes, was estimated to be the eleventh leading cause of death for children under five, the eighth leading cause of death before adulthood, and the deadliest of all vaccine-preventable diseases[24–26]. However, like so many aspects of whooping cough, even its contemporary role in childhood mortality remains uncertain and controversial. Frustratingly, the uncertainty in whooping cough's health burden largely stems from the recent decisions from the WHO and the UN Inter-agency Group for Child Mortality Estimation to no longer include a separate measure of deaths due to *B. pertussis* in their reports on childhood mortality[25,27].

The ongoing rise in whooping cough incidence in some countries has coincided with the proliferation of hypotheses to explain its resurgence (or lack thereof)[18,21]. Indeed even the global patterns of resurgence, persistence, and decline remain enigmatic, with few general mechanisms able to account for the observed patterns[18]. Interestingly, nearly all of the hypotheses and/or the confounding issues surrounding *B. pertussis* and whooping cough epidemiology have been with the disease since its earliest days in the medical and public health literature. For example, a detailed study of whooping cough in New York City was motivated by the following observation: 'during the years 1900 to 1903 there were more deaths from pertussis in Manhattan and Bronx than cases of whooping cough reported in the whole city of New York'[28]. One of the primary reasons for *B. pertussis* reporting issues stems from the comparative difficulty of the diagnosis of pertussis (as compared to measles or smallpox). For example, 'Viewed clinically, whooping cough is also a paradoxical disease. The acute febrile illness is not to be identified by clinical means, for the reaction in the catarrhal stage is that of a number of other mild respiratory infections'[11].

Pertussis today

As we hope you have started to appreciate in this brief prologue, whooping cough caused by

B. pertussis is a fascinating disease. Nearly all aspects of its ecology, evolution, immunology, and history are hotly debated and studying them pushes the limits of even modern experimental, laboratory, sequencing, and computational methods.

The earliest medical literature provides evidence for the proliferation of hypotheses about the causes and treatment of whooping cough and the persistent challenges associated with studying this enigmatic disease. Describing his exasperation at what science would come to cause the 'the correlation vs causation fallacy', William Moss (in his book *A Familiar Medical Survey of Liverpool* (1784)) said, as quoted by Still (p. 474)[5]:

Moss is alive to the absurdity of some of the remedies which obtained popular credence. He says of the chin-cough [whooping cough]: 'When the complaint takes a favourable turn it is frequently attributed to the means that were last used: hence that means is ever after recorded as infallible. As it is a disease of the spasmodic of convulsive kind it has been sometimes relieved, or even removed, by a shock or sudden fright: thus riding upon a bear (a frightful mode of traveling no doubt) from the fright it occasions has been said to be serviceable. Giving the patient a part of some disgustful animal, [such] as a mouse, etc., to eat, and afterwards informing him of it; and so forth.'

To date, leading explanations for observed epidemiological patterns of whooping cough infections include vaccine-driven evolution, multipathogen systems (congenerics), vaccine hesitation/refusal, vaccine failure, and a shifting diagnostic (molecular) and surveillance (active) landscape (pertussis is like the stars: the more we look, the more find). In the coming chapters, you will see evidence, in places contradictory, presented for each of these hypotheses. Our sincere hope is that readers, especially young scientists and medical researchers, are inspired to take up the challenge of whooping cough. Our goal in this work is to describe the current knowledge surrounding whooping cough and *B. pertussis*, outline the proposed hypotheses for its resurgence, persistence, or decline, and attempt to reconcile these hypotheses. The book is structured into the following sections, where the authors go into greater detail regarding the ecology, evolution, epidemiology, molecular biology, and immunology of pertussis:

- *Introduction to pertussis*: (Chapter 1) transmission and epidemiological dynamics, (Chapter 2) pathogenesis, (Chapter 3) immunology, and (Chapter 4) epidemiology.
- *Vaccines and immune kinetics*: (Chapter 5) vaccine schedules, (Chapter 6) animal models, and (Chapter 7) human immunity.
- *Evolution in action*: (Chapter 8) expectation from theory, (Chapter 9) genomic patterns, (Chapter 10) vaccine-driven selection, and (Chapter 11) congenerics.
- *Epidemiological data, modelling, and inference*: (Chapter 12) surveillance, (Chapter 13) contrasting ecological and evolutionary signatures, (Chapter 14) model-based inference, and (Chapter 15) public health outcomes.
- *Epilogue*: an attempt at synthesis and a roadmap.

References

1. Aslanabadi A, Ghabili K, Shad K, et al. Emergence of whooping cough: notes from three early epidemics in Persia. *Lancet Infect Dis* 2015;**15**:1480–4.
2. Rz B. *Summary of Experiences*. Tehran: Tehran University Press; 2011.
3. Cherry JD. Historical review of pertussis and the classical vaccine. *J Infect Dis* 1996;**174**:S259–63.
4. Horman W. *Vulgaria*, Vol. 3. London: Richard Pynson; 1519.
5. Still GF. *The History of Paediatrics*. Oxford: Oxford University Press; 1931.
6. Murray J. *English Dictionary on Historical Principles*. Oxford: Oxford University Press; 1928.
7. Mooi FR. Bordetella pertussis and vaccination: the persistence of a genetically monomorphic pathogen. *Infect Genet Evol* 2010;**10**:36–49.
8. Parkhill J, Sebaihia M, Preston A, et al. Comparative analysis of the genome sequences of Bordetella pertussis, Bordetella parapertussis and Bordetella bronchiseptica. *Nat Genet* 2003;**35**:32–40.
9. Bart MJ, Harris SR, Advani A, et al. Global population structure and evolution of Bordetella pertussis and their relationship with vaccination. *MBio* 2014;**5**: e01074–14.
10. Magner LN. In: Kiple KF (ed), *The Cambridge World History of Human Disease*. Cambridge: Cambridge University Press; 1993, pp 392–400.
11. Gordon JE, Hood RI. Whooping cough and its epidemiological anomalies. *Am J Med Sci* 1951;**222**:333–61.
12. Roush SW, Murphy TV, Vaccine-Preventable Disease Table Working Group. Historical comparisons of mor-

bidity and mortality for vaccine-preventable diseases in the United States. *JAMA* 2007;**298**:2155–63.

13. Kendrick PL. Can whooping cough be eradicated? *J Infect Dis* 1975;**132**:707–12.

14. Linder FE, Grove RD. *Vital Statistics Rates in the United States, 1900–1940*. Washington, DC: US Government Printing Office; 1943.

15. Lapin L. *Whooping Cough*. Springfield, IL: Charles C. Thomas; 1943.

16. Broder KR, Cortese MM, Iskander JK, et al. Preventing tetanus, diphtheria, and pertussis among adolescents: use of tetanus toxoid, reduced diphtheria toxoid and acellular pertussis vaccines recommendations of the Advisory Committee on Immunization Practices (ACIP). *MMWR Recomm Rep* 2006;**55**:1–34.

17. World Health Organization. Pertussis vaccines: WHO position paper – September 2015. *Wkly Epidemiol Rec* 2015;**90**:433–58.

18. de Cellès MD, Magpantay FM, King AA, et al. The pertussis enigma: reconciling epidemiology, immunology and evolution. *Proc R Soc B* 2016;**283**:20152309.

19. Rohani P, Drake JM. The decline and resurgence of pertussis in the US. *Epidemics* 2011;**3**:183–8.

20. Varughese P Incidence of pertussis in Canada. *Can Med Assoc J* 1985;**132**:1041–2.

21. Jackson D, Rohani P. Perplexities of pertussis: recent global epidemiological trends and their potential causes. *Epidemiol Infect* 2014;**142**:672–84.

22. Guris D, Strebel PM, Jafari H, et al. Pertussis vaccination: use of acellular pertussis vaccines among infants and young children: recommendations of the advisory committee on immunization practices (ACIP). *MMWR Recomm Rep* 1997;**46**:1–25.

23. Crowcroft N, Britto J. Whooping cough – a continuing problem. *BMJ* 2002;**324**:1537.

24. World Health Organization. *Pertussis Global Annual Reported Cases and Dtp3 Coverage, 1980–2016*. Geneva: World Health Organization; 2017.

25. Hogan D. *MCEE-WHO Methods and Data Sources for Child Causes of Death 2000–2016*. Geneva: World Health Organization; 2018.

26. Yeung KHT, Duclos P, Nelson EAS, et al. An update of the global burden of pertussis in children younger than 5 years: a modelling study. *Lancet Infect Dis* 2017;**17**:974–80.

27. Hug L, Sharrow D, You D. *Levels and Trends in Child Mortality: Report 2017. Estimates Developed by the UN Inter-agency Group for Child Mortality Estimation*. New York: United Nations Children's Fund; 2017.

28. Luttinger P. The epidemiology of pertussis. *Am J Dis Child* 1916;**12**:290–315.

CHAPTER 1

Introduction to pertussis transmission and epidemiological dynamics

Pejman Rohani and Samuel V. Scarpino

Abstract

Resolving the long-term, population-level consequences of changes in pertussis epidemiology, arising from bacterial evolution, shifts in vaccine-induced immunity, or changes in surveillance, are key challenges for devising effective control strategies. This chapter reviews some of the essential features of pertussis epidemiology, together with the underlying epidemiological principles that set the context for their interpretation. These features include the relationship between the age distribution of cases and pertussis transmission potential, the impact of vaccine uptake on incidence, periodicity and age incidence, as well as spatially explicit recurrent pertussis epidemics and associated extinction frequency. This review highlights some of the predictable and consistent aspects of pertussis epidemiology (e.g. the systematic increase in the inter-epidemic period with the introduction of whole-cell vaccines) and a number of important heterogeneities, including variations in contemporary patterns of incidence and geographic spread.

1.1 Introduction

Pertussis is a respiratory disease that is also known as whooping cough and is caused primarily by the Gram-negative coccobacillus *Bordetella pertussis*[1,2]. Prior to the development and widespread administration of vaccines, pertussis prevalence was high. In the United States, for example, throughout the 1920s, 1930s, and 1940s, annual pertussis incidence ranged from 100 to 200 cases per 100,000, with a concomitant annual fatality incidence of 1–10 per 100,000[3]. It was recognized early on that pertussis exclusively infects humans, with no animal reservoir[3,4]. Earlier epidemiological investigations[5,6] also led researchers to conclude that 'there are no proven chronic carriers of *B. pertussis*'[6]. Thus, the perceived

relative simplicity of this host–pathogen association increased optimism for the prospect of effectively controlling pertussis with vaccines[7,8] and even chemoprophylaxis, such as antibiotics[4]. The pioneering work of Madsen[9], Kendrick and Eldering[10], and others led to the development of so-called whole-cell pertussis (wP) vaccines, containing concentrated bacteria killed by heat or a chemical agent[2]. Subsequent large-scale vaccination campaigns and the roll-out of routine infant immunization throughout much of the industrialized world in the 1940s and 1950s were associated with marked declines in pertussis incidence[3,11,12]. It appeared that the vaccine was working as hoped for, protecting individuals from disease and the population from spread of infection. Beginning in the 1970s, however, some regions began to see a rise in cases[13].

Rohani, P. and Scarpino, S. V., *Introduction to pertussis transmission and epidemiological dynamics*. In: *Pertussis: epidemiology, immunology, and evolution*. Edited by Pejman Rohani and Samuel V. Scarpino: Oxford University Press (2019). © Oxford University Press. DOI: 10.1093/oso/9780198811879.003.0001

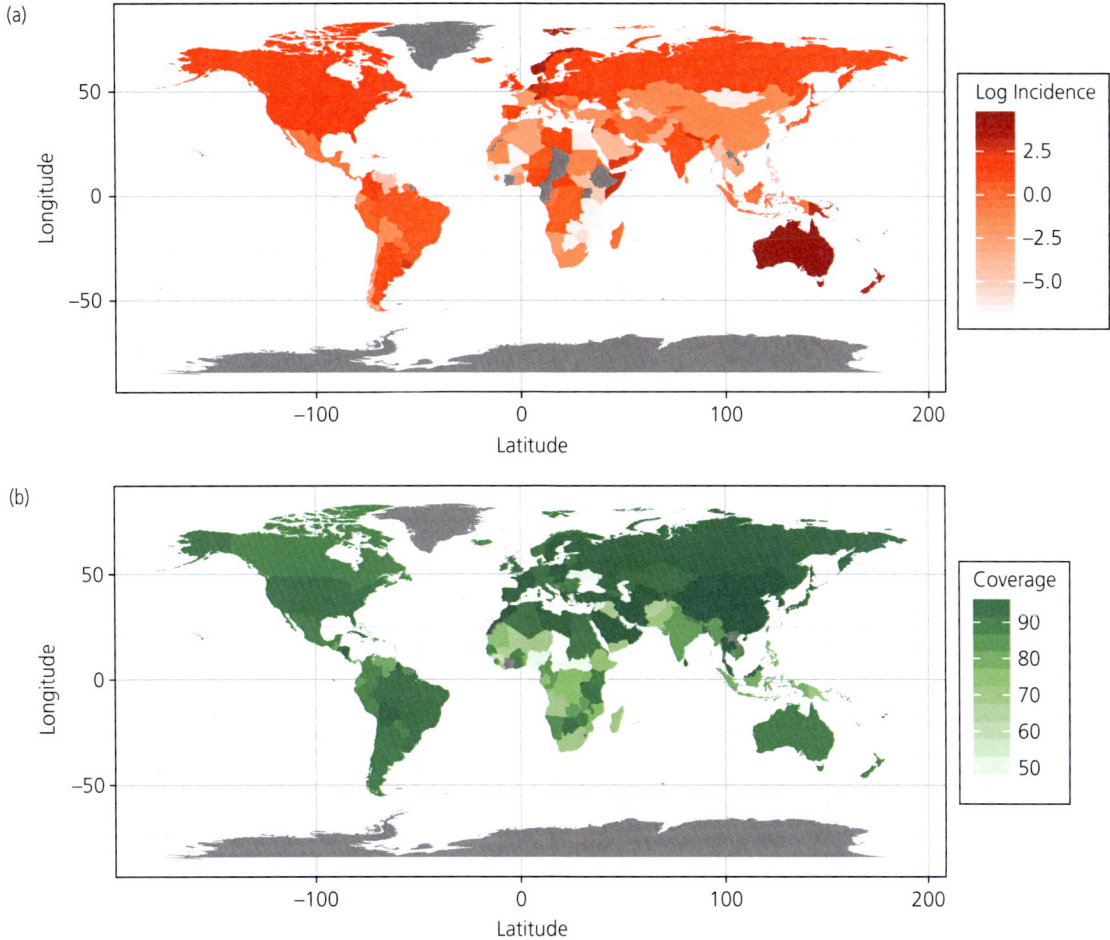

Figure 1.1 Global mean annual pertussis (a) log(incidence per 100,000) and (b) three doses of diphtheria–tetanus–pertussis vaccine (DTP3) coverage from 2011 to 2016. Grey regions represent countries for which data were not available. Population demographic information from[23] and pertussis incidence and vaccine coverage data from[24]. Maps courtesy of Ana Bento.

In some locations, such as the United Kingdom and Sweden, this coincided with a major vaccine scare[14–16]. However, reported incidences increased despite sustained high coverage in many parts of the world including parts of Europe, the United States, Australia, and Taiwan[13,17–21]. Recent epidemiological data indicate that the control of pertussis transmission remains incomplete and problematic; pertussis continues to exact a heavy toll worldwide, with an estimated 24.1 (95 per cent confidence interval (CI): 7–40) million cases and 161,000 (95 per cent CI: 38,000–671,000) deaths in 2014 in children younger than 5 years, for the most part in low-income countries[22].

In Figure 1.1a, we present annual pertussis incidence by country, averaged from 2011 to 2016. The figure demonstrates substantial geographic variability in pertussis burden. In Figure 1.1b, we depict estimated mean annual coverage of at least three doses of pertussis vaccines. This figure also underlines the widespread variation in vaccine coverage, with notably low uptake in much of Africa, the Indian subcontinent, and parts of Southeast Asia. Not surprisingly, the regions with low vaccine coverage tend to be associated with high reported pertussis incidence. Surprisingly, however, some countries experiencing high per capita incidence, such as Australia and the United States (Figure 1.1a), are among those countries that boast high routine vaccine coverage (Figure 1.1b). Thus, diagnosing the reasons underlying the tremendous heterogeneity in pertussis epidemiology, including its resurgence in some settings and the absence of resurgence in other settings, is clearly an important scientific problem as well as a major public health issue—it forms one of the key motivations for this book. In this chapter, we wish to set the scene by presenting key epidemiological features of such infectious disease systems, especially as they pertain to the interpretation of time series of disease incidence.

1.2 Transmission models

Insights into the determinants of epidemiological traits, such as transmission dynamics, pathogen invasion, and long-term disease prevalence (e.g. Figure 1.1a), can be gained by considering simple mathematical models of pertussis transmission[25–28]. In the context of immunizing pathogens, the simplest model is the so-called *SIR* model, which categorizes individuals according to their infection history as Susceptible (those previously not infected), Infectious (individuals currently infectious), or Recovered (previously infected and now immune to reinfection). As we discuss below and in Chapter 14, this base model can be extended and modified to account for a diversity of epidemiologically relevant specifics of pertussis.

A central quantity of interest in transmission models is the *basic reproduction number*, or R_0 [26,28,29]. It quantifies the transmission potential of an infectious disease and a generally accepted definition of R_0 is the number of secondary cases that result from a typical infected case when the entire population is susceptible. Classic work by Anderson and May[30] used data on the age at infection in order to estimate R_0 for pertussis, in the range of 16–18 in urban settings. Though this figure reflects the acknowledged transmissibility of pertussis[3,31,32], care needs to be taken with estimates derived from age-incidence patterns as their interpretation requires consideration of the underlying age-specific contact patterns[33] and infection-derived immunity[34]. That said, subsequent model-based estimates from

pre-vaccine era incidence data are reasonably consistent with those of Anderson and May and indicate values ranging from 11 to 15 in England and Wales[34], approximately 16 in Copenhagen[35], and approximately 17 during the vaccine hiatus period (1986–1996) in Sweden[36]. On the other hand, using serological data and assuming a very short duration of immunity (lasting, on average, 1.1 years), Kretzschmar and colleagues arrived at much lower estimates of R_0 in a number of European countries, with figures ranging from 5.3 to 5.9[37].

1.3 Disease prevalence

In the *SIR* model, assuming constant demographic and epidemiological conditions, prevalence is determined by a combination of population demography (specifically, the per capita birth and death rates, v and μ, respectively), local transmission rate (β), and the ratio of transmission rate to the recovery or clearance rate of infected individuals (R_0). In particular, the relationship between these quantities and prevalence, denoted by I^*, is given by:

$$I^* = \frac{\mu}{\beta}(R_0 - 1). \tag{1}$$

In this model, the basic reproduction number, R_0, is given by $\dfrac{v}{\mu}\dfrac{\beta}{(\mu+\gamma)}$. Thus, according to equation (1), both the per capita birth rate (v) and R_0 are positively associated with disease prevalence. The inclusion of routine infant immunization, whereby a fraction p of newborns are protected by vaccination, modifies this expression, so that:

$$I^* = \frac{\mu}{\beta}(R_0(1-p) - 1). \tag{2}$$

Hence, prevalence linearly decays with vaccine uptake, p. Given that disease prevalence is determined by $R_0(1-p)$ and that R_0 takes into account the per capita birth rate (v), often epidemiologists resort to exploring the association between unvaccinated births (i.e. $v \times (1-p)$) and pertussis incidence. An example of such an exploration is provided in Figure 1.2, where pertussis incidence per 100,000 is plotted against per capita susceptible births[38]. Again, the figure demonstrates substantial variation in this relationship, ranging from a

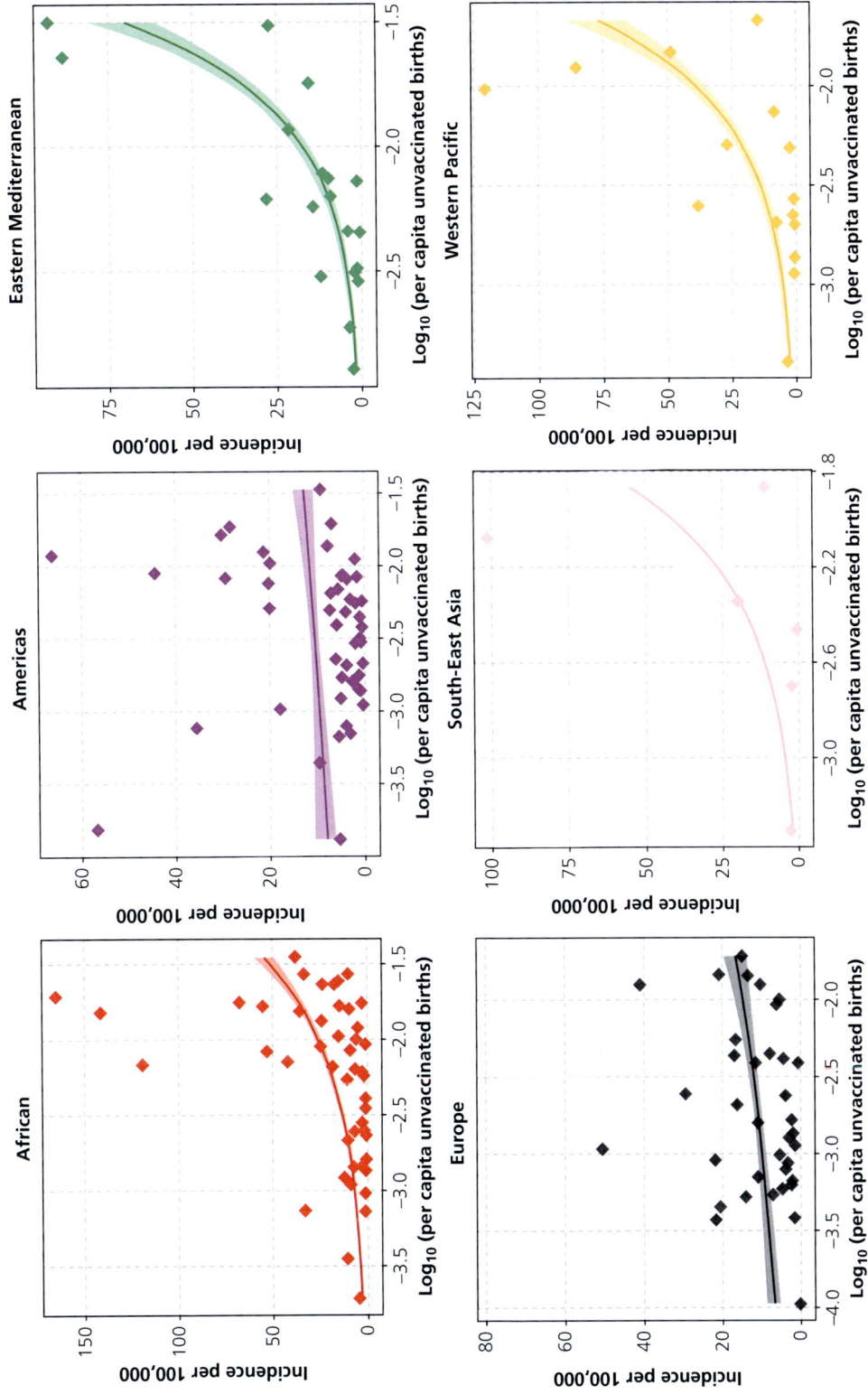

Figure 1.2 Relationship between per capita unvaccinated births and pertussis incidence in different regions, as categorized by the World Health Organization. Population demographic information from[23] and pertussis incidence and vaccine coverage data from[24]. Figure courtesy of Navideh Noori.

reasonably robust association (e.g. Eastern Mediterranean or Western Pacific regions) to apparently an absence of any relationship (e.g. the Americas or Europe regions).

1.3.1 Relationship with changing vaccines

Some authors have claimed that in a number of more economically advantaged countries, the rise in pertussis incidence appears to have followed the switch from whole-cell to acellular pertussis (aP) vaccines[39-43]. The rationale is that acellular vaccines generate an immune response profile that is distinct from both natural infection and whole-cell vaccines[44,45]. In animal studies, aP vaccines have been shown to ameliorate pertussis severity, but not transmission[44,46], leading some to believe the resurgence may be underpinned by largely asymptomatic infections among individuals who had previously received aP vaccines[42,43]. In addition, recent epidemiological studies have concluded that aP-derived immunity is short-lived, lasting perhaps as little as 3 years[47,48]. While these arguments are intuitively appealing, the epidemiological association between vaccine composition and pertussis transmission is not straightforward or geographically consistent[12].

In the study by Domenech de Cellès et al.[49], the authors explored trends in annual incidence reports for pertussis across the globe. They reported substantial temporal and spatial variability in pertussis incidence. Of the 63 countries they examined, there were 31 nations that did not have any period of significant increase (Africa: eight; Americas: eight; Eastern Mediterranean: five; Europe: four; Southeast Asia: three; Western Pacific: three). However, 32 had at least one period of increasing pertussis incidence during 1980–2012, comprising zero in Africa, eight in the Americas, four in Eastern Mediterranean, 11 in Europe, five in southeast Asia, and four in Western Pacific. Of these countries, 28 also experienced at least one period marked by a decrease, with only four—Australia, Israel, the Netherlands, and the United States—exhibiting no period of decrease in pertussis incidence during 1980–2012.

To assess the relationship between these trends and the switch to aP vaccines, Domenech de Cellès et al.[49] reviewed the literature to identify the date of the switch in their data. For those 20 countries, the authors examined the primary immunization schedule and the timing of any paediatric and/or adolescent booster dose. As depicted in Figure 1.3, although the switch to aP coincided with resurgence in some countries (e.g. Spain and the United Kingdom), in others the resurgence occurred much earlier, notably in Australia, the United States, Israel, the Netherlands, Finland, and Bulgaria. In addition, the switch to aP, after a period of high, low, or no vaccination, led to significant decreases in incidence in Finland, Italy, and Sweden, respectively. Trends were similarly variable for the 43 countries that used wP vaccines for primary immunization; for example, Domenech de Cellès et al.[49] reported that pertussis incidence increased with increasing vaccine coverage in Brazil and Columbia, but decreased with increasing coverage in Bolivia, Thailand, and Vietnam. Considering only contemporary data, these trends did not differ between countries that used wP or aP vaccines for primary immunization (Fisher's exact test, $p = 0.22$). Because the number of aP components[50] and timing of primary immunization[51] may affect pertussis epidemiology, these authors also examined this possibility, but found no consistent association across countries.

Under-reporting and incompleteness in notification data are ubiquitous sources of concern. That said, the analysis of overall trends by Domenech de Cellès et al.[49] identified several patterns which they believed to be robust. First, contrary to some claims[52,53], pertussis resurgence is not global: a majority of countries examined experienced sustained periods of decrease in incidence over the last 30 years. Second, except for countries in Africa, for which no significant periods of increase were detected, there appeared to be no consistent geographical pattern in the data. Finally, these authors were unable to identify any association between epidemiological trends and country-specific differences in vaccination (i.e. type of vaccine, vaccine composition, or vaccination schedule), although more complex interactions may be at work.

1.3.2 Relationship with immune waning

There has long been uncertainty around the durability of immunity to pertussis, both in the context of

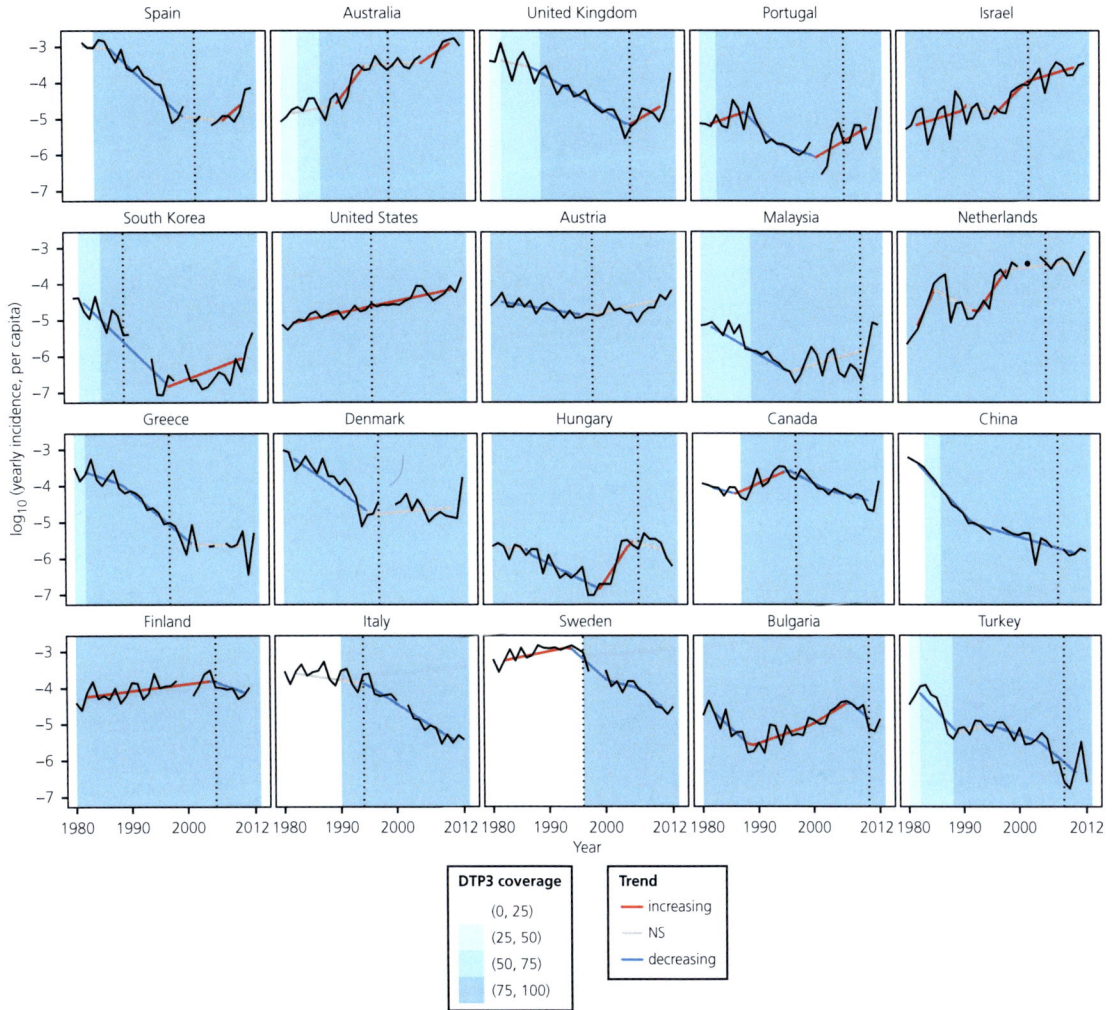

Figure 1.3 Annual incidence data in 20 countries that switched to acellular pertussis (aP) vaccines for primary immunization. Yearly case counts and vaccine coverage estimates from 1980 to 2012 were obtained from the World Health Organization database[24]. The analysis was restricted to countries with more than 80% complete case counts and more than 5 million inhabitants. Incidence data were $\log_{10}$-transformed and a 5-year moving average was applied to remove temporal periodicity[38]. To detect long-term trends in pertussis incidence, first a series of segmented regression models with 0–3 breakpoints and segments longer than 5 years were applied for each country[13,54]; the most parsimonious model was selected according to the Bayesian information criterion. Next, to account for autocorrelation, we used generalized least-squares on each time segment, assuming the residuals autocorrelation structure followed an autoregressive process of order 1. For each country, the time segments were then classified according to their slope, as increasing (significantly positive slope), decreasing (significantly negative slope), or not significant. For each country, we represent the annual incidence (black solid lines), the fitted values from segmented regression, coloured according to the trend (red lines: significantly increasing; grey lines: no significant trend; blue lines: significantly decreasing), and the date of switch to aP vaccination (black vertical dotted lines). Coloured blue areas indicate the vaccine coverage for the third dose of the diphtheria, tetanus, and acellular pertussis (DTaP) vaccine. From left to right and top to bottom, countries are ranked by decreasing value of the last slope. Figure from Domenech de Cellès et al. 2016[49].

vaccine-derived protection but also immunity following natural infection[44,55]. Here, we wish to highlight the changes to equation (1) that result from the assumption that previously infected and immune individuals reactivate susceptibility at rate ω, such that the average duration spent in the recovered and immune class is given by $1/\omega$. In this revised model, referred to as *SIRS*, assuming the infectious period is much greater than the average lifespan ($\gamma \gg \mu$), pertussis prevalence is approximately given by:

$$I^* \approx \frac{\mu}{\beta}(R_0 - 1) \times \left(1 + \frac{\omega}{\mu}\right). \tag{3}$$

Hence, even if infection-derived immunity lasts, on average, for a lifetime ($\omega \sim \mu$), disease prevalence is predicted to double compared to the *SIR* model. It is important to note that, in this simple model, subsequent infections are assumed to have identical epidemiological characteristics (in terms of transmissibility and duration) as a first episode and that this model ignores severity of symptoms and impacts on reporting probability. Other studies have developed more complex models of immune waning, permitting differential transmissibility[34,56,57], immune boosting[34,35,58], and alternative mechanisms of immune failure, including immune 'leakiness'[59–61]. In Chapter 14, King et al. describe two contrasting methods for identifying the mechanism of immune failure together with estimating associated parameters from longitudinal, age-stratified incidence reports.

1.4 Under-reporting

Pertussis infection produces a spectrum of clinical symptoms, ranging from the classical cough to mild, largely asymptomatic infection, especially among adolescents and adults[62]. Consequently, incidence reports represent an incomplete sample of the total transmission events in the reporting interval. This under-reporting of pertussis incidence is considered so problematic by some researchers that they seem to question the reliability of any conclusions drawn on the basis of epidemiological data[63,64]. Attempts to quantify the reporting fidelity of pertussis data have generated reasonably consistent estimates. In the United States, Sutter and Cochi documented substantial age specificity in reporting accuracy, with an overall estimate of 11.6 per cent[65]. As shown in Table 1.1, estimates of reporting efficiency appear to be in the range of 10–32 per cent. We note that epidemiological models routinely specify a so-called observation process, through which the reporting probability can be estimated. The estimates of such studies are consistent with those obtained via routine epidemiological methods.

Table 1.1 Estimates of pertussis reporting efficiency

Estimate (%)	Age cohort	Vaccination	Location	Reference
15	Overall	Pre-vaccine	Denmark	35
11.6	Overall	wP	USA	65
12–32	Adolescents	wP	USA	66
12–32	Adolescents	wP	USA	66
15	Infants (< 1 year)	wP and aP	MA, USA	61
4	Infants (> 5 years)	wP and aP	MA, USA	61
23–28	Adolescents and adults	aP	UK	67
13–20	Adolescents and adults	wP	USA	68
17	Children (< 5 years)	aP	New Zealand	69
7	Older (> 5 years)	aP	New Zealand	69
15	Children (< 12 years)	wP	Uganda	70

1.4.1 Relationship with sources of infection

Complicating our understanding of whooping cough transmission dynamics is the observation that across a broad range of countries, the source of infection for infant *B. pertussis* infections could only be identified in 42 per cent of household transmission studies (24–66 per cent averages reported across studies)[71–77]. As with all aspects of whooping cough, there exist competing explanations for the low rate of whooping cough contact tracing success, ranging from a 2014 household pertussis study in Brazil, which found evidence that a large fraction of *B. pertussis* infections in children likely came from asymptomatically infectious adults[78] to another 2014 study that found over 70 per cent of patients—determined from a sample of 245,249 individuals—infected with *B. pertussis* were not identified by standard hospital procedures[79]. Developing a more complete understanding of factors contributing to this observed aspect of household transmission remains an active area of whooping cough research and may help reconcile the importance of adult under-reporting and asymptotic transmission in whooping cough epidemiology.

1.5 Periodicity

A key characteristic of infectious disease time series is the inter-epidemic period, or the time between cyclic outbreaks[26,80,81]. Infectious diseases that are directly transmitted, highly infectious, and immunizing tend to produce recurrent multi-annual epidemics resulting from the inherently non-linear transmission process, coupled with seasonality. The intuition behind this idea is as follows: epidemiological theory explains that a threshold fraction of susceptibles are necessary to generate an outbreak[26,82]. Once an outbreak occurs, however, it depletes the population of susceptible hosts, which retards the chain of transmission until the susceptible population has been replenished (e.g. following naïve births), thus generating the conditions necessary for a subsequent outbreak. Therefore, the speed of susceptible replenishment is key in determining the inter-epidemic period[38,83]. For a lifetime immunizing infection, this is determined by the rate of influx of susceptible newborns, which is determined by a combination of population birth rate and the effective immunization coverage. In the absence of seasonal variation in transmission, the expression describing the relationship is given by:

$$T = 2\pi \sqrt{AG}. \tag{4}$$

where T is the inter-epidemic period, A is the mean age at infection (given by $1/\{\mu(R_0 - 1)\}$), and G is the generation length of the infectious disease (given by $1/(\mu + \gamma)$), defined as the infectious (and latent) period[26,28,84]. Thus, in general, slow susceptible recruitment, either due to low birth rates or high effective vaccine coverage, leads to a higher mean age at infection, thus generating a longer inter-epidemic period. In particular, if a fraction p of newborns are vaccinated, then the expression for the mean age at infection is modified to $A = 1/\{\mu(R_0(1-p)-1)\}$. This is demonstrated in Figure 1.4a, with an inter-epidemic period of almost 2.8 years in the absence of immunization, increasing to ~4 years with 40 per cent vaccine uptake. Importantly, this result only holds when vaccination successfully blocks transmission. If vaccination only prevents disease, the number of *observed* cases will decrease with higher vaccine coverage, but the inter-epidemic period will remain the same since the underlying transmission has remained unchanged[85]. Crucially, periodicity is insensitive to the extent of under-reporting or in changes in reporting over time so long as they are not directly tied to incidence (i.e. when reporting probability itself scales with disease incidence). This is an important feature for a robust metric given the notoriously low, age-specific, and variable reporting rates for pertussis[65].

If the vaccine provides immunity that is transient and wanes after some period (denoted by $1/\omega$), then the inter-epidemic period is inevitably impacted[28]. Specifically, the mean age at infection is now determined by

$$A = \frac{1}{\mu(R_0(1-p+\omega p/(\mu+\omega))-1)}. \tag{5}$$

Thus, even if the vaccine affords protection for the average duration of natural life ($\omega = \mu$), the expected age at infection is substantially lower than would be predicted for a perfect vaccine. To illustrate, if we assume $R_0 = 15$, an infectious period of 10 days, a mean lifespan of 70 years, and 50 per cent vaccine

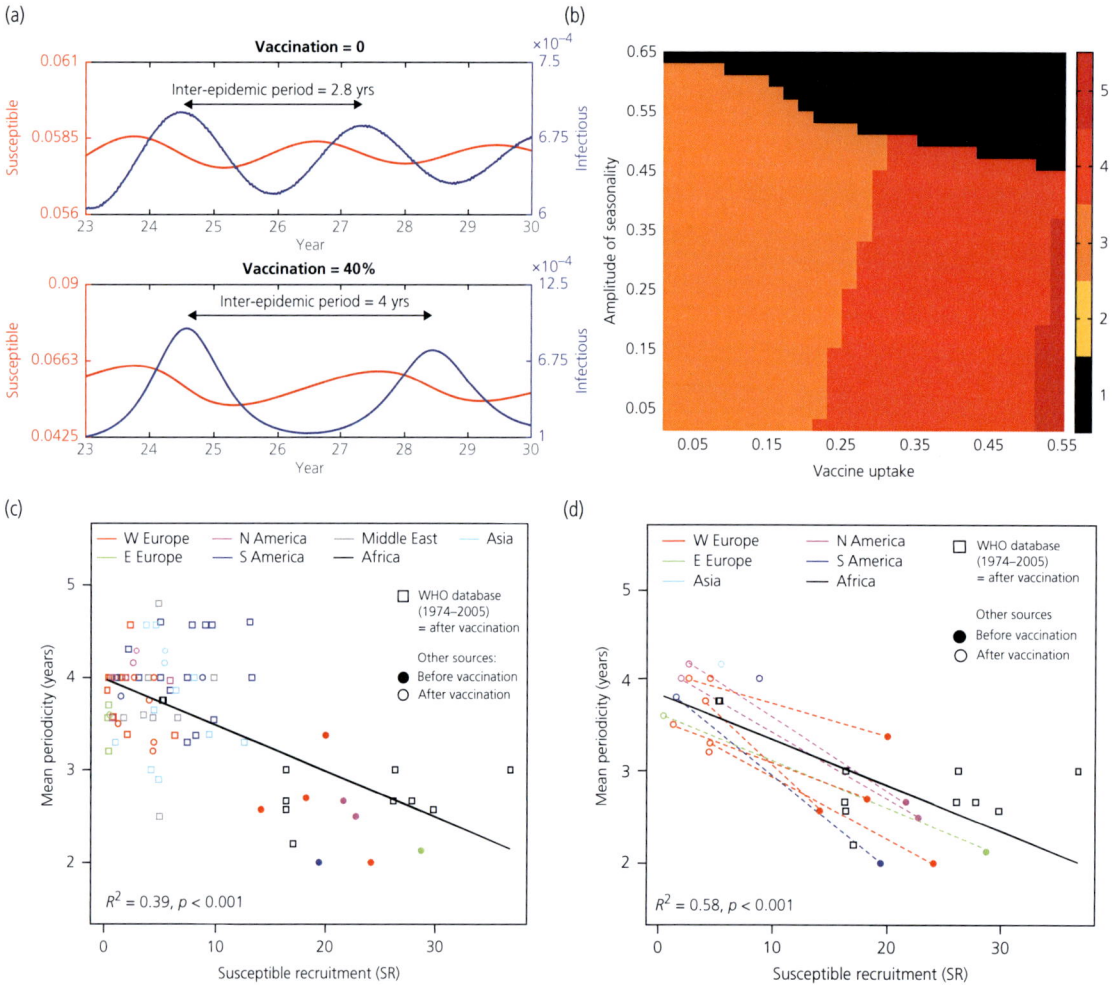

Figure 1.4   Relationship between susceptible recruitment and the inter-epidemic period. (a) Demonstrates an inter-epidemic period of 2.8 years in the absence of vaccination (top panel), which lengthens to 4 years as 40% of newborns are immunized (bottom panel). (b) Illustrates the relationship between periodicity, immunization coverage, and the strength of seasonality in transmission. From Blackwood et al.[91]. (c) Documents the association between periodicity and susceptible recruitment for 64 countries ($R^2 = 0.39, p < 0.001$). Colours represent geographical regions (red, western Europe; pink, North America; grey, Middle East; cyan, Asia; green, eastern Europe; blue, South America; black, Africa). Squares correspond to the World Health Organization (WHO) database and circles to non-WHO data. Open symbols represent vaccine era data and filled symbols represent pre-vaccination era. (d) As for (c) except with WHO database African countries (black squares) and non-WHO data in pre-vaccination (filled circles) and post-vaccine (open circles) periods ($R^2 = 0.58, p < 0.001$). The dashed lines link the pre- and post-vaccination eras for the same country. From Broutin et al.[38].

coverage, this translates into a mean age at infection of almost 7 years for the waning vaccine, rather than ~11 years for the perfect vaccine. Concomitantly, the inter-epidemic period is reduced from 3.4 to 2.7 years.

A number of studies have fruitfully used this underlying theory to gauge the effectiveness of pertussis vaccines in blocking transmission. The first study to do so was by Fine and Clarkson[80], published in 1982, which used aggregated data from England and Wales and documented increases in the observed inter-epidemic period of pertussis following the introduction of national immunization programme using wP vaccines. However, without providing any firm theoretical reasoning, the authors deemed the increase to be less than they had expected and hence interpreted their findings as evidence for the wP vaccine only protecting against disease, not transmission[80]. Since then, numerous authors have cited this study, which has assumed folkloric status within the field of pertussis epidemiology (e.g.[39,64,86–88]). It is routinely referenced as the central evidence against the protective effects of whole-cell vaccine against infection.

From our perspective, there are at least two problems with this claim. First, the periodicity of a host–pathogen system such as pertussis is shaped both by susceptible recruitment and by seasonality[83,89,90], as depicted in Figure 1.4b. Increases in the inter-epidemic period with vaccine uptake are most likely when seasonal forcing is moderate. In settings where the amplitude of seasonal variation in transmission is large, reductions in the replenishment rate of the susceptible pool may have no effect on periodicity (Figure 1.4b). This idea was illustrated by Blackwood et al.[91] who showed no change in the inter-epidemic period of pertussis in Thailand despite a substantial rise in vaccine uptake in the mid 1980s. Examination of pertussis incidence, especially among infants, however, indicated a notable reduction in pertussis circulation[58]. This apparent discrepancy was explained by the authors as arising from a substantial amplitude of seasonal forcing in Thailand, as illustrated using time-series methods[58].

The second reason to question the conclusions of Fine and Clarkson is the number of studies that have documented a positive association between the inter-epidemic period and wP vaccine coverage (e.g.[92,93]). For instance, a more recent and detailed study on pertussis in England and Wales[93], which used city-level incidence data, provided evidence for a systematic increase in the inter-epidemic period following the rollout of vaccination. Similarly, the study by Broutin et al.[38] examined pertussis periodicity across 64 countries and identified an association between the inter-epidemic period and susceptible births (Figure 1.4c). In particular, when considering those countries for which incidence records spanned the introduction of wP vaccines, these authors noted that the inter-epidemic period increased, on average, by 1.27 years (Figure 1.4d). Taken together, these studies lend strong credence to the theory that

historical wP vaccines indeed provided some protection against transmissible disease.

1.6 Extinction profile

The frequency and duration of local pathogen extinction events also hold information regarding immunity in a population[34]. Although pertussis is far from eradication, in sufficiently small populations, weeks or months may pass with no case notifications[43,85,93], suggesting that pertussis has become locally absent (termed a fade-out[94,95]) and is eventually reintroduced from an outside source. As shown in Figure 1.5b, a number of countries around the globe fall below the critical community size (CCS); that is, they are smaller than the threshold population size necessary to ensure sustained pertussis transmission[43,85]. A similar relationship is found when extinction frequency is compared with unvaccinated births (Figure 1.5a).

If a vaccine blocks transmission, then its introduction is predicted to lead to an increase in the CCS; a larger population will be necessary to prevent local extinction. This result holds even if vaccine-derived immunity is not lifelong; however, a short duration of protection leads to a reduced increase in the CCS following vaccination[34] since the susceptible pool is replenished not only by births but also loss of immunity.

There have been fewer studies of the pertussis CCS compared with periodicity analysis perhaps in part because it requires granular spatial and temporal resolution, such as cases aggregated weekly from locations with a range of population sizes. Data from cities in England and Wales demonstrated a systematic increase in the number of weeks with zero cases and the duration of each individual fade-out from the pre-vaccine to the vaccine era, apparently consistent with the predictions of a transmission-blocking vaccine[93]. A similar study using data from small communities in Senegal also showed an increase in the duration of fade-outs with the introduction of vaccination[96]. It is important to point out, however, that, as recently demonstrated out by Althouse and Scarpino[43], if vaccination protects against symptomatic disease but not infection, then interpretation of extinction frequencies in the vaccine era may be biased. This is because the

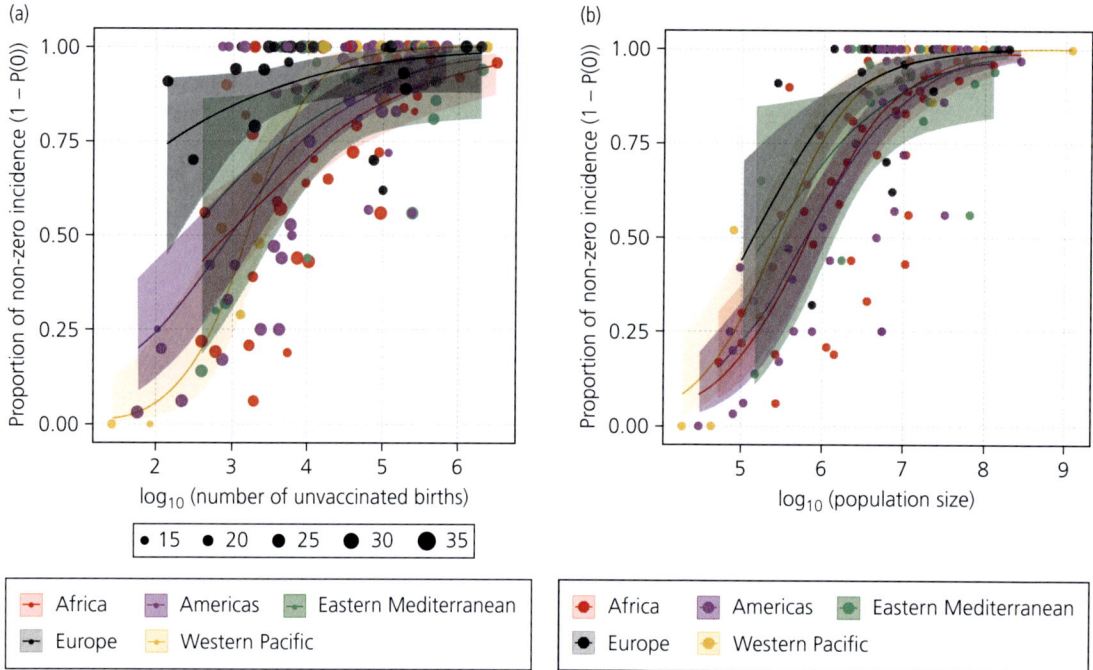

Figure 1.5 Relationship between extinction frequency and (a) susceptible births and (b) population size. Population demographic information from[23] and pertussis incidence and vaccine coverage data from[24]. Figure courtesy of Navideh Noori.

observed incidence would be an unreliable indicator of underlying transmission. Therefore, by itself, a shift in the extinction threshold in a population following changes in vaccination policy ought not to be invoked as evidence for changing transmission.

1.7 Age structure

A growing body of theory and data have demonstrated how changes in transmission resulting from vaccine coverage[18,33,36] or demographic trends[98] can affect age-stratified disease incidence. The key feature that is often focused on is the age-specific force of infection[26], which quantifies the per capita transmission hazard experienced by susceptible individuals in a specified age group. In dynamical models, λ_i, the force of infection for individuals in age group i, is represented as:

$$\lambda_i = q_i \sum_{j=1}^{n} c_{ij} \frac{I_j}{N_j},$$ (6)

where n is the number of age groups, c_{ij} quantifies the rate of contacts between individuals in age groups

i and j, I_j/N_j is the proportion of individuals of age j infected, and q_i is the age-specific susceptibility of individuals in age group i (alternatively, can be thought of as the probability of transmission given contact)[26,27,36,97]. A critical component of such models, therefore, is the specification of the pattern of age-specific contacts (the matrix with entries c_{ij}), often called the Who Acquires Infection From Whom (WAIFW) matrix[26,99]. The most naïve assumption would be homogeneous mixing. That is, the rate of contacts (as related to transmission) among individuals is independent of their respective ages. It has been shown that relaxing this simplifying assumption can have important consequences for interpreting epidemiological data[33]. Furthermore, since the pioneering work of Edmunds[100] and Mossong[101], we now have access to estimates for age-specific contact patterns for a number of countries across the globe[102–106].

One of the most widely noted changes in pertussis epidemiology in the past few decades has been the shift to cases in older individuals[18,61,107–110]. In particular, since the beginning of the resurgence in some countries, an increase in cases among teenagers has been noted[111], along with outbreaks in middle schools and high schools with highly vaccinated student populations[112]. Initially, the increase in teenage and adult cases led many researchers to speculate that immunity, at least to disease, is being lost increasingly rapidly, perhaps due to new bacterial strains that elicit shorter-lasting immunity[113,114] or reduced natural immune boosting arising from a reduction in pathogen circulation[97]. An alternative explanation for these patterns is that they may be the inevitable dynamical consequence of vaccination. As previously described, and illustrated in Figure 1.6a, theory predicts that a transmission-blocking vaccine will lead to a decrease in pathogen circulation, which reduces the force of infection and acts to increase the proportion of infections occurring in older individuals, as evident from equation (5). Thus, the proportion of cases in teenagers should increase simply due to a vaccine-induced reduction in transmission. Evidence in support of this interpretation was provided by Rohani et al.[36], who showed that the documented high contact rates among teenagers, especially with each other[101], can explain the recent increase in teenage cases in Sweden, where pertussis vaccination with aP vaccines was reintroduced in 1996 after a 17-year hiatus[115]. The changes in the age distribution of pertussis incidence in the vaccine-free and the vaccine eras are presented in Figure 1.6b.

The rich information contained in age-stratified incidence records is increasingly harnessed by modellers to better understand pertussis epidemiology and vaccine traits[61,108,116]. In Chapter 14, King et al. utilize dynamic transmission models to quantify long-term protective impacts of infection and vaccination. There is, however, notable variation in the conclusions of these studies and their proposed explanations of contemporary pertussis epidemiology. This arises from the use of different model structures, fitted to data from different countries using different statistical approaches. While it is plausible that pertussis epidemiology is country specific and that aspects of immunity and vaccine effectiveness vary from location to location, it would be insightful to explore whether any consensus will emerge if all models attempt to explain all data. Such an exercise may take the form of recent 'prediction challenges', where multiple groups have independently forecast the epidemiology of influenza[117] or Ebola[118], among others. The critical dimension of these efforts is the focus on the same geographic location for predictions. Alternatively, attempts at finding a unified explanation may require the simultaneous fitting of multiple (competing) models to the same set of incidence data using the same statistical inference algorithm, as was achieved for models of rotavirus vaccines[119].

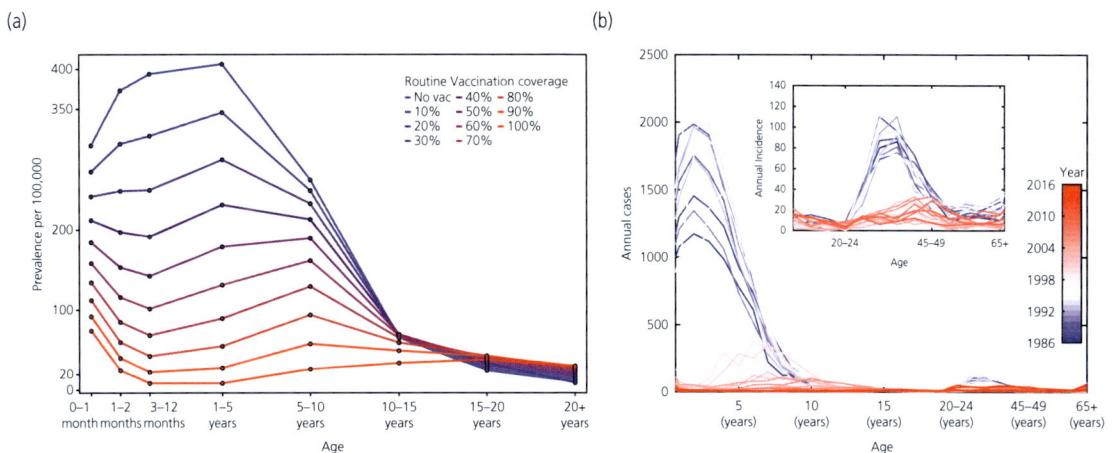

Figure 1.6 (a) Relationship between immunization coverage and the age distribution of pertussis prevalence, assuming vaccination at 3 months. Adapted from Bento and Rohani[120]. (b) Shifting age distribution of pertussis in Sweden during the immunization hiatus (1986–1996; blue lines) and the vaccine era (1996–2015; red lines). Updated from Rohani et al.[36].

1.8 Seasonality

Many infectious diseases of childhood show seasonal variation in incidence[26,121]. The seasonality in pertussis incidence is, however, quite enigmatic. Kilgore et al.[122] stated that unlike 'other respiratory pathogens, *B. pertussis*-associated outbreaks do not uniformly show a distinct seasonality'. This is evident when we examine data from the six largest cities in England and Wales during the pre-vaccine era (Figure 1.7)[123]. In a number of cities (Birmingham, Sheffield, and Leeds), there is a general rise in incidence during the late spring and early summer months, while other cities (London and Manchester) show no clear seasonal peak. Five out of the six cities show a marked reduction in incidence during October, November, and December. In contrast, incidence in Liverpool is highly variable through time without much discernible and consistent yearly signal.

In other countries, however, a strong seasonal peak has been reported, including China[124], Australia[125], and the United States[126,127]. It is interesting to note that the studies by Kaczmarek et al.[125] and Bhatti et al.[127] both used polymerase chain reaction testing data and identified a strong summer spike in positivity rates, yet they pointed out that the highest number of pertussis tests are usually carried out outside the summer months. This clearly points to ways in which pertussis surveillance may be improved.

In countries such as Senegal and Kenya, more complex patterns involving bimodal seasonal peaks have been recorded[128,129]. Such patterns may be the result of climate mediation of disease transmission (e.g. measles in Niger driven by migration during the rainy season[130]) or may result from differential seasonality across age groups. For example, in the Netherlands, from 1996 to 2006, the peak incidence of pertussis among infants and young children was documented in August while the incidence among adolescents peaked in November[88]. As shown in Chapter 14 (Figure 1.7), a similar age-specific pattern of seasonality is at work in Massachusetts,

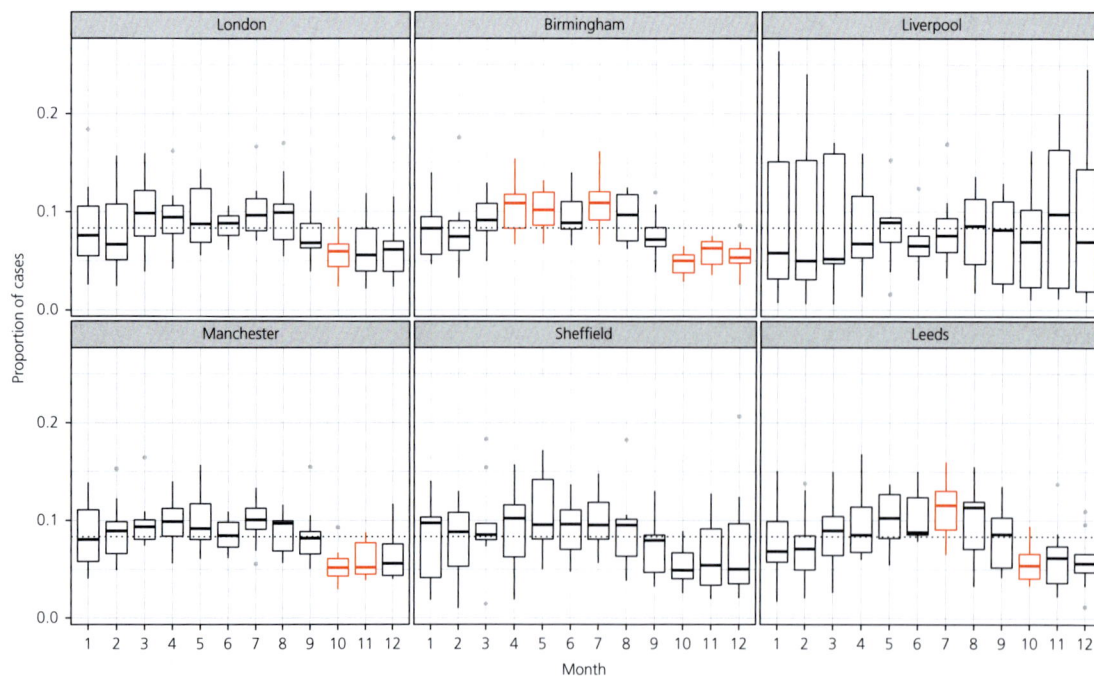

Figure 1.7 Pertussis seasonality in reported cases in the six largest cities in England and Wales. The figure represents, for each city, the year-to-year variability in the proportion of reported cases during each month. The colours indicate the significance of each month, which was assessed by generating 1000 synthetic time series based on a Markov process resampling scheme[61]. Figure by Matthieu Domenech de Cellès.

United States. The peak incidence in infants and 1–4-year-olds occurs in July and August, while cases in 10–20-year-olds peak in October, November, and December. Interestingly, 5–9-year-olds exhibit two peaks, one in July and another in November. Those aged 20 years and older show a general rise in incidence throughout the late summer and early autumn months.

The heterogeneity in the documented seasonal patterns of pertussis incidence highlights our surprising level of ignorance regarding the mechanisms responsible. At present, despite some promising studies[131], it is unclear in general how factors that affect patterns of contact (e.g. school terms and day care attendance), susceptible population size (e.g. births and seasonal immunity), and those that

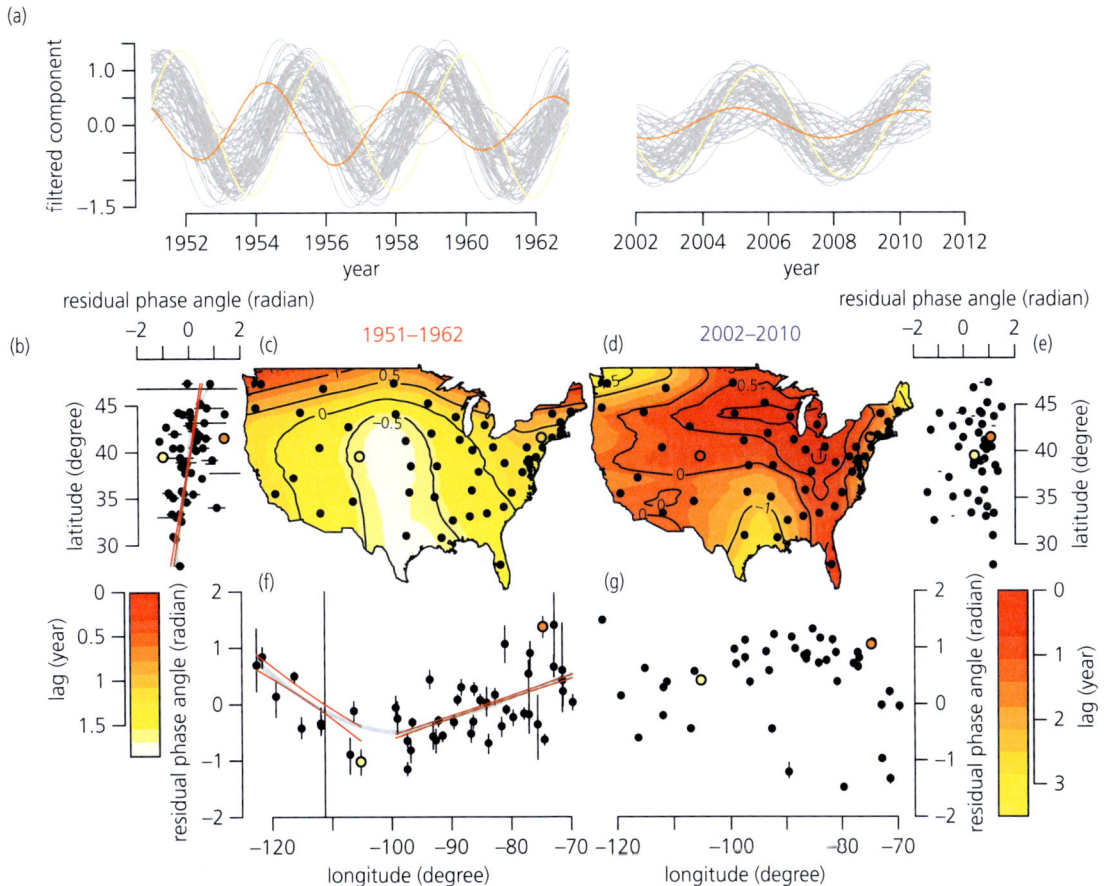

Figure 1.8 Travelling waves of pertussis across the continental US during the 1950s and early 1960s (1951–1962) and in recent years (2002–2010). For illustrative purpose, the states of New York and Colorado are highlighted in orange and yellow, respectively. (a) Time series of pertussis cases for the 49 states filtered between periods of 3.5 and 4.5 years (early era) and 5 and 6 years (recent era). (b and e–g) Residual phase angles of each filtered time series plotted against the latitude (b and e) and the longitude (f and g) of the centres of population of each state for the early era (b and f) and the recent era (e and g). Each small dot represents the value of the residual phase angle for a given month and the big dots show their mean for a given state. Note that, for visual clarity, residual phase angles were deliberately plotted between −2 and +2 radians instead of $-\pi$ and $+\pi$ radians. As a result, some dots are not visible on the graph. The grey areas are the 99% confidence intervals of loess regressions and the coloured straight lines are the 99% confidence intervals of piecewise linear regressions. The cut-off years were obtained via maximum likelihood of the piecewise regressions. (c and d) Colour-coded loess regressions (and their isoclines) of phases angles against longitudes and latitudes (and interaction) of the centres of population of each state (dots). The interpolation values are colour-coded according to the scales on the bottom left (c) and right (d) corners. Figure from Choisy and Rohani[136].

determine the persistence of the bacterium in aerosol form (e.g. climatic variables such as temperature and humidity) affect variation in pertussis transmission.

1.9 Spatial dynamics

Understanding the spatial structure of epidemics and their consequences for possible control strategies has become increasingly appreciated. These include spatially organized seasonal outbreaks of rotavirus in the United States that have been explained in terms of state-specific per capita birth rates[132], the latitudinal gradient in the timing of polio and non-polio enteroviruses[133,134], and the hierarchical spread of seasonal influenza epidemics, driven by differential patterns of mobility[135]. Analyses of pertussis epidemics in the United States since 1951 have revealed a transition from geographically organized epidemics to largely asynchronous outbreaks. Specifically, Choisy and Rohani[136] used wavelet decomposition to document 4-year travelling waves of pertussis epidemics throughout the 1950s (Figure 1.8a–c, f). There were two epicentres for these waves, one in the Northeastern states and the other on the West Coast. This translates into a longitudinal progression speed of 323 km per month (95 per cent CI: 311–336 km per month) westward from the northeast that is almost threefold faster than the eastward wave from the northwest, which travelled at 113 km per month (95 per cent CI: 106–120 km per month). In contrast, the analyses of contemporary data uncovered variable periodicity, with outbreaks occurring approximately every 5.5 years, with little spatial structure (Figure 1.8a, d, e, g). The mechanisms underlying the shift from periodic epidemics spreading across the continents to seemingly spatially unorganized dynamics is not well understood[131]. Intuitively, changes in human mobility, together with heterogeneity in state-specific vaccination policies (e.g. personal belief exemptions)[137] and coverage likely play a major role.

As with other aspects of pertussis, its spatial epidemiology also exhibits substantial variation from place to place. In England and Wales, for example, pre-vaccine epidemics were shown to be inconsistent in their periodicity and largely spatially asynchronous[11]. Throughout the 1960s, 1970s, and 1980s, however, pertussis outbreaks in England and Wales were both periodic (inter-epidemic period of nearly 4 years) and highly synchronized[11]. These authors did not identify any wave-like spread of pertussis in England and Wales. Similarly, in their study of pertussis in the provinces of Thailand, Blackwood et al. found no evidence for any spatial organization of annual epidemics, which they attributed to the large amplitude of seasonal forcing[91].

1.10 Phylodynamics

The field of phylodynamics integrates epidemiology, evolutionary biology, and immunology to study the transmission of pathogenic microparasites[138]. Recent efforts to apply phylodynamic methods to the study of B. pertussis have uncovered evidence for a change in bacterial population diversity associated with the vaccine switch in the United States[43]. However, the analysis by Althouse and Scarpino[43] only focused on the United States and contained a fairly limited number of B. pertussis genome sequences. Additionally, emerging evidence from B. pertussis population genomic studies suggests that a primary source of variation might be physical rearrangements (e.g. inversions or duplications), as opposed to the type of genetic variability most commonly studied in phylodynamics (e.g. single nucleotide polymorphisms)[139]. It is important to extend such analyses to encompass a greater number of countries, to include a more representative geographic span and with respect to further evaluating the role of hidden (i.e. subclinical) infection and transmission in the ongoing whooping cough resurgence. There is also an urgent need to develop methodologies capable of taking into consideration genomic rearrangements in phylodynamic analyses. For instance, one could focus efforts on a set of strategic countries with respect to prevailing hypotheses around modern whooping cough dynamics, for example: (1) the United States, which switched vaccines and is experiencing a resurgence[13]; (2) Australia, which is experiencing a resurgence due to vaccine failure[12]; (3) Italy, which is not experiencing a resurgence[49]; and (4) Uganda, which still uses the wP vaccine[140].

1.11 Summary

Despite the early development, widespread deployment, and initial success of pertussis vaccines, whooping cough still plagues even highly vaccinated populations (Figure 1.1a). In some settings, the burden of pertussis can be straightforwardly explained by the relatively modest immunization coverage (Figure 1.1b). In a number of high income countries with high estimated vaccine coverage (e.g. the United States, United Kingdom, and Australia), the underlying causes for the absence of control remain debated. This is in part because of concerns regarding the fidelity of notification data (Table 1.1) and the fact that key aspects of protective immunity and vaccine efficacy are poorly understood and challenging to study[141]. There are aspects of pertussis epidemiology, however, that appear to conform to our expectations for an immunizing infection and vaccine. These include the association between incidence and susceptible recruitment in many parts of the globe (Figure 1.2), the relationship between population size and extinction risk (Figure 1.5), the predicted increase in the inter-epidemic period in response to wP vaccination (Figure 1.4), and the shifting age distribution of incidence in response to aP vaccination (Figure 1.6). At the same time, in this chapter we have reviewed many inconsistencies in the patterns of pertussis epidemiology, such as the geographic variations in its seasonality (Figure 1.7), contrasting patterns of spatial spread, and differential trends in incidence following the switch to aP vaccines (Figure 1.3). Ultimately, reconciling these differences will likely require us to develop novel integrative methods that bring all available sources of data to bear, including large-scale dynamical time-series data, age-stratified serology, and whole genome sequences.

References

1. Cravitz L, Williams J. A comparative study of the immune response to various pertussis antigens and the disease. *J Pediatr* 1946;**28**:172.
2. Preston N. Pertussis today. In: Wardlaw AC, Parton R (eds), *Pathogenesis and Immunity in Pertussis*. Chichester: John Wiley and Sons Ltd; 1988, pp 1–19.
3. Linnemann CC Jr. Host–parasite interactions in pertussis. In: Manclark CR, Hill JC (eds), *International Symposium on Pertussis*. Washington, DC: US Department of Health; 1978, pp 3–18.
4. Lambert HP. The carrier state: Bordetella pertussis. *J Antimicrob Chemother* 1986;**18** Suppl A;13–6.
5. Kristensen B. Occurrence of the Bordet-Gengou bacillus. *JAMA* 1933;**101**:204–6.
6. Linnemann CC Jr, Bass JW, Smith MDH. The carrier state in pertussis. *J Ecology* 1968;**88**:422–7.
7. Preston NW, Stanbridge TN. Efficacy of pertussis vaccines: a brighter horizon. *Br Med J* 1972;**3**:448–51.
8. Kendrick PL. Can whooping cough be eradicated? *J Infect Dis* 1975;**132**:707–12.
9. Madsen T. Vaccination against whooping cough. *JAMA* 1933;**101**:187–8.
10. Kendrick P, Eldering G. Progress report on pertussis immunization. *Am J Public Health* 1936;**26**:8–12.
11. Rohani P, Earn DJ, Grenfell BT. Opposite patterns of synchrony in sympatric disease metapopulations. *Science* 1999;**286**:968–71.
12. Jackson DW, Rohani P. Perplexities of pertussis: recent global epidemiological trends and their potential causes. *Epidemiol Infect* 2014;**142**:672–84.
13. Rohani P, Drake JM. The decline and resurgence of pertussis in the US. *Epidemics* 2011;**3**:183–8.
14. Gangarosa E, Galazka A, Wolfe C, et al. Impact of anti-vaccine movements on pertussis control: the untold story. *Lancet* 1998;**351**:356–61.
15. Baker JP. The pertussis vaccine controversy in Great Britain, 1974–1986. *Vaccine* 2003;**21**:4003–10.
16. Romanus V, Jonsell R, Bergquist SO. Pertussis in Sweden after the cessation of general immunization in 1979. *Pediatr Infect Dis J* 1987;**6**:364–71.
17. De Melker HE, Schellekens JF, Neppelenbroek SE, et al. Reemergence of pertussis in the highly vaccinated population of the Netherlands: observations on surveillance data. *Emerg Infect Dis* 2000;**6**:348–57.
18. Skowronski DM, De Serres G, MacDonald D, et al. The changing age and seasonal profile of pertussis in Canada. *J Infect Dis* 2002;**185**:1448–53.
19. Pebody RG, Gay NJ, Giammanco A, et al. The sero-epidemiology of Bordetella pertussis infection in Western Europe. *Epidemiol Infect* 1999;**133**:159–71.
20. Celentano LP, Massari M, Paramatti D, et al. Resurgence of pertussis in Europe. *Pediatr Infect Dis J* 2005;**24**:761–5.
21. Lin YC, Yao SM, Yan JJ, et al. Epidemiological shift in the prevalence of pertussis in Taiwan: implications for pertussis vaccination. *J Med Microbiol* 2007;**56**:533–7.
22. Yeung KHT, Duclos P, Nelson EAS, et al. An update of the global burden of pertussis in children younger than 5 years: a modelling study. *Lancet Infect Dis* 2017;**17**:974–80.

23. The World Bank. *World Development Indicators*. 2016. https://data.worldbank.org/products/wdi.

24. World Health Organization. *Immunization Surveillance, Assessment and Monitoring*. 2016. http://www.who.int/immunization/monitoring_surveillance/en/.

25. Grenfell BT, Anderson RM. Pertussis in England and Wales: an investigation of transmission dynamics and control by mass vaccination. *Proc R Soc Lond B Biol Sci* 1989;**236**:213–52.

26. Anderson RM, May RM. *Infectious Diseases of Humans*. Oxford: Oxford University Press; 1991.

27. Hethcote HW. The mathematics of infectious diseases. *SIAM Rev* 2000;**42**:599–653.

28. Keeling MJ, Rohani P. *Modelling Infectious Diseases in Humans and Animals*. Princeton, NJ: Princeton University Press; 2008.

29. Diekmann O, Heesterbeek JAP, Roberts MG. The construction of next-generation matrices for compartmental epidemic models. *J R Soc Interface* 2010;**7**:873–85.

30. Anderson RM, May RM. Directly transmitted infections diseases: control by vaccination. *Science* 1982; **215**:1053–60.

31. Fales WT. The age distribution of whooping cough, measles, chicken pox, scarlet fever and diphtheria in various areas in the United States. *Am J Epidemiol* 1928;**8**:759–99.

32. Gordon JE, Hood RI. Whooping cough and its epidemiological anomalies. *Am J Med Sci* 1951;**222**:333–61.

33. Edmunds WJ, Gay NJ, Kretzschmar M, Pebody RG, et al. The pre-vaccination epidemiology of measles, mumps and rubella in Europe: implications for modelling studies. *Epidemiol Infect* 2000;**125**:635–50.

34. Wearing HJ, Rohani P. Estimating the duration of pertussis immunity using epidemiological signatures. *PLOS Pathog* 2009;**5**:e1000647.

35. Lavine JS, King AA, Andreasen V, et al. Immune boosting explains regime-shifts in prevaccine-era pertussis dynamics. *PLoS One* 2013;**8**:e72086.

36. Rohani P, Zhong X, King AA. Contact network structure explains the changing epidemiology of pertussis. *Science* 2010;**330**:982–5.

37. Kretzschmar M, Teunis PFM, Pebody RG. Incidence and reproduction numbers of pertussis: estimates from serological and social contact data in five European countries. *PLOS Med* 2010;**7**:e1000291.

38. Broutin H, Viboud C, Grenfell BT, et al. Impact of vaccination and birth rate on the epidemiology of pertussis: a comparative study in 64 countries. *Proc Biol Sci* 2010;**277**:3239–45.

39. Amirthalingam G, Gupta S, Campbell H. Pertussis immunisation and control in England and Wales, 1957 to 2012: a historical review. *Euro Surveill* 2013;**18**:20587.

40. Warfel JM, Beren J, Merkel TJ. Airborne transmission of Bordetella pertussis. *J Infect Dis* 2012;**206**:902–6.

41. Ausiello CM, Cassone A. Acellular pertussis vaccines and pertussis resurgence: revise or replace? *MBio* 2014;**5**:e01339-14

42. Edwards KM. Unraveling the challenges of pertussis. *Proc Natl Acad Sci U S A* 2014;**111**:575–6.

43. Althouse BM, Scarpino SV. Asymptomatic transmission and the resurgence of Bordetella pertussis. *BMC Med* 2015;**13**:146.

44. Warfel JM, Zimmerman LI, Merkel TJ. Acellular pertussis vaccines protect against disease but fail to prevent infection and transmission in a nonhuman primate model. *Proc Natl Acad Sci U S A* 2014;**111**:787–92.

45. Mills KHG, Ross PJ, Allen AC, et al. Do we need a new vaccine to control the re-emergence of pertussis? *Trends Microbiol* 2014;**22**:49–52.

46. Smallridge WE, Rolin OY, Jacobs NT, et al. Different effects of whole-cell and acellular vaccines on Bordetella transmission. *J Infect Dis* 2014;**209**:1981–8.

47. Klein NP, Bartlett J, Rowhani-Rahbar A, et al. Waning protection after fifth dose of acellular pertussis vaccine in children. *N Engl J Med* 2012;**367**:1012–19.

48. McGirr A, Fisman DN. Duration of pertussis immunity after DTaP immunization: a meta-analysis. *Pediatrics* 2015;**135**:331–43.

49. Domenech de Cellès M, Magpantay FMG, King AA, et al. The pertussis enigma: reconciling epidemiology, immunology and evolution. *Proc R Soc B* 2016; **283**:20152309.

50. Zhang L, Prietsch S, Axelsson I, et al. *Cochrane Database Syst* 2012;**3**:CD001478.

51. Campbell H, Amirthalingam G, Andrews N, et al. Accelerating control of pertussis in England and Wales. *Emerg Infect Dis* 2012;**18**:38–47.

52. Libster R, Edwards KM. Re-emergence of pertussis: what are the solutions? *Expert Rev Vaccines* 2012; **11**:1346.

53. Plotkin SA. The pertussis problem. *Clin Infect Dis* 2014;**58**:830–3.

54. Muggeo V. Segmented: an R package to fit regression models with broken-line relationships. *Res News* 2008;**8**:20–5.

55. Wendelboe AM, Van Rie A, Salmaso S, et al. Duration of immunity against pertussis after natural infection or vaccination. *Pediatr Infect Dis J* 2005;**24**:S58–61.

56. van Boven M, de Melker HE, Schellekens JF, et al. A model based evaluation of the 1996–7 pertussis epidemic in the Netherlands. *Epidemiol Infect* 2001; **127**:73–85.

57. Magpantay FMG, Domenech de Cellès M, Rohani P, et al. Pertussis immunity and epidemiology: mode and duration of vaccine-induced immunity. *Parasitology* 2016;**143**:835–49.

58. Blackwood JC, Cummings DAT, Broutin H, et al. Deciphering the impacts of vaccination and immunity

on pertussis epidemiology in Thailand. *Proc Natl Acad Sci U S A* 2013;**110**:9595–600.

59. Águas R, Gonçalves G, Gomes MG. Pertussis: increasing disease as a consequence of reducing transmission. *Lancet Infect Dis* 2006;**6**:112–7.

60. Magpantay FMG, Riolo MA, de Cellès MD, et al. Epidemiological consequences of imperfect vaccines for immunizing infections. *SIAM J Appl Math* 2014;**74**:1810–30.

61. Domenech de Cellès M, Magpantay FMG, King AA, et al. The impact of past vaccination coverage and immunity on pertussis resurgence. *Sci Transl Med* 2018;**10**:eaaj1748.

62. Guiso N, Wirsing von König CH, Forsyth K, et al. The Global Pertussis Initiative: report from a round table meeting to discuss the epidemiology and detection of pertussis, Paris, France, 11–12 January 2010. *Vaccine* 2011;**29**:1115–21.

63. Cherry JD. The science and fiction of the "resurgence" of pertussis. *Pediatrics* 2003;**112**:405–6.

64. Cherry JD, Baraff LJ, Hewlett E. The past, present, and future of pertussis. The role of adults in epidemiology and future control. *West J Med* 1989;**150**:319–28.

65. Sutter RW, Cochi SL. Pertussis hospitalizations and mortality in the United States, 1985–1988. Evaluation of the completeness of national reporting. *JAMA* 1992;**267**:386–91.

66. Cherry JD, Grimprel E, Guiso N, et al. Defining pertussis epidemiology: clinical, microbiologic and serologic perspectives. *Pediatr Infect Dis J* 2005;**24**:S25–34.

67. Wang K, Birring SS, Taylor K, et al. Montelukast for postinfectious cough in adults: a double-blind randomised placebo-controlled trial. *Lancet Respir* 2013;**2**:35–43.

68. Cherry JD. The epidemiology of pertussis: a comparison of the epidemiology of the disease pertussis with the epidemiology of Bordetella pertussis infection. *Pediatrics* 2005;**115**:1422–7.

69. Philipson K, Goodyear-Smith F, Grant CC, et al. When is acute persistent cough in school-age children and adults whooping cough? *Br J Gen Pract* 2013;**63**:573–9.

70. Kayina V, Kyobe S, Katabazi FA, et al. Pertussis prevalence and its determinants among children with persistent cough in urban Uganda. *PLoS One* 2015;**10**:e0123240.

71. Baron S, Njamkepo E, Grimprel E, et al. Epidemiology of pertussis in French hospitals in 1993 and 1994: thirty years after a routine use of vaccination. *Pediatr Infect Dis J* 1998;**17**:412–18.

72. Crowcroft NS, Booy R, Harrison T, et al. Severe and unrecognised: pertussis in UK infants. *Arch Dis Child* 2003;**88**:802–6.

73. Bisgard KM, Pascual FB, Ehresmann KR, et al. Infant pertussis: who was the source? *Pediatr Infect Dis J* 2004;**23**:985–9.

74. Kowalzik F, Barbosa AP, Fernandes VR, et al. Prospective multinational study of pertussis infection in hospitalized infants and their household contacts. *Pediatr Infect Dis J* 2007;**26**:238–42.

75. Wendelboe AM, Njamkepo E, Bourillon A, et al. Transmission of Bordetella pertussis to young infants. *Pediatr Infect Dis J* 2007;**26**:293–9.

76. Skoff TH, Kenyon C, Cocoros N, et al. Sources of infant pertussis infection in the United States. *Pediatrics* 2015;**136**:635–41.

77. Staudt A, Mangla AT, Alamgir H. Investigation of pertussis cases in a Texas county, 2008–2012. *South Med J* 2015;**108**:452–7.

78. Berezin EN, de Moraes JC, Leite D, et al. Sources of pertussis infection in young babies from São Paulo State, Brazil. *Pediatr Infect Dis J* 2014;**33**:1289–91.

79. Moore HC, Lehmann D, de Klerk N, et al. How accurate are International Classification of Diseases-10 diagnosis codes in detecting influenza and pertussis hospitalizations in children? *J Pediatr Infect Dis Soc* 2014;**3**:255–60.

80. Fine PE, Clarkson JA. The recurrence of whooping cough: possible implications for assessment of vaccine efficacy. *Lancet* 1982;**1**:666–9.

81. Magpantay FMG, Rohani P. Dynamics of pertussis transmission in the United States. *Am J Epidemiol* 2015;**181**:921–31.

82. Hamer W. The Milroy Lectures on Epidemic Disease in England—the evidence of variability and persistence of type. *Lancet* 1906;**ii**:733–9.

83. Earn DJ, Rohani P, Bolker B, et al. A simple model for complex dynamical transitions in epidemics. *Science* 2000;**287**:667–70.

84. Rohani P, Keeling MJ, Grenfell BT. The interplay between determinism and stochasticity in childhood diseases. *Am Nat* 2002;**159**:469–81.

85. Lavine JS, Rohani P. Resolving pertussis immunity and vaccine effectiveness using incidence time series. *Expert Rev Vaccines* 2012;**11**:1319–29.

86. Cherry JD. Historical review of pertussis and the classical vaccine. *J Infect Dis* 1996;**174**:S259–63.

87. Pebody RG, Gay NJ, Giammanco A, et al. The sero-epidemiology of Bordetella pertussis infection in Western Europe. *Epidemiol Infect* 2005;**133**:159–71.

88. de Greeff SC, Dekkers AL, Teunis P, et al. Seasonal patterns in time series of pertussis. *Epidemiol Infect* 2009;**137**:1388–95.

89. Bauch CT, Earn DJD. Transients and attractors in epidemics. *Proc Biol Sci* 2003;**270**:1573–8.

90. Nguyen HTH, Rohani P. Noise, nonlinearity and seasonality: the epidemics of whooping cough revisited. *J R Soc Interface* 2008;**5**:403–13.

91. Blackwood JC, Cummings DAT, Broutin H, et al. The population ecology of infectious diseases: pertussis in Thailand as a case study. *Parasitology* 2012;**139**:1888–98.

92. Korobeinikov A, Maini PK, Walker WJ. Estimation of effective vaccination rate: pertussis in New Zealand as a case study. *J Theor Biol* 2003;**224**:269–75.

93. Gay NJ, Miller E. Pertussis transmission in England and Wales. *Lancet* 2000;**355**:1553–4.

94. Bartlett M. Measles periodicity and community size. *J R Stat Soc S A Gen* 1957;**120**:48–70.

95. Conlan AJK, Rohani P, Lloyd AL, et al. Resolving the impact of waiting time distributions on the persistence of measles. *J R Soc Interface* 2010;**7**:623–40.

96. Broutin H, Simondon F, Guégan JF. Whooping cough metapopulation dynamics in tropical conditions: disease persistence and impact of vaccination. *Proc Biol Sci* 2004;**271**:S302–5.

97. Lavine JS, King AA, Bjørnstad ON. Natural immune boosting in pertussis dynamics and the potential for long-term vaccine failure. *Proc Natl Acad Sci U S A* 2011;**108**:7259–64.

98. Geard N, Glass K, McCaw JM, et al. The effects of demographic change on disease transmission and vaccine impact in a household structured population. *Epidemics* 2015;**13**:56–64.

99. Wallinga J, Edmunds WJ, Kretzschmar M. Perspective: human contact patterns and the spread of airborne infectious diseases. *Trends Microbiol* 1999; **7**:372–7.

100. Edmunds WJ, O'Callaghan CJ, Nokes DJ. Who mixes with whom? A method to determine the contact patterns of adults that may lead to the spread of airborne infections. *Proc Biol Sci* 1997;**264**:949–57.

101. Mossong J, Hens N, Jit M, et al. Social contacts and mixing patterns relevant to the spread of infectious diseases. *PLOS Med* 2008;**5**:e74.

102. Horby P, Pham QT, Hens N, et al. Social contact patterns in Vietnam and implications for the control of infectious diseases. *PLoS One* 2011;**6**:e16965.

103. Fumanelli L, Ajelli M, Manfredi P, et al. Inferring the structure of social contacts from demographic data in the analysis of infectious diseases spread. *PLoS Comput Biol* 2012;**8**:e1002673.

104. Read JM, Lessler J, Riley S, et al. Social mixing patterns in rural and urban areas of southern China. *Proc Biol Sci* 2014;**281**:20140268.

105. Grijalva CG, Goeyvaerts N, Verastegui H, et al. A household-based study of contact networks relevant for the spread of infectious diseases in the highlands of Peru. *PLoS One* 2015;**10**:e0118457.

106. Nishita M, Park SY, Nishio T et al. Ror2 signaling regulates Golgi structure and transport through IFT20 for tumor invasiveness. *Sci Rep* 2017;**7**:1–12.

107. Clark TA. Changing pertussis epidemiology: everything old is new again. *J Infect Dis* 2014;**209**:978–81.

108. Choi YH, Campbell H, Amirthalingam G, et al. Investigating the pertussis resurgence in England and Wales, and options for future control. *BMC Med* 2016;**14**:121.

109. Bento AI, Riolo MA, Choi YH, et al. Core pertussis transmission groups in England and Wales: a tale of two eras. *Vaccine* 2018;**36**:1160–6.

110. Blackwood JC, Cummings DAT, Iamsirithaworn S, et al. Using age-stratified incidence data to examine the transmission consequences of pertussis vaccination. *Epidemics* 2016;**16**:1–7.

111. Güris, D, Strebel PM, Bardenheier B, et al. Changing epidemiology of pertussis in the United States: increasing reported incidence among adolescents and adults, 1990–1996. *Clin Infect Dis* 1999;**28**:1230–7.

112. van Boven M, De Melker HE, Schellekens JF, et al. Waning immunity and sub-clinical infection in an epidemic model: implications for pertussis in the Netherlands. *Math Biosci* 2000;**164**:161–82.

113. Hozbor D, Mooi F, Flores D, et al. Pertussis epidemiology in Argentina: trends over 2004–2007. *J Infect* 2009;**59**:225–31.

114. Mooi FR. Bordetella pertussis and vaccination: the persistence of a genetically monomorphic pathogen. *Infect Genet Evol* 2010;**10**:36–49.

115. Carlsson RM, Trollfors B. Control of pertussis – lessons learnt from a 10–year surveillance programme in Sweden. *Vaccine* 2009;**27**:5709–18.

116. Gambhir M, Clark TA, Cauchemez S, et al. A change in vaccine efficacy and duration of protection explains recent rises in pertussis incidence in the United States. *PLoS Comput Biol* 2015;**11**:e1004138.

117. Biggerstaff M, Alper D, Dredze M, et al. Results from the Centers for Disease Control and Prevention predict the 2013–2014 influenza season challenge. *BMC Infect Dis* 2016;**16**:357.

118. Viboud C, Sun K, Gaffey R, et al. The RAPIDD Ebola forecasting challenge: synthesis and lessons learnt. *Epidemics* 2018;**22**:13–21.

119. Pitzer VE, Atkins KE, de Blasio BF, et al. Direct and indirect effects of rotavirus vaccination: comparing predictions from transmission dynamic models. *PLoS One* 2012;**7**:e42320.

120. Bento AI, Rohani P. Forecasting epidemiological consequences of maternal immunization. *Clin Infect Dis* 2016;**63**:S205–12.

121. Metcalf CJ, Bjørnstad ON, Grenfell BT, et al. Seasonality and comparative dynamics of six childhood infections in pre-vaccination Copenhagen. *Proc Biol Sci* 2009;**276**:4111–8.

122. Kilgore PE, Salim AM, Zervos MJ, et al. Pertussis: microbiology, disease, treatment, and prevention. *Clin Microbiol Rev* 2016;**29**:449–86.

123. Fine PE, Clarkson JA. Seasonal influences on pertussis. *Int J Epidemiol* 1986;**15**:237–47.

124. Cappello V, Marchetti L, Parlanti P, et al. Ultrastructural characterization of the lower motor system in a mouse model of Krabbe disease. *Sci Rep* 2016;**6**:1–8.

125. Kaczmarek MC, Ware RS, Nimmo GR, et al. Pertussis seasonality evident in polymerase chain reaction and serological testing data, Queensland, Australia. *J Pediatr Infect Dis Soc* 2016;**5**:214–7.

126. Tanaka M, Vitek CR, Pascual FB, et al. Trends in pertussis among infants in the United States, 1980–1999. *JAMA* 2003;**290**:2968–75.

127. Bhatti MM. Eight-year review of Bordetella pertussis testing reveals seasonal pattern in the United States. *J Pediatr Infect Dis Soc* 2017;**6**:91–3.

128. Préziosi MP, Yam A, Wassilak SG, et al. Epidemiology of pertussis in a West African community before and after introduction of a widespread vaccination program. *Am J Epidemiol* 2002;**155**:891–6.

129. Muller A, Leeuwenburg J, Voorhoeve A. *Bull World Health Organ* 1984;**62**:899–908.

130. Bharti N, Tatem AJ, Ferrari MJ, et al. Explaining seasonal fluctuations of measles in Niger using nighttime lights imagery. *Science* 2011;**334**:1424–7.

131. Penman BS, Gupta S, Shanks GD. Rapid mortality transition of Pacific Islands in the 19th century. *Epidemiol Infect* 2017;**145**:1–10.

132. Pitzer VE, Viboud C, Simonsen L, et al. Demographic variability, vaccination, and the spatiotemporal dynamics of rotavirus epidemics. *Science* 2009;**325**: 290–4.

133. Martinez-Bakker M, Bakker KM, King AA, et al. Human birth seasonality: latitudinal gradient and interplay with childhood disease dynamics. *Proc Biol Sci* 2014;**281**:20132438.

134. Pons-Salort M, Oberste MS, Pallansch MA, et al. The seasonality of nonpolio enteroviruses in the United States: patterns and drivers. *Proc Natl Acad Sci U S A* 2018;**115**:3078–83.

135. Viboud C, Bjornstad ON, Smith DL, et al. Synchrony, waves, and spatial hierarchies in the spread of influenza. *Science* 2006;**312**:447–51.

136. Choisy M, Rohani P. Changing spatial epidemiology of pertussis in continental USA. *Proc R Soc Lond B* 2012;**279**:4574–81.

137. Omer SB, Salmon DA, Orenstein WA, et al. Vaccine refusal, mandatory immunization, and the risks of vaccine-preventable diseases. *N Engl J Med* 2009;**360**:1981–8.

138. Grenfell BT. Unifying the epidemiological and evolutionary dynamics of pathogens. *Science* 2004;**303**:327–32.

139. Weigand MR, Peng Y, Loparev V, et al. The history of Bordetella pertussis genome evolution includes structural rearrangement. *J Bacteriol* 2017;**199**:e00806.

140. Kayina V, Kyobe S, Katabazi FA, et al. Pertussis prevalence and its determinants among children with persistent cough in urban Uganda. *PLoS One* 2015; **10**:e0123240.

141. Hewlett EL, Halperin SA. Serological correlates of immunity to Bordetella pertussis. *Vaccine* 1998; **16**:1899–900.

CHAPTER 2

Basics of pertussis pathogenesis

Amanda L. Skarlupka, Bodo Linz, Jennifer Maynard, and Eric T. Harvill

Abstract

The use of animal models and *in vitro* assays has allowed the identification and functional characterization of an iconic set of *Bordetella pertussis* factors that contribute to its pathogenesis. Much research on *B. pertussis* has been focused on the effects of pertussis toxin and adenylate cyclase toxin on pathogenesis and disease progression, and on the function of adherence factors. However, much larger sets of factors have been identified and proposed to be involved in host–pathogen interactions, including *B. pertussis* manipulating the host's metabolism to its advantage. The identification of a third player, the resident microbiota, reveals a complex picture that has yet to be investigated. The ongoing studies of the molecular components produced by *B. pertussis* and their effect on the host during infection, commonly referred to as 'pathogenesis' research, are crucial to generate understanding of the host–pathogen interactions necessary to design new approaches to prevent and cure disease.

2.1 Introduction

The highly contagious aspect of pertussis that allows it to transmit so effectively is a result of the pathogenesis that the causative agent, *Bordetella pertussis*, inflicts on the host respiratory tract. Symptoms of pertussis typically present in three phases: catarrhal (cold-like symptoms with mild cough), paroxysmal (severe, prolonged coughing spells, followed by whooping and post-tussive vomiting), and convalescence. Due to its initial, non-specific, cold-like symptoms, paediatricians are often unable to symptomatically distinguish and diagnose early cases of pertussis. Pertussis is usually suspected only after it progresses into the paroxysmal stage by which time, *B. pertussis* is no longer reliably detected through culture methods; antibiotic therapy during and after the paroxysmal phase is ineffective since the disease is no longer driven by bacterial infection. The disease symptoms persist and continue to worsen, presumably due to residual toxin and damage of host tissues. The cold-like symptoms and coughing induced by *B. pertussis* during the catarrhal phase and the initial 3 weeks of the paroxysmal phase increase its transmissibility to the next host by producing respiratory secretions and air-borne droplets.

2.2 Animal-based experimental studies of pathogenesis

A variety of non-human animals have been used to study *B. pertussis*–host interactions. This section covers the strengths and limitations of non-human animal models as a means to study pathogenesis.

Skarlupka, A. L., Linz, B., Maynard, J., and Harvill, E. T., *Basics of pertussis pathogenesis*. In: *Pertussis: epidemiology, immunology, and evolution*. Edited by Pejman Rohani and Samuel V. Scarpino: Oxford University Press (2019). © Oxford University Press. DOI: 10.1093/oso/9780198811879.003.0002

2.2.1 Baboon model

Multiple aspects of B. pertussis pathogenesis can be studied within the baboon animal model. From a host pathogenesis perspective, pertussis in baboons mimics the human disease in several ways that include the immune response, mucus production, and lymphocytosis. Baboons are the only non-human animal model to induce the characteristic symptomatic paroxysmal cough. Therefore, from a host-to-host transmission perspective, due to the culmination of these and other symptoms, the transmission of B. pertussis can be studied in both co-housed and spatially separated baboons. The baboon animal model is currently the ideal experimental system to study and predict B. pertussis vaccine and therapeutic performances in humans[1].

However, the baboon model is expensive and only accessible to a few groups who conduct a small number of carefully justified experiments which focus on the highest priority studies testing improved vaccines rather than studies on basic pathogenesis. Although the ideal model for some aspects of pathogenesis, the colonization and initial infection processes are difficult to evaluate in baboons due to the unnaturally high inoculation dose of nearly a billion bacteria necessary for inducing symptoms. The human infectious dose of B. pertussis is currently unknown, but based on its highly contagious nature, it is thought that a much lower dose is required for establishing an infection. These various limitations of the baboon model make it poorly suited for the exploration of molecular mechanisms involved in the pathogenesis of B. pertussis.

2.2.2 Mouse model

The mouse experimental infection system has been the cornerstone for B. pertussis pathogenesis discoveries. In the pre-vaccine era, mice were used to measure the efficacy of prototype vaccines against intracerebral challenges of B. pertussis. More recently, mice are challenged via aerosol or liquid delivery to the respiratory tract with a minimum delivery of 10,000 colony-forming units (CFU) of B. pertussis[2]. Those delivered to the lower respiratory tract colonize and grow efficiently for about a week, after which the bacteria are gradually contained and eventually eliminated by the mouse immune system by 50 days post inoculation[3]. The early aspects of detection and inflammatory response, as well as the adaptive immune response that eventually controls infection, appear to be substantially similar in humans and mice. In both systems, the generated antibodies recognize a specific set of prominent antigens, and a similar distribution of T-cell responses are detected. The functions of the human and mouse immune systems contribute to B. pertussis pathogenesis in a seemingly similar fashion.

The mouse model is currently the most common and widely accepted experimental system for studying the mechanistic details of B. pertussis pathogenesis. The accessibility and the relatively low cost of purchasing and housing laboratory mice allows more research groups to maintain sufficient quantities for robust experimentation. Unique to the mouse model, inbred mouse lines allow for genetic homogeneity, greatly increasing reproducibility over time and between laboratories. Additionally, the mammalian immune response has been extensively experimentally manipulated and studied, producing mice with a wide variety of altered immune components demonstrating the roles of each in control/clearance of bacterial infection and disease. Using available genetic and immunological tools for manipulating both B. pertussis and the mouse provides important detailed information about the complex interactions involved.

Unfortunately, the mouse model is not ideal for studying all aspects of pertussis. Most notably, not all the symptoms are induced in mice, including mucus production and the iconic paroxysmal cough of pertussis. Similar to the baboon model, respiratory system colonization has generally involved a high bacterial load, which is probably not how humans encounter the naturally highly infectious B. pertussis. The combination of the lack of characteristic coughing symptoms and the low infectivity prevents efficient B. pertussis transmission between mice, making it difficult to study this critical aspect of its biology. However, the great versatility of the mouse model system justifies its extensive use in the study of pathogenesis, and most of what is known was initially discovered in mice.

2.2.3 Other *Bordetella* species in the mouse model

Other *Bordetella* species naturally infect mice and can overcome some of the weaknesses presented by the study of *B. pertussis*. Both *Bordetella bronchiseptica* and *Bordetella pseudohinzii* have been isolated from mice. These species are highly infectious, mimicking a realistic *B. pertussis* infection, with less than 5 CFU being sufficient to colonize and indefinitely persist within the upper respiratory tracts of mice.

While these species are not *B. pertussis*, they are very closely related. *Bordetella pertussis* and *B. bronchiseptica* have such a high nucleotide sequence homology that they may be considered subspecies. Due to *B. pertussis* emerging from a *B. bronchiseptica*-like progenitor and being specialized to the human host via genome reduction, *B. bronchiseptica* possesses many of the same genes and regulators necessary for pathogenesis[4]. The genes shared between *B. pertussis* and *B. bronchiseptica* are so similar that it is unlikely that they function differently. Studies comparing the immunological response to *B. bronchiseptica* and *B. pertussis* in human versus mouse pathogenesis found that *B. bronchiseptica* exhibited a more similar pathogenesis to *B. pertussis* in humans than in mice. Similar to when *B. pertussis* infects humans, *B. bronchiseptica* colonizes and persists in the upper respiratory tract and induces immunological responses resembling those of *B. pertussis* in mice[5]. Since *B. bronchiseptica* expresses many of the same factors as *B. pertussis*, the roles of many of those factors in pathogenesis have been best studied in the context of this natural and efficient mouse infection.

2.3 Virulence factors and pathogenesis

Bordetella pertussis induces pathogenesis using different toxins and virulence effectors. Through the use of diverse bacterial secretion systems, these bacterial effectors are anchored to the outer membrane or associated with either the extracellular environment or with the bacterial surface. This collection of virulence factors each contribute to pathogenesis in different ways as outlined in this section.

2.3.1 Toxins

Pertussis toxin

Pertussis toxin (PTx) plays a central role in pertussis disease pathology, and the detoxified toxin is included in all acellular vaccines. The *Bordetella* virulence gene (Bvg)⁺-upregulated 105 kDa holotoxin protein is classified as an AB_5 class toxin comprising one catalytically active A subunit and five receptor-binding B subunits. The AB_5 toxins are common virulence factors in pathogenic bacteria and include other highly human-relevant toxins, such as cholera toxin produced by *Vibrio cholerae*, Shiga toxin produced by *Shigella dysenteriae* and Shiga toxigenic *Escherichia coli*, heat-labile enterotoxin produced by enterotoxigenic *E. coli*, and *Pseudomonas aeruginosa* exotoxin A[6].

PTx exerts a complex range of effects in cell culture and animals. Most notably, when exposed to PTx, CHO-K1 cells exhibit a clustering morphology[7]. This phenotypic change provides a convenient *in vitro* assay for active holotoxin detection. In mice, PTx has immunosuppressive properties by inhibiting neutrophil recruitment[8]. Furthermore, PTx alters mononuclear phagocyte circulation, which plays a role in the evasion of bacterial opsonization by monocytes[9]. Early in infection, *B. pertussis* uses PTx to facilitate colonization. At later stages, PTx targets resident cells in the lung tissue such as alveolar macrophages and epithelial cells[10]. There, PTx reduces proinflammatory chemokines and cytokines and inhibits resident cell chemokine production necessary for the recruitment and activation of neutrophils and macrophages[11,12]. Over time, PTx causes lymphocytosis, an identifiable symptom of pertussis. In addition to promoting bacterial colonization, PTx increases *B. pertussis* resistance to antibody-mediated clearance through suppression of antibody responses[13].

The toxin is encoded by a five-gene operon, *ptxABDEC*, which is located adjacent to the *ptl* locus comprising nine genes (*ptlABCDEFGHI*). The *ptl* locus encodes a type IV secretion system that is utilized for the transport of the inactive holotoxin from the periplasmic space to the extracellular environment. Upon contact with the host cell, the B subunit of PTx binds to glycoproteins or sialo-glycoproteins on the cell surface[14], which subsequently triggers host cell receptor-mediated endocytosis and entry into the host cell. The holotoxin is then delivered from the Golgi apparatus to the endoplasmic reticulum, and the PTx A subunit (S1) is released into the cytoplasm after reduction of a disulphide bond and an adenosine triphosphate-dependent conformational change in the B subunit[15]. The S1 subunit, which has adenosine diphosphate (ADP)-ribosyltransferase activity, catalyses ADP-ribosylation of membrane-associated $G_{i/o}$ proteins in the cytoplasm, which results in disruption of G-protein signalling by arresting the $G_{i/o}$ proteins in their inactive form[16,17]. Since $G_{i/o}$ proteins are important inhibitors of adenylate cyclase, the key enzyme for synthesis of cyclic adenosine monophosphate (cAMP), functional inactivation of $G_{i/o}$ proteins leads to increased cellular levels of cAMP[18,19], an essential mediator in cell signalling, which consequently has systemic effects, including interference with neutrophil and macrophage activities[8,9,20,21].

In addition to direct modulation of the host's immune response, PTx induces metabolic disturbances that affect host–pathogen interactions in favour of the invading bacteria[22]. PTx was originally described as islet-activating protein (IAP) because of its stimulating effects on pancreatic beta cells that resulted in secretion of insulin[23-25]. The toxin's stimulation of insulin release is associated with an increase in cAMP levels. High intracellular cAMP levels trigger activation of the cAMP-dependent protein kinase A, which in turn greatly stimulates insulin secretion[26]. Unbalanced insulin secretion leads to hypoglycaemia (i.e. low blood sugar), which deprives the infected host of circulating blood glucose. In contrast to immune cells, which depend on a constant supply of energy in the form of glucose, particularly during the rapid activation and expansion during infection, B. pertussis is largely unaffected by hypoglycaemia, because bacteria of the genus Bordetella are unable to utilize glucose as a carbon and energy source[27].

Adenylate cyclase toxin

The adenylate cyclase/haemolysin toxin (ACT) is actively secreted during Bordetella spp. pathogenesis in the Bvg+ phase. ACT is a 177kDa monomeric protein, CyaA, composed of five sequential regions: (1) the catalytically active N-terminal adenylate cyclase (CAT) domain, (2) the central hydrophobic (HP) domain, (3) the modification region carrying acylation sites at K860 and K983, (4) the C-terminal repeat-in-toxin (RTX) pore-forming haemolysin domain, and (5) the C-terminal secretion signal (Sec). ACT is the substrate of a type I secretion system encoded by cyaBDE, and is secreted from the cytoplasm to the extracellular environment through a channel created by the CyaBDE proteins[28]. After secretion, CyaC modifies ACT through fatty acylation on the lysin residues on the RTX haemolysin domain[29]. After modification, the majority of the protein remains associated with the B. pertussis outer membrane through interactions with filamentous haemagglutinin. Only low levels of calcium-activated ACT (~15 ng/mL) are present on the respiratory mucosa[30], where the ACT toxin can exhibit toxigenic effects on local phagocytes.

The different domains of the ACT toxin are all necessary for its complete contribution to pathogenesis. The calcium-binding RTX motif includes five tandem Gly–Asp-rich repeats which reversibly form beta rolls in the presence of calcium. The RTX domain contributes to pathogenic cytotoxic effects by mediating cell receptor binding, protein translocation, and plasma membrane permeabilization[31,32]. The calcium-activated toxin can then bind to host cell surface receptors CD11b and CD18[32]. Upon interaction of the RTX domain with the host cell surface receptor $αMβ2$ integrin, the HP domain penetrates the membrane. This allows the CAT domain to translocate into the host cell cytosol. There, calmodulin activates the CAT domain leading to the inhibition of host cell antibacterial activities through the rapid generation of supraphysiological cAMP[28]. An ACT-mediated increase in cAMP levels in macrophages results in inactivation of RhoA, a GTP-binding protein playing a central role in the organization of the actin cytoskeleton and in microtubule dynamics, which leads to membrane ruffling and in blocking of complement-dependent phagocytosis[33]. In addition, elevated cAMP levels inhibit chemotaxis and induce lysis of phagocytic cells, including neutrophils[34] and macrophages[35,36], both in vitro and in vivo. ACT also reduces the innate immune response by inhibiting interleukin (IL)-12 production and by inducing expression of the anti-inflammatory cytokine IL-10[37].

Similar to PTx, ACT likely also indirectly impairs the immune response by manipulating metabolic functions. Adenylate cyclase, which rapidly increases intracellular cAMP levels to toxic levels upon activation of the catalytic function inside neutrophils and macrophages, possibly also plays a role in inducing insulin release by pancreatic cells. Once the toxin enters the pancreatic beta cell, it can exhibit its strong adenylate cyclase activity, bypassing and overwhelming the endogenous regulation of the host adenylate cyclase. Similar to the action of PTx, increased cAMP levels then activate cAMP-dependent protein kinase A, a strong mediator of insulin secretion, which subsequently results in significantly elevated release of insulin and thus hypoglycemia[25]. Since *B. pertussis* is unable to utilize glucose, the bacteria are unaffected by hypoglycaemia, but disruption of the glucose homeostasis by PTx and possibly ACT appears to contribute to the inhibition of the host's immune system, thereby extending and enhancing the period of infection.

Direct manipulation and enhancement of insulin production by the bacterial toxins requires their delivery to the pancreatic beta cells. However, PTx and ACT have not been documented to disseminate widely from the infection site in the respiratory tract. Thus, *B. pertussis* needs to spread to the pancreatic tissue to induce the observed drop in blood sugar levels by enhanced insulin production. Dissemination of *B. pertussis* to systemic organs requires a 'vessel' that both transports and protects the microbe in the bloodstream and/or through the lymphatic system. Upon host infection, *B. pertussis* has been shown to survive inside macrophages, which enables the bacteria to evade the host immunity[38,39]. Up to a quarter of the initially phagocytized bacteria can resist digestion by the macrophages and subsequently begin to multiply inside the eukaryotic cell[38,39]. Inside the macrophages, the bacteria are protected from immune surveillance and can use the phagocytic cells as a 'Trojan Horse' for travelling inside the bloodstream to disseminate to systemic organs, including the pancreas.

Dermonecrotic toxin

Dermonecrotic toxin (DNT) was the first toxin identified in the *Bordetella* genus. The heat-labile toxin is found throughout the classical Bordetellae species with 99 per cent amino acid (AA) homology[40].

This AB toxin is a 160 kDa cytoplasmic polypeptide containing a 54 AA N-terminal receptor binding domain and a 300 AA C-terminal enzymatic domain[41,42].

The specific receptor on the mammalian cells necessary for the cell binding of the 54 AA N-terminal domain is currently unidentified[43]. However, the cell binding domain of the protein has been identified to be between AA 2 and AA 30 at the N-terminal[43]. After recognition and binding to the target cell, the C-terminal end is nicked on the C-terminal side of Arg44 by mammalian furins, or furin-like proteases[43,44]. The target cell internalizes the C-terminal fragment through dynamin-dependent endocytosis[43].

Once inside the cell, the C-terminal fragment modifies Rho-family GTPases by addition of polyamines[45], which decreases their GTPase activity and has downstream effects on its effector, the Rho-associated protein kinase. Polyamination of the Rho-GTPase induces rearrangements in the actin cytoskeleton, including the formation of actin stress fibres[45]. While the exact mechanisms and contributions of DNT to pathogenesis are unknown, intradermal injection into mice and other laboratory animals causes localized necrotic lesions, and small-dose intravenous injections are lethal in mice[46,47]. *Bordetella bronchiseptica* strains with attenuated DNT showed decreased turbinate atrophy in infected pigs[48,49]. The addition of DNT isolated from *B. bronchiseptica* to osteoblastic clone MC 3T3 cells induced morphological changes, stimulated DNA replication, and impaired differentiation and proliferation[50,51]. These effects are due to the Rho GTPase activation[52]. This activation leads to tyrosine phosphorylation of focal adhesion kinase and paxillin[53], which are involved in embryonic development and cell locomotion[54]. Activation of these proteins causes cytoskeleton alterations and focal adhesion assembly[55]. By contrast, *B. pertussis* mutants lacking functional DNT show no reduction in virulence in mice[56].

Tracheal cytotoxin

Tracheal cytotoxin (TCT) is a disaccharide-tetrapeptide monomer created by Gram-negative bacteria during breakdown of the peptidoglycan cell wall during growth[57]. This monomer is generally recycled by the integral cytoplasmic membrane protein AmpG[58,59], but the failure to reclaim this monomer causes

B. pertussis to release TCT into its environment. In human nasal epithelial biopsies, TCT causes loss of ciliated cells, cell blebbing, and mitochondrial damage[60]. TCT has been shown *in vitro* to destroy ciliated cells potentially by increasing nitric oxide levels which ciliated cells are more susceptible to[61]. The cytotoxin also causes mitochondrial bloating, disruption of tight junctions, and extrusion of ciliated cells with little or no damage to non-ciliated cells[61,62].

2.3.2 Adhesins

Filamentous haemagglutinin

Filamentous haemagglutinin (FHA) is an important *B. pertussis* adherence factor, which is efficiently secreted and bound to the bacterial surface[63]. Since it is a major pertussis antigen, FHA is included in almost all currently licensed acellular pertussis vaccines. The mature FHA protein is a 220 kDa protein modified from an even longer 367 kDa precursor protein, FhaB[64]. In *B. pertussis*, the protein passes from the cytoplasm to the periplasm via a SecYEG-dependent bacterial secretion mechanism requiring both SecA and SecB which results in cleavage of the N-terminal[65,66]. FHA passes through the FhaC β-barrel channel located in the outer membrane, to the extracellular environment. There FHA acquires its final tertiary structure and binds to the outer membrane[64,66]. FHA contains an Arg-Gly-Asp (RGD) domain that is characterized as a eukaryotic cell-recognition domain interacting with fibronectin and other eukaryotic extracellular proteins[67].

Since FHA mediates *B. pertussis* adherence to host cells, it is considered an important colonization factor[68]. Besides its function during initial colonization, FHA also appears to be important for persistence in the lower respiratory tract[69]. FHA has been shown to bind to adenylate cyclase, to inhibit biofilm formation *in vitro*[70], and to exhibit strong immunomodulatory effects. FHA suppresses inflammation in the respiratory tract by inducing secretion of the immunosuppressive IL-10[71] and other cytokines. It further shows immune-modulatory functions due to inhibiting CD4+ T-cell proliferation[72] and induction of apoptosis[73].

Fimbriae

The fimbriae (Fim) proteins are filamentous, polymeric proteins located on the cell surface used by the Bordetellae to facilitate attachment to host epithelial cells. Two major fimbrial subunits, Fim2 and Fim3, encoded by the two loci *fim2* and *fim3* respectively[74,75], are included in acellular vaccines. They are expressed in the Bvg+ phase and are critical for bacterial pathogenesis and adhesion[76]. During infection, a combination of one or both serotypes can be expressed[77]. The exact contributions of each Fim subunit and their mechanisms have yet to be elucidated. However, *in vitro* studies have shown that purified Fim binds to sugars found ubiquitously in the mammalian respiratory tract. FimD, the tip adhesion and a part of the fimbrial biogenesis operon *fimBCD*, facilitates binding of *B. pertussis* to monocytes, activates CR3, and enhances binding to FHA[78,79]. Mutations in Fim2, Fim3, and FimD significantly reduced bacterial adherence to human bronchial epithelial cells, and addition of purified fimbrial subunits inhibited bacterial adherence by competing for the receptors on the eukaryotic cell surface[80]. *Bordetella pertussis* strains lacking wild-type Fim are unable to multiply in the mouse nasopharynx and trachea[81], and studies conducted with *B. bronchiseptica* strains lacking Fim indicate that fimbriae contribute to colonization and persistence in rat and mouse trachea and produce different serum antibody profiles[82]. In addition, fimbriae play an immunomodulatory role inducing a Th2-mediated host immune response and inhibit killing of *Bordetella* by mouse lung macrophages[83].

2.3.3 Autotransporters

Autotransporters are multidomain proteins composed of a C-terminal translocator domain and an N-terminal passenger domain. They utilize the type V protein secretion pathway, in which the translocator (or autotransporter) domain forms a pore in the outer membrane through which the passenger (or outer membrane) domain(s) of the protein is transported. Unlike non-classical *Bordetella* species, each of which possesses its own unique set of autotransporter genes[4], the classical Bordetellae share most of the 21 genes identified in *B. bronchiseptica*, with

20 genes in *B. parapertussis* and 16 in *B. pertussis* genomes[58]. Several of the autotransporters in *B. pertussis* appear to be involved in adherence, in resistance against complement-mediated bacterial killing, and in persistence in the respiratory tract.

Pertactin

Pertactin (Prn), a putative adhesin, was the first identified and characterized autotransporter in *Bordetella*. The BvgAS system regulates the expression of the protein, elevating expression levels in the Bvg+ phase[84]. The unprocessed protein is a 93.5 kDa polypeptide with 910 AA residues[85] and is composed of the N-terminal signal peptide (1–34 AA) which is cleaved in the inner membrane after guiding the Prn autotransporter region (35–910 AA) into the periplasm[86]. The C-terminal Prn translocator (632–910 AA) inserts itself into the outer membrane to form a hydrophobic pore allowing the outer membrane protein (35–631 AA) to pass through and then be cleaved. The mature outer membrane protein was originally believed to obtain a final molecular weight of 69 kDa and was denoted P69. However, due to aberrant migration in sodium dodecyl sulphate polyacrylamide gels, this weight is incorrect, and it was determined that the true molecular weight of the polypeptide is 60 kDa[86]. The mature protein creates a 16-strand parallel β-barrel and contains an RGD tripeptide motif at residues 260–262 which has been described to be involved in eukaryotic cell attachment[87–89].

In vitro assays with purified Prn and recombinant expression studies suggest that Prn may play a role in adherence to, and potentially intracellular invasion of, mammalian cell lines[88–90]. However, *in vivo* assays with a *B. pertussis* strain lacking Prn exhibited no reduction in colonization and adherence to the mouse respiratory tract[91]. Expression of Prn is lacking in a number of the resurgent strains of *B. pertussis*[92,93], and several authors speculated that Prn-deficient strains may be arising due to selective pressure from vaccination. The multiple identified mechanisms of Prn inactivation indicate that these strains arose independently of each other and not from a single clone[93]. Supporting this hypothesis, *B. pertussis* lacking Prn expression exhibited a competitive advantage in colonizing vaccinated mice[92]. Thus, while Prn is thought to be an adhesin, it is

apparently not essential for host colonization, neither in mice as shown by experimental data, nor in humans as is implied by the drastic increase in Prn-deficient clinical strains. Along this line, a Prn-deficient mutant of *B. bronchiseptica* exhibited no difference in colonizing the rat respiratory tract compared to the wild-type bacteria, even though it was cleared quicker in a mouse lung inflammation model[89,94]. By contrast, in the swine model, Prn is necessary for optimal colonization and contributes to *Bordetella* resisting neutrophil-mediated clearance[95].

Bordetella resistance to killing protein

The *Bordetella* resistance to killing (*brk*) locus contains two genes, *brkA* and *brkB*[96]. The locus is potentially BvgAS regulated due to putative BvgA binding sites upstream of both open reading frames[96]. Both proteins are necessary for serum resistance in *B. pertussis*, and wild-type *B. pertussis* strains overexpressing the proteins due to possessing multiple copies of the *brk* locus exhibited a two- to fivefold increase in serum resistance[97].

BrkA is a predicted autotransporter and shares 29 per cent protein homology with Prn[96]. The precursor protein is 103 kDa in size and consists of 1010 AA residues. After translocation into the periplasm, the 42 AA residue N-terminal signal peptide is cleaved at a conserved proteolytic cleavage site. The C-terminal BrkA translocator (732–1010 AA) then facilitates the transport through the outer membrane through pore formation by the intramolecular folding, followed by cleavage into the mature BrkA protein (43–731 AA; 73 kDa)[98]. After secretion, BrkA remains tightly associated with the bacterial surface[99]. The presence of two RGD tripeptide motifs in its mature form suggests that the protein is involved in adherence to and invasion of epithelial target cells[96]. While BrkB is less well characterized than BrkA, it has homology to other known transporter proteins, and its hydropathy profile predicts that it is a cytoplasmic membrane protein[96]. Both BrkA and BrkB play a critically important role in serum resistance[96,97]. In *B. pertussis*, BrkA contributes to adherence and attachment to host cells, protects against antimicrobial peptides by preventing cell lysis, and is expressed in all prevalent clinical strains[97,100,101]. However, in *B. bronchiseptica* BrkA is not necessary for serum resistance, and the contribution of BrkB is unknown[102].

Vir-activated gene 8

The vir-activated gene 8 (Vag8) protein is another autotransporter and, therefore, shares homology with the C-terminal end of Prn, BrkA, TcfA, and SphB1. The 94.9 kDa outer membrane protein contains 915 amino acid residues with a RGD motif at 151–153 AA. The protein remains associated with and anchored to the outer membrane due to the lack of a C-terminal α-domain cleavage site[103]. *Bordetella pertussis* Vag8 protein is central to one of several bacterial strategies for resistance against killing via complement. It binds the C1 esterase inhibitor, a complement regulatory protein, to the bacterial cell surface[104]. Vag8 expression increases serum resistance in a Bvg-regulated manner[102,104], and this resistance to complement-mediated killing does not require the expression of BrkA or FHA. Vag8 knock-out mutants were more susceptible to killing compared to the isogenic wild-type strains[104], and further research unravelled the molecular mechanisms by which Vag8 inhibits complement deposition on the bacterial surface[105]. Vag8 binding to human C1-inhibitor (C1-inh) was shown to interfere with the binding of C1-inh to C1s, C1r, and MASP-2. This results in the release of active proteases that subsequently cleave C2 and C4 away from the bacterial surface. The depletion of these components leads to decreased complement deposition on *B. pertussis* and thus to less complement-mediated bacterial killing[105]. When used as an antigen, mice vaccinated with Vag8 protein experienced less *B. pertussis* colonization and persistence in the lungs; however, bacterial numbers in the nasopharynx were not affected[106].

Tracheal colonization factor

The Bvg⁺-regulated tracheal colonization factor (TcfA) is unique in that it is the only autotransporter not associated with the cell surface after processing[107]. The 90 kDa cell-associated precursor protein is processed into a 60 kDa mature protein that is released freely into the supernatant after secretion. The N-terminus of the mature protein contains one RGD tripeptide motif. *Bordetella pertussis* strains lacking TcfA exhibited a decreased ability to colonize the mouse trachea indicating that, similar to Prn, TcfA may be involved in adherence to epithelial cells[107]. TcfA is only expressed in *B. pertussis*, but not in *B. parapertussis*, which does not possess the gene, or in *B. bronchiseptica*, because *tcfA* is present as a pseudogene[4,58].

SphB1

SphB1 is a subtilisin-like Ser protease/lipoprotein autotransporter. The N-terminal passenger domain contains a conserved bacterial lipoprotein motif, as well as a subtilisin-like protease protein characteristic, an Asp–His–Ser catalytic triad[108]. The C-terminal ends with the common autotransporter β-barrel domain. After translocation, the protein remains tightly associated with the outer surface of the bacteria[108]. The mature SphB1 interacts and is necessary for the function of another virulence factor, FHA. The protease function cleaves the C-terminal end of FHA, and it was the first ever described autotransporter whose passenger protein is essential for the maturation of another unrelated secreted protein[108]. Thus, the observed reduction in bacterial colonization and persistence in murine lungs following vaccination with the SphB1 antigen in mice challenged with *B. pertussis*[106] may have been, in fact, caused by impaired function of FHA.

2.3.4 *Bordetella* polysaccharide

Bordetella polysaccharide (Bps) is a polymer constructed by the *bpsABCD* operon and negatively regulated by BpsR which binds and represses the promoter of the operon. The BvgAS system does not regulate Bps[109].

Bps confers complement resistance and inhibits deposition of complement proteins during mouse infections[110,111]. Bps is an organ-specific factor involved in the colonization of the mouse nose and trachea but is apparently not needed for colonization of the lungs following high-dose innoculation[109,111]. Furthermore, Bps promotes adherence to the human nasal epithelia but not the lungs[109]. Analysis of human serum indicates that *B. pertussis* expresses Bps during infection. Bps also contributes to the formation of biofilms *in vivo*[109,110,111]. Through immunofluorescence and scanning electron microscopy, *B. bronchiseptica* biofilms have been visualized on the mouse nasal epithelium[112].

Bps is required to stabilize and maintain these nasal biofilms which are necessary for colonization[112].

2.3.5 Host factor–bacteriophage Qβ

Post-transcriptional regulation in *B. pertussis* is just barely being elucidated. Host factor–bacteriophage Qβ (Hfq) is an 8.8 kDa RNA chaperone protein that binds to small non-coding RNAs (sRNA) and messenger RNAs (mRNA) as a means of gene expression regulation. Although understudied, sRNAs have been identified in *B. pertussis*[113] and have been shown in other pathogens to contribute to virulence regulation[114,115]. In its hexameric form, Hfq binds to mRNAs and sRNAs inducing conformational changes, modifying their stability, and positively or negatively affecting expression[116]. Hfq also facilitates the interaction between sRNAs and their target mRNAs which can lead to either activation or repression of translation[117]. Hfq homologues play a known role in virulence and stress tolerance in other pathogenic bacteria, such as in *V. cholerae* and *Legionella pneumophila*[118–120].

Specifically, in *B. pertussis*, Hfq is involved in the expression of numerous genes and several described virulence factors including the type III secretion system (T3SS), Vag8, and TcfA[121]. A *B. pertussis* Hfq deletion mutant exhibited a decrease in production of adenylate cyclase toxin, PTx, and filamentous haemagglutinin, as well as a reduced ability to survive within macrophages[117]. Furthermore, the deletion mutant showed attenuated virulence in the mouse infection model with reduced lung colonization and an increase of the 50% infectious dose value by an order of magnitude[117]. Hence, in addition to the relatively heavily studied transcriptional regulation, *B. pertussis* also utilizes post-transcriptional means to control virulence.

2.3.6 Type III secretion system

The *Bordetella* T3SS locus is regulated by the BvgAS system and is highly conserved among the classical Bordetellae[122]. All three of the classical Bordetellae possess the gene encoding the *Bordetella* T3SS effector A (*bteA*) toxin[123]. However, the expression of T3SS proteins depends on the environmental growth conditions of the pathogen[123]. After successive *in vitro* passages, expression is lost in *B. pertussis*, but can be reactivated by *in vivo* growth within a host[124]. The T3SS is involved in persistence and pathogenesis, but it is not required for transmission[125].

In vitro studies have shown that the *B. bronchiseptica* T3SS induces necrotic cytotoxicity in mammalian cell lines[122,126], dephosphorylation of host cell proteins[122], and activation of mitogen-activated protein kinases[127]. However, *B. pertussis* and human isolates of *B. parapertussis* exhibit no cytotoxicity to mammalian cell lines, possibly due to a post-translational block during the *in vitro* expression of T3SS proteins[94].

In vivo, the *B. bronchiseptica* T3SS contributes to persistent colonization of the trachea in both rats and mice and is required for persistence in the mouse's lower respiratory tract[122,127,128]. The T3SS possibly increases persistence by inducing an interleukin cell-mediated response in the host and stopping the formation of anti-*Bordetella* antibody levels[125]. Therefore, the T3SS may be primarily important for chronic colonization of the host, but does not necessarily cause pathologies. However, the T3SS also increases disease severity by contributing to the development of nasal lesions and pneumonia in swine[125].

2.3.7 Type VI secretion system

The *Bordetella* genus encodes a bacterial type VI secretion system (T6SS) that injects effectors into either eukaryotic host cells or bacterial competitors[4]. The T6SS system is present in both classical and non-classical Bordetellae and shows variability among species which may allow for it to perform roles tailored to different hosts or against different bacterial competitors[4,129]. Genomic comparisons suggested that the T6SS was initially present in the ancestor of the *Bordetella* genus but was lost over time in some species and became divergent in others[4]. While *B. bronchiseptica* has a functional T6SS, the T6SS in *B. parapertussis* is likely non-functional because of a missing subset of genes and/or the presence of pseudogenes in the two lineages[4,129]. *B. pertussis* has lost the entire T6SS locus[4,129].

The T6SS in *B. bronchiseptica* may contribute to immunomodulation and pathogenesis by altering cytokine production, delaying lower respiratory tract clearance, and increasing long-term persistence in the nasal cavity[130]. *In vitro* macrophage studies have shown that the T6SS induces interleukin cytokine production and contributes to cytotoxicity[130]. *In vivo*, *B. bronchiseptica* deficient in a T6SS had a decreased survival rate in the lungs of wild-type mice[131]. However, in immunocompromised mice, the T6SS mutant had enhanced intracellular survival allowing the bacterium to spread to systemic organs and contribute to early host death[131]. Hence, the T6SS is necessary for containment of the pathogen to the respiratory tract, increasing the likelihood of transmission[131].

2.3.8 Lipooligosaccharides

Lipopolysaccharides (LPSs) are pyrogenic and mitogenic endotoxins displayed on the outer membrane of the Bordetellae which provide cell envelope structural integrity and adhesion functions[132,133]. In Gram-negative bacteria, including *B. bronchiseptica* and *B. parapertussis*, LPSs are composed of a lipid A outer membrane anchor and an oligosaccharide core ending with a repetitive O-antigen attachment and are considered smooth LPS. LPS synthesis requires two genetic loci, *wlb* and *wbm.* The *wlb* locus is conserved among the classical Bordetellae and is required for the biosynthesis and assembly of the lipid A and oligosaccharide core[134], while the *wbm* locus is required for the production and assembly of the O-antigen. *B. pertussis* lacks the *wbm* locus and, therefore, *B. pertussis* LPSs are deficient in O-antigen. The LPSs contain only the lipid A anchor and the oligosaccharide core and, thus, represent a lower-molecular-weight rough-type lipooligosaccharide (LOS). LOSs, along with LPSs, can be recognized by the host immune system and elicit a response.

In combination with BrkA, LOSs mediate resistance of *B. pertussis* to serum[94]. Bactericidal antibodies have been isolated from children diagnosed with pertussis indicating that LOSs are expressed during pathogenesis[135]. The LOSs bind to CD14/TLR4/MD2 complexes, which in turn activate the immune response, including the recruitment and activation of neutrophils[136]. The lack of the O-antigen and the decreased recognition by TLR-4 immune components may contribute to the ability of *B. pertussis* to delay host recognition and subsequent clearing[137]. Furthermore, LOSs cause monocyte-derived dendritic cells to produce IL-23, IL-6, and IL-1β, which elicit naïve T-cells to produce IL-17.

2.4 Microbiota

Until the widespread use of deep sequencing approaches there was relatively little known about the microbiota present within the mammalian respiratory tract. Recent 16S-directed and metagenomic sequencing projects revealed the presence of a complex assortment of native organisms with health consequences that we are only beginning to understand. The strengths of the mouse experimental infection model have allowed the interactions between *B. pertussis* and resident microbiota to begin to be probed.

Bordetella pertussis has long been known to colonize and grow within the respiratory tracts of mice when delivered in large numbers (>10,000 CFU). But *B. pertussis* is highly infectious among humans, suggesting that much smaller numbers of bacteria are necessary to establish bacterial colonization. Furthermore, *B. bronchiseptica* colonizes mice efficiently with an effective infectious dose of less than 10 CFU. This disparity is generally attributed to *B. pertussis* specialization to its human host, for example, via the receptor–ligand type specificity which is seen in other pathogens. On the other hand, the variance between the two species may be due to *B. pertussis* being unable to compete with the resident microbiota in the mice nasal cavity. Indeed, during colonization, *B. bronchiseptica* displaces the resident microbiota, but *B. pertussis* does not[2], suggesting *B. pertussis* may have lost the ability to compete with the microbes in the mouse nose. The infectious dose of *B. pertussis* decreased by orders of magnitude in mice treated with intranasally administered topic antibiotics. Hence, *B. pertussis* maintains the ability to efficiently colonize the nasal cavity of mice, but fails to efficiently compete with the resident microbiota in these mice.

The interactions between *B. pertussis* and each individual competitor in the resident microbiota are an active area of study and may reveal new susceptibilities of *B. pertussis*. The effect of resident microbes on *B. pertussis* colonization and pathogenesis also reveal interbacterial competition as a new and important aspect of the study of its pathogenesis.

Variation in the diversity and distribution in host-resident microbiota could explain some of the observed clinical variability of *B. pertussis* pathogenesis in humans. There does not appear to be only two distinct states of immunity to *B. pertussis*, naïve and immune. In fact, knowing the exposure status of the host provides only a limited ability to predict the outcome of infection; those more recently immunized tend to have less severe disease. But most acellular vaccines are estimated to be 80–90 per cent effective, indicating that the vaccines have little effect in a substantial proportion of the population. Could variable host microbiota contribute to these variable outcomes? Does resident microbiota within the respiratory tract, or the gut, at the time of first exposure affect the initial immune response that predisposes subsequent responses (original antigenic sin)?

In addition to a direct impact on variability of pathogenesis within individuals, the effects of microbiota may be observable at the population level. Between different populations there are different distributions of each of several *B. pertussis* lineages, and other *Bordetella* species, that could be affected by the different respiratory microbiota prevalent in those populations. Even within a population there is a wide range of severity of symptoms associated with *B. pertussis* colonization. Since diagnosis, treatment, and containment are critical aspects of the control of the spread of disease, variability in presentation can have a significant effect on the onward spread of infection. Exposures that fail to lead to infections, and infections that go undiagnosed, could have profound effects on the epidemiology of *B. pertussis* within and between populations. Could a more resistant microbiota, such as that apparently present in the mouse nose, be harnessed to protect against *B. pertussis* infection and/or disease?

2.5 Conclusion

In conclusion, now classical bacterial pathogenesis work has allowed for the discovery of *B. pertussis* bacterial toxins and numerous other virulence effectors that contribute to its virulence, and identified protective antigens that prevent the most severe forms of disease. But the ongoing resurgence of

B. pertussis underscores the importance of broader research assays and approaches to understand the mechanisms by which *B. pertussis* achieves initial colonization of the human host and subsequent transmission to the next, to guide strategies to prevent the ongoing spread of pertussis disease. The use of natural host infection systems and disruption of host microbiota to allow *B. pertussis* to efficiently colonize mice are two examples of such approaches.

Future studies of *B. pertussis* pathogenesis are likely to include further optimization of viable animal models to replicate human infection, identification, and detailed characterization of a longer list of factors involved in not just virulence, but other aspects of host–pathogen interactions. Most complex, and therefore potentially most ambitious, is untangling the interactions between the host, pathogen, and resident microbiota. Understanding at a mechanistic level the complex set of factors *B. pertussis* uses to mediate its remarkably efficient transmission and infection will guide us to a more efficient mitigation strategy for this resurging disease.

References

1. Warfel JM, Beren J, Kelly VK, et al. Nonhuman primate model of pertussis. *Infect Immun* 2012;**80**:1530–6.
2. Weyrich LS, Feaga HA, Park J, et al. Resident microbiota affect Bordetella pertussis infectious dose and host specificity. *J Infect Dis* 2014;**209**:913–21.
3. Zhang X, Goel T, Goodfield LL, et al. Decreased leukocyte accumulation and delayed Bordetella pertussis clearance in IL-6-/- mice. *J Immunol* 2011;**186**:4895–904.
4. Linz B, Ivanov YV, Preston A, et al. Acquisition and loss of virulence-associated factors during genome evolution and speciation in three clades of Bordetella species. *BMC Genomics* 2016;**17**.
5. Harvill ET, Cotter PA, Miller JF. Pregenomic comparative analysis between Bordetella bronchiseptica RB50 and Bordetella pertussis Tohama I in murine models of respiratory tract infection. *Infect Immun* 1999;**67**:6109–18.
6. Beddoe T, Paton AW, Le Nours J, et al. Structure, biological functions and applications of the AB5 toxins. *Trends Biochem Sci* 2010;**35**:411–8.
7. Hewlett EL, Sauer KT, Myers GA, et al. Induction of a novel morphological response in Chinese hamster ovary cells by pertussis toxin. *Infect Immun* 1983;**40**:1198–203.
8. Kirimanjeswara GS, Agosto LM, Kennett MJ, et al. Pertussis toxin inhibits neutrophil recruitment to delay

antibody-mediated clearance of Bordetella pertussis. *J Clin Invest* 2005;**115**:3594–601.

9. Meade BD, Kind PD, Manclark CR. Lymphocytosis-promoting factor of Bordetella pertussis alters mononuclear phagocyte circulation and response to inflammation. *Infect Immun* 1984;**46**:733–9.

10. Carbonetti NH, Artamonova GV, Mays RM, et al. Pertussis toxin plays an early role in respiratory tract colonization by Bordetella pertussis. *Infect Immun* 2003;**71**:6358–66.

11. Andreasen C, Carbonetti NH. Pertussis toxin inhibits early chemokine production to delay neutrophil recruitment in response to Bordetella pertussis respiratory tract infection in mice. *Infect Immun* 2008;**76**:5139–48.

12. Rollins BJ. Chemokines. *Blood* 1997;**90**:909–28.

13. Vogel FR, Klein TW, Stewart WE,2nd, et al. Immune suppression and induction of gamma interferon by pertussis toxin. *Infect Immun* 1985;**49**:90–7.

14. Armstrong GD, Howard LA, Peppler MS. Use of glycosyltransferases to restore pertussis toxin receptor activity to asialoagalactofetuin. *J Biol Chem* 1988;**263**:8677–84.

15. Hazes B, Boodhoo A, Cockle SA, et al. Crystal structure of the pertussis toxin-ATP complex: a molecular sensor. *J Mol Biol* 1996;**258**:661–71.

16. Katada T, Tamura M, Ui M. The A protomer of islet-activating protein, pertussis toxin, as an active peptide catalyzing ADP-ribosylation of a membrane protein. *Arch Biochem Biophys* 1983;**224**:290–8.

17. Katada T, Ui M. Direct modification of the membrane adenylate cyclase system by islet-activating protein due to ADP-ribosylation of a membrane protein. *Proc Natl Acad Sci U S A* 1982;**79**:3129–33.

18. Katada T, Ui M. ADP ribosylation of the specific membrane-protein of C6 Cells by islet-activating protein associated with modification of adenylate-cyclase activity. *J Biol Chem* 1982;**257**:7210–6.

19. Katada T, Ui M. Direct modification of the membrane adenylate-cyclase system by islet-activating protein due to ADP-ribosylation of a membrane-protein. *Proc Natl Acad Sci U S A* 1982;**79**:3129–33.

20. Meade BD, Kind PD, Ewell JB, et al. In vitro inhibition of murine macrophage migration by Bordetella pertussis lymphocytosis-promoting factor. *Infect Immun* 1984;**45**:718–25.

21. Schaeffer LM, Weiss AA. Pertussis toxin and lipopolysaccharide influence phagocytosis of Bordetella pertussis by human monocytes. *Infect Immun* 2001;**69**:7635–41.

22. Freyberg Z, Harvill ET. Pathogen manipulation of host metabolism: a common strategy for immune evasion. *PLoS Pathog* 2017;**13**:e1006669.

23. Yajima M, Hosoda K, Kanbayashi Y, et al. Islets-activating protein (IAP) in Bordetella pertussis that potentiates insulin secretory responses of rats—purification and characterization. *J Biochem* 1978;**83**:295–303.

24. Yajima M, Hosoda K, Kanbayashi Y, et al. Biological properties of islets-activating protein (IAP) purified from culture medium of Bordetella pertussis. *J Biochem* 1978;**83**:305–12.

25. Furman BL, Sidey FM, Wardlaw AC. Role of insulin in the hypoglycemic effect of sublethal Bordetella pertussis infection in mice. *Br J Exp Pathol* 1986;**67**:305–12.

26. Simpson N, Maffei A, Freeby M, et al. Dopamine-mediated autocrine inhibitory circuit regulating human insulin secretion in vitro. *Mol Endocrinol* 2012;**26**:1757–72.

27. Ivanov YV, Linz B, Register KB, et al. Identification and taxonomic characterization of Bordetella pseudohinzii sp. nov. isolated from laboratory-raised mice. *Int J Syst Evol Microbiol* 2016;**66**:5452–9.

28. Shrivastava R, Miller JF. Virulence factor secretion and translocation by Bordetella species. *Curr Opin Microbiol* 2009;**12**:88–93.

29. Vojtova J, Kamanova J, Sebo P. Bordetella adenylate cyclase toxin: a swift saboteur of host defense. *Curr Opin Microbiol* 2006;**9**:69–75.

30. Eby JC, Gray MC, Warfel JM, et al. Quantification of the adenylate cyclase toxin of Bordetella pertussis in vitro and during respiratory infection. *Infect Immun* 2013;**81**:1390–8.

31. Bauche C, Chenal A, Knapp O, et al. Structural and functional characterization of an essential RTX subdomain of Bordetella pertussis adenylate cyclase toxin. *J Biol Chem* 2006;**281**:16914–26.

32. El-Azami-El-Idrissi M, Bauche C, Loucka J, et al. Interaction of Bordetella pertussis adenylate cyclase with CD11b/CD18: role of toxin acylation and identification of the main integrin interaction domain. *J Biol Chem* 2003;**278**:38514–21.

33. Kamanova J, Kofronova O, Masin J, et al. Adenylate cyclase toxin subverts phagocyte function by RhoA inhibition and unproductive ruffling. *J Immunol* 2008;**181**:5587–97.

34. Harvill ET, Cotter P, Yuk MH, et al. Probing the function of Bordetella bronchiseptica adenylate cyclase toxin by manipulating host immunity. *Infect Immun* 1999;**67**:1493–500.

35. Khelef N, Zychlinsky A, Guiso N. Bordetella pertussis induces apoptosis in macrophages: role of adenylate cyclase-hemolysin. *Infect Immun* 1993;**61**:4064–71.

36. Khelef N, Guiso N. Induction of macrophage apoptosis by Bordetella pertussis adenylate cyclase-hemolysin. *FEMS Microbiol Lett* 1995;**134**:27–32.

37. Hickey FB, Brereton CF, Mills KH. Adenylate cyclase toxin of Bordetella pertussis inhibits TLR-induced IRF-1 and IRF-8 activation and IL-12 production and enhances IL-10 through MAPK activation in dendritic cells. *J Leukoc Biol* 2008;**84**:234–43.

38. Lamberti Y, Gorgojo J, Massillo C, et al. Bordetella pertussis entry into respiratory epithelial cells and intracellular survival. *Pathog Dis* 2013;**69**:194–204.

39. Lamberti YA, Hayes JA, Perez Vidakovics ML, et al. Intracellular trafficking of Bordetella pertussis in human macrophages. *Infect Immun* 2010;**78**:907–13.

40. Walker KE, Weiss AA. Characterization of the dermonecrotic toxin in members of the genus Bordetella. *Infect Immun* 1994;**62**:3817–28.

41. Horiguchi Y, Nakai T, Kume K. Purification and characterization of Bordetella bronchiseptica dermonecrotic toxin. *Microb Pathog* 1989;**6**:361–8.

42. Cowell JL, Hewlett EL, Manclark CR. Intracellular localization of the dermonecrotic toxin of Bordetella pertussis. *Infect Immun* 1979;**25**:896–901.

43. Fukui-Miyazaki A, Ohnishi S, Kamitani S, et al. Bordetella dermonecrotic toxin binds to target cells via the N-terminal 30 amino acids. *Microbiol Immunol* 2011;**55**:154–9.

44. Matsuzawa T, Fukui A, Kashimoto T, et al. Bordetella dermonecrotic toxin undergoes proteolytic processing to be translocated from a dynamin-related endosome into the cytoplasm in an acidification-independent manner. *J Biol Chem* 2004;**279**:2866–72.

45. Masuda M, Betancourt L, Matsuzawa T, et al. Activation of rho through a cross-link with polyamines catalyzed by Bordetella dermonecrotizing toxin. *EMBO J* 2000;**19**:521–30.

46. Parton R. Effect of prednisolone on the toxicity of Bordetella pertussis for mice. *J Med Microbiol* 1985;**19**:391–400.

47. Iida T, Okonogi T. Lienotoxicity of Bordetella pertussis in mice. *J Microbiol* 1971;**4**:51–61.

48. Roop RM, Veit HP, Sinsky RJ, et al. Virulence factors of Bordetella bronchiseptica associated with the production of infectious atrophic rhinitis and pneumonia in experimentally infected neonatal swine. *Infect Immun* 1987;**55**:217–22.

49. Magyar T, Chanter N, Lax AJ, et al. The pathogenesis of turbinate atrophy in pigs caused by Bordetella bronchiseptica. *Vet Microbiol* 1988;**18**:135–46.

50. Horiguchi Y, Nakai T, Kume K. Effects of Bordetella bronchiseptica dermonecrotic toxin on the structure and function of osteoblastic clone MC3T3-E1 cells. *Infect Immun* 1991;**59**:1112–6.

51. Horiguchi Y, Sugimoto N, Matsuda M. Stimulation of DNA synthesis in osteoblast-like MC3T3-E1 cells by Bordetella bronchiseptica dermonecrotic toxin. *Infect Immun* 1993;**61**:3611–5.

52. Horiguchi Y, Inoue N, Masuda M, et al. Bordetella bronchiseptica dermonecrotizing toxin induces reorganization of actin stress fibers through deamidation of Gln-63 of the GTP-binding protein Rho. *Microbiology* 1997;**94**:11623–6.

53. Lacerda HM, Pullinger GD, Lax AJ, et al. Cytotoxic necrotizing factor 1 from Escherichia coli and dermonecrotic toxin from Bordetella bronchiseptica induce p21rho-dependent tyrosine phosphorylation of focal adhesion kinase and paxillin in Swiss 3T3 cells. *J Biol Chem* 1997;**272**:9587–96.

54. Llic D, Furuta Y, Kanazawa S, et al. Reduced cell motility and enhanced focal adhesion contact formation in cells from FAK-deficient mice. *Nature* 1995;**337**:539–44.

55. Seufferlein T, Rozengurt E. Sphingosylphosphorylcholine rapidly induces tyrosine phosphorylation of p125fak and paxillin, rearrangement of the actin cytoskeleton and focal contact assembly. *J Biol Chem* 1995;**270**:24343–51.

56. Weiss AA, Goodwin M. Lethal infection by Bordetella pertussis mutants in the infant mouse model. *Infect Immun* 1989;**57**:3757–64.

57. Cookson BT, Tyler AN, Goldman WE. Primary structure of the peptidoglycan-derived tracheal cytotoxin of Bordetella pertussis. *Biochemistry* 1989;**28**:1744–9.

58. Parkhill J, Sebaihia M, Preston A, et al. Comparative analysis of the genome sequences of Bordetella pertussis, Bordetella parapertussis and Bordetella bronchiseptica. *Nat Genet* 2003;**35**:32–40.

59. Rosenthal RS, Nogami W, Cookson BT, et al. Major fragment of soluble peptidoglycan released from growing Bordetella pertussis is tracheal cytotoxin. *Infect Immun* 1987;**55**:2117–20.

60. Wilson R, Read R, Thomas M, et al. Effects of Bordetella pertussis infection on human respiratory epithelium in vivo and in vitro. *Infect Immun* 1991;**59**:337–45.

61. Goldman WE, Klapper DG, Baseman JB. Detection, isolation, and analysis of a released Bordetella pertussis product toxic to cultured tracheal cells. *Infect Immun* 1982;**36**:782–94.

62. Cookson BT, Cho HL, Herwaldt LA, et al. Biological activities and chemical composition of purified tracheal cytotoxin of Bordetella pertussis. *Infect Immun* 1989;**57**:2223–9.

63. Arai H, Sato Y. Separation and characterization of two distinct hemagglutinins contained in purified leukocytosis-promoting factor from Bordetella pertussis. *Biochim Biophys Acta* 1976;**444**:765–82.

64. Lambert-Buisine C, Willery E, Locht C, et al. N-terminal characterization of the Bordetella pertussis filamentous haemagglutinin. *Mol Microbiol* 1998;**28**:1283–93.

65. Chevalier N, Moser M, Koch HG, et al. Membrane targeting of a bacterial virulence factor harbouring an extended signal peptide. *J Mol Microbiol Biotechnol* 2004;**8**:7–18.

66. Jacob-Dubuisson F, El-Hamel C, Saint N, et al. Channel formation by FhaC, the outer membrane protein involved in the secretion of the Bordetella pertussis filamentous hemagglutinin. *J Biol Chem* 1999;**274**:37731–5.

67. Relman DA, Domenighini M, Tuomanen E, et al. Filamentous hemagglutinin of Bordetella pertussis: nucleotide sequence and crucial role in adherence. *Biochemistry* 1989;**86**:2637–41.

68. Ishibashi Y, Relman DA, Nishikawa A. Invasion of human respiratory epithelial cells by Bordetella pertussis: possible role for a filamentous hemagglutinin Arg-Gly-Asp sequence and alpha5beta1 integrin. *Microb Pathog* 2001;**30**:279–88.

69. Melvin JA, Scheller EV, Noel CR, et al. New insight into filamentous hemagglutinin secretion reveals a role for full-length FhaB in Bordetella virulence. *MBio* 2015;**6**:e01189-15.

70. Hoffman C, Eby J, Gray M, et al. Bordetella adenylate cyclase toxin interacts with filamentous haemagglutinin to inhibit biofilm formation in vitro. *Mol Microbiol* 2017;**103**:214–28.

71. Dirix V, Mielcarek N, Debrie AS, et al. Human dendritic cell maturation and cytokine secretion upon stimulation with Bordetella pertussis filamentous haemagglutinin. *Microbes Infect* 2014;**16**:562–70.

72. Boschwitz JS, Batanghari JW, Kedem H, et al. Bordetella pertussis infection of human monocytes inhibits antigen-dependent CD4 T cell proliferation. *J Infect Dis* 1997;**176**:678–86.

73. Abramson T, Kedem H, Relman DA. Proinflammatory and proapoptotic activities associated with Bordetella pertussis filamentous hemagglutinin. *Infect Immun* 2001;**69**:2650–8.

74. Mooi FR, van der Heide HG, ter Avest AR, et al. Characterization of fimbrial subunits from Bordetella species. *Microb Pathog* 1987;**2**:473–84.

75. Livey I, Duggleby CJ, Robinson A. Cloning and nucleotide sequence analysis of the serotype 2 fimbrial subunit gene of Bordetella pertussis. *Mol Microbiol* 1987;**1**:203.

76. Willems R, Paul A, van de Winkel JGJ, et al. Fimbrial phase variation in Bordetella pertussis: a novel mechanism for transcriptional regulation. *EMBO J* 1990;**9**:2803–9.

77. Vaughan TE, Pratt CB, Sealy K, et al. Plasticity of fimbrial genotype and serotype within populations of Bordetella pertussis: analysis by paired flow cytometry and genome sequencing. *Microbiology* 2014;**160**: 2030–44.

78. Hazenbos WLW, van den Berg BM, Geuijen CAW, et al. Binding of FimD on Bordetella pertussis to very late antigen-5 on monocytes activates complement receptor type 3 via protein tyrosine kinases. *J Immunol* 1995;**155**:3972–8.

79. Hazenbos WLW, Geuijen CAW, van den Berg BM, et al. Bordetella pertussis fimbriae bind to human monocytes via the minor fimbrial subunit FimD. *J Infect Dis* 1995;**171**:924–9.

80. Guevara C, Zhang C, Gaddy JA, et al. Highly differentiated human airway epithelial cells: a model to study host cell-parasite interactions in pertussis. *Infect Dis (Lond)* 2016;**48**:177–88.

81. Geuijen CAW, Willems R, Bongaerts M, et al. Role of the Bordetella pertussis minor fimbrial subunit, FimD, in colonization of the mouse respiratory tract. *Infect Immun* 1997;**65**:4222–8.

82. Mattoo S, Miller JF, Cotter P. Role of Bordetella bronchiseptica fimbriae in tracheal colonization and development of a humoral immune response. *Infect Immun* 2000;**68**:2024–33.

83. Vandebriel RJ, Hellwig SMM, Vermeulen JP, et al. Association of Bordetella pertussis with host immune cells in the mouse lung. *Microb Pathog* 2003;**35**:19–29.

84. Kinnear SM, Boucher PE, Stibitz S, Carbonetti NH. Analysis of BvgA activation of the pertactin gene promoter in Bordetella pertussis. *J Bacterio* 1999;**181**: 5234–41.

85. Charles IG, Dougan G, Pickard D, et al. Molecular cloning and characterization of protective outer membrane protein P.69 from Bordetella pertussis. *Biochemistry* 1989;**86**:3554–8.

86. Makoff AJ, Oxer MD, Ballantine SP, et al. Protective surface antigen P69 of Bordetella pertussis: its characterization and very high level expression in Escherichia coli. *Biotechnology* 1990;**8**:1030–3.

87. Emsley P, McDermott G, Charles IG, et al. Crystallographic characterization of pertactin, a membrane-associated protein from Bordetella pertussis. *J Mol Biol* 1994;**235**:772–3.

88. Leininger E, Roberts M, Kenimer JG, et al. Pertactin, an Arg-Gly-Asp-containing Bordetella pertussis surface protein that promotes adherence of mammalian cells. *Biochemistry* 1991;**88**:345–9.

89. Inatsuka CS, Xu Q, Vujkovic-Cvijin I, et al. Pertactin is required for Bordetella species to resist neutrophil-mediated clearance. *Infect Immun* 2010;**78**:2901–9.

90. Edwards JA, Groathouse NA, Boitano S. Bordetella bronchiseptica adherence to cilia is mediated by multiple adhesin factors and blocked by surfactant protein A. *Infect Immun* 2005;**73**:3618–26.

91. Roberts M, Fairweather NF, Leininger E, et al. Construction and characterization of Bordetella pertussis mutants lacking the vir-regulated P.69 outer membrane protein. *Mol Microbiol* 1991;**5**:1393–404.

92. Safarchi A, Octavia S, Luu LD, et al. Pertactin negative Bordetella pertussis demonstrates higher fitness under vaccine selection pressure in a mixed infection model. *Vaccine* 2015;**33**:6277–81.

93. Lam C, Octavia S, Ricafort L, et al. Rapid increase in pertactin-deficient Bordetella pertussis isolates, Australia. *Emerg Infect Dis* 2014;**20**:626–33.

94. Mattoo S, Cherry JD. Molecular pathogenesis, epidemiology, and clinical manifestations of respiratory infections due to Bordetella pertussis and other Bordetella subspecies. *Clin Microbiol Rev* 2005;**18**:326–82.

95. Nicholson TL, Brockmeier SL, Loving CL. Contribution of Bordetella bronchiseptica filamentous hemagglutinin and pertactin to respiratory disease in swine. *Infect Immun* 2009;**77**:2136–46.

96. Fernandez R, Weiss AA. Cloning and sequencing of a Bordetella pertussis serum resistance locus. *Infect Immun* 1994;**62**:4727–38.

97. Fernandez R, Weiss AA. Serum resistance in bvg-regulated mutants of Bordetella pertussis. *FEMS Microbiol Lett* 1998;**163**:57–63.

98. Shannon JL, Fernandez R. The C-terminal domain of the Bordetella pertussis autotransporter BrkA forms a pore in lipid bilayer membranes. *J Bacteriol* 1999;**181**:5838–42.

99. Jain S, van Ulsen P, Benz I, et al. Polar localization of the autotransporter family of large bacterial virulence proteins. *J Bacteriol* 2006;**188**:4841–50.

100. Fernandez RC, Weiss AA. Susceptibilities of Bordetella pertussis strains to antimicrobial peptides. *Antimicrob Agents Chemother* 1996;**40**:1041–3.

101. Marr N, Oliver DC, Laurent V, et al. Protective activity of the Bordetella pertussis BrkA autotransporter in the murine lung colonization model. *Vaccine* 2008;**26**:4306–11.

102. Rambow AA, Fernandez R, Weiss AA. Characterization of BrkA expression in Bordetella bronchiseptica. *Infect Immun* 1998;**66**:3978–80.

103. Finn TM, Amsbaugh DF. Vag8, a Bordetella pertussis bvg-regulated protein. *Infect Immun* 1998;**66**:3985–9.

104. Marr N, Shah NR, Lee R, et al. Bordetella pertussis autotransporter Vag8 binds human C1 esterase inhibitor and confers serum resistance. *PloS One* 2011;**6**:e20585.

105. Hovingh ES, van den Broek B, Kuipers B, Pinelli E, Rooijakkers SHM, Jongerius I. Acquisition of C1 inhibitor by Bordetella pertussis virulence associated gene 8 results in C2 and C4 consumption away from the bacterial surface. *PLoS Pathog* 2017;**13**:e1006531.

106. de Gouw D, de Jonge MI, Hermans PWM, et al. Proteomics-identified Bvg-activated autotransporters protect against Bordetella pertussis in a mouse model. *PloS One* 2014;**9**:e105011.

107. Finn TM, Stevens LA. Tracheal colonization factor: a Bordetella pertussis secreted virulence determinant. *Mol Microbiol* 1995;**16**:625–34.

108. Coutte L, Antoine R, Drobecq H, et al. Subtilisin-like autotransporter serves as maturation protease in a bacterial secretion pathway. *EMBO J* 2001;**20**:5040–8.

109. Conover MS, Redfern CJ, Ganguly T, et al. BpsR modulates Bordetella biofilm formation by negatively regulating the expression of the Bps polysaccharide. *J Bacteriol* 2011;**194**:233–42.

110. Ganguly T, Johnson JB, Kock ND, et al. The Bordetella pertussis Bps polysaccharide enhances lung colonization by conferring protection from complement-mediated killing. *Cell Microbiol* 2014;**16**:1105–18.

111. Little DJ, Milek S, Bamford NC, et al. The protein BpsB is a poly-beta-1,6-N-acetyl-D-glucosamine deacetylase required for biofilm formation in Bordetella bronchiseptica. *J Biol Chem* 2015;**290**:22827–40.

112. Sloan GP, Love CF, Sukumar N, et al. The Bordetella Bps polysaccharide is critical for biofilm development in the mouse respiratory tract. *J Bacteriol* 2007;**189**:8270–6.

113. Hot D, Slupek S, Wulbrecht B, et al. Detection of small RNAs in Bordetella pertussis and identification of a novel repeated genetic element. *BMC Genomics* 2011;**12**:1–13.

114. Johansson J, Mandin P, Renzoni A, et al. An RNA thermosensor controls expression of virulence genes in Listeria monocytogenes. *Cell* 2002;**110**:551–61.

115. Toledo-Arana A, Repoila F, Cossart P. Small noncoding RNAs controlling pathogenesis. *Curr Opin Microbiol* 2007;**10**:182–8.

116. Link TM, Valentin-Hansen P, Brennan RG. Structure of Escherichia coli Hfq bound to polyriboadenylate RNA. *Proc Natl Acad Sci U S A* 2009;**106**:19292–7.

117. Bibova I, Skopova K, Masin J, et al. The RNA chaperone Hfq is required for virulence of Bordetella pertussis. *Infect Immun* 2013;**81**:4081–90.

118. Ding Y, Davis BM, Waldor MK. Hfq is essential for Vibrio cholerae virulence and downregulates sigma expression. *Mol Microbiol* 2004;**53**:345–54.

119. McNealy TL, Forsbach-Birk V, Shi C, et al. The Hfq homolog in Legionella pneumophila demonstrates regulation by LetA and RpoS and interacts with the global regulator CsrA. *J Bacteriol* 2005;**187**:1527–32.

120. Sittka A, Pfeiffer V, Tedin K, et al. The RNA chaperone Hfq is essential for the virulence of Salmonella typhimurium. *Mol Microbiol* 2007;**63**:193–217.

121. Bibova I, Hot D, Keidel K, et al. Transcriptional profiling of Bordetella pertussis reveals requirement of RNA chaperone Hfq for type III secretion system functionality. *RNA Biol* 2015;**12**:175–85.

122. Yuk MH, Harvill ET, Miller JF. The BvgAS virulence control system regulates type III secretion in Bordetella bronchiseptica. *Mol Microbiol* 1998;**28**:945–59.

123. Hegerle N, Rayat L, Dore G, et al. In-vitro and in-vivo analysis of the production of the Bordtella type three secretion system effector A in Bordetella pertussis, Bordetella parapertussis and Bordetella bronchiseptica. *Microbes Infect* 2013;**15**:399–408.

124. Gaillard ME, Bottero D, Castuma CE, et al. Laboratory adaptation of Bordetella pertussis is associated with

the loss of type three secretion system functionality. *Infect Immun* 2011;**79**:3677–82.

125. Nicholson TL, Brockmeier SL, Loving CL, et al. The Bordetella bronchiseptica type III secretion system is required for persistence and disease severity but not transmission in swine. *Infect Immun* 2014;**82**:1092–103.

126. van den Akker WMR. Bordetella bronchiseptica has a BvgAS-controlled cytotoxic effect upon interaction with epithelial cells. *FEMS Microbiol Lett* 1997;**156**:239–44.

127. Yuk MH, Harvill ET, Cotter P, et al. Modulation of host immune responses, induction of apoptosis and inhibition of NF-kB activation by the Bordetella type III secretion system. *Mol Microbiol* 2000;**35**:991–1004.

128. Pilione MR, Harvill ET. The Bordetella bronchiseptica type III secretion system inhibits gamma interferon production that is required for efficient antibody-mediated bacterial clearance. *Infect Immun* 2006;**74**:1043–9.

129. Park J, Zhang Y, Buboltz AM, et al. Comparative genomics of the classical Bordetella subspecies: the evolution and exchange of virulence-associated diversity amongst closely related pathogens. *BMC Genomics* 2012;**13**:1–17.

130. Weyrich LS, Rolin OY, Muse SJ, et al. A Type VI secretion system encoding locus is required for Bordetella bronchiseptica immunomodulation and persistence in vivo. *PloS One* 2012;**7**:e45892.

131. Bendor L, Weyrich LS, Linz B, et al. Type six secretion system of Bordetella bronchiseptica and adaptive immune components limit intracellular survival during infection. *PloS One* 2015;**10**:e0140743.

132. Ayme G, Caroff M, Chaby R, et al. Biological activities of fragments derived from Bordetella pertussis endotoxin: isolation of a nontoxic, Shwartzman-negative lipid A possessing high adjuvant properties. *Infect Immun* 1980;**27**:739–45.

133. Nakase Y, Tateisi M, Sekiya K, et al. Chemical and biological properties of the purified O antigen of Bordetella pertussis. *Jpn J Microbiol* 1970;**14**:1–8.

134. Preston A, Thomas R, Maskell D. Mutational analysis of the Bordetella pertussis wlb LPS biosynthesis locus. *Microb Pathog* 2002;**33**:91–5.

135. Robbins JB, Schneerson R, Kubler-Kielb J, et al. Toward a new vaccine for pertussis. *Proc Natl Acad Sci U S A* 2014;**111**:3213–6.

136. Eby JC, Hoffman CL, Gonyar LA, et al. Review of the neutrophil response to Bordetella pertussis infection. *Pathog Dis* 2015;**73**:ftv081.

137. Brummelman J, Veerman RE, Hamstra HJ, et al. Bordetella pertussis naturally occurring isolates with altered lipooligosaccharide structure fail to fully mature human dendritic cells. *Infect Immun* 2015;**83**:227–38.

CHAPTER 3

The immunology of *Bordetella pertussis* infection and vaccination

Mieszko M. Wilk, Aideen C. Allen, Alicja Misiak, Lisa Borkner, and Kingston H.G. Mills

Abstract

Bordetella pertussis causes whooping cough (pertussis), a severe and sometimes fatal respiratory infectious disease, especially in young infants. Pertussis can be prevented in infants and children by immunization with either whole-cell pertussis (wP) or acellular pertussis (aP) vaccines; however, its incidence is increasing in many countries despite high vaccine coverage. This resurgence in populations immunized with aP vaccines has been attributed to (1) genetic changes in circulating strains of *B. pertussis* resulting from vaccine-driven immune selection, (2) waning protective immunity due to poor induction of immunological memory, or (3) a failure of aP vaccines to induce the appropriate arm(s) of the cellular immune responses required to prevent infection. Studies in a baboon model have suggested that previous infection prevents reinfection as well as disease, whereas aP vaccines fail to prevent nasal colonization and transmission of *B. pertussis*. Studies in the mouse model have demonstrated that immunization with wP vaccines induces Th1 and Th17 responses, whereas aP vaccines promote Th2-skewed responses and high antibody titres. Thus, while aP vaccine-induced antibodies may prevent pertussis, they may not prevent nasal colonization or transmission. Emerging data have suggested that replacing alum with novel adjuvants based on pathogen-associated molecular patterns has the capacity to switch the responses induced with aP vaccines to the more protective Th1/Th17 responses and may also enhance immunological memory. It is likely that third-generation pertussis vaccines will be based on live attenuated bacteria or aP formulations with novel adjuvants, which prevent nasal and lung infection and induce sustained immunity through induction of memory T cells.

3.1 Introduction

Bordetella pertussis causes the acute respiratory disease whooping cough (pertussis) that is very severe in infants and young children, lasting for weeks to months, sometimes with a fatal outcome. However, the majority of infected individuals clear the infection and develop immunity to subsequent infection, at least in the short term. Studies in animal models have suggested that innate immune responses, including antimicrobial peptides (AMPs), complement, macrophages, neutrophils, natural killer (NK) cells, and $\gamma\delta$ T cells control the early infection in naïve individuals, but full clearance of the bacteria is dependent on adaptive immune responses, mediated by *B. pertussis*-specific T and B cells, that develop several weeks into the

Wilk, M. M., Allen, A. C., Misiak, A., Borkner, L., and Mills, K. H. G., *The immunology of* Bordetella pertussis *infection and vaccination*. In: *Pertussis: epidemiology, immunology, and evolution*. Edited by Pejman Rohani and Samuel V. Scarpino: Oxford University Press (2019). © Oxford University Press. DOI: 10.1093/oso/9780198811879.003.0003

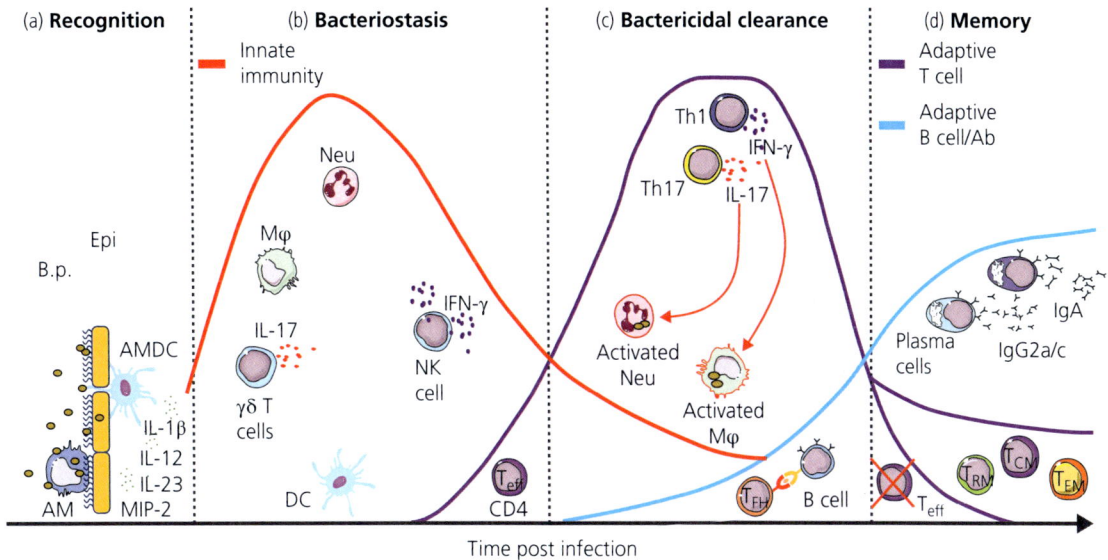

Figure 3.1 The kinetics of protective innate and adaptive immune responses that mediate clearance of a primary infection with *Bordetella pertussis*. (a) *Bordetella pertussis* (B.p.) binds to epithelial cells (Epi) in the respiratory tract which can secrete antimicrobial peptides in response to the binding. *Bordetella pertussis* can also be taken up by resident alveolar macrophages (AM) and airway mucosal dendritic cells (AMDCs) that migrate to lymph nodes and prime naïve T cells to become effector T cells. The complement system enhances the clearance of the pathogen. (b) $\gamma\delta$ T cells that secrete IL-17 and inflammatory macrophages (Mφ) are recruited to the lungs early in infection, followed by neutrophils (Neu) and natural killer (NK) cells, that secrete IFN-γ. (c) Effector T cells (T$_{eff}$) migrate from draining lymph nodes around week 2–3 of infection. Effector Th1 and Th17 cells activate macrophages and neutrophils to kill intracellular *B. pertussis*. T follicular helper (T$_{FH}$) cells stimulate B cells and plasma cells, which secrete antibodies (IgA and IgG) that help to opsonize the bacteria. (d) Central memory T (T$_{CM}$), effector memory T (T$_{EM}$), and tissue-resident memory T (T$_{RM}$) cells together with memory B cells develop and help to protect against re-infection. Modified and re-drawn with permission from the authors, Figure 1. Brummelman et al *Pathogens and Disease* 2015 Nov; 73(8): ftv067, doi: 10.1093/femspd/ftv067.

infection[1] (see Figure 3.1). The persistence of the infection and the delay in induction of adaptive immunity is largely a reflection of the sophisticated immune subversion strategies evolved by the bacteria and is mediated through an armour of bacterial virulence factors and immunomodulatory molecules.

The induction of protective immunity by vaccination is the best approach for controlling this pathogen and indeed current vaccines are able to prevent severe disease; however, it is less clear that they can prevent respiratory infection, especially nasal colonization[2]. Whole-cell pertussis (wP) vaccines, developed in the 1940s, have variable efficacy against whooping cough, although the best wP vaccines have been reported to have up to 95 per cent efficacy against severe disease. Unfortunately, wP vaccines have been associated with side effects including fevers and rare cases of febrile seizures and were replaced in most industrialized countries in the late 1990s and early 2000s with acellular pertussis (aP) vaccines, composed

of one to five antigens formulated with alum as the adjuvant. Although the newer aP vaccines have a significantly improved safety profile over the wP vaccines, they are not as effective, only conferring 84–85 per cent efficacy against typical whooping cough and 71–78 per cent efficacy against mild pertussis disease[3–5]. In recent years, the incidence of whooping cough has increased in many countries with high vaccination rates, including Ireland[6], Australia[7], the United States,[8,9] and England and Wales[10].

From an immunological perspective, there is a clear dichotomy in the profile of immune response induced by the two vaccines, with aP vaccines inducing strong antibody responses (quantified by enzyme-linked immunosorbent assay) to vaccine antigen and T helper (Th) 2-dominated T-cell responses, while wP vaccines induce more potent cellular immune responses mediated by Th1 cells, with a smaller contribution from Th17 cells, as well as humoral immune responses[11,12]. The profile of adaptive immune responses induced by wP vaccines is more closely aligned with that induced by previous infection, and this is consistent with the view that natural infection or immunization with wP vaccines confers more potent and persistent protective immunity than current aP vaccines. Furthermore, a published report from the baboon model[2], as well as emerging evidence from the mouse model (Wilk MM, Borkner L, and Mills KH, unpublished data), have suggested that previous infection or immunization with a wP vaccine prevents nasal colonization as well as lung infection, whereas immunization with a current aP vaccine fails to prevent nasal colonization.

3.2 Innate immunity

The innate immune system is the first line of defence against infection and reacts within minutes to invading microorganisms. It is comprised of AMPs, components of the complement system, and different effector cells, such as macrophages, dendritic cells (DCs), neutrophils, and NK cells. These innate cells also secrete cytokines that recruit and activate cells of the adaptive immune system, thereby inducing specific immune responses.

3.2.1 Antimicrobial peptides

AMPs are mostly small cationic peptides secreted by the mucosal epithelia. AMPs can disrupt the outer membrane of Gram-negative bacteria by interacting with the negatively charged lipopolysaccharide (LPS) lipid A, forming pores in the cytoplasmic membrane leading to cell death[13]. Studies in a mouse *B. pertussis* infection model have shown that interleukin (IL)-17 production by $\gamma\delta$ T cells early in infection promotes the expression of the AMPs, siderophore-binding AMP Lcn2, and the cathelicidin CRAMP, which have both been shown to be crucial in the mucosal defence against Gram-negative bacteria[14]. It has been established from *in vitro* assays that *B. pertussis* is susceptible to killing by the AMPs of the cecropin family, while members of the magainin family and α-defensin were less effective[13].

In canines and pigs, AMPs of the β-defensin family have antimicrobial activity against *Bordetella bronchiseptica* and *B. pertussis* respectively[15,16]. However, *Bordetella* species have developed several strategies

to evade AMPs. The type III secretion system (T3SS) of *B. bronchiseptica* can block the expression of the β-defensin tracheal AMP by inhibiting NFκB signalling, thereby improving colonization and survival[17]. It has also been shown that modification of lipid A, the membrane-proximal region of LPS, with glucosamine residues mediated by LgmA, LgmB, and LgmC of *B. pertussis* increases resistance against cationic AMPs, such as human LL-37, by reducing the perturbation of the outer membrane[18].

3.2.2 Complement

Bordetella pertussis has developed mechanisms of evading killing by each of all three activation pathways of the complement system, thereby escaping the first line of host immune defence. It has been shown in the mouse model that mutants lacking the *bpsABCD* locus encoding Bps polysaccharides are more susceptible than wild-type bacteria to complement-mediated killing due to increased deposition of complement proteins at their surface[19]. *Bordetella* resistance to killing A (BrkA) protein is also a key virulence factor that inhibits the bactericidal activity of complement. Experiments with mutants either lacking or overexpressing BrkA revealed that BrkA reduces C3 and C4 attachment to the surface of the bacteria and the production of the membrane attack complex[20]; however, the exact mechanism of complement evasion via BrkA has not yet been elucidated[21]. Furthermore, *Bordetella* autotransporter protein-C (BapC) has been identified as a factor promoting serum resistance of *B. pertussis*, but again the mechanism remains elusive[22].

Many bacteria recruit host regulatory factors to resist complement-mediated killing and *B. pertussis* is no exception[23]. The complement regulator C1 esterase inhibitor (C1-inh) inactivates proteases involved in the classical and lectin pathway of complement activation[24]. *Bordetella pertussis*, but not the related strains *B. bronchiseptica*, *Bordetella parapertussis*, *Bordetella holmesii*, and *Bordetella avium*, has been shown to bind C1-inh via the passenger domain of the autotransporter Vag8, which is also a correlate of serum resistance[25]. C4b-binding protein (C4BP), another regulator of the classical and the lectin pathway[24], is recruited to the surface of *B. pertussis* via factors expressed by the *Bordetella* virulence gene (*bvg*) gene locus, one of which has been identified as the protein filamentous haemagglutinin (FHA)[26]. *Fha* mutants, however, retain a residual binding capacity for C4BP, indicating that at least one additional component is involved in C4BP recruitment[26]. Surface-bound C4BP is still functional[27], but whether it protects *B. pertussis* against complement-mediated lysis remains to be seen[21]. Finally, *B. pertussis* also engages factor H (FH) and FH family proteins[28], which are inhibitory factors of the alternative pathway of complement activation[24]. Surface-bound FH remains functionally active and has been shown to facilitate survival of *Bordetella* strains in human medium[28].

3.2.3 Macrophages

Macrophages are long-lived phagocytic cells which reside in the tissues, especially in areas vulnerable to infections such as the lung and the gut. They are among the first cells of the immune system to encounter microbes that managed to break through the epithelial layer. Depletion of airway macrophages during *B. pertussis* infection enhanced the infection suggesting that macrophages contribute to immunity against the pathogen[29]. Experiments in inducible nitric oxide (NO) synthase-defective mice showed increased susceptibility of those animals to *B. pertussis* infection, suggesting a role for NO as an effector mechanism[30]. Interferon (IFN)-γ and IL-17 secreted by CD4 T cells during *B. pertussis* infection promote macrophage-mediated killing of the bacteria[31,32]; however, only IFN-γ treatment induced NO production, suggesting that IL-17 might induce bacterial killing via a different mechanism[31].

While a high percentage of *B. pertussis* die shortly after phagocytosis in phagolysosomes, significant numbers of engulfed *B. pertussis* survive and even proliferate in non-acidic compartments within macrophages[33]. Studies in the monocyte cell line THP-1, which differentiate into macrophages *in vitro*, showed that inflammatory and bactericidal responses were downregulated after *B. pertussis* uptake, suggesting that the bacteria manipulate the macrophages to create a favourable environment for survival[34]. Survival of *B. pertussis* strains lacking either adenylate cyclase toxin (ACT) or pertussis toxin (PT) was reduced compared to the wild-type bacteria,

indicating that both toxins might play a role in bacterial survival in macrophages[34,35]. The demonstration that *B. pertussis* has the capacity to survive within macrophages[36–38] suggests that the bacteria may create an intracellular niche that may serve to prolong the period of infection and increase the chance of the bacteria being passed to a new host.

3.2.4 Neutrophils

Neutrophils are short-lived phagocytes that circulate in the blood and are recruited towards the site of infection by chemokines released by activated macrophages as well as by microbial peptides. It has been demonstrated that human neutrophils can take up and kill *B. pertussis*[39]. Although a fraction of the phagocytosed bacteria can survive within the neutrophils[40], it is unlikely that neutrophils serve as a niche of survival for *B. pertussis* due to the short lifespan of the cells[1].

While depletion of neutrophils in *B. pertussis* infection had no effect on the severity of the disease in naïve mice, neutrophil depletion led to a higher bacterial load in the lower respiratory tract in immune mice or mice that had been passively immunized with serum[1,41]. This suggests that neutrophils play a protective role against *B. pertussis* via antibody-mediated phagocytosis.

Bordetella pertussis virulence factors PT and ACT have an inhibitory effect on neutrophil recruitment and function. PT has been shown to delay neutrophil recruitment early in infection by inhibiting the expression of keratinocyte-derived chemokine (KC) and macrophage inflammatory protein 2 (MIP-2) as well as LPS-induced CXC chemokine (LIX)[42]. PT also acts directly on neutrophils by ADP ribosylation of the G-protein $G_{i\alpha}$ and thereby disrupts signalling of $G_{i\alpha}$-dependent chemokine receptors[43]. ACT inhibits neutrophil chemotaxis, phagocytosis, and neutrophil extracellular trap formation by increasing cAMP levels within host cells[43].

3.2.5 Dendritic cells

DCs are found in lymphoid organs, epithelia of the skin, the gut, and the respiratory tract, as well as in the interstitium of most organs. While DCs are phagocytes, their main task is not the immediate clearance of pathogens, but the processing and presentation of pathogen-derived antigens to T cells. LPS produced by *B. pertussis* activates DCs via toll-like receptor (TLR)-4 and stimulates them to release proinflammatory cytokines (tumour necrosis factor (TNF), IL-12p40, IL-12p70, IL-6, IL-1β) as well as inflammatory chemokines (macrophage inflammatory protein (MIP)-1α, MIP-1β, MIP-2)[44]. IL-12 directs the induction of IFN-γ-secreting CD4 T cells (Th1 cells), which play a key role in protective immunity to *B. pertussis*.

A study on the role of DC subtypes in a mouse model showed a strong increase in the numbers of CD11c$^+$CD8α^+ and CD11c$^+$CD8α^+CD103$^+$ DCs in the cervical lymph nodes shortly after *B. pertussis* infection that migrate to the lungs and produce IFN-γ, and also support the induction of Th1 cells[45]. Depletion of CD8α^+ cells or blocking of CD103 early in infection attenuated the Th1 response and delayed bacterial clearance[45]. Adoptive transfer of CD11c$^+$CD8α^+ DCs into *B. pertussis*-infected mice shortly after infection enhanced clearance of the bacteria from the respiratory tract[45].

DCs activated with the *B. pertussis* toxin ACT induce Th17 cells, which together with Th1 cells help to clear bacteria from the lungs of infected mice[12,46]. ACT activates caspase-1 and the NOD-like receptor (NLR) family, pyrin domain-containing 3 (NLRP3) inflammasome, which leads to the release of IL-1β and then together with IL-23 expands Th17 cells[46]. Mice defective in IL-1 signalling have reduced IL-17 production and more prolonged infection with *B. pertussis*[46]. Human DCs have also been shown to express IL-1β and IL-23 after infection with *B. pertussis*[47].

Plasmacytoid DCs (pDCs), a subset of DCs with the main function of secreting type I interferon, produce IFN-α in the lungs of mice early in *B. pertussis* infection, which delays the induction of Th17 responses in mice[48]. Blocking of IFN-α during infection increased the frequency of Th17 cells early in infection and reduced the bacterial load in lungs. Migration of pDCs into the lymph nodes is dependent on G-protein-coupled receptors, which are suppressed by PT. Therefore, it is possible that PT obstructs pDC migration from the lungs to the lymphatic system, thereby creating an IFN-α-rich environment that delays the Th17 response[48].

3.2.6 NK cells

NK cells are lymphocytes of the innate immune system which can detect and kill cells infected by viruses or intracellular bacteria, as well as tumour cells, and they also secrete IFN-γ which activates macrophages. In *B. pertussis* infection, NK cells are a key initial source of IFN-γ[49]. In IFN-γ receptor-deficient mice, *B. pertussis* disseminates from the lungs to the liver early after infection before the adaptive response to the infection is established, emphasizing the critical protective role of early IFN-γ production[50]. Depletion of NK cells had a similar effect, with *B. pertussis* disseminating into the liver via the blood and resulting in the induction of Th2 rather than Th1 responses, which reduced the rate of bacterial clearance[49]. It has also been demonstrated that PT can inhibit NK cell chemotaxis *in vitro*[51], providing further indirect evidence that NK cells play a protective role during *B. pertussis* infection.

3.3 Adaptive immunity

3.3.1 B cells and antibody

A significant number of studies in animal models and humans have demonstrated that infection with *B. pertussis* induces antigen-specific immunoglobulin (Ig)-G and IgA antibodies (reviewed by Higgs et al.[1]). IgA is detected in mucosal secretions earlier than IgG in the serum and plays a role in the clearance of a primary infection with *B. pertussis* from the respiratory tract[52]. IgG antibodies may play a role in the later stage of *B. pertussis* infection and in adaptive immunity induced by previous infection and vaccination. In contrast to infection, parenteral immunization with pertussis vaccines does not induce secretory IgA; however, *B. pertussis*-specific IgG can be detected in the serum[11,12]. Although, the antigens in aP vaccines were selected based on an ability to induce robust antibody responses, aP vaccines do not confer as high a level of protection against pertussis disease as high-efficacy wP vaccines[5], and importantly, studies in baboons have suggested that current aP vaccines may not prevent nasal colonization and transmission of *B. pertussis*[2].

Evidence of a protective role for antibodies in humans was presented in a study which demonstrated that passive immunization with antisera against *B. pertussis* reduced the severity of disease in infected individuals[53]. However, antibody titres, especially against PT, decrease rapidly after immunization with aP vaccines[54]. Consequently, cell-mediated immunity may be more important for long-term protection against *B. pertussis* infection.

The production of high-affinity antibodies is dependent on help from a subtype of CD4+ T cells called follicular helper T cells (T_{FH}). Natural infection of mice with *B. pertussis* or immunization with wP vaccines induces T_{FH} cells (Wilk MM, Allen AA, and Mills KH, unpublished data). Furthermore, T-cell-derived cytokines, together with contact-dependent interactions between T and B cells, promote Ig class switching. Th1-type cytokines (IFN-γ) are associated with secretion of murine IgG2a/IgG2c antibodies by mature B cells[55-57], whereas Th2-type immune responses favour IgG1 production[58,59]. Consistent with this, wP vaccines predominantly induced antigen-specific IgG2a, whereas aP vaccines induced potent IgG1 antibody responses in mice and the corresponding IgG subclasses in humans.

Interestingly, B cells can play a role in protective immunity against *B. pertussis* independently of antibody production. Immunization of mice lacking B cells resulted only in partial protection that was completely restored after transfer of *B. pertussis*-specific B cells; furthermore, the reconstituted B-cell-defective mice had little, if any, detectable *B. pertussis*-specific antibodies[60].

3.3.2 CD4 T cells (Th1 and Th17 cells)

While antibodies are an important element of the adaptive immune response to *B. pertussis*, their induction is dependent on help from CD4 T cells. CD4 T cells also play a crucial role in protective cellular immunity to *B. pertussis*, largely through the recruitment and activation of phagocytic cells, macrophages and neutrophils, at the site of infection[12]. Studies in animal models of *B. pertussis* infection, as well as in humans naturally infected or immunized with wP vaccines, have provided convincing evidence of a role for Th1 and Th17 cells in bacterial clearance. In mice, *B. pertussis* infection results in the induction of mixed Th1 and Th17 responses[12,61,62]. A similar T-cell profile is observed

in infected baboons[63]. Polarized Th1 responses are also seen in *B. pertussis*-infected or convalescent children[64,65]. To date, there is a paucity of data on *B. pertussis*-specific Th17 responses in humans. However, the evidence from baboon studies as well as *in vitro* data from human cells suggest that Th17 cells may also contribute to local protective immunity to *B. pertussis* in humans[47,63].

Initial studies on the role of CD4 T cells in immunity to *B. pertussis* in mice highlighted the importance of Th1 responses in mediating bacterial clearance and containment in the lung[50,61,66]. IFN-$\gamma^{-/-}$ mice develop disseminating infections, which can be lethal in a proportion of animals[50]. In contrast, bacterial clearance is not impaired in IL-4$^{-/-}$ mice, suggesting that Th2 cells are dispensable for the control of a primary infection with *B. pertussis*[62]. Furthermore, mice deficient in IL-17 have reduced neutrophil recruitment to the lung and delayed bacterial clearance[12]. In addition, *B. pertussis* clearance is delayed in convalescent IL-17$^{-/-}$ mice suggesting that IL-17 is involved in protection against reinfection and may contribute to development of immune memory (Ross PR, Borkner L and Mills KH, unpublished observations). Transfer of *B. pertussis*-specific Th1 or Th17 cells into naïve recipients significantly enhances bacterial clearance from the lung[12]. However, the protection is greatest following transfer of Th1 and Th17 cells, highlighting the importance of both subsets in protective immunity against *B. pertussis*. Studies in baboons have confirmed findings in mice and complemented them by showing that *B. pertussis* infection induces long-lived memory Th1 and Th17 cells[63]. Since IL-17 is emerging as an important player in mucosal immunity, strong local Th17 responses are likely to make a significant contribution to sterilizing immunity seen in convalescent baboons[2,67]. Taken together, the studies from animal models, supported by more limited data from humans, suggest that protection against *B. pertussis* requires strong cellular immunity mediated by Th1 and Th17 cells.

3.3.3 Memory T cells

T and B cells of the adaptive immune system, which mediate cellular and humoral immunity respectively, have unique T cell receptors (TCRs) and B cell receptors (BCRs) capable of recognizing antigens, which confers specificity and allows them to recall an encounter with the antigen following subsequent exposure. This leads to more efficient elimination of invading organisms from the body and helps to prevent reinfection. Typically, naïve T cells encounter antigens in secondary lymphoid organs such as the lymph nodes. Upon activation, they rapidly proliferate and differentiate into either effector or memory T cells. Effector T cells promptly migrate to inflamed tissues to assist with pathogen elimination. In contrast, memory T cells do not have an immediate effector function, but persist after the infection has been cleared and recirculate in the bloodstream, acting as a surveillance system in case of secondary challenge with the same pathogen, when they can provide immediate protection[68].

Initially, the memory T-cell population was divided into two subtypes of cells, central memory T (T_{CM}) and effector memory T (T_{EM}) cells, which recirculate between the blood, lymph nodes, and tissue after pathogen clearance. They are distinguished by the expression of CCR7, a chemokine receptor which directs cell homing to secondary lymphoid organs[69]. Resting T_{CM} cells express high levels of CCR7 and CD62L, and are more sensitive to antigenic stimulation than naïve T cells. Upon encountering cognate antigen, they rapidly proliferate to repopulate the memory T-cell population, and can then differentiate into T_{EM} cells. In contrast, resting T_{EM} cells lack CCR7 and express CD44, in addition to an array of surface markers and chemokine receptors which allow them to migrate to inflamed tissues during infection[69,70].

In recent years, a third subtype of memory T cell has been described. Tissue-resident memory T (T_{RM}) cells remain in the tissues after infection has been cleared and do not recirculate to the bloodstream and secondary lymphoid organs. Although T_{RM} cells are distinct from T_{EM} cells, they do share a number of features, including high expression of CD44, low expression of CD62L and CCR7, and the ability to produce cytokines[71]. Additionally, they express the marker CD69 with or without CD103 on the cell surface. CD8+ and CD4+ T_{RM} cells are protective against viral infections in various tissues, including the lungs and skin of mice[72,73].

Infection of mice with *B. pertussis* induces the development of T_{RM} cells in the lungs, and these cells play a role in protective immunity against secondary infection with this pathogen[74] (see Figure 3.2). Treatment of convalescent mice with FTY720, which inhibits lymphocyte egress from the draining lymph nodes, did not affect clearance of a secondary infection with *B. pertussis*. Furthermore, T_{RM} cells rapidly expanded locally in the FTY720-treated and untreated mice following secondary challenge. Moreover, adoptive transfer of CD4+ T_{RM} cells from convalescent mice into naïve mice conferred protection against *B. pertussis* infection[74]. We have also found that immunization of mice with a wP, but not an aP vaccine, induced CD4+ T_{RM} cells in the lungs that expanded rapidly after *B. pertussis* infection despite treatment with FTY720 (Wilk MM, Borkner L, and Mills KH, unpublished data). Collectively, these findings suggest that induction of T_{RM} cells may explain the superior immunological memory induced by previous infection or immunization with a wP vaccine and the more limited memory induced with an aP vaccine. These preliminary data underscore the importance of finding new aP vaccine-adjuvant combinations and immunization approaches that can induce T_{RM} cells Superscript: 188.

3.3.4 Regulatory T cells

Although pertussis is an acute respiratory disease, the infection can persist for many weeks. In the mouse model, complete clearance of the bacteria from the respiratory tract can take 7–8 weeks in normal immunocompetent animals. Furthermore, adaptive immune responses are only detectable 3–4 weeks into the infection in naïve mice[61], whereas an adaptive immune response to a pertussis vaccine can be detected after 7 days. This is in part a reflection of the various immune subversion strategies evolved by *B. pertussis*. For example, FHA[75,76], ACT[77], T3SS[78], and LPS[44] all induced IL-10 production by macrophages and/or DCs and this can promote the development of inducible IL-10-secreting regulatory T (Treg) cells. The combined effect of innate IL-10 and induced Treg cells suppressed effector Th1 and Th7 cells, thereby slowing clearance of the bacteria from the respiratory tract. There is a high frequency of FoxP3+ Treg cells in the lungs throughout the course of infection with *B. pertussis*,

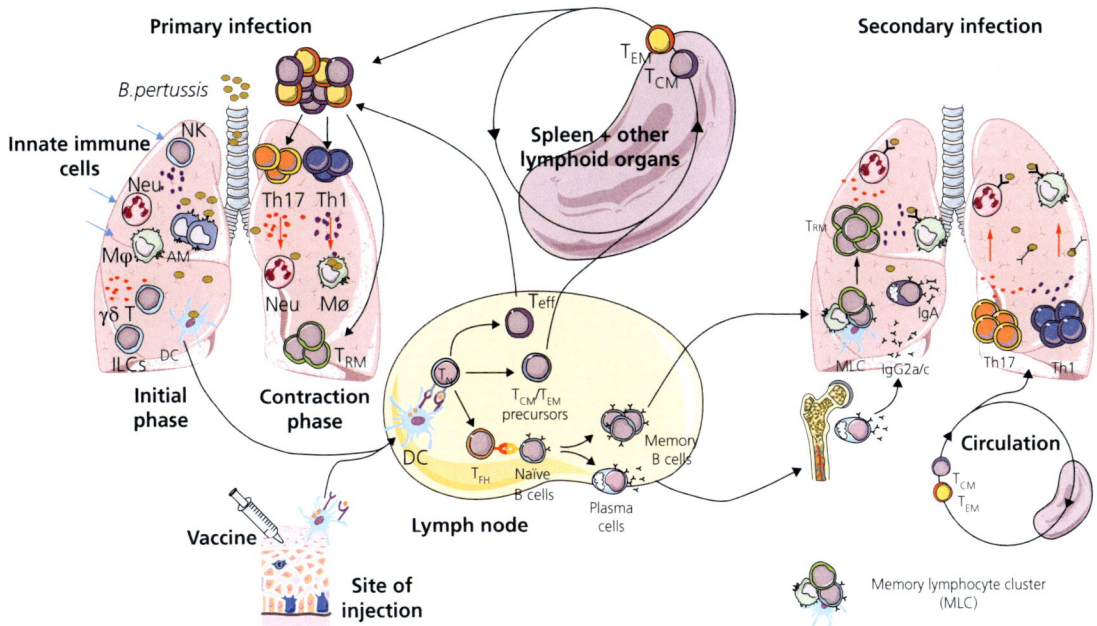

Figure 3.2 Effector and memory immune responses that mediate protection against primary infection and re-infection with *Bordetella pertussis*. During primary infection with *B. pertussis*, various innate immune cells are recruited and activated in the lungs. Activated dendritic cells (DCs) migrate to draining lymph nodes, where they present the antigen and prime naïve T cells to become effector (T_{eff}) Th1, Th17, and T_{FH} cells. T_{FH} cells activate B cells and promote the development of plasma cells that produce *B. pertussis*-specific antibodies. Effector Th1 and Th17 cells migrate via the circulation to the lungs where they help to recruit and activate macrophages ($M\varphi$) and neutrophils (Neu), which with the help of opsonizing antibodies take up and kill *B. pertussis*. Memory T cells (T_{EM} and T_{CM} cells) are also generated in the lymph nodes and migrate via the circulation and secondary lymphoid tissues to the lungs. Some of the memory T cells become tissue-resident memory T (T_{RM}) cells and maintain their residency and function in the lung parenchyma in long-lived structures called memory lymphocyte clusters (MLCs). These T_{RM} cells are poised and ready to immediately proliferate and respond to re-infection, where they can become Th1 and Th17 cells, thereby limiting the extent of the secondary infection.

and these Treg cells also help subvert bacterial clearance by modulating protective Th1 cells[79].

Alveolar macrophages, which play a role in maintaining tolerance in the lungs, help to promote the expansion of Treg cells during *B. pertussis* infection[80]. The majority of natural Treg cells in the lungs of *B. pertussis*-infected mice, which suppressed Th1 responses via IL-10, are CD4+FoxP3+CD25− T cells[79]. While innate and IL-10-secreting Treg cells may be exploited by the bacteria to persist in the host, there is also evidence that these responses can be beneficial to the host by limiting immunopathology caused by lung inflammation during *B. pertussis* infection[44].

3.3.5 $\gamma\delta$ T cells

$\gamma\delta$ T cells constitute 1–5 per cent of the peripheral T cell population, but they are enriched in tissues,

where in mice they can account for up to 50 per cent. Up until recently, $\gamma\delta$ T cells had been thought to act solely as innate-like lymphocytes by responding directly to cytokines or damage-associated molecules. However, a number of studies have shown that they can also mount antigen-specific memory T-cell responses[81]. Zachariadis et al. showed that $\gamma\delta$ T cells may act to limit early inflammatory responses during *B. pertussis* infection: TCRδ−/− mice had higher inflammation in the lung[82]. More recently, $\gamma\delta$ T cells have been shown to be an early source of IL-17 in the lungs of *B. pertussis*-infected mice, thus helping to control the initial bacterial replication by helping to recruit neutrophils[14]. The initial wave of $\gamma\delta$ T cells are innate-like cells, secreting IL-17 in response to IL-1 and IL-23 without TCR engagement. Furthermore, the Vγ4 $\gamma\delta$ T cell subset was found to produce IL-17 in an antigen-specific fashion and to establish a pool of lung T_{RM} $\gamma\delta$ T cells, which were activated rapidly after rechallenge[14]. Depletion of Vγ4 $\gamma\delta$ T cells from convalescent animals resulted in a significant increase in bacterial counts on day 2 post reinfection, suggesting that $\gamma\delta$ T cells contribute to protective memory against *B. pertussis* in mice. The contribution of $\gamma\delta$ T cells to immunity against *B. pertussis* in humans is currently unknown. However, $\gamma\delta$ T cells were reported to decrease in the blood of children with whooping cough, suggesting that they are recruited to the site of infection[83].

3.4 Local immunity in respiratory tract

3.4.1 Immunoglobulin A

IgA is the predominant antibody isotype at mucosal surfaces where it is present in its dimeric form as secretory IgA[84]. It is produced by mucosal plasma cells and transported across the epithelial barrier via the polymeric Ig receptor. Serum IgA facilitates receptor-mediated phagocytosis while secretory IgA functions in neutralization of toxins and blocking of microbial adhesion. Despite the abundance of IgA on mucosal surfaces, selective IgA immunodeficiencies are relatively common and are not associated with a strong phenotype[85], but there is evidence of an increased incidence of respiratory tract infections[86,87].

Bordetella pertussis infection induces IgG and IgA production in mice and humans, and since infection but not parenteral vaccination induces IgA production, detection of *B. pertussis*-specific IgA in serum can be used to confirm a recent infection[88–90]. Since natural infection results in better protection than that induced by any of the currently licensed vaccines, it is likely that secretory IgA has at least some role to play in natural immunity to *B. pertussis*. However, there have been a limited number of studies on the role of IgA in clearance of *B. pertussis* and these did not directly address the contribution of secretory IgA[52,91]. A study by Wolfe et al. showed that IgA$^{-/-}$ mice clear primary and secondary *B. pertussis* infection with similar kinetics to wild-type mice, suggesting that IgA does not play a significant role[91]. The clearance of *B. bronchiseptica*, on the other hand, was significantly affected. The difference in involvement of IgA in protection against these two

Bordetella species in the mouse model may be due to the difference in the natural host range of these bacteria: *B. pertussis* is a human-adapted pathogen while *B. bronchiseptica* infects mostly other mammals[92].

IgA purified from the sera of patients with pertussis was shown to enhance phagocytosis and killing of *B. pertussis* by human polymorphonuclear leucocytes[52]. The same study showed that infection of transgenic mice expressing human IgA receptor with IgA-opsonized *B. pertussis* resulted in significantly lower bacterial counts compared with wild-type mice. Secretory IgA is present in breast milk and maternal vaccination against pertussis has been shown to lead to a significant increase in antigen-specific IgA titres in colostrum and milk of humans and animals[93,94]. IgA was also significantly increased in serum and bronchoalveolar lavage fluid of piglets born to vaccinated sows and this was associated with significant protection against *B. pertussis* challenge[94]. Given the association of IL-17 with IgA class switching[95], and in light of the findings from baboon studies[2], it is likely that secretory IgA plays an important role in prevention of *B. pertussis* infection in humans.

3.4.2 Cellular immunity in lungs

Effective immunity against *B. pertussis* is dependent on both local and systemic immune responses. The lung-resident innate immune cells together with ciliated epithelial cells provide the first line of defence. During infection with *B. pertussis*, bacteria bind to epithelial cells of the respiratory tract. Alternatively, it can also be taken up by DCs and macrophages, and the DCs migrate to the draining lymph nodes where they present antigen to naïve T cells. Further innate cells including monocytes, neutrophils, NK cells, and $\gamma\delta$ T cells are recruited to the lungs. Innate and antigen-specific $\gamma\delta$ T cells are expanded in the lung in the first few hours of infection and 7–14 days after challenge, respectively, and these cells produce IL-17 which helps to recruit neutrophils[14]. IFN-γ-secreting NK cells are also recruited to the lungs during infection with *B. pertussis* and play a role in directing the induction of antigen-specific Th1 cells[49].

Bordetella pertussis-specific CD4 T cells (Th1 and Th17 cells) activated in the lymph nodes enter the circulation and migrate in response to inflammatory signals to the infected lungs where they mediate effector functions, helping to recruit and activate neutrophils and macrophages that kill *B. pertussis*. The protective role of Th1 and Th17 cells in immunity to *B. pertussis* is well established, but this evidence is based largely on peripheral T cells. It was shown that lung and splenic T cells respond differently to purified bacterial antigens suggesting a compartmentalization of T cell responses between the lung and the periphery during *B. pertussis* infection[96]. This is associated with strong suppressive properties of lung macrophages on local T-cell populations in the lungs of infected mice[96]. Recent studies demonstrated that CD4[+] T$_{RM}$ cells accumulate in the lung during *B. pertussis* infection and significantly expand through local proliferation following reinfection[74]. These lung T$_{RM}$ cells are antigen specific and secrete IL-17, or IL-17 and IFN-γ, and may be crucial for long-term immunity against *B. pertussis*.

3.5 Immune modulation by *B. pertussis* virulence factors

Bordetella pertussis expresses many virulence factors which enhance bacterial survival and modulate immune responses of the host. Many of these factors, including FHA, pertactin (Prn), and ACT are common to different *Bordetella* species.

3.5.1 Pertussis toxin

PT is expressed exclusively by *B. pertussis*[97]. It is a secreted ADP-ribosylating toxin consisting of one active subunit and five binding subunits (AB$_5$ toxin). It binds to glycosylated molecules and thus can potentially intoxicate any mammalian cell[98]. Upon entering the cytosol, PT interferes with G-protein-coupled receptor signalling[99]. Purified PT has been shown to enhance bacterial colonization when administered intranasally even 14 days prior to infection, suggesting that it acts mainly on lung-resident cells and not on recruited cells[100]. PT has been shown to impair the function of airway macrophages and to interfere with neutrophil recruitment through inhibition of chemokine

production[29,42]. Although PT functions mainly to inhibit host immune responses, it was also found to promote IL-17 production at the later stages of the infection[42,101].

3.5.2 Filamentous haemagglutinin

FHA is the major attachment factor of *B. pertussis* and can be expressed on the cell surface or secreted into the extracellular milieu[102]. FHA binds to CD11b/CD18 on human neutrophils and enhances phagocytosis of *B. pertussis*, most likely as a mechanism of gaining access to the cell interior rather than unintentional opsonization of bacteria[103]. Soluble FHA has been shown to inhibit IL-12 but enhance IL-10 production by macrophages and DCs, which in turn leads to the generation of Treg cells that suppress Th1 responses[75,76]. Recently, FHA was shown to inhibit the induction of IL-17 responses in the lung[104]. Infection of mice with FHA-deficient *B. bronchiseptica* resulted in greater inflammatory responses at the early stages of infection with a steady influx of IL-17[+] neutrophils and macrophages.

3.5.3 Fimbriae

Fimbriae are thin, hair-like structures composed of pilin subunits[105]. They belong to the type I pili family and are thought to be involved in mediating adhesion of *B. pertussis* to ciliated epithelium. Antibodies specific to fimbriae have been shown to inhibit *B. pertussis* binding to respiratory epithelium *in vitro* and fimbriae protein (Fim)-deficient *B. bronchiseptica* was unable to effectively colonize the lower respiratory tract of mice[106,107]. Furthermore, a Fim-deficient mutant was found to induce a stronger inflammatory response than the wild-type bacteria, suggesting that Fim promotes anti-inflammatory responses in the host.

3.5.4 Adenylate cyclase toxin

ACT, similar to FHA, is able to bind to the CD11b/CD18 integrin expressed by macrophages and neutrophils. However, it is also able to intoxicate CD11b[−]/CD18[−] cells[108]. ACT is a bipartite toxin with adenylate cyclase and pore-forming activities[109].

Enzymatic activity of the toxin leads to deregulation of cellular signalling through stimulation of uncontrolled ATP to cAMP conversion, while its pore-forming ability induces innate IL-1β production in macrophages and DCs[46]. The adenylate cyclase activity of ACT has been shown to inhibit activation of antigen-presenting cells and induce their apoptosis[110–112]. Furthermore, it was found to inhibit IL-12 production by DCs but enhance LPS-induced IL-10 secretion by these cells, thus promoting induction of Tr1 cells and inhibiting generation of Th1 responses[113–115].

3.5.5 Pertactin

Prn is a 69kDa autotransporter protein that has been implicated in mediating adherence of *B. pertussis* to mammalian cells[116]. Prn has been shown to be required by *B. bronchiseptica* to resist neutrophil-mediated clearance[117]. However, Prn-deficient *B. pertussis* strains show no impairment of colonization of murine respiratory tract compared to wild-type bacteria[118–121]. Nonetheless, high anti-Prn antibody titres were found to correlate with protection against *B. pertussis* and to be crucial for its phagocytosis[122,123]. An increase in the number of Prn⁻ strains of *B. pertussis* has been reported in a number of countries and it is thought to be one of the reasons for the increasing incidence of pertussis, especially in vaccinated individuals[124,125].

3.5.6 Dermonecrotic toxin

Dermonecrotic toxin (DNT), also termed heat-labile toxin due to its temperature sensitivity, is a single-chain polypeptide with an N-terminal receptor-binding domain and a C-terminal enzymatic domain[126,127]. DNT is activated by proteolytic cleavage by furin-type proteases inside the endosomal compartment[127]. It is then translocated into the cytosol where it catalyses polyamination or deamidation of intracellular Rho GTPases, resulting in their constitutive activation[128]. DNT stimulates DNA and protein synthesis and alters cell morphology through reorganization of actin stress fibres[129]. *Bordetella bronchiseptica* DNT has been shown to be the principal mediator of turbinate atrophy in pigs[130]. DNT is lethal to mice when injected intravenously and causes necrotic skin lesions upon subcutaneous injection[126]. A mutant of *B. bronchiseptica* lacking DNT has been shown to be as capable as the wild-type strain at colonizing the lungs of mice[131]. However, DNT-sufficient bacteria cause more severe lung damage with characteristic necrotizing lesions.

3.5.7 Tracheal cytotoxin

Tracheal cytotoxin (TCT) is a 9.2 kDa muramyl peptide derived from the *B. pertussis* cell wall component peptidoglycan. It is released during the log phase of bacterial growth and causes damage to the respiratory epithelium and ciliostasis, thus impairing clearance of bacteria through normal ciliary motion[132,133]. It also inhibits DNA synthesis in the epithelium and prevents cilia regeneration[134]. TCT was shown to induce IL-1 production in hamster trachea epithelium *in vitro*[135]. Treatment of respiratory epithelium with IL-1, but not TNF or IL-6, reproduced toxic effects of TCT implicating IL-1 as a mediator of TCT-induced tissue damage[133]. TCT was shown to strongly synergize with LPS for induction of IL-1α, NO production, and inhibition of DNA synthesis[135]. Furthermore, it was shown to be toxic to neutrophils and inhibit their function[136]. TCT is structurally related to other muramyl peptides which activate the intracellular NLRs, NOD-1 and NOD-2. TCT has been shown to strongly stimulate murine NOD-1 but not its human orthologue[137].

3.5.8 Type III secretion system

The T3SS is a sophisticated toxin delivery apparatus which is expressed by a variety of Gram-negative bacteria[138]. It is composed of a number of structural and effector proteins. T3SS allows for direct injection of bacterial toxins into the target cell via a needle-like structure. The T3SS has been most extensively studied in *B. bronchiseptica*, but the gene locus is highly conserved in *B. pertussis* and *B. parapertussis*[139]. Studies using laboratory-adapted strains of *B. pertussis* had suggested that the T3SS is not functional in *B. pertussis*[140]. However, the components of *B. pertussis* T3SS are readily detectable in recent clinical isolates but are lost in laboratory-adapted strains after prolonged *in vitro* culture[78]. Infection of mice with a mutant *B. pertussis* strain that does not express

T3SS was shown to result in significantly lower colonization of the respiratory tract compared with the wild-type strain[78]. Furthermore, T3SS-defective mutant strains induced significantly greater antigen-specific IL-17 and IFN-γ production suggesting that T3SS promotes bacterial growth by dampening innate and adaptive immune response.

3.5.9 Lipopolysaccharide

LPS is a component of the cell wall of most Gram-negative bacteria. It is composed of lipid A, a core oligosaccharide, and polysaccharide O-antigen. *B. pertussis* LPS lacks the O-antigen and therefore is often referred to as lipooligosaccharide (LOS)[126]. Similar to LPS from other Gram-negative bacteria, *B. pertussis* LPS has mitogenic and pyrogenic properties[141]. LPS binds to TLR4 on innate immune cells such as DCs inducing their maturation and stimulating the production of a number of innate cytokines. *Bordetella pertussis* LPS has been shown to induce the production of IL-1β, IL-23, IL-12, and IL-10 by mouse DCs and human monocyte-derived DCs[31, 142]. C3H/HeJ mice, which have defective TLR4 signalling, have more severe infection, reduced IL-10 production, and increased pathology compared to wild-type mice[44,143]. C3H/HeJ mice also have significantly lower antigen-specific IL-17 and IFN-γ responses following wP vaccination[31]. TLR4-defective spleen cells or DCs produce significantly lower quantities of a number of cytokines, including IL-1β,

TNF, IL-12, IL-23, and IL-10[31,44,143]. Consequently, supernatants from TLR4-defective murine DCs stimulated with *B. pertussis* LPS fail to induce IL-17 production from T cells. TLR4 signalling has been shown to be crucial for neutrophil infiltration into the lung during the first 24 hours of *B. pertussis* infection[144]. Recently, it has been shown that some clinical *B. pertussis* isolates express lipid A with modifications that preclude it from activating TLR4 signalling[145]. These isolates show impaired activation of DC maturation and cytokine secretion, highlighting the importance of TLR4 signalling in activation of innate immune responses against *B. pertussis* in humans as well as in mice.

3.6 Vaccine-induced immunity

3.6.1 Whole-cell and acellular pertussis vaccines

The wP vaccines comprising killed whole *B. pertussis* organisms are highly immunogenic and confer a high level of protection against pertussis, but they are also reactogenic. The reactogenicity of wP vaccines is largely due to the presence of LPS in the vaccine[31,146]. wP vaccines are more effective than aP vaccines and one explanation for this is that they mimic natural infection by inducing a mixed Th1 and Th17 cell response[12], whereas the aP vaccines elicit strong Th2, but weak Th1 responses[12,62,147,148] (see Table 3.1). The polarization of the response to Th2 over Th1 with the aP vaccines probably reflects

Table 3.1 Summary of immune response induced by previous infection and immunization with pertussis vaccines

Response (species)	Natural infection	Whole-cell pertussis vaccine	Acellular pertussis vaccine
Side effects (humans)	Severe disease	Fevers (common), seizures (rare)	Few; rare cases of leg swelling after boosters
Antibody response (humans and mice)	Strong	Moderate	High (to vaccine antigens)
T-cell response (humans and mice)	Th1 (strong) Th17 (strong)	Th1 (strong), Th17 (moderate)	Th2 (strong), Th17 (weak)
Protection against disease (children)	High	>90% efficacy	70–85% efficacy
Protection against colonization/transmission (baboons and mice)	High	Moderate	None
Persistence of protection (humans and mice)	Good	Moderate	Poor
T-cell memory (mice)	Very good	Moderate	Poor

The summary of responses is a consensus view by the authors of this chapter, based on numerous reports from studies in humans and animal models cited in this review.

the absence of a number of pathogen-associated molecular patterns (PAMPs), including LPS and TLR2 ligands, produced by the live bacteria and present in the wP vaccine, and the use of alum as the adjuvant[12,31,149].

3.6.2 Vaccine antigens

A number of *B. pertussis* virulence factors target the host, which enhances bacterial survival and modulates host immune responses. Many of these factors, including FHA, Prn, and ACT, are common to different *Bordetella* species. Different combinations of *B. pertussis* antigens, including PT, FHA, Prn, and Fim are components of current aP vaccines, selected on the basis of their virulence and bacterial attachment, and their ability to generate antibody responses when used as immunogens in mice and humans. All licensed aP vaccines include detoxified PT (dPT) and most also include FHA. The consensus view is that the efficacy of aP vaccines reflects the number of component antigens, with five- and three-component vaccines having higher efficacy than two- or one-component vaccines[5,150].

3.6.3 Adjuvants for pertussis vaccines: old and new

Purified antigens are not very immunostimulatory and adjuvants are added to subunit vaccines to enhance the immunogenicity of the antigens[151]. The wP vaccine contains several intrinsic adjuvants including the TLR4 ligand LPS[152] and a number of TLR2 lipoproteins[149]. However, these PAMPs are absent or at very low levels in the purified antigens included in current aP vaccines, necessitating the addition of an exogenous adjuvant.

Alum

Alum, which was first used as a vaccine adjuvant in the 1940s, was chosen as the adjuvant for aP vaccines largely based on its good safety profile[151]. However, while alum is effective at promoting robust antibody and Th2-type responses, it is inefficient at inducing Th1 cells, which are important for *B. pertussis* clearance[12]. Indeed, alum has been shown to impair the induction of Th1 responses by directly inhibiting IL-12p35 expression and, therefore, IL-12

production by DCs[153]. Although aP vaccines also induce Th17 cells, the polarization to Th2 over Th1 responses is in our view one of the shortcomings of alum as an adjuvant for aP vaccines. This has motivated the search for alternative adjuvants for pertussis vaccines that induce cellular as well as humoral immunity.

TLR agonists

Several studies in animal models have shown that substituting alum for TLR agonists enhances the efficacy of experimental aP vaccines. Indeed, the Mills' laboratory has shown that replacing alum with the TLR2 agonist LP1569[149] or the TLR9 agonist CpG[12] in a laboratory-prepared aP vaccine enhanced protection against *B. pertussis* challenge in a mouse model by inducing potent Th1 and Th17 cell responses. Kindrachuk and colleagues found that the combination of CpG and the synthetic innate defence regulator peptide HH2 (CpG-HH2) was an effective adjuvant for an aP vaccine consisting of dPT[154]. Indeed, mice immunized intranasally with this laboratory-prepared aP vaccine formulated with CpG-HH2 produced higher levels of PT-specific IgG2a and IgG1 antibodies than mice immunized with the aP vaccines formulated with either CpG or HH2 alone[154]. These findings indicated that the adjuvant combination CpG-HH2 is capable of inducing a mixed Th1/Th2 response. Asokanathan et al. also showed that addition of CpG to an aP vaccine already formulated with alum accelerated *B. pertussis* clearance from the lungs, enhanced IFN-γ production by spleen cells, and strongly activated peritoneal macrophages, compared with mice immunized with an aP vaccine formulated with alum only[155].

Other studies have focused on generating less toxic versions of LPS as adjuvants. Brummelman and colleagues demonstrated that addition of a meningococcal LPS derivative, LpxL1 to a commercial alum-adjuvanted aP vaccine enhanced the antigen-specific IFN-γ response and IgG2a to IgG1 ratio in mice, compared with mice immunized with the commercial alum-adjuvanted aP vaccine alone, suggesting enhancement of Th1 responses[156]. Furthermore, the LPS analogue monophosphoryl lipid A (MPL) has been shown to be a suitable replacement for alum in an aP vaccine; Geurtsen and colleagues found that mice immunized with an MPL-adjuvanted aP vaccine had lower bacterial burden in the lungs post challenge than mice immunized with an alum-adjuvanted aP vaccine[157]. Moreover, antigen-specific IL-5 production by spleen cells restimulated with *B. pertussis* was lower in those obtained from mice immunized with the aP vaccine formulated with MPL compared with those from mice immunized with the alum-adjuvanted aP vaccine, indicating that MPL downregulated the Th2 response[157]. The adjuvant ASO4, which consists of MPL co-formulated with alum, has already been approved for human use and therefore it could be a promising alternative to alum[158]. Indeed, Agnolon et al. demonstrated that addition of MPL to an alum-adjuvanted aP vaccine induced high antigen-specific serum IgG2a titres in immunized mice, indicating a Th1-biased response, whereas mice immunized with the alum-adjuvanted aP vaccine only produced IgG1, suggesting a Th2 profile[159].

It has also been demonstrated that addition of a small molecule TLR7 agonist to an alum-adjuvanted aP vaccine enhanced the efficacy of the vaccine to the level seen with the wP vaccine Superscript: 160. Furthermore, it redirected the immune responses from predominantly Th2 and IgG1 antibodies induced with alum, to Th1 and Th17 IgG2a/b antibodies[160]. Thus, several TLR agonists have been shown to be more effective adjuvants for aP vaccines than alum alone. Given the low costs involved in the manufacture of these molecules and their chemical stability, they are promising adjuvants for next-generation pertussis vaccines.

We have recently demonstrated that the adjuvant activity of the TLR agonist LP1569 can be enhanced when used in combination with c-di-GMP, an agonist for intracellular receptor stimulator of interferon genes (STING) Superscript: 188. Immunization of mice with an experimental aP vaccine formulated with LP1569 and c-di-GMP induced potent Th1 and Th17 responses, and conferred protection against lung infection and nasal colonization with *B. pertussis*, whereas the same vaccine formulated with alum failed to protect against nasal colonization. Furthermore, immunization with an aP vaccine formulated with LP1569 and c-di-GMP by the intranasal route induced potent respiratory T_{RM} CD4 T cells that sustained protective immunity against nasal colonization, as

well as lung infection, for at least 10 months Superscript: 188.

Cytokines and emulsions

Several studies in cancer and infectious disease models have demonstrated that T-cell-polarizing cytokines can act as potent adjuvants. Indeed IL-12, which is the primary cytokine involved in promoting Th1 cell differentiation[161], has been shown to be an effective adjuvant for an aP vaccine; Mahon et al. demonstrated that mice immunized with an experimental aP vaccine consisting of FHA, dPT, and Prn formulated with IL-12 rapidly cleared *B. pertussis* infection and were as protected from aerosol challenge as mice immunized with a wP vaccine[162].

MF59 is an oil-in-water emulsion adjuvant which has been licensed for use in human vaccines since 1997[163]. Currently, MF59 is used as an adjuvant in pandemic and seasonal influenza vaccines[164,165]. MF59 has been shown to induce more potent T-cell and antibody responses than alum[166] and indeed, a study by Agnolon et al. demonstrated that an aP vaccine formulated with MF59 induced a Th1-biased (IgG2a) profile in mice[159].

Particulate antigen delivery systems

Incorporating soluble antigens into particles is an effective method of enhancing the efficacy of subunit vaccines, including the aP vaccine. Garlapati et al. have shown that incorporation of an aP vaccine adjuvanted with CpG and the innate defence regulator peptide 1002 (IDR) into polyphosphazene (PCEP) microparticles enhanced vaccine efficacy; mice immunized with the PCEP vaccine were more protected from intranasal *B. pertussis* challenge than mice immunized with a soluble formulation of the vaccine, and had similar bacterial counts in the lungs on day 7 post infection as mice immunized with a licensed alum-adjuvanted aP vaccine[166]. Furthermore, spleen cells from mice immunized with the PCEP vaccine had a high IFN-γ/IL-4 ratio, indicating a Th1-biased response, whereas mice immunized with an alum-adjuvanted aP vaccine had a low one, which is indicative of a Th2-type response[167].

Further evidence of the benefit of delivering soluble antigens encapsulated within a particle was provided by Li et al., who showed that spleen cells from mice immunized with dPT encapsulated in biodegradable polylactic-co-glycolic acid (PLGA)-based nano/microparticles produced significantly more *B. pertussis*-specific IFN-γ and IL-17 (Th1/Th17) than spleen cells from mice immunized with the soluble antigens[168]. Furthermore, mice immunized with the particulate vaccine formulation were more protected from *B. pertussis* challenge than mice immunized with the soluble vaccine formulation[168].

Conway et al. have also demonstrated that immunizing mice twice with dPT and FHA formulated in PLGA microparticles induced Th1 responses and conferred a high level of protection against *B. pertussis* infection[169]. Furthermore, a single intraperitoneal administration of *B. pertussis* Fim entrapped within PLGA microparticles induced potent antigen-specific antibody production in mice and protected them from infection as efficiently as a single intraperitoneal immunization with alum-adjuvanted Fim[170]. The PLG polymer has already been approved for human use by the US Food and Drug Administration[171]. These findings suggest that microparticles may be safe and effective delivery systems for enhancing immune responses to *B. pertussis* antigens.

3.6.4 Outer membrane vesicles

Outer membrane vesicles (OMVs) are naturally secreted from the surface of most Gram-negative bacteria during normal cell processes and include antigens in their native conformation[172]. *B. pertussis* growing in liquid culture has been shown to release OMVs containing multiple *B. pertussis* antigens including LOS, Prn, and PT[173]. As these OMVs have inherent adjuvanticity and immunogenic properties (they contain some of the main immunogens included in aP vaccines), they could be a valuable asset in the development of third-generation aP vaccines. Indeed, administering OMVs by the intranasal route has been shown to confer protection against *B. pertussis* intranasal challenge, as has administration by the intraperitoneal route when the OMVs are adjuvanted with alum[174]. Immunizing mice with *B. pertussis* OMVs formulated with diphtheria and tetanus toxoids have been shown to induce as potent Th1 and Th17 responses as a wP vaccine, and a greater IgG2a/IgG1 ratio than seen

with an aP vaccine[175]. Furthermore, an OMV-based pertussis vaccine is more effective than a commercial aP vaccine in protecting against recent clinical isolates of *B. pertussis*[176]. In a different study, an OMV-based vaccine was found to induce milder inflammatory responses than wP vaccines without compromising the efficacy[177].

3.6.5 Live attenuated *B. pertussis* vaccines

Genetically attenuated bacteria are still capable of colonizing the host and inducing an immune response without causing severe disease, and this could potentially be a very effective method of inducing protective immunity to *B. pertussis*. Indeed, a live attenuated strain of *B. pertussis*, BPZE1, has been developed, that lacks key virulence factors and does not cause disease in humans[178]. A single intranasal immunization with BPZE1 was shown to induce a potent Th1 response and to protect infant (3-week-old) mice significantly more than two doses of an aP vaccine from *B. pertussis* and *B. parapertussis* infection[178]. Furthermore, immunization with BPZE1 has been shown to promote robust Th17 cell responses. CD4+ but not CD8+ T cells have been shown to be crucial for BPZE1-mediated protection, presumably through the production of IFN-γ and IL-17[179]. BPZE1 has also been shown to induce systemic *B. pertussis*-specific IgG[178] and local IgA responses[180] in immunized mice. Long-term protection induced with BPZE1 has also been shown to be substantially better than that conferred with an aP vaccine; 12 months after immunization, infant and adult mice immunized intranasally once with BPZE1 were significantly more protected from intranasal challenge with *B. pertussis* than mice immunized twice intraperitoneally with an aP vaccine[181].

BPZE1 could be administered to infants as a single-dose intranasal vaccine prior to the scheduled course of diphtheria, tetanus, and pertussis vaccinations, and indeed, studies in mice investigating the efficacy of this option have been promising. Neonatal mice intranasally primed with BPZE1 maintain their Th1 and Th17 cell profile, and a greater IgG2a/IgG1 antibody ratio, despite two subsequent intraperitoneal booster immunizations with aP vaccines[182]. Furthermore, mice intranasally immunized with

BPZE1 and boosted once with an aP vaccine were significantly more protected from intranasal challenge with *B. pertussis* than mice that were intranasally administered BPZE1 once, or intraperitoneally immunized with an aP vaccine once[182].

The safety of BPZE1 was evaluated using immunodeficient mouse strains in advance of clinical studies in humans. Unlike wild-type bacteria, BPZE1 did not cause a disseminating infection in IFN-γR$^{-/-}$ mice[183]. Furthermore, infection with live attenuated BPZE1 was well tolerated in neonatal mice, whereas infection with virulent *B. pertussis* resulted in significant mortality[183]. Finally, this vaccine has been evaluated in phase I clinical trials in humans and has been shown to be immunogenic and well tolerated, indicating it has substantial potential as an alternative vaccine for pertussis[184,185].

3.7 Conclusions and future directions

Since the introduction of aP vaccines into paediatric immunization programmes in most high income countries in the late 1990s and early 2000s, our knowledge on the mechanism of natural and vaccine-induced immunity to *B. pertussis* has increased significantly. Prior to that, most of the focus had been on the role of antibodies in protective immunity. Since most current prophylactic vaccines against infectious diseases prevent infection by inducing antibodies and memory B cells, the focus on antibodies was understandable, but this may also have resulted in the development of vaccines that were suboptimal. While aP vaccines are clearly safer than the wP vaccines they replaced and while they do prevent death from severe diseases, the emerging view, based largely on animal models, is that current aP vaccines are not as effective as wP or natural immunity because they fail to induce the appropriate cellular immune responses, in particular memory T cells.

Although there had been a number of papers describing *B. pertussis*-specific T cells in previously infected humans[186,187], the first papers demonstrating protective T cells were not published until after the large phase 3 clinical trials in Sweden and Italy in the early 1990s. These early studies, together with a significant number of reports over the last two decades, have now provided a more comprehensive picture of the role of innate and adaptive immune systems in the clearance of a primary infection and the complementary role of antibodies and T cells in natural and vaccine-induced acquired immunity. There is very convincing data from mouse models, supported by indirect evidence from human and baboon studies, to suggest that Th1 cells play a critical role in adaptive immunity to *B. pertussis*, with growing evidence of a role for Th17 cells. Furthermore IgA and T$_{RM}$ cells appear to be critical for local immunity and memory in the respiratory tract and are induced by previous infection, but less effectively by parenteral vaccination.

Studies on the immune responses to infection and vaccination have provided significant insight into the mechanism of immunity to *B. pertussis*, however there are several unanswered questions that still need to be resolved in order to design more effective vaccines:

- Why do previous infection and wP, but not aP, vaccination prevent nasal colonization?
- How can sterilizing immunity be generated by vaccination?
- What immunization approaches can enhance the persistence of protective immunity and immunological memory?
- What vaccine adjuvants are most effective for induction of Th1/Th17 cells and T$_{RM}$ cells with aP vaccines, and will these be safe and acceptable for use in infants?
- Will it be possible to switch Th2-polarized responses in aP vaccinated individuals to Th1/Th17 responses by booster immunization with a new aP vaccine formulated with novel adjuvants?
- What is the mechanism of protective immunity against nasal colonization?
- Can we identify correlates or biomarkers of protective immunity in humans?

The much articulated notion that research will reveal a single correlate or surrogate of protective immunity against *B. pertussis* that will facilitate the development of a new vaccine does not give credit to the complexity of the host immune system, or indeed the bacterium. There are examples of

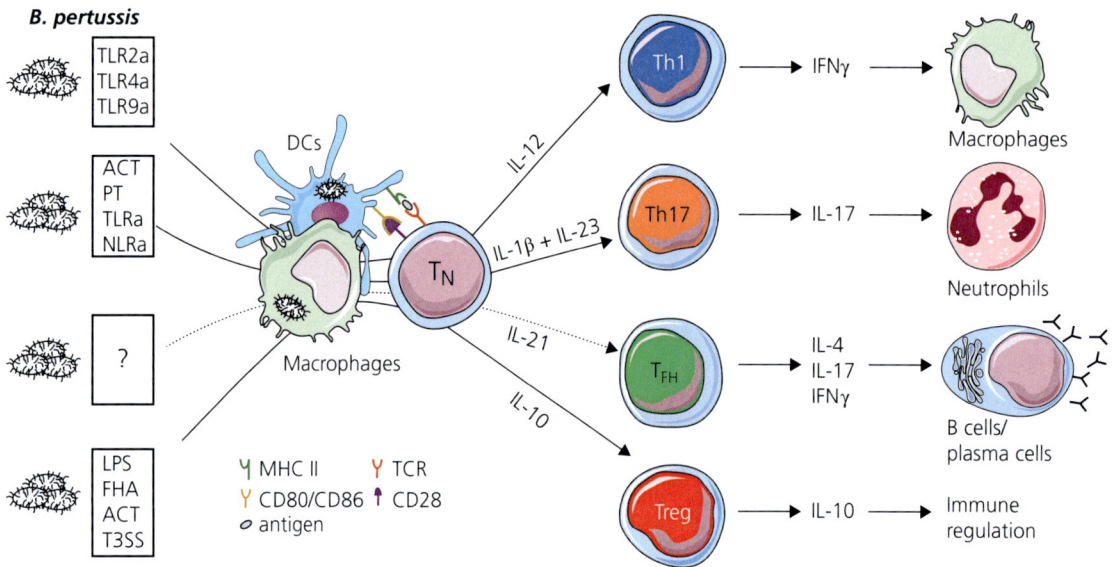

Figure 3.3 *Bordetella pertussis* virulence factors promote the development of multiple subtypes of CD4 T cells. The priming of naïve T cells to become effector T cells requires three signals: (1) TCR interaction with MHC antigen peptide (*B. pertussis* provides a wide range of antigens for signal 1), (2) CD28 interaction with CD80/CD86 (expression of co-stimulatory molecules on dendritic cells (DCs) or macrophages (Mφ) is promoted by LPS and other TLR ligands), and (3) T-cell polarization cytokines secreted by DCs, Mφ, and other innate cells in response to *B. pertussis* virulence factors. The TLR4 ligand LPS, TLR2 lipopeptide ligands, and bacterial DNA binding to TLR9 promote production of IL-12. Adenylate cyclase toxin (ACT) and pertussis toxin (PT) together with various TLR ligands promote NLRP3 inflammasome activation, leading to IL-1 induction, which together with IL-23 enhances induction of Th17 cells. LPS, FHA, ACT, and T3SS effector molecules induce IL-10 production by DCs and Mφ, which promotes induction of Treg cells. T$_{FH}$ cells are also induced during *B. pertussis* infection, and while this may involve IL-21, the virulence factors involved have not yet been identified. IFN-γ secreted by Th1 cells activates Mφ and IL-17 secreted by Th17 cells helps to recruit neutrophils. IL-4, IL-17, and IFN-γ secretion by T$_{FH}$ cells help to activate B cells and plasma cells leading to antibody production and immunoglobulin class switching. Treg cells help to regulate effector T-cell responses and prevent immunopathology in the lungs from excessive inflammation during *B. pertussis* infection.

diseases where the serum antibody titres against a single toxin can predict protection. Indeed, antibodies that neutralize PT may be sufficient to prevent deaths from pertussis in infants. However, *B. pertussis* produces multiple toxins and virulence factors that mediate bacterial attachment to host cells and modulate immune responses, and there are a range of host immune responses that help to prevent or clear the infection (see Figure 3.1 and Figure 3.3). Therefore, the identification of a correlate of protection, where protection means the prevention of nasal infection/carriage and transmission of *B. pertussis*, will not be so straightforward.

The introduction of a new vaccine against pertussis will face many logistic obstacles, especially in relation to clinical trials, regulatory approval, and assays that can be used to assess vaccine efficacy or potency. Nevertheless, it is imperative that the latest research in immunology, in particular strategies for inducing Th1/Th17 cells and T$_{RM}$ cells, is exploited to its fullest to facilitate the development of more effective third-generation vaccines against pertussis.

Acknowledgements

KHGM is supported by research grants from Science Foundation Ireland (11/PI/1036 and 16/IA/4468) and the PERISCOPE project, which has received funding from the Innovative Medicines Initiative 2 Joint Undertaking under grant agreement No 115910. This Joint Undertaking receives support from the European Union's Horizon 2020 research and innovation programme, the European Federation of Pharmaceutical Industries and Associations, and the Bill & Melinda Gates Foundation.

References

1. Higgs R, Higgins SC, Ross PJ, et al. Immunity to the respiratory pathogen Bordetella pertussis. *Mucosal Immunol* 2012;**5**:485–500.

2. Warfel JM, Zimmerman LI, Merkel TJ. Acellular pertussis vaccines protect against disease but fail to prevent infection and transmission in a nonhuman primate model. *Proc Natl Acad Sci U S A* 2014;**111**:787–92.

3. Greco D, Salmaso S, Mastrantonio P, et al. A controlled trial of two acellular vaccines and one whole-cell vaccine against pertussis. Progetto Pertosse Working Group. *N Engl J Med* 1996;**334**:341–8.

4. Gustafsson L, Hallander HO, Olin P, et al. A controlled trial of a two-component acellular, a five-component acellular, and a whole-cell pertussis vaccine. *N Engl J Med* 1996;**334**:349–55.

5. Zhang L, Prietsch SO, Axelsson I, et al. Acellular vaccines for preventing whooping cough in children. *Cochrane Database Syst Rev* 2012:CD001478.

6. Barret AS, Ryan A, Breslin A, et al. Pertussis outbreak in northwest Ireland, January–June 2010. *Euro Surveill* 2010;**15**:19654.

7. Octavia S, Sintchenko V, Gilbert GL, et al. Newly emerging clones of Bordetella pertussis carrying prn2 and ptxP3 alleles implicated in Australian pertussis epidemic in 2008–2010. *J Infect Dis* 2012;**205**:1220–4.

8. Winter K, Harriman K, Zipprich J, et al. California pertussis epidemic, 2010. *J Pediatr* 2012;**161**:1091–6.

9. Cherry JD. Epidemic pertussis in 2012—the resurgence of a vaccine-preventable disease. *N Engl J Med* 2012; **367**:785–7.

10. Amirthalingam G, Gupta S, Campbell H. Pertussis immunisation and control in England and Wales, 1957 to 2012: a historical review. *Euro Surveill* 2013;**18**:20587.

11. Redhead K, Watkins J, Barnard A, et al. Effective immunization against Bordetella pertussis respiratory infection in mice is dependent on induction of cell-mediated immunity. *Infect Immun* 1993;**61**:3190–8.

12. Ross PJ, Sutton CE, Higgins S, et al. Relative contribution of Th1 and Th17 cells in adaptive immunity to Bordetella pertussis: towards the rational design of an improved acellular pertussis vaccine. *PLoS Pathog* 2013;**9**:e1003264.

13. Fernandez RC, Weiss AA. Susceptibilities of Bordetella pertussis strains to antimicrobial peptides. *Antimicrob Agents Chemother* 1996;**40**:1041–3.

14. Misiak A, Wilk MM, Raverdeau M, et al. IL-17-producing innate and pathogen-specific tissue resident memory gammadelta T Cells expand in the lungs of Bordetella pertussis-infected mice. *J Immunol* 2017;**198**: 363–74.

15. Elahi S, Buchanan RM, Attah-Poku S, et al. The host defense peptide beta-defensin 1 confers protection against Bordetella pertussis in newborn piglets. *Infect Immun* 2006;**74**:2338–52.

16. Erles K, Brownlie J. Expression of beta-defensins in the canine respiratory tract and antimicrobial activity against Bordetella bronchiseptica. *Vet Immunol Immunopathol* 2010;**135**:12–9.

17. Legarda D, Klein-Patel ME, Yim S, et al. Suppression of NF-kappaB-mediated beta-defensin gene expression in the mammalian airway by the Bordetella type III secretion system. *Cell Microbiol* 2005;**7**:489–97.

18. Shah NR, Hancock RE, Fernandez RC. Bordetella pertussis lipid A glucosamine modification confers resistance to cationic antimicrobial peptides and increases resistance to outer membrane perturbation. *Antimicrob Agents Chemother* 2014;**58**:4931–4.

19. Ganguly T, Johnson JB, Kock ND, et al. The Bordetella pertussis Bps polysaccharide enhances lung colonization by conferring protection from complement-mediated killing. *Cell Microbiol* 2014;**16**:1105–18.

20. Barnes MG, Weiss AA. BrkA protein of Bordetella pertussis inhibits the classical pathway of complement after C1 deposition. *Infect Immun* 2001;**69**:3067–72.

21. Jongerius I, Schuijt TJ, Mooi FR, et al. Complement evasion by Bordetella pertussis: implications for improving current vaccines. *J Mol Med (Berl)* 2015;**93**:395–402.

22. Noofeli M, Bokhari H, Blackburn P, et al. BapC autotransporter protein is a virulence determinant of Bordetella pertussis. *Microb Pathog* 2011;**51**:169–77.

23. Zipfel PF, Hallstrom T, Riesbeck K. Human complement control and complement evasion by pathogenic microbes—tipping the balance. *Mol Immunol* 2013;**56**: 152–60.

24. Zipfel PF, Skerka C. Complement regulators and inhibitory proteins. *Nat Rev Immunol* 2009;**9**:729–40.

25. Marr N, Shah NR, Lee R, et al. Bordetella pertussis autotransporter Vag8 binds human C1 esterase inhibitor and confers serum resistance. *PLoS One* 2011;**6**:e20585.

26. Berggard K, Johnsson E, Mooi FR, et al. Bordetella pertussis binds the human complement regulator C4BP: role of filamentous hemagglutinin. *Infect Immun* 1997;**65**:3638–43.

27. Berggard K, Lindahl G, Dahlback B, et al. Bordetella pertussis binds to human C4b-binding protein (C4BP) at a site similar to that used by the natural ligand C4b. *Eur J Immunol* 2001;**31**:2771–80.

28. Amdahl H, Jarva H, Haanpera M, et al. Interactions between Bordetella pertussis and the complement inhibitor factor H. *Mol Immunol* 2011;**48**:697–705.

29. Carbonetti NH, Artamonova GV, Van Rooijen N, et al. Pertussis toxin targets airway macrophages to promote Bordetella pertussis infection of the respiratory tract. *Infect Immun* 2007;**75**:1713–20.

30. Canthaboo C, Xing D, Wei XQ, et al. Investigation of role of nitric oxide in protection from Bordetella pertussis respiratory challenge. *Infect Immun* 2002;**70**: 679–84.

31. Higgins SC, Jarnicki AG, Lavelle EC, et al. TLR4 mediates vaccine-induced protective cellular immunity to Bordetella pertussis: role of IL-17-producing T cells. *J Immunol* 2006;**177**:7980–9.

32. Mahon BP, Mills KH. Interferon-gamma mediated immune effector mechanisms against Bordetella pertussis. *Immunol Lett* 1999;**68**:213–7.

33. Lamberti YA, Hayes JA, Perez Vidakovics ML, et al. Intracellular trafficking of Bordetella pertussis in human macrophages. *Infect Immun* 2010;**78**:907–13.

34. Valdez HA, Oviedo JM, Gorgojo JP, et al. Bordetella pertussis modulates human macrophage defense gene expression. *Pathog Dis* 2016;**74**:ftw073.

35. Masure HR. The adenylate cyclase toxin contributes to the survival of Bordetella pertussis within human macrophages. *Microb Pathog* 1993;**14**:253–60.

36. Friedman RL, Nordensson K, Wilson L, et al. Uptake and intracellular survival of Bordetella pertussis in human macrophages. *Infect Immun* 1992;**60**:4578–85.

37. Saukkonen K, Cabellos C, Burroughs M, et al. Integrin-mediated localization of Bordetella pertussis within macrophages: role in pulmonary colonization. *J Exp Med* 1991;**173**:1143–9.

38. Hellwig SM, Hazenbos WL, van de Winkel JG, et al. Evidence for an intracellular niche for Bordetella pertussis in broncho-alveolar lavage cells of mice. *FEMS Immunol Med Microbiol* 1999;**26**:203–7.

39. Lenz DH, Weingart CL, Weiss AA. Phagocytosed Bordetella pertussis fails to survive in human neutrophils. *Infect Immun* 2000;**68**:956–9.

40. Lamberti Y, Perez Vidakovics ML, van der Pol LW, et al. Cholesterol-rich domains are involved in Bordetella pertussis phagocytosis and intracellular survival in neutrophils. *Microb Pathog* 2008;**44**:501–11.

41. Andreasen C, Carbonetti NH. Role of neutrophils in response to Bordetella pertussis infection in mice. *Infect Immun* 2009;**77**:1182–8.

42. Andreasen C, Carbonetti NH. Pertussis toxin inhibits early chemokine production to delay neutrophil recruitment in response to Bordetella pertussis respiratory tract infection in mice. *Infect Immun* 2008;**76**:5139–48.

43. Eby JC, Hoffman CL, Gonyar LA, et al. Review of the neutrophil response to Bordetella pertussis infection. *Pathog Dis* 2015;**73**:ftv081.

44. Higgins SC, Lavelle EC, McCann C, et al. Toll-like receptor 4-mediated innate IL-10 activates antigen-specific regulatory T cells and confers resistance to Bordetella pertussis by inhibiting inflammatory pathology. *J Immunol* 2003;**171**:3119–27.

45. Dunne PJ, Moran B, Cummins RC, et al. CD11c+ CD8alpha+ dendritic cells promote protective immunity to respiratory infection with Bordetella pertussis. *J Immunol* 2009;**183**:400–10.

46. Dunne A, Ross PJ, Pospisilova E, et al. Inflammasome activation by adenylate cyclase toxin directs Th17 responses and protection against Bordetella pertussis. *J Immunol* 2010;**185**:1711–9.

47. Fedele G, Spensieri F, Palazzo R, et al. Bordetella pertussis commits human dendritic cells to promote a Th1/Th17 response through the activity of adenylate cyclase toxin and MAPK-pathways. *PLoS One* 2010;**5**:e8734.

48. Wu V, Smith AA, You H, et al. Plasmacytoid dendritic cell-derived IFNalpha modulates Th17 differentiation during early Bordetella pertussis infection in mice. *Mucosal Immunol* 2016;**9**:777–86.

49. Byrne P, McGuirk P, Todryk S, et al. Depletion of NK cells results in disseminating lethal infection with Bordetella pertussis associated with a reduction of antigen-specific Th1 and enhancement of Th2, but not Tr1 cells. *Eur J Immunol* 2004;**34**:2579–88.

50. Mahon BP, Sheahan BJ, Griffin F, et al. Atypical disease after Bordetella pertussis respiratory infection of mice with targeted disruptions of interferon-gamma receptor or immunoglobulin mu chain genes. *J Exp Med* 1997;**186**:1843–51.

51. Sozzani S, Locati M, Zhou D, et al. Receptors, signal transduction, and spectrum of action of monocyte chemotactic protein-1 and related chemokines. *J Leukoc Biol* 1995;**57**:788–94.

52. Hellwig SM, van Spriel AB, Schellekens JF, et al. Immunoglobulin A-mediated protection against Bordetella pertussis infection. *Infect Immun* 2001;**69**:4846–50.

53. Granstrom M, Olinder-Nielsen AM, Holmblad P, et al. Specific immunoglobulin for treatment of whooping cough. *Lancet* 1991;**338**:1230–3.

54. Aase A, Herstad TK, Jorgensen SB, et al. Anti-pertussis antibody kinetics following DTaP-IPV booster vaccination in Norwegian children 7–8 years of age. *Vaccine* 2014;**32**:5931–6.

55. Snapper CM, Peschel C, Paul WE. IFN-gamma stimulates IgG2a secretion by murine B cells stimulated with bacterial lipopolysaccharide. *J Immunol* 1988;**140**:2121–7.

56. Bossie A, Vitetta ES. IFN-gamma enhances secretion of IgG2a from IgG2a-committed LPS-stimulated murine B cells: implications for the role of IFN-gamma in class switching. *Cell Immunol* 1991;**135**:95–104.

57. Zhang Z, Goldschmidt T, Salter H. Possible allelic structure of IgG2a and IgG2c in mice. *Mol Immunol* 2012;**50**:169–71.

58. Coffman RL, Seymour BW, Lebman DA, et al. The role of helper T cell products in mouse B cell differentiation and isotype regulation. *Immunol Rev* 1988;**102**:5–28.

59. Moon HB, Severinson E, Heusser C, et al. Regulation of IgG1 and IgE synthesis by interleukin 4 in mouse B cells. *Scand J Immunol* 1989;**30**:355–61.

60. Leef M, Elkins KL, Barbic J, et al. Protective immunity to Bordetella pertussis requires both B cells and CD4(+) T cells for key functions other than specific antibody production. *J Exp Med* 2000;**191**:1841–52.

61. Mills KH, Barnard A, Watkins J, et al. Cell-mediated immunity to Bordetella pertussis: role of Th1 cells in bacterial clearance in a murine respiratory infection model. *Infect Immun* 1993;**61**:399–410.

62. Mills KH, Ryan M, Ryan E, et al. A murine model in which protection correlates with pertussis vaccine efficacy in children reveals complementary roles for humoral and cell-mediated immunity in protection against Bordetella pertussis. *Infect Immun* 1998;**66**:594–602.

63. Warfel JM, Merkel TJ. Bordetella pertussis infection induces a mucosal IL-17 response and long-lived Th17 and Th1 immune memory cells in nonhuman primates. *Mucosal Immunol* 2013;**6**:787–96.

64. Ryan M, Murphy G, Gothefors L, et al. Bordetella pertussis respiratory infection in children is associated with preferential activation of type 1 T helper cells. *J Infect Dis* 1997;**175**:1246–50.

65. Hafler JP, Pohl-Koppe A. The cellular immune response to Bordetella pertussis in two children with whooping cough. *Eur J Med Res* 1998;**3**:523–6.

66. Barbic J, Leef MF, Burns DL, et al. Role of gamma interferon in natural clearance of Bordetella pertussis infection. *Infect Immun* 1997;**65**:4904–8.

67. Jaffar Z, Ferrini ME, Herritt LA, et al. Cutting edge: lung mucosal Th17-mediated responses induce polymeric Ig receptor expression by the airway epithelium and elevate secretory IgA levels. *J Immunol* 2009;**182**:4507–11.

68. Murphy K, Travers P, Walport M. *Janeway's Immunobiology*. 7th ed. New York: Garland Science; 2008.

69. Sallusto F, Lenig D, Forster R, et al. Two subsets of memory T lymphocytes with distinct homing potentials and effector functions. *Nature* 1999;**401**:708–12.

70. Masopust D, Vezys V, Marzo AL, et al. Preferential localization of effector memory cells in nonlymphoid tissue. *Science* 2001;**291**:2413–7.

71. Mueller SN, Mackay LK. Tissue-resident memory T cells: local specialists in immune defence. *Nat Rev Immunol* 2016;**16**:79–89.

72. Wu T, Hu Y, Lee YT, et al. Lung-resident memory CD8 T cells (TRM) are indispensable for optimal cross-protection against pulmonary virus infection. *J Leukoc Biol* 2014;**95**:215–24.

73. Gebhardt T, Wakim LM, Eidsmo L, et al. Memory T cells in nonlymphoid tissue that provide enhanced local immunity during infection with herpes simplex virus. *Nat Immunol* 2009;**10**:524–30.

74. Wilk MM, Misiak A, McManus RM, et al. Lung CD4 tissue-resident memory T cells mediate adaptive immunity induced by previous infection of mice with Bordetella pertussis. *J Immunol* 2017;**199**:233–43.

75. McGuirk P, McCann C, Mills KH. Pathogen-specific T regulatory 1 cells induced in the respiratory tract by a bacterial molecule that stimulates interleukin 10 production by dendritic cells: a novel strategy for evasion of protective T helper type 1 responses by Bordetella pertussis. *J Exp Med* 2002;**195**:221–31.

76. McGuirk P, Mills KH. Direct anti-inflammatory effect of a bacterial virulence factor: IL-10-dependent suppression of IL-12 production by filamentous hemagglutinin from Bordetella pertussis. *Eur J Immunol* 2000;**30**:415–22.

77. Ross PJ, Lavelle EC, Mills KH, et al. Adenylate cyclase toxin from Bordetella pertussis synergizes with lipopolysaccharide to promote innate interleukin-10 production and enhances the induction of Th2 and regulatory T cells. *Infect Immun* 2004;**72**:1568–79.

78. Fennelly NK, Sisti F, Higgins SC, et al. Bordetella pertussis expresses a functional type III secretion system that subverts protective innate and adaptive immune responses. *Infect Immun* 2008;**76**:1257–66.

79. Coleman MM, Finlay CM, Moran B, et al. The immunoregulatory role of CD4(+) FoxP3(+) CD25(-) regulatory T cells in lungs of mice infected with Bordetella pertussis. *FEMS Immunol Med Microbiol* 2012;**64**:413–24.

80. Coleman MM, Ruane D, Moran B, et al. Alveolar macrophages contribute to respiratory tolerance by inducing FoxP3 expression in naive T cells. *Am J Respir Cell Mol Biol* 2013;**48**:773–80.

81. Lalor SJ, McLoughlin RM. Memory γδ T cells-newly appreciated protagonists in infection and immunity. *Trends Immunol* 2016;**37**:690–702.

82. Zachariadis O, Cassidy J, Brady J, Mahon B. Gammadelta T cells regulate the early inflammatory response to Bordetella pertussis infection in the murine respiratory tract. *Infect Immun* 2006;**74**:1837–45.

83. Bertotto A, De Benedictis FM, Vagliasindi C, et al. Gamma delta T cells are decreased in the blood of children with Bordetella pertussis infection. *Acta Paediatr* 1997;**86**:114–5.

84. Brandtzaeg P. Secretory IgA: designed for anti-microbial defense. *Front Immunol* 2013;**4**:222.

85. Stiehm ER. The four most common pediatric immunodeficiencies. *J Immunotoxicol* 2008;**5**:227–34.

86. Jorgensen GH, Gardulf A, Sigurdsson MI, et al. Clinical symptoms in adults with selective IgA deficiency: a case-control study. *J Clin Immunol* 2013;**33**:742–7.

87. Aghamohammadi A, Cheraghi T, Gharagozlou M, et al. IgA deficiency: correlation between clinical and immunological phenotypes. *J Clin Immunol* 2009;**29**:130–6.

88. von Linstow ML, Pontoppidan PL, von König CH, Evidence of Bordetella pertussis infection in vaccinated 1-year-old Danish children. *Eur J Pediatr* 2010;**169**: 1119–22.

89. Poynten M, Hanlon M, Irwig L, et al. Serological diagnosis of pertussis: evaluation of IgA against whole cell and specific Bordetella pertussis antigens as markers of recent infection. *Epidemiol Infect* 2002;**128**:161–7.

90. Hendrikx LH, Öztürk K, de Rond LG, et al. Serum IgA responses against pertussis proteins in infected and Dutch wP or aP vaccinated children: an additional role in pertussis diagnostics. *PLoS One* 2011;**6**:e27681.

91. Wolfe DN, Kirimanjeswara GS, Goebel EM, et al. Comparative role of immunoglobulin A in protective immunity against the Bordetellae. *Infect Immun* 2007;**75**: 4416–22.

92. Mattoo S, Cherry J. Molecular pathogenesis, epidemiology, and clinical manifestations of respiratory infections due to Bordetella pertussis and other Bordetella subspecies. *Clin Microbiol Rev* 2005;**18**:326–82.

93. Abu Raya B, Srugo I, Kessel A, et al. The induction of breast milk pertussis specific antibodies following gestational tetanus-diphtheria-acellular pertussis vaccination. *Vaccine* 2014;**32**:5632–7.

94. Elahi S, Buchanan RM, Babiuk LA, et al. Maternal immunity provides protection against pertussis in newborn piglets. *Infect Immun* 2006;**74**:2619–27.

95. Christensen D, Mortensen R, Rosenkrands I, et al. Vaccine-induced Th17 cells are established as resident memory cells in the lung and promote local IgA responses. *Mucos Immunol* 2017;**10**:260–70.

96. McGuirk P, Mahon BP, Griffin F, et al. Compartmentalization of T cell responses following respiratory infection with Bordetella pertussis: hyporesponsiveness of lung T cells is associated with modulated expression of the co-stimulatory molecule CD28. *Eur J Immunol* 1998;**28**:153–63.

97. Parkhill J, Sebaihia M, Preston A, et al. Comparative analysis of the genome sequences of Bordetella pertussis, Bordetella parapertussis and Bordetella bronchiseptica. *Nat Genet* 2003;**35**:32–40.

98. Witvliet MH, Burns DL, Brennan MJ, et al. Binding of pertussis toxin to eucaryotic cells and glycoproteins. *Infect Immun* 1989;**57**:3324–30.

99. Katada T, Tamura M, Ui M. The A protomer of islet-activating protein, pertussis toxin, as an active peptide catalyzing ADP-ribosylation of a membrane protein. *Arch Biochem Biophys* 1983;**224**:290–8.

100. Carbonetti NH, Artamonova GV, Mays RM, et al. Pertussis toxin plays an early role in respiratory tract colonization by Bordetella pertussis. *Infect Immun* 2003;**71**:6358–66.

101. Carbonetti NH. Pertussis toxin and adenylate cyclase toxin: key virulence factors of Bordetella pertussis and cell biology tools. *Future Microbiol* 2010;**5**:455–69.

102. Locht C, Bertin P, Menozzi FD, et al. The filamentous haemagglutinin, a multifaceted adhesion produced by virulent Bordetella spp. *Mol Microbiol* 1993;**9**:653–60.

103. Mobberley-Schuman PS, Weiss AA. Influence of CR3 (CD11b/CD18) expression on phagocytosis of Bordetella pertussis by human neutrophils. *Infect Immun* 2005;**73**:7317–23.

104. Henderson MW, Inatsuka CS, Sheets AJ, et al. Contribution of Bordetella filamentous hemagglutinin and adenylate cyclase toxin to suppression and evasion of interleukin-17-mediated inflammation. *Infect Immun* 2012;**80**:2061–75.

105. Scheller EV, Cotter PA. Bordetella filamentous hemagglutinin and fimbriae: critical adhesins with unrealized vaccine potential. *Pathog Dis* 2015;**73**:ftv079.

106. Rodríguez ME, Hellwig SM, Pérez Vidakovics ML, et al. Bordetella pertussis attachment to respiratory epithelial cells can be impaired by fimbriae-specific antibodies. *FEMS Immunol Med Microbiol* 2006;**46**:39–47.

107. Scheller EV, Melvin JA, Sheets AJ, et al. Cooperative roles for fimbria and filamentous hemagglutinin in Bordetella adherence and immune modulation. *MBio* 2015;**6**:e00500–15.

108. Eby JC, Gray MC, Mangan AR, et al. Role of CD11b/CD18 in the process of intoxication by the adenylate cyclase toxin of Bordetella pertussis. *Infect Immun* 2012;**80**:850–9.

109. Benz R, Maier E, Ladant D, et al. Adenylate cyclase toxin (CyaA) of Bordetella pertussis. Evidence for the formation of small ion-permeable channels and comparison with HlyA of Escherichia coli. *J Biol Chem* 1994;**269**:27231–9.

110. Khelef N, Guiso N. Induction of macrophage apoptosis by Bordetella pertussis adenylate cyclase-hemolysin. *FEMS Microbiol Lett* 1995;**134**:27–32.

111. Kamanova J, Kofronova O, Masin J, et al. Adenylate cyclase toxin subverts phagocyte function by RhoA inhibition and unproductive ruffling. *J Immunol* 2008;**181**:5587–97.

112. Skopova K, Tomalova B, Kanchev I, et al. cAMP-elevating capacity of the adenylate cyclase toxin-hemolysin is sufficient for lung infection but not for full virulence of Bordetella pertussis. *Infect Immun* 2017;**85**:e00937-16.

113. Ross P, Lavelle E, Mills K, et al. Adenylate cyclase toxin from Bordetella pertussis synergizes with lipopolysaccharide to promote innate interleukin-10 production and enhances the induction of Th2 and regulatory T cells. *Infect Immun* 2004;**72**:1568–79.

114. Boyd AP, Ross PJ, Conroy H, et al. Bordetella pertussis adenylate cyclase toxin modulates innate and

adaptive immune responses: distinct roles for acylation and enzymatic activity in immunomodulation and cell death. *J Immunol* 2005;**175**:730–8.

115. Hickey FB, Brereton CF, Mills KH. Adenylate cyclase toxin of Bordetella pertussis inhibits TLR-induced IRF-1 and IRF-8 activation and IL-12 production and enhances IL-10 through MAPK activation in dendritic cells. *J Leukoc Biol* 2008;**84**:234–43.

116. Leininger E, Roberts M, Kenimer JG, et al. Pertactin, an Arg-Gly-Asp-containing Bordetella pertussis surface protein that promotes adherence of mammalian cells. *Proc Natl Acad Sci U S A* 1991;**88**:345–9.

117. Inatsuka CS, Xu Q, Vujkovic-Cvijin I, et al. Pertactin is required for Bordetella species to resist neutrophil-mediated clearance. *Infect Immun* 2010;**78**:2901–9.

118. Khelef N, Bachelet CM, Vargaftig BB, et al. Characterization of murine lung inflammation after infection with parental Bordetella pertussis and mutants deficient in adhesins or toxins. *Infect Immun* 1994;**62**:2893–900.

119. Roberts M, Fairweather NF, Leininger E, et al. Construction and characterization of Bordetella pertussis mutants lacking the vir-regulated P.69 outer membrane protein. *Mol Microbiol* 1991;**5**:1393–404.

120. van den Berg BM, Beekhuizen H, Willems RJ, et al. Role of Bordetella pertussis virulence factors in adherence to epithelial cell lines derived from the human respiratory tract. *Infect Immun* 1999;**67**:1056–62.

121. Everest P, Li J, Douce G, et al. Role of the Bordetella pertussis P.69/pertactin protein and the P.69/pertactin RGD motif in the adherence to and invasion of mammalian cells. *Microbiology* 1996;**142**:3261–8.

122. Storsaeter J, Hallander HO, Gustafsson L, et al. Levels of anti-pertussis antibodies related to protection after household exposure to Bordetella pertussis. *Vaccine* 1998;**16**:1907–16.

123. Hellwig SM, Rodriguez ME, Berbers GA, et al. Crucial role of antibodies to pertactin in Bordetella pertussis immunity. *J Infect Dis* 2003;**188**:738–42.

124. Martin SW, Pawloski L, Williams M, et al. Pertactin-negative Bordetella pertussis strains: evidence for a possible selective advantage. *Clin Infect Dis* 2015;**60**:223–7.

125. Poolman JT. Shortcomings of pertussis vaccines: why we need a third generation vaccine. *Expert Rev Vaccines* 2014;**13**:1159–62.

126. Smith AM, Guzmán CA, Walker MJ. The virulence factors of Bordetella pertussis: a matter of control. *FEMS Microbiol Rev* 2001;**25**:309–33.

127. Matsuzawa T, Fukui A, Kashimoto T, et al. Bordetella dermonecrotic toxin undergoes proteolytic processing to be translocated from a dynamin-related endosome into the cytoplasm in an acidification-independent manner. *J Biol Chem* 2004;**279**:2866–72.

128. Fukui A, Horiguchi Y. Bordetella dermonecrotic toxin exerting toxicity through activation of the small GTPase Rho. *J Biochem* 2004;**136**:415–9.

129. Horiguchi Y, Inoue N, Masuda M, et al. Bordetella bronchiseptica dermonecrotizing toxin induces reorganization of actin stress fibers through deamidation of Gln-63 of the GTP-binding protein Rho. *Proc Natl Acad Sci U S A* 1997;**94**:11623–6.

130. Brockmeier SL, Register KB, Magyar T, et al. Role of the dermonecrotic toxin of Bordetella bronchiseptica in the pathogenesis of respiratory disease in swine. *Infect Immun* 2002;**70**:481–90.

131. Magyar T, Glávits R, Pullinger GD, et al. The pathological effect of the Bordetella dermonecrotic toxin in mice. *Acta Vet Hung* 2000;**48**:397–406.

132. Luker KE, Collier JL, Kolodziej EW, et al. Bordetella pertussis tracheal cytotoxin and other muramyl peptides: distinct structure-activity relationships for respiratory epithelial cytopathology. *Proc Natl Acad Sci U S A* 1993;**90**:2365–9.

133. Heiss LN, Moser SA, Unanue ER, et al. Interleukin-1 is linked to the respiratory epithelial cytopathology of pertussis. *Infect Immun* 1993;**61**:3123–8.

134. Goldman WE, Cookson BT. Structure and functions of the Bordetella tracheal cytotoxin. *Tokai J Exp Clin Med* 1988;**13** Suppl:187–91.

135. Flak TA, Heiss LN, Engle JT, et al. Synergistic epithelial responses to endotoxin and a naturally occurring muramyl peptide. *Infect Immun* 2000;**68**:1235–42.

136. Cundell DR, Kanthakumar K, Taylor GW, et al. Effect of tracheal cytotoxin from Bordetella pertussis on human neutrophil function in vitro. *Infect Immun* 1994;**62**:639–43.

137. Magalhaes JG, Philpott DJ, Nahori MA, et al. Murine Nod1 but not its human orthologue mediates innate immune detection of tracheal cytotoxin. *EMBO Rep* 2005;**6**:1201–7.

138. Notti RQ, Stebbins CE. The structure and function of type III secretion systems. *Microbiol Spectr* 2016;**4**:VMBF-0004-2015.

139. Yuk MH, Harvill ET, Miller JF. The BvgAS virulence control system regulates type III secretion in Bordetella bronchiseptica. *Mol Microbiol* 1998;**28**:945–59.

140. Mattoo S, Yuk MH, Huang LL, et al. Regulation of type III secretion in Bordetella. *Mol Microbiol* 2004;**52**:1201–14.

141. Watanabe M, Takimoto H, Kumazawa Y, et al. Biological properties of lipopolysaccharides from Bordetella species. *J Gen Microbiol* 1990;**136**:489–93.

142. Fedele G, Nasso M, Spensieri F, et al. Lipopolysaccharides from Bordetella pertussis and Bordetella parapertussis differently modulate human dendritic cell functions resulting in divergent prevalence of Th17-polarized responses. *J Immunol* 2008;**181**:208–16.

143. Banus HA, Vandebriel RJ, de Ruiter H, et al. Host genetics of Bordetella pertussis infection in mice: significance of Toll-like receptor 4 in genetic susceptibility and pathobiology. *Infect Immun* 2006;**74**:2596–605.

144. Moreno G, Errea A, Van Maele L, et al. Toll-like receptor 4 orchestrates neutrophil recruitment into airways during the first hours of Bordetella pertussis infection. *Microbes Infect* 2013;**15**:708–18.

145. Brummelman J, Veerman RE, Hamstra HJ, et al. Bordetella pertussis naturally occurring isolates with altered lipooligosaccharide structure fail to fully mature human dendritic cells. *Infect Immun* 2015;**83**:227–38.

146. Donnelly S, Loscher CE, Lynch MA, et al. Whole-cell but not acellular pertussis vaccines induce convulsive activity in mice: evidence of a role for toxin-induced interleukin-1beta in a new murine model for analysis of neuronal side effects of vaccination. *Infect Immun* 2001;**69**:4217–23.

147. Ausiello CM, Lande R, Urbani F, et al. Cell-mediated immune responses in four-year-old children after primary immunization with acellular pertussis vaccines. *Infect Immun* 1999;**67**:4064–71.

148. Ryan M, Murphy G, Ryan E, et al. Distinct T-cell subtypes induced with whole cell and acellular pertussis vaccines in children. *Immunology* 1998;**93**:1–10.

149. Dunne A, Mielke LA, Allen AC, et al. A novel TLR2 agonist from Bordetella pertussis is a potent adjuvant that promotes protective immunity with an acellular pertussis vaccine. *Mucosal Immunol* 2015;**8**:607–17.

150. Olin P, Rasmussen F, Gustafsson L, et al. Randomised controlled trial of two-component, three-component, and five-component acellular pertussis vaccines compared with whole-cell pertussis vaccine. Ad Hoc Group for the Study of Pertussis Vaccines. *Lancet* 1997;**350**:1569–77.

151. Bomford R. Will adjuvants be needed for vaccines of the future? *Dev Biol Stand* 1998;**92**:13–7.

152. Johnson AG, Gaines S, Landy M. Studies on the O antigen of Salmonella typhosa. V. Enhancement of antibody response to protein antigens by the purified lipopolysaccharide. *J Exp Med* 1956;**103**:225–46.

153. Mori A, Oleszycka E, Sharp FA, et al. The vaccine adjuvant alum inhibits IL-12 by promoting PI3 kinase signaling while chitosan does not inhibit IL-12 and enhances Th1 and Th17 responses. *Eur J Immunol* 2012;**42**:2709–19.

154. Kindrachuk J, Jenssen H, Elliott M, et al. A novel vaccine adjuvant comprised of a synthetic innate defence regulator peptide and CpG oligonucleotide links innate and adaptive immunity. *Vaccine* 2009;**27**:4662–71.

155. Asokanathan C, Corbel M, Xing D. A CpG-containing oligodeoxynucleotide adjuvant for acellular pertussis vaccine improves the protective response against

Bordetella pertussis. *Hum Vaccin Immunother* 2013;**9**:325–31.

156. Brummelman J, Helm K, Hamstra HJ, et al. Modulation of the CD4(+) T cell response after acellular pertussis vaccination in the presence of TLR4 ligation. *Vaccine* 2015;**33**:1483–91.

157. Geurtsen J, Banus HA, Gremmer ER, et al. Lipopolysaccharide analogs improve efficacy of acellular pertussis vaccine and reduce type I hypersensitivity in mice. *Clin Vaccine Immunol* 2007;**14**:821–9.

158. Boland G, Beran J, Lievens M, et al. Safety and immunogenicity profile of an experimental hepatitis B vaccine adjuvanted with AS04. *Vaccine* 2004;**23**:316–20.

159. Agnolon V, Bruno C, Leuzzi R, et al. The potential of adjuvants to improve immune responses against TdaP vaccines: a preclinical evaluation of MF59 and monophosphoryl lipid A. *Int J Pharm* 2015;**492**:169–76.

160. Misiak A, Leuzzi R, Allen AC, et al. Addition of a TLR7 agonist to an acellular pertussis vaccine enhances Th1 and Th17 responses and protective immunity in a mouse model. *Vaccine* 2017;**35**:5256–63.

161. Mullen AC, High FA, Hutchins AS, et al. Role of T-bet in commitment of TH1 cells before IL-12-dependent selection. *Science* 2001;**292**:1907–10.

162. Mahon BP, Ryan MS, Griffin F, et al. Interleukin-12 is produced by macrophages in response to live or killed Bordetella pertussis and enhances the efficacy of an acellular pertussis vaccine by promoting induction of Th1 cells. *Infect Immun* 1996;**64**:5295–301.

163. O'Hagan DT, Ott GS, Nest GV, et al. The history of MF59((R)) adjuvant: a phoenix that arose from the ashes. *Expert Rev Vaccines* 2013;**12**:13–30.

164. Podda A. The adjuvanted influenza vaccines with novel adjuvants: experience with the MF59-adjuvanted vaccine. *Vaccine* 2001;**19**:2673–80.

165. Gasparini R, Schioppa F, Lattanzi M, et al. Impact of prior or concomitant seasonal influenza vaccination on MF59-adjuvanted H1N1v vaccine (Focetria) in adult and elderly subjects. *Int J Clin Pract* 2010;**64**:432–8.

166. Seubert A, Monaci E, Pizza M, et al. The adjuvants aluminum hydroxide and MF59 induce monocyte and granulocyte chemoattractants and enhance monocyte differentiation toward dendritic cells. *J Immunol* 2008;**180**:5402–12.

167. Garlapati S, Eng NF, Kiros TG, et al. Immunization with PCEP microparticles containing pertussis toxoid, CpG ODN and a synthetic innate defense regulator peptide induces protective immunity against pertussis. *Vaccine* 2011;**29**:6540–8.

168. Li P, Asokanathan C, Liu F, et al. PLGA nano/micro particles encapsulated with pertussis toxoid (PTd) enhances Th1/Th17 immune response in a murine model. *Int J Pharm* 2016;**513**:183–90.

169. Conway MA, Madrigal-Estebas L, McClean S, et al. Protection against Bordetella pertussis infection following parenteral or oral immunization with antigens entrapped in biodegradable particles: effect of formulation and route of immunization on induction of Th1 and Th2 cells. *Vaccine* 2001;**19**:1940–50.

170. Jones DH, McBride BW, Jeffery H, et al. Protection of mice from Bordetella pertussis respiratory infection using microencapsulated pertussis fimbriae. *Vaccine* 1995;**13**:675–81.

171. Jain RA. The manufacturing techniques of various drug loaded biodegradable poly(lactide-co-glycolide) (PLGA) devices. *Biomaterials* 2000;**21**: 2475–90.

172. Schwechheimer C, Kuehn MJ. Outer-membrane vesicles from Gram-negative bacteria: biogenesis and functions. *Nat Rev Microbiol* 2015;**13**:605–19.

173. Hozbor D, Rodriguez ME, Fernandez J, et al. Release of outer membrane vesicles from Bordetella pertussis. *Curr Microbiol* 1999;**38**:273–8.

174. Roberts R, Moreno G, Bottero D, et al. Outer membrane vesicles as acellular vaccine against pertussis. *Vaccine* 2008;**26**:4639–46.

175. Bottero D, Gaillard ME, Zurita E, et al. Characterization of the immune response induced by pertussis OMVs-based vaccine. *Vaccine* 2016;**34**:3303–9.

176. Gaillard ME, Bottero D, Errea A, et al. Acellular pertussis vaccine based on outer membrane vesicles capable of conferring both long-lasting immunity and protection against different strain genotypes. *Vaccine* 2014;**32**:931–7.

177. Raeven RH, Brummelman J, Pennings JL, et al. Bordetella pertussis outer membrane vesicle vaccine confers equal efficacy in mice with milder inflammatory responses compared to a whole-cell vaccine. *Sci Rep* 2016;**6**:38240.

178. Mielcarek N, Debrie AS, Raze D, et al. Live attenuated B. pertussis as a single-dose nasal vaccine against whooping cough. *PLoS Pathog* 2006;**2**:e65.

179. Feunou PF, Bertout J, Locht C. T- and B-cell-mediated protection induced by novel, live attenuated pertussis vaccine in mice. Cross protection against parapertussis. *PLoS One* 2010;**5**:e10178.

180. Mielcarek N, Debrie AS, Mahieux S, et al. Dose response of attenuated Bordetella pertussis BPZE1-induced protection in mice. *Clin Vaccine Immunol* 2010;**17**:317–24.

181. Feunou PF, Kammoun H, Debrie AS, et al. Long-term immunity against pertussis induced by a single nasal administration of live attenuated B. pertussis BPZE1. *Vaccine* 2010;**28**:7047–53.

182. Feunou PF, Kammoun H, Debrie AS, et al. Heterologous prime-boost immunization with live attenuated B. pertussis BPZE1 followed by acellular pertussis vaccine in mice. *Vaccine* 2014;**32**:4281–8.

183. Skerry CM, Cassidy JP, English K, et al. A live attenuated Bordetella pertussis candidate vaccine does not cause disseminating infection in gamma interferon receptor knockout mice. *Clin Vaccine Immunol* 2009;**16**:1344–51.

184. Thorstensson R, Trollfors B, Al-Tawil N, et al. A phase I clinical study of a live attenuated Bordetella pertussis vaccine—BPZE1; a single centre, double-blind, placebo-controlled, dose-escalating study of BPZE1 given intranasally to healthy adult male volunteers. *PLoS One* 2014;**9**:e83449.

185. Jahnmatz M, Amu S, Ljungman M, et al. B-cell responses after intranasal vaccination with the novel attenuated Bordetella pertussis vaccine strain BPZE1 in a randomized phase I clinical trial. *Vaccine* 2014;**32**: 3350–6.

186. Peppoloni S, Nencioni L, Di Tommaso A, et al. Lymphokine secretion and cytotoxic activity of human CD4+ T-cell clones against Bordetella pertussis. *Infect Immun* 1991;**59**:3768–73.

187. De Magistris MT, Romano M, Bartoloni A, et al. Human T cell clones define S1 subunit as the most immunogenic moiety of pertussis toxin and determine its epitope map. *J Exp Med* 1989;**169**:1519–32.

188. Allen AC, Wilk MM, Misiak A, Borkner L, Murphy D, Mills KHG. Sustained protective immunity against Bordetella pertussis nasal colonization by intranasal immunization with a vaccine-adjuvant combination that induces IL-17-secreting TRM cells. *Mucosal Immunol.* 2018. Aug 20. doi: 10.1038/s41385-018-0080-x.

CHAPTER 4

Pertussis epidemiology

Natasha Crowcroft and Elizabeth Miller

Abstract

Bordetella pertussis is an exclusively human pathogen found worldwide in all populations. Complications of pertussis including bronchopneumonia, failure to thrive from post-tussive vomiting, cerebral hypoxia leading to brain damage, and death are strongly concentrated in infants. The similarity in mortality profile with age between countries that are markedly different with respect to demography, health systems, socioeconomic status, and surveillance systems is striking. It is because infants are most likely to suffer serious complications of infection or die that prevention of pertussis in infants is the primary goal of immunization programmes. Of all the vaccine-preventable infections, pertussis remains the most epidemiologically challenging to understand. This is partly because of surveillance issues but also the lack of an established correlate of protection that allows susceptible and immune individuals to be distinguished. These factors are compounded by our imperfect understanding of the mechanism of protection from acellular and whole-cell vaccines, and the product-specific differences in efficacy and effectiveness. It can therefore be difficult for policymakers to be confident about the optimal number and timing of primary doses and how many and at what age booster doses should be given in their setting, as reflected in the plethora of schedules in use throughout the world. There are even greater challenges for the World Health Organization when attempting to make global policy as lessons learned in one setting may not appear to apply in others. What is clear is that high coverage early in infancy with three doses of an effective vaccine greatly reduces pertussis mortality and severe morbidity in all settings and that many infants in resource-poor settings are still deprived of that benefit.

4.1 Introduction

At a global level, the epidemiology of *Bordetella pertussis* is that of an exclusively human pathogen found worldwide in all populations. Complications of pertussis including bronchopneumonia, failure to thrive from post-tussive vomiting, cerebral hypoxia leading to brain damage, and death are strongly concentrated in infants. Nearly all deaths from pertussis are of infants infected in the first 6 months to 1 year of life as shown in a review conducted in 2014 by the London School of Hygiene and Tropical Medicine for the World Health Organization (WHO) Strategic Advisory Group of Experts (see Figure 4.1)[1]. The similarity in mortality profile with age between countries that are markedly different with respect to demography, health systems, socioeconomic status, and surveillance systems is striking. It is because infants are most likely to suffer serious complications of infection or die that prevention of pertussis in infants is the primary goal of

Crowcroft, N. and Miller, E., *Pertussis epidemiology*. In: *Pertussis: epidemiology, immunology, and evolution*. Edited by Pejman Rohani and Samuel V. Scarpino: Oxford University Press (2019). © Crown copyright is held by Public Health Ontario. DOI: 10.1093/oso/9780198811879.003.0004

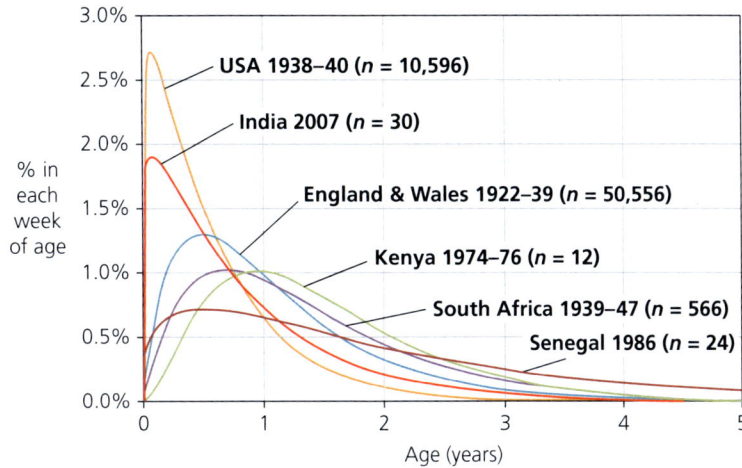

Figure 4.1 Age distribution of pertussis deaths in children under 5 years of age in the pre-vaccine era: curves fitted to available datasets. Reproduced with permission of Prof Colin Sanderson.

immunization programmes[2]. For the same reason, epidemiological data tend to be biased towards ascertainment in infants.

In the pre-vaccine era, pertussis was predominantly a disease of young children. The infection was typically recognized and reported in children less than 10 years old with most reported cases occurring between 3 and 6 years of age[3]. In the past, pertussis was even thought to occur exclusively in childhood, but it is now well understood that the infection can occur at any age and reinfection is also recognized[4-8]. Pertussis may be mild in older children and adults, particularly if vaccinated, and the diagnosis may not be suspected, leading to under-diagnosis and under-reporting. Case definitions that focus on severe disease under-detect pertussis and overestimate vaccine effectiveness. Note that the WHO clinical case definition designed for ensuring specificity of diagnosis in clinical trials is different from the more sensitive WHO surveillance case definition in requiring three or more weeks of coughing rather than two or more, leading to higher estimates of vaccine efficacy from clinical trials than vaccine effectiveness estimated from surveillance data[9,10].

In the post-vaccine era, under-detection and under-reporting of pertussis and poor surveillance data quality in much of the world make it hard to compare age-specific incidence and mortality by setting. Many aspects of how pertussis vaccination programmes affect the immunity of populations, transmission of infection, and persistence of the

organism in human populations are poorly defined. These limitations present a challenge to research and policy development.

The epidemiology of pertussis has been extensively examined in recent years. WHO reviewed pertussis epidemiology in 2014[11] and summarized the findings in its position paper in 2015[2]. This chapter will build on what has been learnt since the review of 2014, particularly in some of the countries identified as experiencing resurgences of disease (United Kingdom, Australia, Chile, Portugal, United States), compared with some countries that have not observed a resurgence (e.g. Canada), and examine new evidence emerging from low- and middle-income countries (LMIC). The objectives of this chapter are to introduce the reader to the basics of pertussis epidemiology, compare and contrast the epidemiology of pertussis in different settings, and reflect on what the similarities and differences tell us about the pathogen, the disease, and the vaccine.

4.2 Basic epidemiological parameters applied in understanding pertussis epidemiology and control

4.2.1 Key epidemiological parameters: mode of transmission, infectious period, and role of disease severity in transmission

Pertussis is spread through droplet transmission and is very infectious; secondary attack rates in households are very high, up to 80 per cent in susceptible contacts[12–14]. The infectious period is from the onset of a catarrhal prodrome to 3 weeks after the onset of illness, or 5 days if the individual is on antibiotic treatment[15]. An initial coryzal stage is followed by development of a cough which is most typically paroxysmal, followed by a whoop or vomiting. Younger children may not develop paroxysms or a whoop and may present with apnoea alone[16]. The clinical severity varies widely, with milder cases occurring particularly in older children and adults and in vaccinated individuals. Although for many years clinicians recognized that pertussis may present with mild symptoms, only more recently has asymptomatic infection been acknowledged.

Asymptomatic carriage (defined as prolonged occupation of an ecological niche in the respiratory tract without any detriment to the host but as an important source of transmission to others) is a characteristic feature of bacteria such as *Neisseria meningitidis* and *Streptococcus pneumoniae* but has not been definitively demonstrated for *B. pertussis*. Asymptomatic carriage is distinct from immune boosting, in which an individual shows signs that their immune system has responded to an infectious agent without completing the full manifestation of disease by becoming infectious to others. This distinction is important, because individuals who undergo boosting do not contribute to transmission but do contribute to sustaining herd immunity.

Studying these phenomena is challenging both because long-term prospective studies would need to be very large to identify sufficient pertussis cases, which makes them very expensive, and because of the limits of available laboratory methods. Culture may be the best method of determining that an individual is infectious, but is too insensitive to be informative. The polymerase chain reaction (PCR) technique is more sensitive but cannot necessarily determine whether an individual is actually infectious. Serological methods cannot distinguish infection from boosting, limiting many of the published studies[17]. From a design perspective, in household transmission studies it is often hard to know who infected whom without onset dates which are generally missing for asymptomatic infections. The best available observational method for studying infectivity is probably through human challenge studies[18]. These are likely to be extremely informative, but

unlikely to include immunologically naïve individuals, and will not (for ethical reasons) include families with young infants.

The precise degree to which asymptomatic infections contribute to pertussis transmission is uncertain, but milder and atypical infections particularly in older children and adults certainly have an important and substantial role to play in maintaining pertussis in populations. Routine surveillance is biased towards detecting younger and more severe cases. In order to understand the epidemiology of disease, studies of older children and adults with prolonged coughing illness have been conducted in a range of different study designs, settings, and epidemic periods. These uncover an important burden of undetected pertussis whenever and wherever they are conducted. Although findings seem similar, some caution is required in their interpretation. Earlier studies used serological and PCR laboratory diagnostic methods that have needed to be refined to be more specific over time. In the United Kingdom, two studies in school-aged children used the same approach and oral fluid antibody detection for diagnosis and are therefore comparable. A preschool booster was implemented in 2001. In 2001–2005, 38 per cent of children with cough aged 5–16 years had evidence of pertussis, compared with 20 per cent in 2010–2012[19,20]. Recent epidemics and past epidemics may strongly influence the age distribution of cases found, depending on which age groups are currently immune. In France during a resurgence of pertussis that followed a prolonged period of good control, 32 per cent of adults with prolonged cough were found to have pertussis; furthermore, some of the patients reported having also had pertussis as a child[7]. In Canada in the late 1990s, during a highly epidemic period following the use of a low-effectiveness whole-cell pertussis (wP) vaccine, 33 percent of coughing adolescents and 16–19 per cent of coughing adult cases were found to have pertussis[5]. In contrast, the proportion with pertussis in a study in New Zealand using oral fluid antibody detection was only 7 per cent in coughing adults, significantly lower than the 17 per cent of undiagnosed pertussis in children ($p = 0.003$)[6]. Although most of the research has been conducted in high-income countries (HICs), when investigations have been conducted in children and adults with prolonged cough in low-income countries, pertussis has been found[21,22].

4.2.2 Pertussis incubation period

The incubation period of pertussis, like other infectious diseases, varies in the form of a log normal distribution[23]. Incubation periods cited in the literature usually take the onset of cough as the onset of disease. The onset of cough is usually preceded by a non-specific prodromal coryzal illness during which the individual is infectious. As the prodrome involves non-specific upper respiratory symptoms, it can be missed as being of no significance. The range of the incubation period of pertussis is often quoted at 6–20 days[24] although a recent systematic review stated that it could be as short as 3 days[25]. Data from 1953 indicated that the incubation period ranged between 5 and 21 days, usually 7 and rarely more than 10 days[3]. Two detailed analyses describe patterns of spread prior to vaccination[23,26]. The first, published in 1933, describes the intervals between cases in households as the incubation period, but this is actually the serial interval which is shorter in an infectious disease with an infectious prodrome. It found that serial intervals may be as short as 3 days, can often be 4–6 days, but is most frequently 1 week. The corresponding incubation periods for these serial intervals depend upon the distribution and duration of the prodromal period. The second study gives the incubation period for 11/26 cases described, but does not explain how the incubation periods were defined. They ranged from 2 to 3 days to possibly 17 days with a median of around 5 days.

Only occasionally is the precise timing of exposure known for pertussis because most cases are spread in households, where individuals are in continuous contact with each other. Experimental pertussis in two unvaccinated brothers aged 6 and 8 years occurred following an incubation period of 7 days before coughing started: the timing of the onset of coryza was not recorded[27]. Two infants infected at birth by a midwife started to cough at 8 and 10 days; again, a prodromal illness was not reported[28]. It is difficult to draw conclusions from case reports and small studies, and challenge studies in humans may be needed to obtain definitive data[18].

4.2.3 Who infects whom?

A number of studies and a review have been conducted of the source of infection and these have been useful for informing immunization policy[29,30]. Although there is heterogeneity in sources of infection, reflecting variation in local epidemiology and sociodemographic factors, in general there is a surprisingly consistent pattern. Most often, when the source of infection for infants can be identified, it is someone within the household, mostly parents, usually the mother, followed by siblings[31]. This pattern reflects the contact patterns of young infants. Evidence in the United States shows a recent trend away from mothers towards siblings as the most frequent source of infection[32]. In addition to the source of infection for the infant, the question of who initially brings pertussis into the household is slightly different. In a recent study of household transmission in Spain, more than 80 per cent of the primary cases were children, and further transmission to contacts occurred within 16.1 per cent of households[33].

For a substantial proportion of cases the source of infection is unknown, which may indicate someone outside the household, or potentially transmission from an asymptomatic household contact[34]. As is true for pertussis studies in general, findings are strongly influenced by the local epidemiology of disease both at the time and in the past. In addition, studies of the source of infection vary with regards to whether cases were laboratory-confirmed, the laboratory diagnostic methods used, and the degree to which contacts were followed up and tested, which is another source of any variation in the proportion for whom the source is unknown. Such variation means that similarities of study findings are often more remarkable than their differences[35]. In France in 1993–1994, 20 per cent of the sources could not be identified using culture and serology compared with 30 per cent in the United Kingdom in 1998–2000 using culture, PCR, and serology[29,36]. Both countries were experiencing relatively good control of pertussis during these time periods. In the United States, a study in the 1990s used solely clinical symptoms to identify sources of infection, and these could not be identified for 57 per cent of cases[37]. More recently, studies from 2006 to 2013 and from 2008 to 2012 in the United States also used clinical case definitions for the source of infection and these could not be identified for 44 per cent and 76 per cent of cases, respectively[32,38]. Heterogeneity between studies that do or do not use laboratory confirmation to identify sources probably means they cannot be combined to generate overall estimates. The range of estimates (20–76 per cent just in the two studies previously cited from the United States) also indicates that this would be of limited utility.

4.2.4 Basic reproduction number and average age at first infection

Some basic parameters that are used to describe the behaviour of infectious diseases include the basic reproduction number (R_0) and the average age at first infection. R_0 is a theoretical concept defined as the average number of people who are infected by each infectious case of pertussis in a totally susceptible population. It is a summary epidemiological measure that is used as a parameter in mathematical models and for estimating the herd immunity threshold. R_0 varies locally according to factors that affect the likelihood of transmission including birth rate and population density. Estimates have varied quite widely, beyond the range of 10–18 that is based on historical reported data[39,40]. The variation is unsurprising given the range of contexts, the rarity of opportunity to observe directly the spread of pertussis introduced into a totally susceptible population, and the variety of methods of estimation.

The average age at first infection with pertussis is linked to R_0 because the age at which people become infected is an indicator of the infectiousness of the organism in a population without a vaccination programme. The younger the average age of acquisition, the higher the values of R_0. Once a programme is implemented, the average age at first infection is linked to the effective reproduction number (R_e, the average number of cases that are infected by an infectious case in a population with some level of existing immunity) since indirect effects change the average age of infection due to herd immunity that slows the accumulation of susceptible individuals. A landmark paper in the understanding of pertussis transmission highlighted that pertussis vaccination seemed to have a greater impact on disease than on transmission[41]. It took many years for the implications of the observation to be fully appreciated by the scientific community. More recent literature points to this finding as contentious, when it might be fairer to see it as opening the door to a less binary and more nuanced discussion of the extent to which both wP and acellular pertussis (aP) vaccines change pertussis transmission dynamics[42]. Evidence that pertussis vaccines do not produce the sterilizing immunity of vaccines such as those for measles and rubella is plentiful, for example, from the continuing epidemic cycles observed in surveillance data (see Figure 4.2).

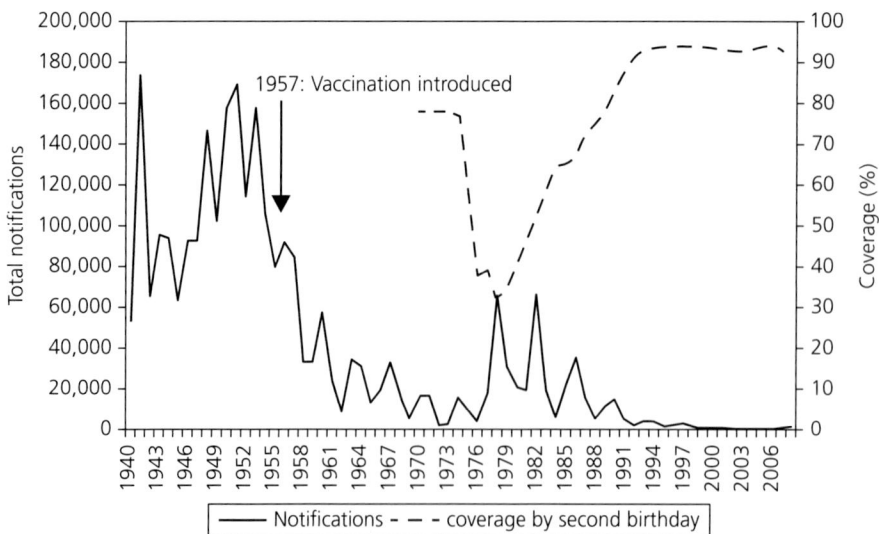

Figure 4.2 Number of annual notifications between 1940 and 2006 in England and Wales and coverage of the primary pertussis vaccination course by 2 years of age.

Equally, evidence that pertussis vaccines have to some significant degree an impact on transmission is plentiful from the lengthening of inter-epidemic cycles and the increased average age at infection[43,44]. A unique illustration of the impact of immunization increasing the age distribution is given by the young age distribution in outbreaks of pertussis occurring in communities that do not vaccinate occurring at the same time as cases with an older age distribution who live in surrounding highly vaccinated communities, as seen in Canada (see 'Countries without widespread resurgences of pertussis').

In the context of other highly effective childhood vaccines, the most important question now is not whether pertussis vaccines interrupt transmission, but how can their impact on transmission be enhanced? The current solution within the choice of existing vaccines is to vaccinate women in pregnancy. Future solutions likely need to draw on interdisciplinary research to tease out the basic biological, human, and social factors that affect pertussis transmission which we need to understand in order to make better vaccines and more effective vaccination programmes.

4.3 Burden of disease and variation by sociodemographic and economic factors

4.3.1 Sources of data, bias, and surveillance

Pertussis is notifiable by law in most countries, and data are shared with the WHO. Case definitions including requirements for laboratory testing and reporting vary. Under-detection and under-reporting can be a major challenge for surveillance. In Spain, more than 90 per cent of cases of children with clinically compatible respiratory symptoms (reportable by law) are not reported[45]. The prism of the surveillance system introduces bias into what is reported and observed, which needs to be taken into consideration when evaluating studies that rely on routine surveillance data to model pertussis[46,47]. Considerable uncertainty remains about the true underlying distribution of infection and disease, leading to wide ranges in derived estimates of the total burden of infection[48].

Surveillance of pertussis in most countries is dominated by reports of cases in young children in whom the highest incidence is generally observed. This pattern is strongly influenced by the age-specific variation in severity and risk of hospitalization.

Infants are the age group in which clinicians are most likely to think of pertussis and carry out investigations. Infants may also be more likely to have positive laboratory results because they present earlier in the illness and may remain positive for longer. Conversely, very young infants may present without prominent typical features of pertussis, increasing the risk of under-diagnosis. Deaths in infants may also be under-reported when they occur rapidly and without typical symptoms and signs[49,50]. Such variation in severity of disease and test likelihood affects the probability of presenting for healthcare, investigation, diagnosis, and reporting. It introduces bias into routinely available data on pertussis incidence. Changes in diagnostic methods (see Chapter 12) have also changed the ascertainment rate and bias in what is reported. PCR has increased test diagnostic sensitivity dramatically compared with traditional culture. Traditional serological diagnostic methods as used for other pathogens (e.g. immunoglobulin (Ig)-M detection in acute phase sera or demonstration of a fourfold rise in titre between acute and convalescent sera) are more difficult to apply to pertussis as cases often do not present until late in the course of the illness. Instead, demonstration of a high IgG titre to pertussis toxin (PT)—a pertussis-specific antigen—in late sera has been used as laboratory evidence of pertussis in those with a chronic cough. Such serological methods of diagnosis enable better characterization of the burden of disease in older children and adults who generally present later with prolonged cough[20]. Nevertheless, even where serological diagnosis is available, there is important under-reporting of disease in older children and adults[47,51].

Because of the difficulties in comparing surveillance data between countries, or within a country over time, attempts have been made to use seroepidemiological methods to assess age-specific pertussis incidence. Most vaccine-preventable diseases have an established correlate of protection which can be used to derive a population immunity profile by age and from this the rate at which susceptible individuals in a specific age group acquire infection, that is, the age-specific force of infection (see Chapter 1). However, no correlate of protection has been established for pertussis so an alternative approach to deriving incidence has been based on the population prevalence of high titres of IgG antibodies to PT. This has been shown to be a sensitive and specific indicator of an acute pertussis infection, at least in those presenting with prolonged cough. With knowledge of the kinetics of decay of PT IgG following infection, the proportion of individuals in an age group with titres above a certain threshold can be used as a measure of the incidence of infection (e.g. within the previous year)[52]. This method was used to conduct a comparative study of pertussis incidence in six Western European countries which concluded that the incidence of infection was highest in 10–19-year-olds in high-coverage countries and in 0–9-year-olds in low-coverage countries[53]. However, use of newer pertussis vaccines for priming or boosting that generate high PT antibody levels can limit the value of this method for estimating incidence in vaccine-eligible cohorts.

4.3.2 Global burden of disease

The WHO commissioned an exhaustive review of the epidemiology of pertussis in 2000 as part of the Global Burden of Disease (GBD) study methods development, searching grey and published literature, but data from LMIC were scarce[54]. Based on what data were available and historical data from HICs, the methods to estimate burden assumed that everyone would get pertussis by the age of 15 years if unvaccinated, higher levels of disease in younger children in LMIC than HICs (to account for herd effects and larger family size), and a higher case fatality ratio (CFR) in infants of 3.7 per cent versus 0.2 per cent (Table 4.1). Based on this method, global deaths from pertussis in children in 2000–2013 were estimated to comprise 1 per cent of deaths under 5 years[55], most of which would be expected to have occurred in the neonatal period. The WHO has now updated the approach, with the most recent estimates for 2014 of 24.1 million pertussis cases and 160,700 deaths from pertussis in children below 5 years[56]. Uncertainties about older children led to estimates

Table 4.1 Case fatality ratios used for WHO GBD estimates in 2003 and 2017 (%)

High, middle, or low income countries	<1 year	1–4 years
High-income	0.2	0.04
Low- and middle-income	3.7	1.0

no longer being included in the revised model. Overall pertussis has been falling as immunization coverage has been increasing globally; the African region contributes the largest proportion of cases but the country with the highest number of cases and deaths is India, followed by Nigeria. Uncertainties about a few key parameters continue to result in wide ranges in sensitivity analyses.

Given the paucity of data underpinning the models, efforts have been made to conduct studies in LMIC to get better estimates of pertussis mortality and morbidity. Nearly two decades after the first GBD study, data are starting to accumulate on the burden of pertussis in infants, but data remain of uneven quality particularly on the cause of neonatal deaths in the community[57]. A recent excellent systematic review found only 17 studies, including studies that were used to develop the WHO GBD methods, highlighting the ongoing need for better data[50]. The review reiterated the findings of the previous work: good evidence that mortality from pertussis was substantial in the pre-vaccine era, deaths were not restricted to infants, and pertussis mortality could be substantial in settings where infant mortality is generally high.

4.3.3 Burden of pertussis in low- and middle-income countries

Several studies of pertussis burden have been conducted in South Africa in recent years, both hospital and community based. Among LMIC, South Africa is unique in having introduced aP vaccination in 2009. A 1-year study of infants with respiratory illness hospitalized in one hospital in Soweto was conducted in 2015[58]. Overall, 2.3 per cent (42/1257) of infants with respiratory illness had pertussis, 86 per cent of which occurred in infants less than 3 months of age, and higher in HIV-exposed (2.7 per cent) than HIV-unexposed infants. Pertussis-associated hospitalization occurred in 2.9/1000 infants per year. The in-hospital CFR was 4.8 per cent, based on two deaths. Another study in Soweto recruited 1254 women during pregnancy, leveraging a randomized controlled trial of influenza vaccination in pregnancy that enrolled women between March and August 2011, and followed up the mother and their infants from birth to 24 weeks of age for respiratory illness[59]. A total of 37 infant pertussis cases were identified, with an incidence of 7.4 and 5.5 per 1000 infant-months in HIV-exposed and HIV-unexposed infants, respectively. Six cases were hospitalized and one died, a total CFR of 2.7 per cent and an in-hospital CFR of 16.7 per cent, but the numbers are small, making the estimates imprecise. Infections were also identified among mothers of the infants. A community-based study in Zambia in 2015 recruited 1981 mother–infant pairs who were healthy at recruitment, excluding underweight, and premature infants who were followed up every 2–3 weeks to 14 weeks of age. Pertussis was identified in ten infants, an incidence of 2.4 per 1000 infant-months or 5.2 cases per 1000 infants. Only one case was severe, and, although infants were hospitalized and died during the follow-up, no hospitalizations or deaths were attributable to pertussis[60]. A convenience sample of 449 children in Uganda who had been coughing for two or more weeks and were aged 3 months to 12 years, found by PCR that 15 per cent had pertussis, and cases in household contacts including adults were epidemiologically linked[21].

An analysis of infants recruited into the Pneumonia Etiology Research for Child Health (PERCH) study found significant heterogeneity by region with more pertussis detected in participating sites in Africa than in South East Asia[60]. Enrolment was of hospitalized infants aged 1–59 months of age during 2 years between August 2011 and January 2014. Community controls were also recruited. Overall, pertussis was detected in 1.3 per cent (53/4200) of cases, ranging by site from 0 to 2.5 per cent, and in 0.2 per cent (11/5196) of controls, ranging by site from 0 to 0.6 per cent. Of 300 deaths in hospital across all sites, eight were confirmed with pertussis, but given the heterogeneity between regions it is probably only appropriate to look at the African sites together where five pertussis-associated deaths occurred among 137 deaths, an aetiological fraction of 3.7 per cent. The overall CFR in the African sites was 9.6 per cent but numbers were small (5/52)[60].

The PERCH study found little evidence that pertussis carried a significant burden in its sites in Bangladesh and Thailand[60]. A community-based study in Pakistan enrolled 2021 infants from birth up to 10 weeks of age and followed them to 18 weeks with intensive follow-up at home[61]. A total of eight cases were identified

giving an incidence of 3.96 per 1000 infants or 1.14 per 1000 infant months, comparable to some of the West African studies. Three cases were classified as severe. One death occurred in a case that had been classified as non-severe (CFR = 12.5 per cent). Similar to Pakistan, the incidence of PCR-confirmed pertussis in a well-designed, nested, population-based study in Nepal was only 13.3 cases per 1000 infant-years (95 per cent confidence interval (CI) 7.7–21.3)[62].

The apparent relatively low contribution of pertussis to the overall burden of mortality and morbidity in various studies in West Africa has unsurprisingly led to conclusions that pertussis may not currently be an important pathogen in Africa[63]. However, the evidence is insufficient to reject the hypothesis that pertussis could be an important pathogen in West Africa. Surveillance remains poor, and some of the best sources of data such as the PERCH study do not start at birth and/or do not follow up for the whole first year of life[59,60]. Studies generally have not been run for at least one whole epidemic cycle, so there is no way of knowing if the peak or trough has been measured. Given that peaks occur every 3–5 years, studies that last for a year or less are over two to four times more likely to encompass a low- than a high-incidence period. Hospital-based studies will miss deaths in the community that don't reach healthcare facilities, which are most likely to happen in the neonatal period. Furthermore, very young infants frequently present with atypical symptoms so that clinicians miss the diagnosis. It is hard to draw many conclusions beyond the need for more data. Better evidence is needed and may be forthcoming from the Child Health and Mortality Prevention (CHAMPS) study[64].

4.3.4 Burden of disease in high-income countries

In HICs with good immunization programmes and healthcare systems, pertussis continues to cause severe infections and epidemic cycles. The CFR estimates used in the WHO burden of disease estimates may not be reliable since these vary considerably. For example, in England and Wales, the CFR among hospitalized infants in the 9 months before the introduction of the maternal pertussis immunization programme in October 2012 was 1.7 per cent (ten deaths among 578 hospitalized cases under a year)[65]. Similarly, in the Californian outbreak in 2010, there were ten deaths among 584 (1.7 per cent) of hospitalized infants under 6 months of age[66]. In a recent case series from Australia and New Zealand from 2002 to 2014, pertussis was found to cause 1 per cent of admissions to paediatric intensive care units (PICUs), of whom 4.8 per cent died, with the highest incidence in 2009–2012[67]. Another Australian-led study over 17 years found that more than half of all cases occurred during that epidemic period of 2009–2012. PICU admissions were most likely in young infants before the age of first immunization, and 6.3 per cent of children admitted to PICUs with pertussis died[68]. In Israel, retrospective testing by PCR of stored nasopharyngeal washes showed that pertussis was the cause of 11 of 72 (15 per cent) PICU admissions in infants aged under a year with a lower respiratory tract infection, none of whom was diagnosed clinically showing that the burden of severe pertussis-attributable disease may be underestimated unless appropriate laboratory investigations are requested.

4.3.5 Impact of socioeconomic factors on disease

As with any infectious disease, a range of demographic and social determinants are important drivers of pertussis transmission. These include population birth rates, family size, contact patterns, and school attendance[69,70].

In relation to equity, perhaps the best way to compare countries is to consider vaccination coverage as a measure of access to protection from pertussis among children. Time series analysis indicates that coverage is rising globally but 53 countries have been flagged as missing the Global Vaccine Action Plan target of 90 per cent coverage of three doses of diphtheria, tetanus, and pertussis (DTP3) vaccines by 2015[71]. Most of these countries are in sub-Saharan Africa and south Asia. Immunization coverage is clearly related to social determinants of health as well as access to and quality of healthcare. Low coverage correlates with healthcare provision such as the proportion of births attended by skilled health staff, and with socioeconomic factors such as primary school completion in Africa and government health spending in the Eastern Mediterranean. Such correlations may help predict countries at ongoing risk of failing to reach targets and highlight the broader context in which immunization programmes are situated.

In HICs, social determinants such as socioeconomic status may also influence the risk of pertussis by being associated with barriers to accessing timely immunization and also by affecting the outcome of disease. For example, in one study in New Zealand, where coverage of pertussis immunization is high, immunization series were less frequently completed on time in the most deprived fifth of families[72], who together accounted for more than half of all cases. In England and Wales, in the whooping cough resurgence in the 1970s and 1980s, not only were children from lower-income families more likely to be unvaccinated against pertussis but among those who got pertussis, hospital admission rates were highest in the lowest-income families[73]. In addition to poverty, outbreaks occur in unvaccinated groups who are not necessarily poor but differ with respect to culture or religious beliefs. Such groups place the broader population at risk, bring pertussis into communities, and are among the early cases in outbreaks[74]. In Canada, failure to be vaccinated is associated with low income, single parents, and less parental education[75].

4.4 Impact of vaccination on disease epidemiology

4.4.1 Overall trends

The most effective control measure for pertussis is mass vaccination which has a marked impact on the epidemiology of the disease. Other public health interventions such as isolation of cases or antimicrobial prophylaxis of vulnerable contacts may reduce the risk of spread to immediate contacts but have no impact on transmission at the population level and are discussed in more detail in Chapter 15.

In a WHO review of schedules in use globally in 2014, of 194 member states 32 per cent had a three-dose schedule, 39 per cent a four-dose schedule, and 29 per cent countries had a five-dose schedule for children under 7 years of age[76]. Global coverage for the three-dose primary course as reported to

WHO in 2015 averaged at 86 per cent with 129 countries reporting a coverage of at least 90 per cent. However, six countries reported a coverage of less than 50 per cent and 14 per cent a coverage of between 50 and less than 70 per cent in 2015[77]. High coverage with an effective vaccine substantially reduces the overall number of cases in the population by rendering susceptible individuals immune before they can become infected. Any infectious person therefore has less opportunity to meet and infect a susceptible individual which reduces overall transmission in the population and generates herd immunity. To be maximally effective in directly protecting the most vulnerable and in reducing transmission, vaccination needs to be targeted at the youngest infants in the population, usually from 6 to 8 weeks of age. The currently available vaccines, whether wP or aP vaccines, do not produce long-lasting sterilizing immunity and so pertussis elimination cannot be achieved despite sustained high coverage. The persistence of pertussis even in high-coverage settings means that herd immunity is incomplete and the protection of infants too young

to be vaccinated requires additional strategies (discussed in Chapter 15). Continued transmission also ensures that if vaccine coverage declines, disease control is quickly lost[78] (see Figure 4.2).

Reduced transmission as a result of mass vaccination results in an upward shift in the age distribution of cases as infection, when it does occur, is on average deferred until an older age. This is nicely shown by the experience in England and Wales when coverage was restored following its collapse in the 1970s and 1980s; the increasing coverage from the 1990s onwards reduced transmission in the population and was associated with a progressive increase in the average age of notified cases (see Figure 4.3). Similar changes of increased average age at infection were seen following the re-introduction of a pertussis vaccination programme in Sweden following a prolonged hiatus with no programme from 1979 to 1996[79]. In Canada, an increase in average age at infection occurred in a very different context in which coverage had remained stable and high but protection had dropped due to the use of a low-effectiveness wP vaccine. In this situation, the

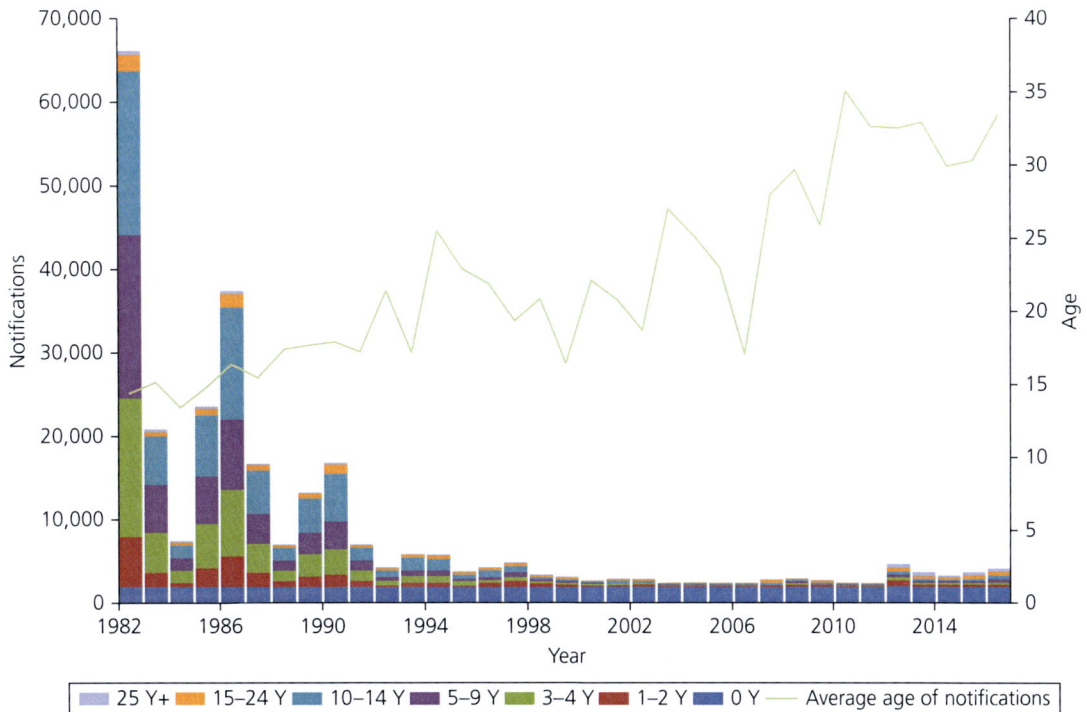

Figure 4.3 Number of notifications by age group in England and Wales 1982–2013 and average age of notifications. Y, years.

pattern of a right shift in the age distribution followed the introduction of a highly effective aP vaccine[80]. An increase in the average age at infection after the introduction of vaccination has also been demonstrated in Thailand and in an African setting[81,82]. Studies of changes in age distribution in larger populations are limited by the availability of consistent high-quality data at sufficiently granular levels on both disease incidence and coverage. The implications of changes in diagnostic methods and the myriad challenges of obtaining good coverage data may not be readily apparent[83]. The error margin for many coverage surveys encompasses the herd immunity threshold; hence, local increases in pertussis can reflect hidden failure to vaccinate as much as vaccine failure. At suboptimal levels of coverage, a resurgence is inevitable sooner or later. In addition, local variation in coverage can be an important driver of heterogeneity in pertussis control, and this may be particularly evident in federalized countries with large populations, decentralized immunization programme delivery, and significant inequity. Marginalized populations that have high birth rates and poor access to healthcare are doubly at risk of outbreaks.

Interestingly, high coverage did not appear to affect the epidemic cycle in the United Kingdom as much as other vaccines such as measles vaccine, which, although increased, has remained at about 3–4 years throughout periods of low and high coverage[70,84,85]. Some greater lengthening of the inter-epidemic might be expected based on epidemiological theory (discussed in Chapter 1). This suggests that the vaccine provides better protection against clinically typical pertussis than against infection so that the impact on transmission is less than on notified cases, as suggested by a recent modelling study in the United Kingdom[99].

4.4.2 Impact on pertussis mortality

Studies on trends in pertussis mortality have been available in some countries for more than a century[50]. A historical review shows declining trends in pertussis mortality that predated the introduction of pertussis immunization programmes and coincided with better living conditions and smaller family sizes. Although pertussis mortality declines started before immunization was introduced, these trends were accelerated by vaccination[86]. However, because of the incomplete herd immunity, pertussis mortality remains at around 1–2 per 100,000, even in HICs with sustained high coverage[11]. Despite the increase in the average age of infection following the introduction of vaccination, deaths remain concentrated in the first year of life due to the very high CFR associated with infant disease[75] (see Figure 4.4).

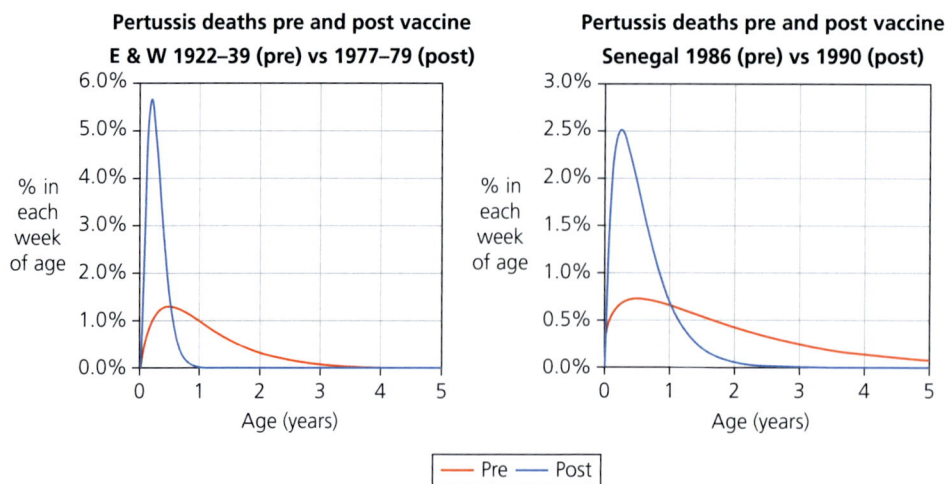

Figure 4.4 Age of children at time of death from pertussis in the pre- and post-vaccine eras: England and Wales (E & W) and Senegal. Reproduced with permission of Prof Colin Sanderson.

4.4.3 Impact of whole-cell versus acellular vaccines

Both wP and aP vaccines are effective at controlling pertussis if high coverage is achieved with effective products. In Sweden, where wP vaccination was abandoned in 1979, the resulting national resurgence of pertussis was effectively brought under control by the introduction of aP vaccines in 1996 with evidence of herd immunity among unvaccinated cohorts[44]. Both wP and aP vaccines are highly effective in providing direct protection to infants after one or two doses[11]. All wP and aP vaccines are not the same, however. While wP vaccines are regarded as a challenge from a regulatory perspective because the content is hard to standardize, the components included in the aP are standardized but vary between products used at different ages both within countries and between countries and change with time. The effectiveness of both wP and aP vaccines therefore varies[87]. This can be a challenge for the interpretation of epidemiological data and for comparison of trends.

Most of the world still uses wP vaccines that have high effectiveness. However, many HICs have switched to aP vaccines over the last two decades to reduce the local reactions and fever that can be associated with wP vaccines. This change may have come at the cost of reduced protection. A meta-analysis conducted in 2016 of clinical trial efficacy and post-licensure effectiveness data for available aP and wP vaccines showed an overall efficacy estimate within 3 years of primary vaccination against clinically typical pertussis of 84 per cent (95 per cent CI 81–87 per cent) for aP vaccines compared with 94 per cent (95 per cent CI 88–97 per cent) for wP vaccines[87]. Moreover, priming with wP leads to a longer duration of protection[88,89] even when priming is with a wP that is not highly efficacious[90]. Use of aP boosters at school entry or in adolescence can extend the duration of protection but protection is not long lasting[91,92]. Also, there is no evidence that adolescent boosters materially reduce the risk of pertussis in infants too young to be vaccinated[2].

4.4.4 Countries with resurgences of pertussis

A number of countries with established vaccination programmes have recently reported a resurgence in pertussis that is not associated with a drop in coverage. In an attempt to understand the reasons for this, in 2014 the WHO Strategic Advisory Group of Experts (SAGE) conducted a review of pertussis epidemiology in selected HICs and LMIC with and without a reported resurgence, using ether aP or wP vaccines and with surveillance data that was considered sufficiently good to allow disease trends to be studied[11]. While WHO does collect, where available, annual pertussis data from every country via the Joint Reporting Form that is used for all vaccine-preventable infections[93], it recognized that such data are insufficiently reliable to make detailed comparisons between countries and to tease out the factors driving the transmission dynamics of pertussis in different settings. The WHO review therefore aimed to obtain data from selected countries that, while not globally representative, did have long-standing high vaccine coverage rates with evidence of periods of effective disease control, and were able to provide high-quality data on vaccine coverage and trends in pertussis disease burden over time. The countries provided surveillance data according to a detailed format that allowed the effect of changes over time in diagnostic methods, vaccines used and their effectiveness, the number and timing of booster doses, and changes in surveillance methodology to be taken into account. For the purposes of the WHO review, a resurgence was defined as a larger number of cases than expected, given the periodic variability of naturally recurring pertussis disease, when compared to previous cycles in the same setting[2]. One key question that SAGE sought to understand was the potential role of aP vaccines in the generation of the reported resurgences.

Detailed epidemiological and vaccine coverage data up to 2012 were obtained from 19 countries of which only five were LMICs, reflecting the dearth of reliable pertussis surveillance data in these settings. In some countries that had reported a resurgence, it was unclear whether the change to more sensitive laboratory diagnostic methods had generated the increase in reported cases, reflecting the difficulty in interpreting trends in disease incidence from routine surveillance data. Comparisons between countries were also complicated by differences in the wP and aP vaccines used, and the number and timing of doses offered in the primary and booster schedules. The WHO review concluded that definitive evidence for a resurgence based on an increase in infant mortality or hospitalizations unexplained by other factors could be found in only four countries: Australia, England and Wales, the United States, and Portugal. All four countries had changed to aP vaccines 6–10 years before the resurgence[11]. However, no common pattern in terms of primary schedule, timing, and number of booster doses was evident.

In Australia, which uses a 2-, 4-, and 6-month primary schedule, aP replaced wP vaccines for primary immunization in 1999. An 18-month aP booster was introduced in 1997 which was discontinued in 2003 and replaced by an adolescent booster. A large increase in cases occurred in 2009 in all age groups which continued until 2012. Despite the increase in hospitalized cases in infants under 3 months of age, no increase in pertussis deaths was recorded. Since 2012, disease incidence declined and by 2014 was at pre-resurgence levels, although pertussis increased again in subsequent years, in line with epidemic cycles[94,95]. A study of culture-positive pertussis (to control for the ascertainment increase due to PCR introduction) found an increase in fully immunized children more than 6 months old when the 18-month booster was discontinued, contributing to the impact of replacing wP by aP vaccines with its more rapidly waning protection[96]. A mathematical model of pertussis infection was developed and calibrated using broad epidemiological trends, namely the 2–4.5-year inter-epidemic period, the fourfold reduction in infections in infants under 6 months after introduction of vaccination, seroepidemiological data suggesting high immunity levels in adults, and the resurgence. The model only reproduced these four epidemiological features if the duration of natural immunity was decades longer than vaccine-induced immunity[97]. The model did not suggest that there were significant differences in duration of immunity between aP and wP vaccines which does not accord with studies from Australia and the United States[88]. The Australian model did not allow for differences in the ability of natural or vaccine-induced immunity to protect against infection and disease transmission which may have obscured some key differences between aP and wP vaccines. It did, however, confirm the importance of the 18-month booster in Australia which was reinstated in 2016[98].

In England and Wales, aP vaccine replaced wP vaccine in 2004, with a single aP booster at 3–4 years of age introduced in 2001. An accelerated 2-, 3-, and 4-month primary schedule has been used since 1990. A resurgence of pertussis evidenced by an increase in pertussis deaths and hospitalizations in young infants occurred in 2012[58]. Cases also increased in other age groups. To protect infants who were too young to be vaccinated, a maternal pertussis immunization programme was introduced in October 2012 which proved highly effective and reduced incidence in infants to pre-resurgence levels. In other age groups, however, disease incidence has remained elevated throughout the period to the end of 2016. A mathematical model of pertussis transmission was developed to investigate the possible reasons for the resurgence and the likely impact of introducing an 18-month and/or adolescent booster[99]. The model was fitted to 60 years of notification data for England and Wales and allowed natural and vaccine-induced immunity to wane and for aP and wP vaccines to differ with respect to their ability to protect against infection, as suggested by baboon challenge studies[100]. The model parameters that best fitted the observed epidemiological trends were ones in which the duration and/or degree of protection against infection was longer for wP than aP vaccines; the duration of protection from natural immunity was longest. The model predicted that the elevated incidence would continue and that the 2012 resurgence represented a resetting of the endemic level of transmission which increased as a result of the lower level of protection from aP vaccine. Neither an 18-month nor adolescent booster was predicted to have much impact on infant disease.

In the United States, wP vaccine was replaced by aP vaccine in 1997 using a 2-, 4-, and 6-month priming schedule with boosters at 18 months and 5 years. An additional early adolescent booster was recommended in 2005 and maternal immunization in every pregnancy has been recommended since 2011 in response to the overall increasing trend in cases since 2004[11]. Although surveillance and diagnostic methods had improved over this time, this increase was considered real as hospital admissions in young infants increased, though not mortality[11]. Reporting rates peaked in 2012 but have declined in 2013–2015, though this may reflect the usual epidemic cycle[101]. Mathematical models were developed to investigate the potential cause of the resurgence; the model that best fits the observed trends is one in which vaccine effectiveness and duration of protection of the aP vaccine is lower than that of the wP vaccine[102]. Interestingly, some evidence indicates a nationwide resurgence in pertussis in the United States starting sometime after the mid 1970s[103]. However, this may have been due in part to the low effectiveness of the wP vaccines used in the United States at the time; when one such vaccine was evaluated as the comparator vaccine in aP trials in Sweden and Italy in the 1990s, its efficacy was considerably lower than the aP vaccines under study which have been shown in placebo-controlled trials to be highly efficacious[104].

Portugal has a long-standing pertussis programme with aP vaccine replacing wP vaccine in 2006. Five doses are given, the three primary doses at 2, 4, 6 months, an 18-month booster, and another booster at 5–6 years of age[105]. Vaccination coverage

has been sustained at over 90 per cent for DTP3[106]. Portugal continues to observe epidemic cycles of pertussis with a peak in 2012, but is not among the countries with the highest incidence of reported disease in Europe[107]. However, in 2012, the usual epidemic peak was associated with a large increase in hospitalizations and deaths in infants aged less than 1 year[44]. Cases occur in both infants and older children, who are most often infected by adult contacts[108].

Israel was one of the countries where despite a sustained recent increase in reported cases, SAGE took the view that the evidence for a resurgence was not clear. Israel used a four-dose schedule of wP vaccine (2, 4, 6, and 12 months) until 2005 when it was replaced by aP vaccine with two further boosters introduced at 7 and 13 years of age[11]. Coverage is relatively high at over 90 per cent[109]. Israel seems to have had a large increase in incidence among infants in 2010–2011 including fatal cases (see Figure 4.5). The WHO review found that although there was some evidence of increased transmission[110], it was difficult to distinguish resurgence

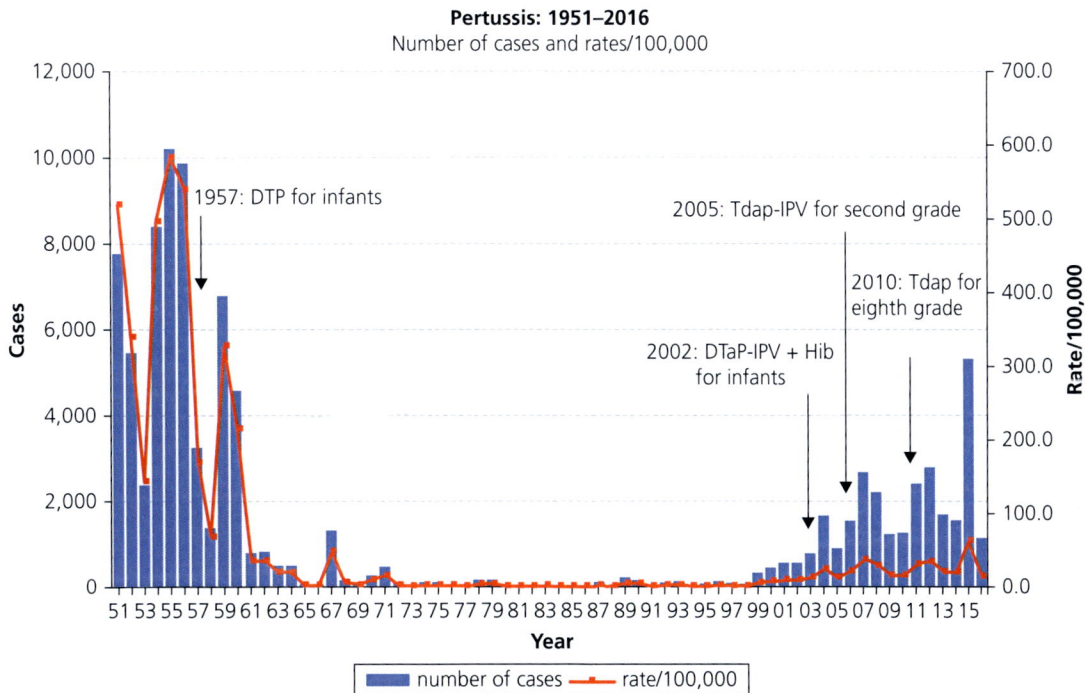

Figure 4.5 Pertussis epidemiology in Israel 1951–2016. Reproduced with permission of Ministry of Health, Israel.

from improved ascertainment associated with the availability of more sensitive laboratory tests. Infant cases in Israel, as in many countries experiencing a resurgence, were found to be more likely to have had a low birth weight, be of high birth order (fourth and higher), be from poorer socioeconomic status, be linked to communities with high rates of immunization exemptions, and be unvaccinated or have delayed vaccination[111].

Two wP vaccine-using countries, Brazil and Chile, reported increases, largely in infant cases and deaths in 2011/2012. Both use a 2-, 4-, and 6-month primary schedule with a wP booster at 15–18 months and again at 4–6 years. However, in both countries coverage had declined prior to the increase and was non-homogeneous across the country. Surveillance also improved around the time of the increase and in Chile there were also concerns about the specificity of the laboratory methods used for confirmation. WHO concluded that evidence of resurgence was limited and that the drop in coverage and improvements in surveillance sensitivity may have caused the increase in reported cases.

4.4.5 Countries without widespread resurgences of pertussis

A number of aP vaccine-using countries in the WHO survey did not report an increase in cases. No resurgences were observed in Sweden, Denmark, and Norway, countries that have used aP vaccines for at least 15 years; all use a 3-, 5-, and 10–12-month schedule. Similarly, Finland, which changed to aP vaccine in 2005 with a 3-, 5-, and 12-month schedule, has not reported a resurgence.

Widespread increases have not been observed in Canada either, where aP vaccine has been in use since the late 1990s. In Canada, immunization programmes policy is determined at provincial or territorial level and programmes vary across the country[112]. Secular trends are dominated by large outbreaks in the late 1980s and 1990s attributable to the use of a wP vaccine with low effectiveness of around 60 per cent, a unique feature of the history of the programme in Canada[113] (see Figure 4.6). A switch to aP vaccine was implemented in 1997–1998. Large sustained national increases have not been observed in the intervening period, but localized outbreaks and epidemic cycles continue to be observed[114].

The epidemiology at subnational level in Canada is heterogeneous. In New Brunswick, a large outbreak of pertussis occurred in 2012[115] and a smaller one in 2015[116] despite high coverage; both affected largely school-aged children[117]. In contrast, in Saskatchewan and Ontario, under-vaccinated communities contribute to the epidemiology[118]. Outbreaks in 2010 and 2015 in Saskatchewan affected younger children including infant deaths[119]. In Ontario, outbreaks in recent years have been concentrated in communities

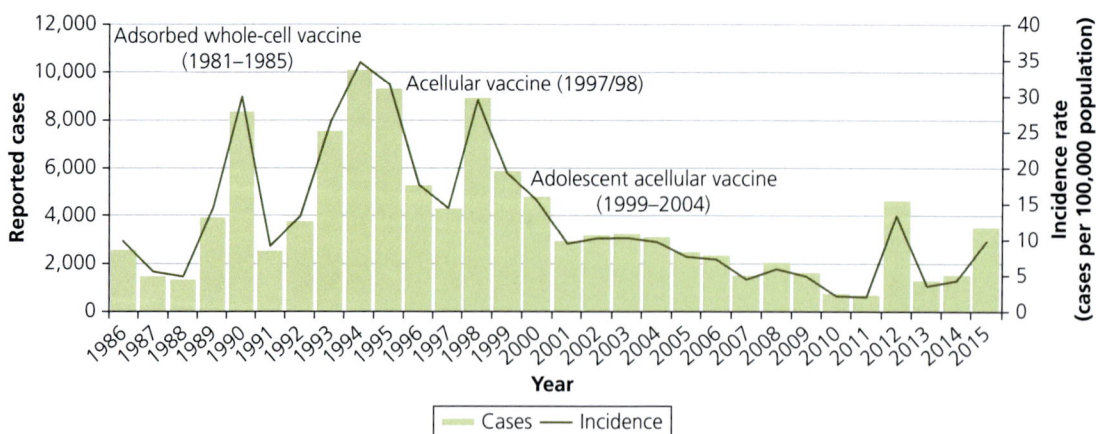

Figure 4.6 Secular trends in Canada. © All rights reserved. *Canada Communicable Disease Report.* 7 February 2014 – Volume 30 – No. 3: Pertussis surveillance in Canada: trends to 2012. Public Health Agency of Canada, 2014. Adapted and reproduced with permission from the Minister of Health, 2017.

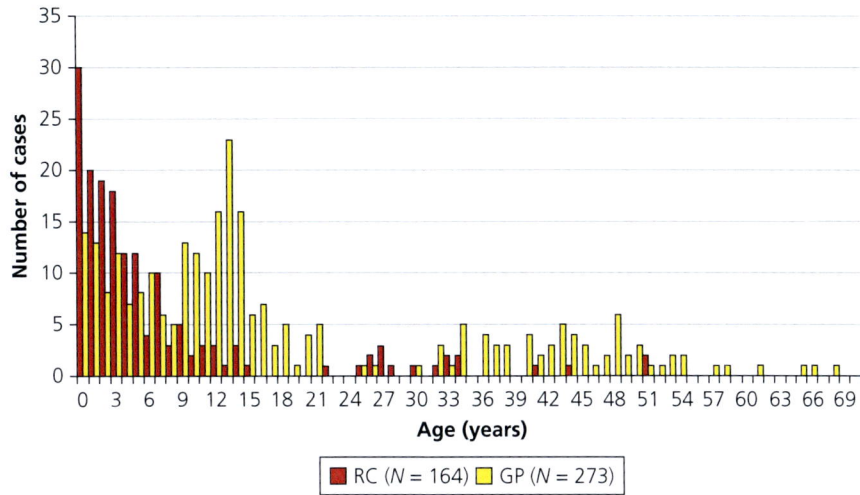

Figure 4.7 Pertussis cases by age and community status: Ontario, 1 November 2011 to 15 April 2013. GP, general population; RC, religious communities. © All rights reserved. *Canada Communicable Disease Report.* 7 February 2014 – Volume 30 – No. 3: Prolonged pertussis outbreak in Ontario originating in an under-immunized religious community. Public Health Agency of Canada, 2014. Adapted and reproduced with permission from the Minister of Health, 2017.

that reject immunization[120]. The age distribution in these communities reflected pre-vaccine epidemiology and affected younger children (see Figure 4.7), but serious outcomes have fortunately not occurred[120]. At the same time, outbreaks in highly immunized communities in the same region have affected older children, consistent with waning immunity and negligible effectiveness from 7 years after last immunization[90]. Outbreaks have also affected Aboriginal communities in Canada. A large outbreak spread across in Nunavut during 2016, including severe illness in infants[121].

4.5 Conclusion

Interpretation of the secular trends in pertussis epidemiology is complex and, as shown by the WHO 2014 review, reports of resurgences need to be interpreted with due account taken of other factors that may have influenced reporting. The conclusion of the WHO review was that there was no evidence of a global resurgence in pertussis but in those countries with definitive evidence of a resurgence, the most likely explanation was the lower and/or shorter duration of protection from aP compared to wP vaccines[2]. This view was supported by additional evidence from baboon challenge studies[79]

and insights from the mathematic models of pertussis transmission (to be discussed in subsequent chapters). SAGE therefore recommended that countries that had not yet changed to an aP vaccine should continue to use wP vaccines[2]. However, it is recognized that attribution of pertussis resurgences solely to the use of aP vaccines is likely to be too simplistic given that there are aP vaccine-using countries with good surveillance that have not experienced a resurgence and that reliable data on the relative duration and degree of protection against disease and infection afforded by different aP and wP vaccines are still lacking[42].

For aP vaccine-using countries, the factors that determine whether or when they experience a resurgence in pertussis are not understood and may differ between countries. The Australian experience suggested that removal of the 18-month booster dose was a driver, but a resurgence occurred in the United States (although not Canada) which has maintained a 15–18-month booster plus a school entry and adolescent booster. Scandinavian countries that use a 3-, 5-, and 10–12-month schedule have not as yet experienced a resurgence, but it seems unlikely that this schedule is superior to the United States or Canadian schedules with its three 2-, 4-, and 6-month priming doses and the early booster in the second year of life. Evolution of bacterial strains deficient in key aP vaccine antigens, such as pertactin, has been documented in Australia and the United States where the prevalence in some regions has reached 80 per cent and may have acted as drivers for the resurgence in these countries[122]. However, the strains circulating during the 2012 outbreak in England and Wales were genetically very similar to those circulating during periods when the incidence of pertussis was low, with only one pertactin-negative strain found in 2012, though the numbers tested were relatively small[123]. The issue of strain evolution under vaccine pressure will be discussed in detail in a later chapter.

Modelling of the resurgence in England and Wales suggested that the resurgence represented a resetting of the endemic transmission level arising from the use of aP vaccines which provide less protection (either by degree or duration) against infection than the wP vaccine formerly used in that country. This would suggest that a resurgence would not be seen in countries with an already high level of endemic transmission such as Sweden where aP vaccine was introduced after a period when no vaccination was given and countries such as Canada where the wP vaccine used was not highly effective. Unfortunately, reliable data are not available to make valid comparisons between countries in absolute disease incidence rates due to differences in surveillance sensitivity, and the completeness and accuracy of reporting.

Of all the vaccine-preventable infections, pertussis remains the most epidemiologically challenging to understand. This is partly because of surveillance issues (discussed in detail in Chapter 12) but also due to the lack of an established correlate of protection that allows susceptible and immune individuals to be distinguished. These factors are compounded by our imperfect understanding of the mechanism of protection from aP and wP vaccines, and the product-specific differences in effectiveness which may exist for aP vaccines and which have existed for wP vaccines used in the past. It can therefore be difficult for policymakers to be confident about the optimal number and timing of primary doses and how many and at what age booster doses should be given in their setting, as reflected in the plethora of schedules in use throughout the world.[1] There are even greater challenges for the WHO when attempting to make global policy as lessons learned in one setting may not appear to apply in others and the logistic challenges of timely vaccine delivery in hard-to-reach populations need to be factored into schedule recommendations. What is clear is that high coverage early in infancy with three doses of an effective vaccine greatly reduces pertussis mortality and severe morbidity in all settings and that many infants in resource-poor settings are still deprived of that benefit.

Acknowledgements

Thanks to Dr Colin Sanderson for the review of the age at death in the pre- and post-vaccine eras (Figures 4.1 and 4.4); to Public Health England for permission to reproduce Figure 4.2; to the Ministry of Health, Israel, for permission to reproduce Figure 4.5; to the Public Health Agency of Canada for permission to reproduce Figures 4.6 and 4.7; and to Delaney Hines and Allison Crehore for assistance with the literature review and citations.

References

1. World Health Organization. *SAGE Whole Cell Pertussis Vaccines: Summary of Evidence Relevant to Schedules.* 2015. http://www.who.int/immunization/sage/meetings/2015/april/2_wP_summary_WG_23Mar2015_submitted.pdf?ua=1.
2. World Health Organization. Pertussis vaccines: WHO position paper—August 2015. Geneva, Switzerland. *WER* 2015;**90**:433–60. http://www.who.int/wer/2015/wer9035.pdf?ua=1.
3. Gordon JE, Hood RI. Whooping cough and its epidemiological anomalies. *Am J Med Sci* 1951;**222**:333–61.
4. Strebel P, Nordin J, Edwards K, et al. Population-based incidence of pertussis among adolescents and adults, Minnesota, 1995–1996. *J Infect Dis* 2001;**183**:1353–9.
5. Senzilet LD, Halperin SA, Spika JS, et al. Pertussis is a frequent cause of prolonged cough illness in adults and adolescents. *Clin Infect Dis* 2001;**32**:1691–7.
6. Philipson K, Goodyear-Smith F, Grant CC, et al. When is acute persistent cough in school-age children and adults whooping cough? A prospective case series study. *Br J Gen Pract* 2013;**63**:e573–9.
7. Gilberg S, Njamkepo E, Du Chatelet IP, et al. Evidence of Bordetella pertussis infection in adults presenting with persistent cough in a French area with very high whole-cell vaccine coverage. *J Infect Dis* 2002;**186**:415–8.

8. Yesmin K, Shamsuzzaman S, Chowdhury A, et al. Isolation of potential pathogenic bacteria from nasopharynx from patients having cough for more than two weeks. *Bangladesh J Med Microbiol* 2010;**4**:13–18.

9. World Health Organization. *WHO-Recommended Surveillance Standard of Pertussis.* http://www.who.int/immunization/monitoring_surveillance/burden/vpd/surveillance_type/passive/pertussis_standards/en/.

10. World Health Organization. *WHO Meeting on Case Definition of Pertussis.* January 10–11, 1991. http://apps.who.int/iris/handle/10665/66921.

11. World Health Organization. *SAGE Pertussis Working Group—Background Paper.* April 2014. http://www.who.int/immunization/sage/meetings/2014/april/1_Pertussis_background_FINAL4_web.pdf.

12. Ontario Agency for Health Protection and Promotion, Provincial Infectious Diseases Advisory Committee. *Routine Practices and Additional Precautions in all Health Care Settings.* November 2012. https://www.publichealthontario.ca/en/eRepository/RPAP_All_HealthCare_Settings_Eng2012.pdf.

13. Heininger U. Pertussis: an old disease that is still with us. *Curr Opin Infect Dis* 2001;**14**:329–35.

14. Muller AS, Leeuwenburg J, Pratt DS. Epidemiology and control of pertussis. *Trop Doct* 1987;**17**:182–90.

15. Public Health England. *Guidance on Infection Control in Schools and Other Childcare Settings.* May 2014. https://www.gov.uk/government/uploads/system/uploads/attachment_data/file/522337/Guidance_on_infection_control_in_schools.pdf.

16. Ranganathan S, Tasker R, Booy R, et al. Pertussis is increasing in unimmunized infants: is a change in policy needed? *Arch Dis Child* 1999;**80**:297–9.

17. Storsaeter J, Hallander HO, Gustafsson L, et al. Low levels of antipertussis antibodies plus lack of history of pertussis correlate with susceptibility after household exposure to Bordetella pertussis. *Vaccine* 2003;**21**:3542–9.

18. Merkel TJ, Halperin SA. Nonhuman primate and human challenge models of pertussis. *J Infect Dis* 2014;**209** Suppl 1:S20–3.

19. Harnden A, Grant C, Harrison T, et al. Whooping cough in school age children with persistent cough: prospective cohort study in primary care. *BMJ* 2006;**333**:174–7.

20. Wang K, Fry NK, Campbell H, et al. Whooping cough in school age children presenting with persistent cough in UK primary care after introduction of the preschool pertussis booster vaccination: prospective cohort study. *BMJ* 2014;**348**:g3668.

21. Kayina V, Kyobe S, Katabazi FA, et al. Pertussis prevalence and its determinants among children with persistent cough in urban Uganda. *PLoS One* 2015;**10**:e0123240.

22. Castillo ME, Bada C, Del Aguila O, et al. Detection of Bordetella pertussis using a PCR test in infants younger than one year old hospitalized with whooping cough in five Peruvian hospitals. *Int J Infect Dis* 2015;**41**:36–41.

23. Stocks P. Some epidemiological features of whooping-cough. *Lancet* 1933;**221**:265–9.

24. Public Health England. Pertussis. In: *The Green Book: Immunisation Against Infectious Disease.* London: Public Health England; 2016, pp 1–26. https://www.gov.uk/government/uploads/system/uploads/attachment_data/file/514363/Pertussis_Green_Book_Chapter_24_Ap2016.pdf.

25. European Centre for Disease Prevention and Control. *Systematic Review on the Incubation and Infectiousness/Shedding Period of Communicable Diseases in Children.* Stockholm, Sweden: ECDC; 2016. http://ecdc.europa.eu/en/publications/Publications/systematic-review-incubation-period-shedding-children.pdf.

26. Smith RE. A review of recent work on whooping-cough. *QJM* 1936;**5**:307–26.

27. MacDonald H, MacDonald EJ. Experimental pertussis. *J Infect Dis* 1933;**53**:328–30.

28. Phillips J. Whooping-cough contracted at the time of birth, with report of two cases. *Am J Medical Sci* 1921;**161**:163–5.

29. Baron S, Njamkepo E, Grimprel E, et al. Epidemiology of pertussis in French hospitals in 1993 and 1994: thirty years after a routine use of vaccination. *Pediatr Infect Dis J* 1998;**17**:412–8.

30. Wiley KE, Zuo Y, Macartney KK, et al. Sources of pertussis infection in young infants: a review of key evidence informing targeting of the cocoon strategy. *Vaccine* 2013;**31**:618–25.

31. Wendelboe AM, Njamkepo E, Bourillon A, et al. Transmission of Bordetella pertussis to young infants. *Pediatr Infect Dis J* 2007;**26**:293–9.

32. Skoff TH, Kenyon C, Cocoros N, et al. Sources of infant pertussis infection in the United States. *Pediatrics* 2015;**136**:635–41.

33. Godoy P, Garcia-Cenoz M, Toledo D, et al. Factors influencing the spread of pertussis in households: a prospective study, Catalonia and Navarre, Spain, 2012 to 2013. *Euro Surveill* 2016;**21**:30393.

34. Wendelboe AM, Hudgens MG, Poole C, et al. Estimating the role of casual contact from the community in transmission of Bordetella pertussis to young infants. *Emerg Themes Epidemiol* 2007;**4**:15.

35. Kowalzik F, Barbosa AP, Fernandes VR, et al. Prospective multinational study of pertussis infection in hospitalized infants and their household contacts. *Pediatr Infect Dis J* 2007;**26**:238–42.

36. Crowcroft NS, Booy R, Harrison T, et al. Severe and unrecognised: pertussis in UK infants. *Arch Dis Child* 2003;**88**:802–6.

37. Bisgard KM, Pascual FB, Ehresmann KR, et al. Infant pertussis: who was the source? *Pediatr Infect Dis J* 2004;**23**:985–9.

38. Staudt A, Mangla AT, Alamgir H. Investigation of pertussis cases in a Texas county, 2008–2012. *South Med J* 2015;**108**:452–7.

39. Anderson RM, May RM. *Infectious Diseases of Humans: Dynamics and Control.* Oxford: Oxford University Press; 1992.

40. Kretzschmar M, Teunis PF, Pebody RG. Incidence and reproduction numbers of pertussis: estimates from serological and social contact data in five European countries. *PLoS Med* 2010;**7**:e1000291.

41. Fine PE, Clarkson JA. The recurrence of whooping cough: possible implications for assessment of vaccine efficacy. *Lancet* 1982;**1**:666–9.

42. Domenech de Celles M, Magpantay FM, King AA, et al. The pertussis enigma: reconciling epidemiology, immunology and evolution. *Proc Biol Sci* 2016; **283**:20152309.

43. Gay NJ, Miller E. Pertussis transmission in England and Wales. *Lancet* 2000;**355**:1553–4.

44. Carlsson RM, von Segebaden K, Bergstrom J, et al. Surveillance of infant pertussis in Sweden 1998–2012; severity of disease in relation to the national vaccination programme. *Euro Surveill* 2015;**20**:21032.

45. Solano R, Crespo I, Fernandez MI, et al. Underdetection and underreporting of pertussis in children attended in primary health care centers: do surveillance systems require improvement? *Am J Infect Control* 2016;**44**:e251–6.

46. Schielke A, Takla A, von Kries R, et al. Marked under-reporting of pertussis requiring hospitalization in infants as estimated by capture-recapture methodology, Germany, 2013–2015. *Pediatr Infect Dis J* 2018; **37**:119–25.

47. van der Maas NAT, Hoes J, Sanders EAM, et al. Severe underestimation of pertussis related hospitalizations and deaths in the Netherlands: a capture-recapture analysis. *Vaccine* 2017;**35**:4162–6.

48. McGirr AA, Tuite AR, Fisman DN. Estimation of the underlying burden of pertussis in adolescents and adults in Southern Ontario, Canada. *PLoS One* 2013;**8**:e83850.

49. Chow MY, Khandaker G, McIntyre P. Global childhood deaths from pertussis: a historical review. *Clin Infect Dis* 2016;**63** Suppl 4:S134–41.

50. Crowcroft NS, Andrews N, Rooney C, et al. Deaths from pertussis are underestimated in England. *Arch Dis Child* 2002;**86**:336–8.

51. Litt DJ, Samuel D, Duncan J, et al. Detection of anti-pertussis toxin IgG in oral fluids for use in diagnosis and surveillance of Bordetella pertussis infection in children and young adults. *J Med Microbiol* 2006;**55**:1223–8.

52. Teunis PF, van der Heijden OG, de Melker HE, et al. Kinetics of the IgG antibody response to pertussis toxin after infection with B. pertussis. *Epidemiol Infect* 2002;**129**:479–89.

53. Pebody RG, Gay NJ, Giammanco A, et al. The seroepidemiology of Bordetella pertussis infection in Western Europe. *Epidemiol Infect* 2005;**133**:159–71.

54. Crowcroft NS, Stein C, Duclos P, et al. How best to estimate the global burden of pertussis? *Lancet Infect Dis* 2003;**3**:413–8.

55. Liu L, Oza S, Hogan D, et al. Global, regional, and national causes of child mortality in 2000–13, with projections to inform post-2015 priorities: an updated systematic analysis. *Lancet* 2015;**385**:430–40.

56. Yeung KHT, Duclos P, Nelson EAS, et al. An update of global burden of pertussis in children aged below 5 years: a modeling study. *Lancet Infect Dis* 2017;**17**:974–80.

57. Sobanjo-Ter Meulen A, Duclos P, McIntyre P, et al. Assessing the evidence for maternal pertussis immunization: a report from the Bill & Melinda Gates Foundation Symposium on Pertussis Infant Disease Burden in Low- and Lower-Middle-Income Countries. *Clin Infect Dis* 2016;**63** Suppl 4:S123–33.

58. Soofie N, Nunes MC, Kgagudi P, et al. The burden of pertussis hospitalization in HIV-exposed and HIV-unexposed South African infants. *Clin Infect Dis* 2016;**63** Suppl 4:S165–73.

59. Nunes MC, Downs S, Jones S, et al. Bordetella pertussis infection in South African HIV-infected and HIV-uninfected mother-infant dyads: a longitudinal cohort study. *Clin Infect Dis* 2016;**63** Suppl 4:S174–80.

60. Barger-Kamate B, Deloria Knoll M, Kagucia EW, et al. Pertussis-associated pneumonia in infants and children from low- and middle-income countries participating in the PERCH study. *Clin Infect Dis* 2016;**63** Suppl 4:S187–96.

61. Omer SB, Kazi AM, Bednarczyk RA, et al. Epidemiology of pertussis among young Pakistani infants: a community-based prospective surveillance study. *Clin Infect Dis* 2016;**63** Suppl 4:S148–53.

62. Hughes MM, Englund JA, Kuypers J, et al. Population-based pertussis incidence and risk factors in infants less than 6 months in Nepal. *J Pediatric Infect Dis Soc* 2017;**6**:33–9.

63. Kampmann B, Mackenzie G. Morbidity and mortality due to Bordetella pertussis: a significant pathogen in West Africa? *Clin Infect Dis* 2016;**63** Suppl 4:S142–7.

64. CHAMPS—Child Health and Mortality Prevention Surveillance. Homepage. https://champshealth.org/.

65. Amirthalingam G, Andrews N, Campbell H, et al. Effectiveness of maternal pertussis vaccination in England: an observational study. *Lancet* 2014;**384**:1521–8.

66. Winter K, Harriman K, Zipprich J, et al. California pertussis epidemic, 2010. *J Pediatr* 2012;**161**:1091–6.

67. Straney L, Schibler A, Ganeshalingham A, et al. Burden and outcomes of severe pertussis infection in critically ill infants. *Pediatr Crit Care Med* 2016;**17**:735–42.

68. Kaczmarek MC, Ware RS, McEniery JA, et al. Epidemiology of pertussis-related paediatric intensive care unit (ICU) admissions in Australia, 1997–2013: an observational study. *BMJ Open* 2016;**6**:e010386.

69. Magpantay FM, Rohani P. Dynamics of pertussis transmission in the United States. *Am J Epidemiol* 2015;**181**:921–31.

70. Broutin H, Viboud C, Grenfell BT, et al. Impact of vaccination and birth rate on the epidemiology of pertussis: a comparative study in 64 countries. *Proc Biol Sci* 2010;**277**:3239–45.

71. de Figueiredo A, Johnston IG, Smith DM, et al. Forecasted trends in vaccination coverage and correlations with socioeconomic factors: a global time-series analysis over 30 years. *Lancet Glob Health* 2016;**4**:e726–35.

72. Macdonald-Laurs E, Ganeshalingham A, Lillie J, et al. Increasing incidence of life-threatening pertussis: a retrospective cohort study in New Zealand. *Pediatr Infect Dis J* 2017;**36**:282–9.

73. Pollock TM, Miller E, Lobb J. Severity of whooping cough in England before and after the decline in pertussis immunisation. *Arch Dis Child* 1984;**59**:162–5.

74. Robison SG, Liko J. The timing of pertussis cases in unvaccinated children in an outbreak year: Oregon 2012. *J Pediatr* 2017;**183**:159–63.

75. Gilbert NL, Gilmour H, Wilson SE, et al. Determinants of non-vaccination and incomplete vaccination in Canadian toddlers. *Hum Vaccin Immunother* 2017;**13**:1–7.

76. Miller E. *Evidence in Support/Against Various Primary Vaccination Schedules* (WHO SAGE meeting presentation). 2015. http://www.who.int/immunization/sage/meetings/2015/april/Miller_Pertussis Review_Evidence_SAGE_April2015.pdf.

77. World Health Organization. *Immunization Coverage with DTP3 Vaccines in Infants (From <50%)*. 2015. http://www.who.int/immunization/monitoring_surveillance/burden/vpd/surveillance_type/passive/big_dtp3_map_global_coverage.jpg.

78. Gangarosa EJ, Galazka AM, Wolfe CR, et al. Impact of anti-vaccine movements on pertussis control: the untold story. *Lancet* 1998;**351**:356–61.

79. Carlsson RM, Trollfors B. Control of pertussis—lessons learnt from a 10-year surveillance programme in Sweden. *Vaccine* 2009;**27**:5709–18.

80. Skowronski DM, De Serres G, MacDonald D, et al. The changing age and seasonal profile of pertussis in Canada. *J Infect Dis* 2002;**185**:1448–53.

81. Blackwood JC, Cummings DA, Iamsirithaworn S, et al. Using age-stratified incidence data to examine the transmission consequences of pertussis vaccination. *Epidemics* 2016;**16**:1–7.

82. Preziosi MP, Yam A, Wassilak SG, et al. Epidemiology of pertussis in a West African community before and after introduction of a widespread vaccination program. *Am J Epidemiol* 2002;**155**:891–6.

83. Cutts FT, Claquin P, Danovaro-Holliday MC, et al. Monitoring vaccination coverage: defining the role of surveys. *Vaccine* 2016;**34**:4103–9.

84. Rohani P, Earn DJ, Grenfell BT. Impact of immunisation on pertussis transmission in England and Wales. *Lancet* 2000;**355**:285–6.

85. Miller E, Gay N. Effect of age on outcome and epidemiology of infectious diseases. *Biologicals* 1997;**25**:137–42.

86. van Wijhe M, McDonald SA, de Melker HE, et al. Effect of vaccination programmes on mortality burden among children and young adults in the Netherlands during the 20th century: a historical analysis. *Lancet Infect Dis* 2016;**16**:592–8.

87. Fulton TR, Phadke VK, Orenstein WA, et al. Protective effect of contemporary pertussis vaccines: a systematic review and meta-analysis. *Clin Infect Dis* 2016;**62**:1100–10.

88. Sheridan SL, Frith K, Snelling TL, et al. Waning vaccine immunity in teenagers primed with whole cell and acellular pertussis vaccine: recent epidemiology. *Expert Rev Vaccines* 2014;**13**:1081–106.

89. Liko J, Robison SG, Cieslak PR. Priming with whole-cell versus acellular pertussis vaccine. *N Engl J Med* 2013;**368**:581–2.

90. Schwartz KL, Kwong JC, Deeks SL, et al. Effectiveness of pertussis vaccination and duration of immunity. *CMAJ* 2016;**188**:E399–406.

91. Misegades LK, Winter K, Harriman K, et al. Association of childhood pertussis with receipt of 5 doses of pertussis vaccine by time since last vaccine dose, California, 2010. *JAMA* 2012;**308**:2126–32.

92. Koepke R, Eickhoff JC, Ayele RA, et al. Estimating the effectiveness of tetanus-diphtheria-acellular pertussis vaccine (Tdap) for preventing pertussis: evidence of rapidly waning immunity and difference in effectiveness by Tdap brand. *J Infect Dis* 2014;**210**:942–53.

93. World Health Organization. *Immunization, Vaccines and Biologicals: Data, Statistics and Graphics*. 2016. http://www.who.int/immunization/monitoring_surveillance/data/en/.

94. Australian Government Department of Health. *Annual Report of the National Notifiable Diseases Surveillance System*. 2014. http://www.who.int/immunization/monitoring_surveillance/data/en/.

95. Australian Government Department of Health. *National Notifiable Diseases Surveillance System. Notifications of Pertussis by Month and Year.* 2017. http://www9. health.gov.au/cda/source/cda-index.cfm.

96. Hale S, Quinn HE, Kesson A, et al. Changing patterns of pertussis in a children's hospital in the polymerase chain reaction diagnostic era. J Pediatr 2016;170:161–5.e1.

97. Campbell PT, McCaw JM, McIntyre P, et al. Defining long-term drivers of pertussis resurgence, and optimal vaccine control strategies. Vaccine 2015;33:5794–800.

98. Australian Government Department of Health. New Whooping *Cough Booster for 18 Month Old Children.* 2016. http://www.immunise.health.gov.au/internet/ immunise/publishing.nsf/content/home.

99. Choi YH, Campbell H, Amirthalingam G, et al. Investigating the pertussis resurgence in England and Wales, and options for future control. *BMC Med* 2016;**14**:121.

100. Warfel JM, Merkel TJ. The baboon model of pertussis: effective use and lessons for pertussis vaccines. *Expert Rev Vaccines* 2014;**13**:1241–52.

101. Center for Disease Control and Prevention. *Pertussis Outbreak Trends.* 2015. https://www.cdc.gov/pertussis/ outbreaks/trends.html.

102. Gambhir M, Clark TA, Cauchemez S, et al. A change in vaccine efficacy and duration of protection explains recent rises in pertussis incidence in the United States. *PLoS Comput Biol* 2015;**11**:e1004138.

103. Rohani P, Drake JM. The decline and resurgence of pertussis in the US. *Epidemics* 2011;**3**:183–8.

104. Miller E. Acellular pertussis vaccines. *Arch Dis Child* 1995;**73**:390–1.

105. Directorate-General of Health (DGS). *Portugal—Portuguese National Vaccination Programme.* http://venice. cineca.org/documents/portugal_ip.pdf.

106. Vaccine European New Integrated Collaboration Effort. *Report on First Survey of Immunisation Programs in Europe: Venice Project.* 2007. http://venice.cineca.org/ Report_II_WP3.pdf.

107. European Centre for Disease Prevention and Control. *Annual Epidemiological Report 2016—Pertussis.* 2016. http://ecdc.europa.eu/en/healthtopics/pertussis/ Pages/Annual-epidemiological-report-2016.aspx.

108. Almeida AF, Flor-de-Lima F, Simoes JS, et al. Pertussis in a Portuguese pediatric tertiary care hospital. *Pediatr Infect Dis J* 2016;**35**:466–7.

109. Moerman L, Leventhal A, Slater PE, et al. The re-emergence of pertussis in Israel. *Isr Med Assoc J* 2006;**8**:308–11.

110. Abu Raya B, Bamberger E, Spiegel G, et al. Two closely related strains associated with pertussis resurgence in Israel. *Pediatr Infect Dis J* 2012;**31**:761–2.

111. Zamir CS, Dahan DB, Shoob H. Pertussis in infants under one year old: risk markers and vaccination status-a case-control study. *Vaccine* 2015;**33**:2073–8.

112. National Advisory Committee on Immunization (NACI). Statement on the recommended use of pentavalent and hexavalent vaccines. *Can Commun Dis Rep* 2007;**33**:1–14.

113. De Serres G, Boulianne N, Duval B, et al. Effectiveness of a whole cell pertussis vaccine in child-care centers and schools. *Pediatr Infect Dis J* 1996;**15**:519–24.

114. Smith T, Rotondo J, Desai S, et al. Pertussis surveillance in Canada: trends to 2012. *Can Commun Dis Rep* 2014;**40**:21–30.

115. Government of New Brunswick. *Weekly Abbreviated Epidemiology Update: New Brunswick Pertussis Outbreak (Data to January 4, 2013).* 2013. http://www2.gnb.ca/ content/dam/gnb/Departments/h-s/pdf/en/ CDC/PertussisOutbreakE.pdf.

116. CBC News. *Whooping Cough Outbreak Jumps to 47 Cases.* Updated 29 October 2015. http://www.cbc. ca/news/canada/new-brunswick/whooping-cough-outbreak-1.3293947.

117. CBC News. *Whooping Cough Outbreak Declared Over.* Updated 1 May 2013. http://www.cbc.ca/news/ canada/new-brunswick/whooping-cough-outbreak-declared-over-1.1341582.

118. Global News. *Number of Whooping Cough Cases on the Rise in Saskatchewan.* Updated April 25, 2016. http:// globalnews.ca/news/2661689/number-of-whooping-cough-cases-on-the-rise-in-saskatchewan/.

119. Government of Saskatchewan. *Pertussis (Whooping Cough).* Updated 2017. https://www.saskatchewan. ca/residents/health/diseases-and-conditions/ pertussis-whooping-cough#top.

120. Deeks S, Lim G, Walton R, et al. Prolonged pertussis outbreak in Ontario originating in an under-immunized religious community. *Can Commun Dis Rep* 2014;**40**:42–9.

121. CBC News. *Whooping Cough Outbreak Has Now Spread Throughout Territory: Nunavut Officials.* 17 October 2017. http://www.cbc.ca/news/canada/north/whooping-cough-outbreak-nunavut-1.3808762.

122. Hegerle N, Guiso N. Bordetella pertussis and pertactin-deficient clinical isolates: lessons for pertussis vaccines. *Expert Rev Vaccines* 2014;**13**:1135–46.

123. Sealey KL, Harris SR, Fry NK, et al. Genomic analysis of isolates from the United Kingdom 2012 pertussis outbreak reveals that vaccine antigen genes are unusually fast evolving. *J Infect Dis* 2015;**212**: 294–301.

CHAPTER 5

Role of vaccine schedules

Jodie McVernon and Hester de Melker

Abstract

Over the past 60 years, pertussis vaccines have been implemented in national immunization programme schedules at a variety of ages. All countries administer the vaccine as an infant primary series, but the number and exact timing of these doses differ, as do the number and timing of booster doses. In consequence, short-term direct protection of age-appropriately immunized children varies by schedule. This variability influences vaccine impact on the burden of disease, because the risk of severe morbidity and mortality is greatest in early life and decreases thereafter. In addition, vaccine-derived immunity wanes over time, meaning that longer intervals between doses are predicted to increase the risk of breakthrough infections throughout childhood and adolescence, fuelling ongoing transmission and cyclic epidemics. The duration of protection following vaccination is substantially shorter than following infection, although absolute estimates vary widely across populations and studies.

5.1 Introduction

Over the past 60 years, pertussis vaccines have been implemented in national immunization programme schedules at a variety of ages. All countries administer the vaccine as an infant primary series, but the number and exact timing of these doses differ, as do the number and timing of booster doses. In consequence, short-term direct protection of age-appropriately immunized children varies by schedule. This variability influences vaccine impact on the burden of disease, because the risk of severe morbidity and mortality is greatest in early life and decreases thereafter. In addition, vaccine-derived immunity wanes over time, meaning that longer intervals between doses are predicted to increase the risk of breakthrough infections throughout childhood and adolescence, fuelling ongoing transmission and cyclic epidemics. The duration of protection following vaccination is substantially shorter than following infection, although absolute estimates vary widely across populations and studies.

In recent decades, in many high-income countries, the number of pertussis vaccines routinely scheduled over the life course has increased from three or four doses to a total of six or seven doses. This increase has been motivated by an observed upward march in the age incidence of disease in populations with high vaccine coverage in early life. Ascertainment of cases in older children and adolescents has been facilitated by a combination of practitioner awareness, and the greater sensitivity and specificity of molecular diagnostics compared with traditional culture methods. Despite an overall increase in pertussis vaccine delivery, a resurgence of disease has recently been observed in a number

McVernon, J. and de Melker, H., *Role of vaccine schedules*. In: *Pertussis: epidemiology, immunology, and evolution.* Edited by Pejman Rohani and Samuel V. Scarpino: Oxford University Press (2019). © Oxford University Press. DOI: 10.1093/oso/9780198811879.003.0005 of these settings, resulting in severe morbidity and death among infants too young to have been fully immunized.

During the same period, there has also been a shift from the use of a variety of reactogenic whole-cell pertussis (wP) vaccines, which demonstrated variable efficacy across populations, to acellular pertussis (aP) vaccines. These less reactogenic acellular formulations were introduced first in many countries as booster doses, and later for the primary course. Changing recommendations for vaccines and schedules over time have resulted in cohorts of children and adolescents with mixed immunization histories. An increasing body of evidence supports the view that children immunized solely with aP vaccines in infancy are less well protected against disease over time than those who received efficacious wP vaccines. Identification of immunologic correlates of protection to inform assessment of population risk remains elusive for pertussis, and is further complicated by differential and in general relatively short-term persistence of antibodies to discrete vaccine antigens[1].

Direct comparison of immunization programmes and schedules is made challenging by differences in case ascertainment and achieved vaccine coverage over time and place[2]. Immunization strategy has also shifted in recent years, from a focus on elimination to mitigation of severe disease in the very young through targeted vaccination of mothers in the postnatal, then antenatal periods (see Chapter 15). This chapter summarizes current understanding of the relative influences of vaccine timing, dosing interval, dose number, and dose formulation on the age incidence and level of endemicity of pertussis infection, and highlights evidence gaps to prompt future research.

5.2 Vaccine timing and dose interval

5.2.1 Primary series

The World Health Organization (WHO) recommends a three-dose infant primary course of pertussis vaccine, based on evidence of increasing protection with each additional dose. Ideally, the series should be completed by 6 months of age, to maximize protection in the early months of life, when disease is most severe. In addition, early primary course completion reduces the observed risk window in the second half of the first year of life associated with 'two-plus-one' schedules where the first dose is administered at 11 or 12 months of age[3].

A recent analysis of vaccine-preventable disease mortality in the Netherlands has confirmed a marked decline in the overall mortality burden attributable to pertussis in individuals aged less than 20 years since vaccines were introduced in 1954. Prior to vaccine availability, pertussis was the cause of 3.8 per cent of deaths before the age of 20, and now accounts for 0.024 per cent of mortality in this age group. Most deaths occurred among infants less than 1 year of age. Over the period 1903–2012, it is estimated that pertussis vaccine has saved 6000 lives and more than 100,000 life years under the age of 20 in the Netherlands[4].

The impact of infant pertussis immunization on disease has been further demonstrated from observations of pertussis vaccine introduction and withdrawal, associated with wP vaccine safety concerns[5]. Sweden reintroduced pertussis immunization as a 3-, 5-, and 11-month schedule in 1996, after 17 years without a pertussis vaccine recommendation. Over the next decade, 1 million infants were immunized and followed for nearly 6 million person years. Mandatory notifications data indicate protection against whooping cough disease on the order of 40 per cent between 3 and 5 months of age following a single vaccine dose, increasing to more than 90 per cent after the second dose. Protection against hospitalization is greater. While the impact of immunization has been substantial at a population level, improved timeliness of coverage is still needed to realize an optimal reduction in the burden of disease[6].

The finding of enhanced vaccine protection against severe disease is consistent with screening method estimates of the effectiveness of the 3-, 5-, and 11-month wP immunization schedule in the United Kingdom in the late 1980s[7]. Unimmunized children under the age of 5 years were significantly more likely to be hospitalized or experience prolonged paroxysmal cough with whoop and vomiting than vaccine failures. This study further demonstrated the importance of epidemiological context to the assessment of vaccine efficacy, which differed significantly between epidemic (87 per cent) and non-epidemic (93 per cent) periods[7]. This apparent reduction in vaccine protection in the face of greater infection pressure is supported by the high secondary-case proportion observed in household follow-up studies of a documented pertussis case[7].

The WHO recommendations state that earlier completion of the primary series is associated with enhanced infant protection. The United Kingdom switched from an expanded 3-, 5-, and 11-month recommendation to an accelerated immunization schedule at 2, 3, and 4 months of age in 1990. This strategy was adopted to maximize early life protection and coverage, while minimizing reactogenicity, and was highly successful in reducing overall incidence in infancy. Improved protection following implementation of this schedule was observed throughout the first year of life, and was particularly notable through a reduction in the proportion of cases presenting between 6 and 11 months of age[8].

Timing of the initiation of the primary series may be less influential than receipt of three primary vaccine doses within the first 6 months of life, compared with only two. In 1999, the Netherlands switched from a 3-, 4-, 5-, and 11-month wP vaccine schedule to an advanced schedule of 2, 3, 4, and 11 months in an attempt to improve protection in very young infants[9]. This measure failed to show a clear impact on disease incidence over the first year of life, and in particular no change was observed among infants aged less than 2 months.

Many other countries administer these three primary doses at 2-month intervals. For example, New Zealand administers a three-dose primary course at 2, 4, and 6 months, and reports 43 per cent efficacy of the first dose against hospitalization between 6 weeks and 2 months of age. Protection increases significantly with subsequent doses, rising to 84 per cent for the second dose and 93 per cent for the third[10]. In Australia, which employs the same schedule, half of all infant (<1 year) hospitalizations occur by 12 weeks of age, with the majority being too young to be vaccinated[11]. Less than 20 per cent of admissions are in children aged 6 months or older, and these cases are often associated with incomplete immunization status[11]. Of note, studies in Australia and the United States show that late receipt of the first vaccine dose makes it less likely that infants will complete all three doses in the series[11].

Ongoing pertussis transmission following vaccine introduction is well documented with persistence of epidemic cycles observed, even in high-uptake settings[12]. However, a reduction in the incidence of disease among infants too young to be immunized provides evidence of additional indirect protection as a determinant of overall programme impact[13].

5.2.2 Second-year booster

The WHO recommends reinforcement of immunity from the primary course, preferably during the second year of life, unless there is strong local evidence against this strategy[3]. New Zealand, for example, continues to observe high vaccine efficacy following a three-dose primary course of aP vaccine, persisting at more than 80 per cent until administration of a preschool booster dose at 4 years[10]. In line with these findings, in the Netherlands, vaccine efficacy estimates using the screening method after primary vaccination (at 2, 3, 4, and 11 months) for 1-, 2-, and 3-year-olds (i.e. before the preschool booster) ranged from 86 to 92 per cent, 75 to 91 per cent, and 65 to 85 per cent, respectively[9].

However, in support of this WHO recommendation, withdrawal of the 18-month booster dose in Australia in 2003 (following a 2-, 4-, and 6-month primary series) is considered to have contributed significantly to pertussis resurgence from 2008 onwards[14]. A period of sustained epidemic activity was experienced between 2008 and 2010, during which time notification rates in 2–3-year-old children were higher than during previous epidemic cycles, when the booster dose had been in place. Age-specific incidence in this group was only surpassed by infants aged less than 6 months[15]. This increase in disease was preceded by a marked reduction in the proportion of children aged less than 4 years with detectable antibodies to pertussis toxin, falling from 75 per cent in 1997 to 38 per cent in 2007[16]. Following evaluation of the drivers of pertussis resurgence, Australia reintroduced a funded 18-month booster dose of pertussis in 2016.

Concordant with the second-year booster recommendation, the National Immunisation Program of Sri Lanka administers four doses of wP vaccine at

2, 4, 6, and 18 months of age. A recent seroepidemiological study was suggestive of ongoing protection from this schedule through to 7 years of age, with high-titre antibodies (a proxy of recent exposure) observed in less than 10 per cent of children aged 8–15 years, rising later to peak at 25 per cent in 16–19-year-olds[17].

5.2.3 Preschool booster

In recognition of the importance of siblings as household introducers of infection, a preschool booster dose was incorporated in the United Kingdom schedule in 2001 to reduce the risk to young infants during ongoing epidemic cycles of infection. This fourth dose was estimated to provide an additional 46 per cent effectiveness compared with the three primary doses[18]. In many other countries, this schedule time point has been introduced in response to an upward age shift in the age of disease associated with high and sustained coverage of infant vaccination, with or without a booster in the second year of life.

In Sweden, where aP vaccine was reintroduced in 1996 as a 3-, 5-, and 12-month schedule after a hiatus of 17 years, the changing age distribution of disease with vaccination has been particularly well observed. An initial gradual decline in disease incidence was seen across the population following vaccine introduction, associated with an increase in the average age of infection from 4 to 10 years of age. A subsequent increase in the absolute incidence of disease in 5–6-year-olds, together with serologic evidence of waning population immunity, prompted introduction of a preschool booster dose[19].

The Israeli pertussis vaccine schedule has been consistent over many decades, administering vaccine at 2, 4, 6, and 12 months. An overall increase in cases was observed from 1998 onwards, and in 2005 was sufficient to prompt the introduction of a booster dose in primary school (7–8 years). Of note, a switch to aP vaccines was only made in 2002. Age-specific incidence rates in the target age group have declined, but indirect effects are less clear. An ongoing increase in incidence among secondary school-aged children prompted introduction of an adolescent dose (13–14 years) in 2008[20].

In the Netherlands, introduction of an acellular preschool booster dose led to about a 50 per cent or greater reduction in pertussis incidence in 4–6-year-olds and somewhat less in 7–9-year-olds. Estimated vaccine effectiveness was about 80 per cent in the target age group but declined to 50 per cent within 5 years[9]. This observation demonstrates that while the introduction of preschool boosters has a demonstrable short-term impact on disease, protection wanes rapidly following administration of aP vaccine as a fifth dose at preschool age.

Similarly, during a period of epidemic activity in California, observed vaccine effectiveness among children aged 8–12 years (a group eligible for receipt of five doses) was only 24 per cent[21]. The risk of pertussis by year since vaccination was assessed in 2010 for children living in the geographically and epidemiologically diverse states of Minnesota and Oregon. A consistent trend for an increasing risk of breakthrough disease over 6 years following immunization was observed in either setting, rising to a risk ratio of 8.9 (incidence rate 140 per 100,000) in Minnesota and 4.0 (incidence rate 20 per 100,000) in Oregon[22].

5.2.4 Adolescent booster

In recognition of increasing case numbers through the second decade of life, reduced-dose formulation adolescent booster doses of pertussis vaccine (diphtheria, tetanus, and acellular pertussis (dTap) vaccine) have been introduced in many countries[3]. The Global Pertussis Initiative, a vaccine advocacy group, further recommended administration at this time point as a means of indirect infant protection[23,24], but anticipated benefits for infant risk reduction have not been realized in practice. For this reason, while the WHO endorses adolescent booster doses for direct protection, it does not recommend vaccination at this time point as a strategy for infant disease control[3].

Early reports of an increase in adolescent and young adult pertussis were attributed to heightened practitioner awareness and ascertainment in many settings. However, comparative studies of pertussis seroepidemiology in Western Europe confirmed a genuine upward age shift in infection incidence, using high-titre pertussis antibodies as a proxy of recent infection exposure[25]. The prevalence of high-titre sera was higher in adolescents than children in countries with sustained high vaccine coverage (Finland, the Netherlands, France, and East Germany), with the reverse trend observed in low-coverage countries (England and Wales, West Germany, and Italy)[25]. A study from the Netherlands[26] has observed a similar trend, in the context of improved control of early life infections following reinforcement of infant immunization.

Adolescent pertussis immunization was introduced in Australia in 2004 and predominantly administered in schools, through existing state and territory delivery systems. Programme implementation varied by state, including the age of routine vaccine administration and extent of catch-up across year levels at the time of introduction[27]. Regardless of approach, ecological evidence of a reduction in disease risk was initially observed among both immunized and unimmunized cohorts between 2005 and 2007. Direct vaccine protection persisted, albeit to a lesser degree over time, through a period of sustained epidemic activity commencing in 2008[27].

Recent data from the United States similarly indicate a short duration of protection following dTap administration in early adolescence, with observed vaccine effectiveness under epidemic conditions declining from 73 per cent within 1 year of vaccination to 34 per cent at 2–4 years post vaccination[28]. Moreover, evaluation of changes in the age incidence of pertussis disease in the United States over time reveals different epochs in effectiveness of the adolescent dose, when applied to cohorts primed either with wP or aP vaccines. Booster immunization was first recommended for adolescents from 2005, and over the next 4 years was associated with a rapid decline in the incidence of disease among this age group. The trend was sharply reversed from 2010 onwards however, as aP vaccine-primed children entered the adolescent age class. From that time, the rate of notifications in this cohort rose more rapidly than in all other age groups combined[29].

5.3 Number of vaccine doses

The number of doses of aP vaccine administered does not appear to substantially reinforce immunity over the course of a series. A recent meta-analysis comparing loss of protection after three or five vaccine doses estimated a 1.33-fold increase in odds of breakthrough disease with each year since schedule completion, independent of the number of completed doses achieved[30]. This absence of immune reinforcement is in contrast to many other routinely administered vaccines, such as tetanus.

Moreover, as evident from the recent resurgence of pertussis in many high-coverage countries, the number of doses administered over the life course does not appear to correlate with the set point endemicity of disease in a given population. This absence of association is likely due to the relatively poor immunogenicity of currently available pertussis vaccines, compared with natural infection. Seroepidemiological studies consistently show that high vaccine coverage is associated with a decline in measurable population immunity over time, with epidemic cycles showing a distinct signature of high-titre antibody production[16].

5.3.1 Pertussis vaccine formulation

Population observations in the United States[31] and Australia[32] have shown reduced vaccine effectiveness against pertussis disease of a four- or five-dose vaccine schedule comprised solely of aP vaccines, compared with a mixed schedule including at least one dose of wP formulation during the primary immunization series. Protection declines more rapidly following acellular priming[28], but can be partially recovered by administration of a sixth vaccine dose in early adolescence[33].

On these grounds, the WHO has advised that countries currently administering wP vaccines should only consider switching to aP vaccine formulations if additional booster immunizations can be reliably sustained, to ensure population protection. Countries currently using aP vaccines are advised to consider reinforcing immunization protection through additional doses, including antenatal vaccination programmes[3].

A modelling study incorporating this knowledge of differential vaccine effectiveness supports the cost-effectiveness of including at least one wP vaccine in the primary infant schedule[34]. It is unlikely, however, that reintroduction of such a reactogenic vaccine would be acceptable in high-income countries.

An increase in the proportion of pertactin-deficient strains of pertussis has been observed in association with the introduction of aP vaccines in some settings. To date, however, there is no evidence that this deficiency compromises vaccine protection. A recent study from Vermont, United States, where greater than 90 per cent of disease-causing pertussis strains were pertactin deficient, found an overall DTaP vaccine efficacy of 84 per cent and dTap vaccine efficacy of 70 per cent. These observations were similar to estimates observed in previous studies[35].

5.4 Case studies: Australia and the Netherlands

To demonstrate the importance of context to vaccine schedule impact, we present here two national case studies—Australia and the Netherlands. These studies outline the sequence of schedule decision-making, in response to epidemiological trends, thereby highlighting some of the commonalities and inconsistencies of vaccine impact in different settings.

5.4.1 Australia

Pertussis first became a nationally notifiable disease in Australia in the 1930s, demonstrating 3–5-yearly epidemic cycles with a peak incidence of 767.3 per 100,000 population. Reporting ceased in 1949 following the success of mass immunization with monovalent pertussis vaccines during the 1940s and did not recommence until 1979, with an all-time low incidence of 1.1 per 100,000 reported in 1988[36]. The current Australian National Notifiable Diseases Surveillance System was established in 1990, and since 1991 has reported on an agreed list of communicable diseases with consistent case definitions, including pertussis.

Since that time, a number of changes have been made to the pertussis immunization schedule in response to observed trends in the age incidence of disease.

Locally manufactured combined diphtheria–tetanus–pertussis vaccine was introduced in Australia in 1953 (Figure 5.1)[36]. States and territories administered the vaccine at different time points until 1975, when a uniform national schedule was agreed at 2, 4, and 6 months of age[37]. A booster dose of vaccine at 18 months (which had previously been included but removed) was reintroduced in 1985[38], and this four-dose schedule was in place at commencement of the period of consistent reporting from 1991 onwards. Reliable estimates of national vaccine coverage were made possible with introduction of the Australian Childhood Immunisation Register in 1996, which was associated with an overall improvement in vaccine uptake. Stable high (>90 per cent) coverage of the primary series has been observed since 2000[16].

A fifth preschool booster dose was added in 1994 for children aged 4–5 years, to reduce transmission among primary school children[38]. Anecdotally, coverage at this age was poor until aP vaccines were funded for this schedule time point in 1997, and for the primary series in 1999, resulting in a rapid transition in vaccine formulation received[39]. Between 1995 and 2005, higher than average pertussis activity was observed nationally in 1997, 2001, and 2005 (Figures 5.2 and 5.3), although this increase was not synchronous across all states and territories[38]. In 1997, peak incidence was observed in the less than 6 months age group, followed by 5–9- and 10–19-year-olds. In 2001, the relative incidence in these latter two age groups was reversed, with very low rates of disease activity among primary

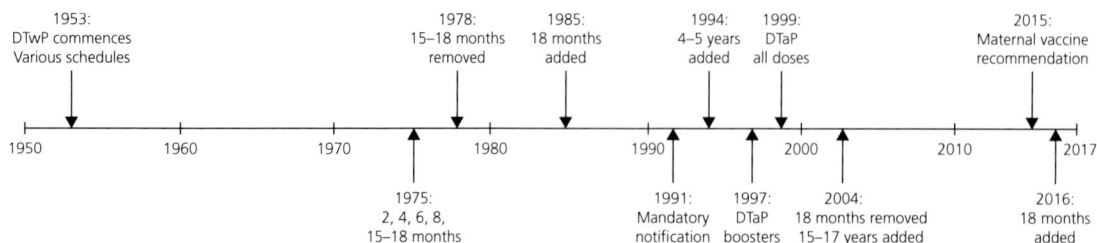

Figure 5.1 Changes to the Australian immunization schedule for pertussis, 1953–2016. Changes to the number, scheduling, and formulation of vaccine doses funded for Australian children are shown on the timeline. Whole-cell and acellular pertussis vaccines are denoted DTwP and DTaP, respectively.

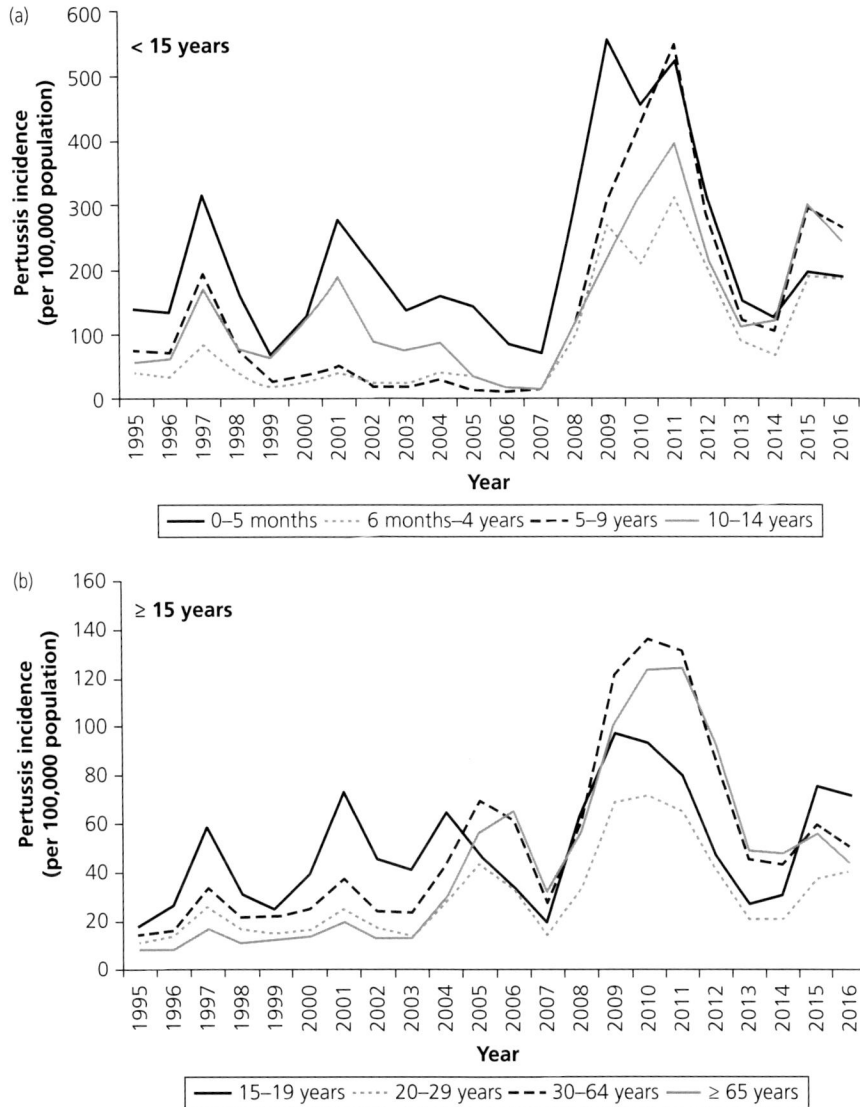

Figure 5.2 Age-specific incidence of pertussis notifications per 100,000 population for groups aged (a) less than 15 years and (b) 15 years or older, 1995–2016. Notifications Data: National Notifiable Diseases Surveillance System, figure prepared by Dr Helen Quinn, National Centre for Immunisation Research and Surveillance of Vaccine Preventable Diseases.

school-aged children attributed to increased uptake of the preschool booster dose[38].

Local and international concern about increasing disease incidence during the second decade of life prompted addition of an adolescent vaccine dose, administered predominantly through school-based campaigns from 2004[27]. A critical trade-off in the National Immunisation Program Schedule occurred

at this time. The 18-month booster dose was withdrawn, supported by evidence from Italy of sustained protection of a three-dose primary course of aP vaccine through to 6 years of age[40]. While this new five-dose schedule initially seemed effective in reducing overall levels of pertussis, a resurgence of disease was observed from 2008–2009 involving all age groups, but led by children under 10 years of age.

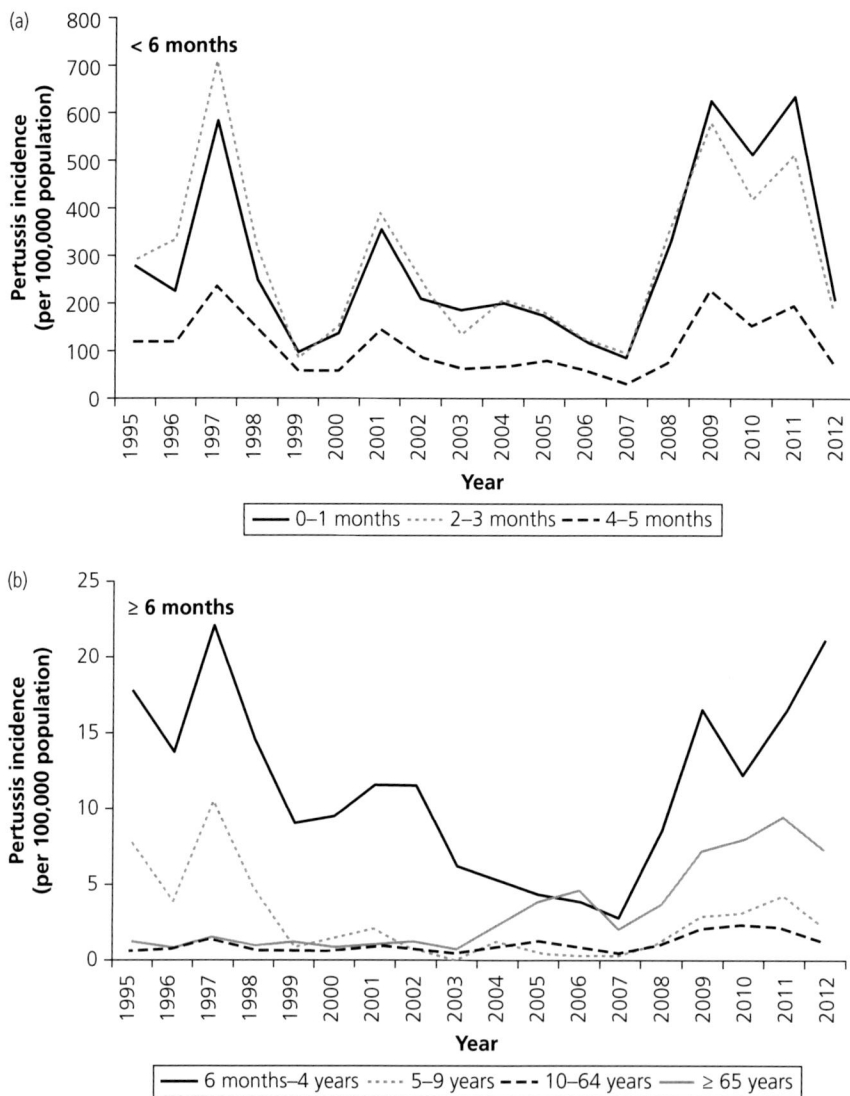

Figure 5.3 Age-specific hospitalization rates of pertussis per 100,000 population for groups aged (a) less than 6 months and (b) 6 months or older, 1995–2012. Hospitalizations data: Australian Institute of Health and Welfare, figure prepared by Dr Helen Quinn, National Centre for Immunisation Research and Surveillance of Vaccine Preventable Diseases.

Consistent with historical trends, peak notifications during the 2008–2011 epidemic were among infants aged less than 6 months, resulting in levels of hospitalization similar to the 1997 outbreak, the most severe previously recorded (Figures 5.2 and 5.3)[41]. Among preschoolers, notifications rose markedly in 3-year-olds to 411 per 100,000 and in 5–9-years-olds to 556 per 100,000 in 2011[41]. Diagnostic practice likely accounted for the magnitude of these peaks; however, associated age-specific hospitalization rates were more consistent with the 2001 epidemic than the more severe experience of 1997, suggesting ongoing disease modification by vaccine[41]. During this period of heightened epidemic activity, incidence rates among immunized adolescents were approximately half those observed in the unimmunized population[27].

Resurgence was attributed to a combination of schedule change, waning immunity, and aP vaccine introduction[41]. A follow-up study of children born in 1998, a year in which both vaccines were available, compared vaccine effectiveness by primary course received during both pre-epidemic (1998–2008) and epidemic (2009–2011) periods[32]. Compared with wP vaccine recipients, children administered aP vaccine had a significantly increased risk of vaccine failure up to 2008 (rate ratio 2.53; 95% confidence interval 1.06–6.07), which was greater during the epidemic phase (rate ratio 3.29; 95% confidence interval 2.44–4.46)[32].

In 2015, pertussis vaccine was recommended for pregnant women between 28 and 32 weeks of gestation by the Australian Technical Advisory Group on Immunisation. The 18-month booster dose was reintroduced into the National Immunisation Program Schedule in 2016, to reduce pertussis disease among children of preschool age, and transmission to infants.

5.4.2 Netherlands

Whole-cell pertussis vaccination introduced as a routine vaccination in the 1950s

The National Institute of Public Health started production of a single wP vaccine in 1949. A few years later (1952), a combination vaccine with diphtheria, pertussis, and tetanus became available. From 1953 onwards, this vaccine was made available free of charge and used in mass vaccination efforts. In 1957 (i.e. after the polio epidemic), the National Immunisation Programme formally started providing diphtheria, pertussis, tetanus, and polio (DTP-IPV) developed by the National Institute of Public Health. While pertussis mortality had already been declining, a further decrease in mortality and case fatality was observed[4] following introduction of the routine schedule.

Children were offered routine DTP-IPV vaccination at 3, 4, 5, and 11 months. The Health Council advised at that time to introduce a booster dose at preschool age with either DT-IPV or DTwP-IPV, given the lack of evidence regarding the need for a booster dose of pertussis. Practice varied throughout the country and over time, with some regions recommending DTwP-IPV to reduce the likelihood of older siblings transmitting pertussis infection to young unvaccinated infants, while others questioned the validity of this indication[42]. From 1965 onwards, a uniform national recommendation was made for booster doses of DT-IPV at 4 and 9 years of age[43]. See Figure 5.4.

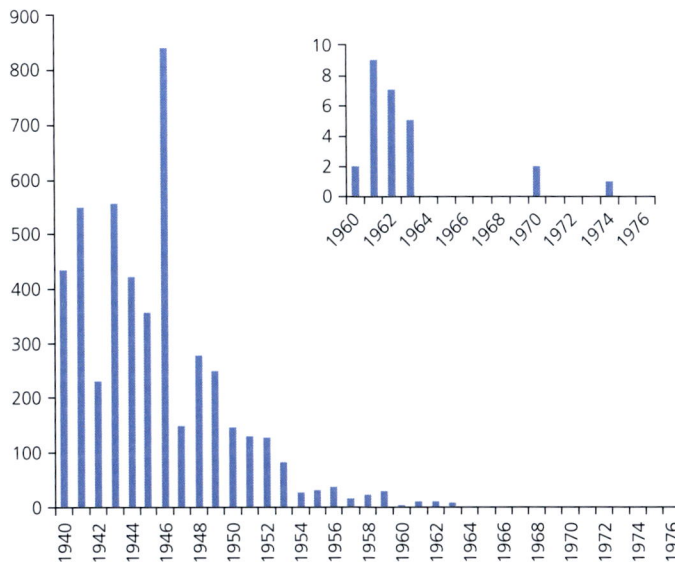

Figure 5.4 Pertussis mortality in the Netherlands including all ages, 1940–1976, demonstrating the impact of vaccine availability (1953) and introduction of the routine vaccination schedule (1957).

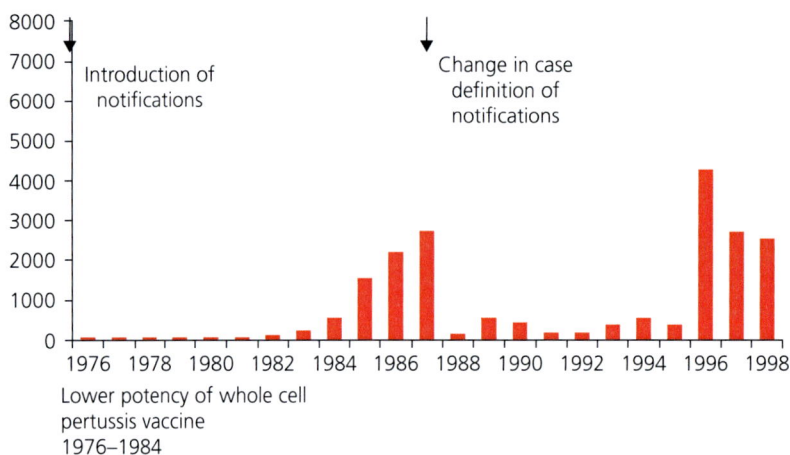

Figure 5.5 Pertussis notifications in the Netherlands in all ages, 1976 (year of introduction of pertussis notifications by law)[a] up to 1999[b]. [a] In the period 1976–1984, a lower potency whole-cell pertussis vaccine was used. [b] In 1999, one-point serology (high titre in one serum) was included in the case definition for notifications and the pertussis primary vaccination schedule changed from 3, 4, 5, and 11 months to 2, 3, 4, and 11 months. OSIRIS, notifications by law, The Netherlands.

Reduction of potency in whole-cell pertussis vaccine: data from 1976 to 1998

Mandatory notification of pertussis disease in the Netherlands commenced in 1976. From this time through to 1984, reports of vaccine adverse events due to the pertussis component raised concerns about vaccine safety. In response to these concerns, the potency of the pertussis component of the vaccine produced in the Netherlands was reduced. A review of notifications through this period reveals an increase in pertussis reports dating back to the early 1980s, ascribed in part to increased awareness but likely also reflecting an underlying trend.

In 1996–1997, a sudden increase in pertussis incidence in the Netherlands was observed among all age groups, but most strongly in vaccinated individuals (see Figure 5.5). Changes in vaccination coverage, diagnostics, notification, or interference with other vaccines could not explain this sudden rise. A mismatch between the vaccine seed strain and circulating *Bordetella pertussis* isolates was found. It was concluded that the low immunogenicity profile of the Dutch vaccine might have resulted in greater vulnerability to antigenic changes in *B. pertussis*. In response to this event, the potency of the wP vaccine produced by the Netherlands Vaccine Institute was enhanced; that is, lots should contain at least 7 IU instead of 4 IU. Despite this change, pertussis

endemicity remained higher than previously, with epidemic peaks every 2–3 years.

Trying to control pertussis resulted in changes in the vaccination schedule in 1999, 2001, and 2005

As a result of this increased incidence, several changes were made to the vaccination schedule. In 1999, the first dose of the primary infant series was brought forward to 2 months of age (i.e. offering vaccination at 2, 3, 4, and 11 months), with the aim of inducing earlier protection of young infants most vulnerable to severe disease. Given the high incidence in school-aged children (5–9 years), a booster dose of aP vaccine was added at 4 years of age in 2001.

Furthermore, the Health Council also concluded that the Dutch wP vaccine was less effective than the aP vaccine. From 2005 onwards, the Dutch vaccine, containing wP vaccine (DTwP-IPV-Hib), was replaced by a combination vaccine including acellular pertussis (i.e. DTaP-IPV-Hib).

Future perspective: control of pertussis among infants, the most vulnerable group

The numbers of infant pertussis cases were not reduced adequately by these measures. Seroepidemiological studies showed that high serum immunoglobulin G-pertussis toxin concentrations (a proxy indicator of exposure) more than doubled in 2006/2007

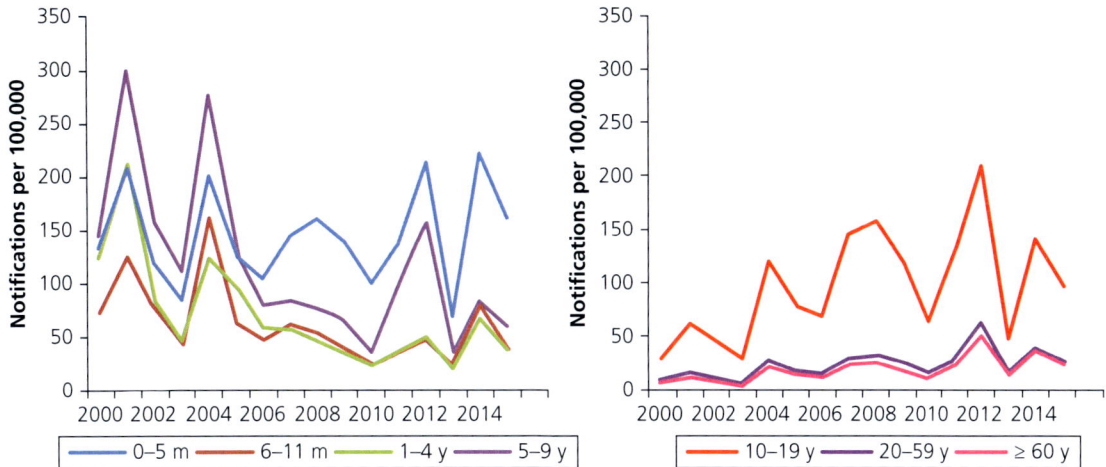

Figure 5.6 Age-specific incidence of pertussis in the Netherlands 2000–2015: inclusion of booster acellular pertussis vaccine for 4-year-olds in 2001, and change to acellular pertussis vaccine in primary series in 2005. m, months; y, years. OSIRIS, notifications by law, The Netherlands.

compared with 1995/1996. High circulation levels of *B. pertussis* limited the impact of paediatric vaccination[26]. In December 2015, the Health Council advised the introduction of maternal pertussis vaccination to reduce the high number of pertussis cases in young infants needing hospitalization.

5.5 Discussion

A robust understanding of the drivers of resurgence is important for identification of optimal strategies for disease control. However, the influence of a given vaccine formulation or schedule on observed pertussis epidemiology is exquisitely context dependent. The resulting heterogeneity of population experience makes direct extrapolation of recommendations from one country to another challenging. In addition, pertussis vaccine impact on disease is often short-lived given changing transmission dynamics.

Given these limitations and the lack of a substantial impact of higher numbers of doses given at various ages, increased emphasis has been applied to control of severe disease among those too young to be vaccinated. Among the various options, such as cocooning and neonatal vaccination, maternal vaccination strategies have recently been shown to be very promising (see Chapter 15). As disease burden remains greatest in most settings over the first year of life, ensuring high infant vaccine coverage

will always be the most highly cost-effective immunization strategy, with or without maternal passive protection. More robust evidence of benefit would be needed for high-income countries to return to the use of reactogenic wP vaccines, particularly given the historical collapse of pertussis immunization programmes in many countries associated with the side effect profile of these formulations.

Waning immunity following the infant primary course and booster doses poses challenges, particularly when either or both of transmission potential or disease burden are appreciable in older age groups. In such instances, modelling studies demonstrate that boosters must be administered at intervals less than the waning rate for vaccine-induced immunity to have a sustained impact[44]. Given the very short duration of protection observed under recent epidemic conditions in many settings, this frequency of booster administration appears an infeasible task.

5.5.1 Key open questions and future studies

Alternative approaches to maintenance of population immunity for pertussis mitigation and control over the longer term are being actively considered and here, epidemiologic models of infection can be useful tools to learn from population experience and/or predict future scenarios. One current open question is whether reducing the number of routinely scheduled vaccine doses after childhood would enable natural circulation of pertussis to more effectively boost immunity through adolescence and adulthood, given the limitations of currently available vaccines. Clearly, maintenance of high vaccine coverage among the very young would be critical to ensure ongoing control of severe disease under such a strategy.

An absence of defined correlates of protection against pertussis following either infection exposure or vaccination is another critical information gap when seeking evidence for the best strategies for pertussis control, including identification and development of more efficacious candidate vaccines for sustained reduction of transmission and disease (see Chapters 3 and 7). The PERISCOPE project, a European study funded by the Innovative Medicines Initiative, is seeking to define more clearly the differences in immune responses and their persistence following each of infection exposure and vaccination with either aP or wP vaccines. Correlation of such differences with observed clinical protection against transmission and disease at individual and population levels will be an important step in identification of vaccine targets for improved vaccines and approaches.

Clearer delineation of natural and vaccine-derived immunity would potentially afford new insights into population-level serosurveillance studies. Such studies have been used in the past to compare trends in the age incidence of disease across Europe, or to understand epidemic cycles in the context of shifting immunologic profiles. Improved methods are needed to tease out key parameters defining population protection from sequential serologic surveys in a range of country settings, to inform the likely impact of current or future vaccine approaches.

The combined experiences reported here further demonstrate that robust prediction of the likely impact of pertussis immunization strategies will be influenced by a population's pertussis vaccine history including vaccine uptake and timeliness. This knowledge supports investment in whole-of-life vaccine registers, alongside epidemiologic surveillance, to determine optimal setting-specific schedules for sustained pertussis control.

References

1. Silfverdal SA, Assudani D, Kuriyakose S, et al. Immunological persistence in 5 y olds previously vaccinated with hexavalent DTPa-HBV-IPV/Hib at 3, 5, and 11 months of age. *Hum Vaccin Immunother* 2014;**10**:2795–8.

2. Heininger U, Andre P, Chlibek R, et al. Comparative epidemiologic characteristics of pertussis in 10 Central and Eastern European countries, 2000–2013. *PLoS One* 2016;**11**:e0155949.

3. World Health Organization. Pertussis vaccines: WHO position paper, August 2015—recommendations. *Vaccine* 2016;**34**:1423–5.

4. van Wijhe M, McDonald SA, de Melker HE, et al. Effect of vaccination programmes on mortality burden among children and young adults in the Netherlands during the 20th century: a historical analysis. *Lancet Infect Dis* 2016;**16**:592–8.

5. Gangarosa E, Galazka A, Wolfe C, et al. Impact of anti-vaccine movements on pertussis control: the untold story. *Lancet* 1998;**351**:356–61.

6. Nilsson L, Lepp T, von Segebaden K, et al. Pertussis vaccination in infancy lowers the incidence of pertussis disease and the rate of hospitalisation after one and two doses: analyses of 10 years of pertussis surveillance. *Vaccine* 2012;**30**:3239–47.

7. Ramsay M, Farrington C, Miller E. Age-specific efficacy of pertussis vaccine during epidemic and non-epidemic periods. *Epidemiol Infect* 1993;**111**:41–8.

8. Amirthalingam G, Gupta S, Campbell H. Pertussis immunisation and control in England and Wales, 1957 to 2012: a historical review. *Euro Surveill* 2013;**18**:20587.

9. van der Maas NA, Mooi FR, de Greeff SC, et al. Pertussis in the Netherlands, is the current vaccination strategy sufficient to reduce disease burden in young infants? *Vaccine* 2013;**31**:4541–7.

10. Radke S, Petousis-Harris H, Watson D, et al. Age-specific effectiveness following each dose of acellular pertussis vaccine among infants and children in New Zealand. *Vaccine* 2017;**35**:177–83.

11. Wood N, Quinn HE, McIntyre P, et al. Pertussis in infants: preventing deaths and hospitalisations in the very young. *J Paediatr Child Health* 2008;**44**:161–5.

12. Fine PE, Clarkson JA. The recurrent of whooping cough: possible implications for assessment of vaccine efficacy. *Lancet* 1982;**1**:666–9.

13. Gay N, Miller E. Pertussis transmission in England and Wales. *Lancet* 2000;**355**:1553–4.

14. Campbell PT, McCaw JM, McIntyre P, et al. Defining long-term drivers of pertussis resurgence, and optimal vaccine control strategies. *Vaccine* 2015;**33**:5794–800.

15. Quinn HE, Snelling TL, Macartney KK, et al. Duration of protection after first dose of acellular pertussis vaccine in infants. *Pediatrics* 2014;**133**:e513–9.

16. Campbell P, McIntyre P, Quinn H, et al. Increased population prevalence of low pertussis toxin antibody

levels in young children preceding a record pertussis epidemic in Australia. *PLoS One* 2012;**7**:e35874.

17. Sigera S, Perera J, Rasarathinam J, et al. Seroprevalence of Bordetella pertussis specific Immunoglobulin G antibody levels among asymptomatic individuals aged 4 to 24 years: a descriptive cross sectional study from Sri Lanka. *BMC Infect Dis* 2016;**16**:729.

18. Campbell H, Amirthalingam G, Andrews N, et al. Accelerating control of pertussis in England and Wales. *Emerg Infect Dis* 2012;**18**:38–47.

19. Hallander HO, Nilsson L, Gustafsson L. Is adolescent pertussis vaccination preferable to natural booster infections? *Expert Rev Clin Pharmacol* 2011;**4**:705–11.

20. Stein-Zamir C, Shoob H, Abramson N, et al. The impact of additional pertussis vaccine doses on disease incidence in children and infants. *Vaccine* 2010;**29**: 207–11.

21. Witt MA, Katz PH, Witt DJ. Unexpectedly limited durability of immunity following acellular pertussis vaccination in preadolescents in a North American outbreak. *Clin Infect Dis* 2012;**54**:1730–5.

22. Tartof SY, Lewis M, Kenyon C, et al. Waning immunity to pertussis following 5 doses of DTaP. *Pediatrics* 2013;**131**:e1047–52.

23. Forsyth KD, Campins-Marti M, Caro J, et al. New pertussis vaccination strategies beyond infancy: recommendations by the global pertussis initiative. *Clin Infect Dis* 2004;**39**:1802–9.

24. Esposito S, Principi N, European Society of Clinical Microbiology and Infectious Diseases (ESCMID) Vaccine Study Group (EVASG). Immunization against pertussis in adolescents and adults. *Clin Microbiol Infect* 2016;**22** Suppl 5:S89–95.

25. Pebody RG, Gay N, Giammanco A, et al. The seroepidemiology of Bordetella pertussis infection in Western Europe. *Epidemiol Infect* 2005;**133**:159–71.

26. de Greeff SC, de Melker HE, van Gageldonk PG, et al. Seroprevalence of pertussis in The Netherlands: evidence for increased circulation of Bordetella pertussis. *PLoS One* 2010;**5**:e14183.

27. Quinn HE, McIntyre PB. The impact of adolescent pertussis immunization, 2004–2009: lessons from Australia. *Bull World Health Organ* 2011;**89**:666–74.

28. Acosta AM, DeBolt C, Tasslimi A, et al. Tdap vaccine effectiveness in adolescents during the 2012 Washington State pertussis epidemic. *Pediatrics* 2015;**135**:981–9.

29. Skoff TH, Martin SW. Impact of tetanus toxoid, reduced diphtheria toxoid, and acellular pertussis vaccinations on reported pertussis cases among those 11 to 18 years of age in an era of waning pertussis immunity: a follow-up analysis. *JAMA Pediatr* 2016;**170**: 453–8.

30. McGirr A, Fisman DN. Duration of pertussis immunity after DTaP immunization: a meta-analysis. *Pediatrics* 2015;**135**:331–43.

31. Liko J, Robison SG, Cieslak PR. Priming with whole-cell versus acellular pertussis vaccine. *N Engl J Med* 2013;**368**:581–2.

32. Sheridan SL, Ware RS, Grimwood K, et al. Number and order of whole cell pertussis vaccines in infancy and disease protection. *JAMA* 2012;**308**:454–6.

33. Witt MA, Arias L, Katz PH, et al. Reduced risk of pertussis among persons ever vaccinated with whole cell pertussis vaccine compared to recipients of acellular pertussis vaccines in a large US cohort. *Clin Infect Dis* 2013;**56**:1248–54.

34. DeAngelis H, Scarpino SV, Fitzpatrick MC, et al. Epidemiological and economic effects of priming with the whole-cell Bordetella pertussis vaccine. *JAMA Pediatr* 2016;**170**:459–65.

35. Breakwell L, Kelso P, Finley C, et al. Pertussis vaccine effectiveness in the setting of pertactin-deficient pertussis. *Pediatrics* 2016;**137**:e20153973.

36. Hall R. Notifiable diseases surveillance, 1917 to 1991. *Comm Dis Intell* 1993;**17**:226–36.

37. Gidding HF, Burgess MA, Kempe AE. A short history of vaccination in Australia. *Med J Aust* 2001;**174**:37–40.

38. Quinn HE, McIntyre PB. Pertussis epidemiology in Australia over the decade 1995–2005—trends by region and age group. *Comm Dis Intell* 2007;**31**:205–15.

39. Torvaldsen S, Hull B, McIntyre P. Using the Australian Childhood Immunisation Register to track the transition from whole-cell to acellular pertussis vaccines. *Comm Dis Intell* 2002;**26**:581–3.

40. Salmaso S, Mastrantonio P, Tozzi AE, et al. Sustained efficacy during the first 6 years of life of 3-component acellular pertussis vaccines administered in infancy: the Italian experience. *Pediatrics* 2001;**108**:E81.

41. Pillsbury A, Quinn HE, McIntyre PB. Australian vaccine preventable disease epidemiological review series: pertussis, 2006–2012. *Commun Dis Intell Q Rep* 2014;**38**:E179–94.

42. Swaak A. A study of children born in 1962 on the relation between local complaints and the type of vaccine used in the immunization for whooping cough, diphtheria, tetanus and poliomyelitis [article in Dutch]. *Ned Tijdschr Geneeskd* 1966;**110**:196–9.

43. National Institute for Public Health and the Environment. *Dutch National Immunisation Programme.* https://rijksvaccinatieprogramma.nl/English https://rijksvaccinatieprogramma.nl/professionals/richtlijn-uitvoering-rvp

44. Riolo MA, Rohani P. Combating pertussis resurgence: one booster vaccination schedule does not fit all. *Proc Natl Acad Sci U S A* 2015;**112**:E472–7.

CHAPTER 6

Animal models

Eric T. Harvill and Tracy Nicholson

Abstract

There is a long history of the study of *Bordetella* species in animal hosts, built on the foundation of Koch's postulates: experimentally inoculating animals with virulent bacteria to define various pathogenic outcomes. Inoculation of mice, rats, pigs, baboons, and humans simulate whooping cough with increasing accuracy, albeit with exponentially increasing costs and difficulties. While most of the basic processes of immune activation and pathogenesis are quite similar from rodents to primates, relative to other pathogen infection systems there are marked differences that are important to consider. While most of this work has involved *Bordetella pertussis*, the closely related species *Bordetella bronchiseptica* naturally, and highly efficiently, infects a variety of animals, allowing aspects of pathogenesis to be examined in the context of natural infections. More recently, the ongoing transmission of *B. pertussis* within highly vaccinated populations has increased interest in understanding the nature of the transmission process. Several innovative animal models have now been established that allow transmission of *B. bronchiseptica* among mice, rabbits, and pigs, and transmission of *B. pertussis* among baboons. Together, these animal model systems have taught us most of what we know of the nature of the complex interactions within an individual host, transmission between hosts, and the past and ongoing evolution of these species. Recent and ongoing improvements of the historical animal infection systems, and the generation of new experimental infection systems to study pathogenesis and transmission, are critical to advance our understanding and control of the resurgence of this highly infectious disease.

6.1 Introduction

A detailed understanding of the events that are triggered in a host throughout a bacterial infection is needed to develop new, or to improve existing, therapies and vaccines. There is no substitute for using animal studies for gaining detailed mechanistic information regarding the dynamics of host–pathogen interactions, the complex interactions between the different cell types and organs involved in a host response to a pathogen, or mechanisms used by a pathogen to manipulate its host. Bordetellae in a variety of animals, including those they naturally infect, present among the best paradigms for exploring these host–pathogen dynamics. The significant impact of the knowledge gained by experimental animal infections on *Bordetella pertussis* pathogenesis and therapeutics, as well as the relative strengths and weaknesses of each animal infection model will be discussed in this chapter.

Given that *B. pertussis* and *Bordetella parapertussis* are believed to have evolved independently from a *Bordetella bronchiseptica*-like ancestor, the 'classical *Bordetella*' species represents current clinically significant human pathogens that evolved from zoonotic sources.

Harvill, E. T. and Nicholson, T., *Animal models*. In: *Pertussis: epidemiology, immunology, and evolution*. Edited by Pejman Rohani and Samuel V. Scarpino: Oxford University Press (2019). © Oxford University Press.
DOI: 10.1093/oso/9780198811879.003.0006

While *B. pertussis* is only known to naturally infect humans, humans make poor experimental subjects for a long and varied list of reasons. The complexity, expense, and ethical concerns of working with humans have limited such work to clinical trials of vaccine formulations and have not allowed for the experimental infections that are feasible in other animals.

Before discussing specific animal infection models, there is value in highlighting the standpoint from which to view experimental infection of animals as 'models'. Although an anthropocentric view that considers primarily human disease (pertussis) is understandably common, the classical Bordetellae (*B. pertussis*, *B. parapertussis*, and *B. bronchiseptica*) infect and cause diseases in a wide variety of animals, and can be important to different stakeholders for different reasons. For example, pet owners are deeply committed to ensuring a high health status for their pets and recognize *B. bronchiseptica* as a significant threat, vaccinating their dogs and cats against it before bringing them to a kennel. Such vaccines for companion animals must be safe not only for the animal, but for the pet owners themselves. For example, the widespread use of attenuated strains could present problems for certain pets as well as other members of their household including infants, elderly people, and/or immunocompromised individuals (e.g. patients with cystic fibrosis). So, attenuated live vaccines must protect against disease by inducing a robust immune response, without the risk of transmitting to or harming other pets or immunocompromised family members. These requirements demonstrate that although the needs of the veterinary medicine community might be different from those of the biomedical community, much of what is learned in these more tractable experimental systems can contribute to advance human medicine.

Bordetella infections impact the health of numerous domestic and wildlife species. The ability of a farmer or rancher to generate an income, and to sustain their family, is linked to the health of the animals they are raising. For both cattle and swine producers in the United States, respiratory disease is the number one health concern. Spillover of respiratory infections (such as *Bordetella* infections) are a top concern for ranchers who raise livestock species such as cattle or bison on open range land where their herds routinely are in contact with wildlife species. Respiratory infections (such as *Bordetella* infections) are a top concern for veterinarians and zoo professionals as well. The 'One Health' approach simultaneously considers all aspects of health for humans, other animals, and the environment. Agriculture is the intersection between public health and veterinary health at local, national, and international levels. Improving animal health positively impacts global food security and economies, which in turn benefits society by decreasing the burden of poverty and illness.

6.2 Baboons

Among the non-human primate species, baboons (*Papio anubis*) are an invaluable model for infectious disease and vaccine research due to the numerous physiological and structural qualities shared with humans[1]. Baboons are genetically very similar to humans with an approximate DNA sequence difference of 4 per cent[1]. For comparison, macaques have over 6 per cent DNA sequence difference compared to humans[1]. Baboons have an immune system similar to humans encompassing homologues of immunoglobulin (Ig) classes and subclasses and cross-reactive CD antigen expression on lymphocytes and other cell types and, unlike macaques, produce all four isotypes of IgG antibody (IgG1–4)[2,3].

Baboons exhibit key characteristic clinical signs of human pertussis, including leucocytosis, high nasopharyngeal colonization levels, moderate rise in body temperature (1°C), and paroxysmal coughing. These features along with an opportunity to investigate neonatal and maternal vaccination and immune responses enable baboons to serve as an ideal animal model for investigating *B. pertussis* pathogenesis and immune responses. The baboon model of pertussis has been successfully used to evaluate key aspects of *B. pertussis* disease dynamics, immune responses to infection, and vaccination. Highlights of these studies will be mentioned below and can also be found in other reviews[4,5].

An additional benefit to using baboons as an animal model for *B. pertussis* research is their core body temperature of 37°C. In contrast, rhesus macaques have a higher body temperature of 39°C. Warfel et al. demonstrated the negative impact of the slightly higher temperature on *B. pertussis* growth and antigen/virulence factor production[6]. The authors demonstrated that *B. pertussis* exhibited a reduced growth rate and reduced production of adenylate cyclase toxin at 39°C compared to levels produced at 37°C. The authors hypothesized that negative impacts such as these resulting from the slightly higher temperature contributed to the reduced disease observed in macaques compared to the disease severity observed in baboons[6].

Using a baboon model of pertussis, Warfel et al. demonstrated that baboons are able to develop an adaptive immune response preventing reinfection after subsequently clearing an initial *B. pertussis* challenge, in which leucocytosis, high nasopharyngeal colonization levels, and paroxysmal coughing were observed[6,7]. Using a baboon model, it was demonstrated that acellular pertussis vaccines prevent pertussis disease symptoms but do not prevent colonization and transmission. More importantly, it was demonstrated that acellular pertussis vaccines induce a suboptimal immune response that differs from the immune response induced by natural infection and whole-cell pertussis vaccines[7,8]. *Bordetella pertussis* requires unbroken and successful transmission among people due to several deficiencies, such as severe host restriction, lack of prolonged persistent carriage state, lack of an animal reservoir, and the inability to survive in the environment. Studies investigating the impact of immune responses to infection and vaccination on transmission are essential to the design of strategies to mitigate and block transmission.

The baboon model of pertussis has been successfully used to investigate vaccine strategies to protect newborns. For example, it was demonstrated that neonatal acellular pertussis vaccination protected baboons[9] and maternal vaccination protected infant baboons from clinical disease[9]. The baboon model has been used to investigate innate and adaptive immune responses to infection and vaccination, revealing the features of the induced immune response that provide protection from colonization and subsequent transmission, such as interleukin-17 memory induction[7,10]. The baboon model also offers the opportunity to identify and quantify virulence factors and antigens produced during a *B. pertussis* infection, via immunological assays such as the toxin neutralization assay[11,12], for example.

Tremendous insights into key aspects of *B. pertussis* disease and immune dynamics have been gained from the baboon model of pertussis. Despite these benefits, there are significant disadvantages to utilizing baboons as a model of pertussis. These disadvantages include a requirement for costly biocontainment facilities and highly trained veterinary staff, as well as voluminous regulatory compliance and approval issues. While the outbred nature of baboons better replicates human vaccination and infection processes, it presents challenges such as variation among or between experimental groups as well as difficulty in obtaining statistical significance. In addition, ethical concerns related to responsible use of non-human primates need to be fully weighed before initiating studies utilizing baboons as a model for pertussis. Together, these substantial challenges make the baboon model of pertussis the most labour- and cost-intensive non-human animal model.

6.3 Pigs

Since pigs are the one domesticated livestock species discussed here, it's worth highlighting the 'One Health' perspective that values such animals as experimental systems in which to study disease and vaccine that can also impact humans. Due to the close association of domesticated livestock with humans, zoonotic pathogens of livestock represent a significant public health concern. Since just 2007, specific examples of zoonotic diseases affecting humans that have been identified in association with livestock, particularly swine, include *Streptococcus suis*, methicillin-resistant *Staphylococcus aureus*, and pandemic H1N1 influenza. Domesticated animals throughout the world, particularly swine, are an important source of emerging zoonotic diseases and provide an invaluable resource for investigating zoonotic agent–host interactions.

From a clinical human disease or pertussis perspective, pigs are a valuable animal model for investigating *B. pertussis* and *B. bronchiseptica* pathogenesis and immune responses. For instance, humans (and non-human primates) and swine share the same anatomical structure of lymphoid tissues in the upper respiratory tract[13,14], while rodents and rabbits lack the pharyngeal and palatine tonsils that are known to function as inductive sites for secretory antibody responses[15–17]. Additionally, the tonsils of humans (and non-human primates) and pigs have deep antigen-retaining crypts, which express germinal centres shortly after birth, whereas the rodent nasal-associated lymphoid tissues have plain surfaces that require external stimuli to induce the expression of germinal centers[18]. These features, along with an opportunity to investigate neonatal and maternal vaccination and immune responses, make pigs a valuable animal model for investigating key aspects of *B. pertussis* and *B. bronchiseptica* disease dynamics, immune responses to infection, vaccination, and transmission.

6.3.1 *Bordetella pertussis* newborn piglet model

The newborn piglet model of pertussis involves challenging newborn (3 days old) up to 4-week-old piglets with relatively high doses of *B. pertussis* by an intrapulmonary route of infection[19]. Infected piglets subsequently exhibit a range of clinical symptoms from nasal discharge and non-paroxysmal coughing, to breathing difficulties arising from bronchopneumonia. Only young pigs (4 weeks old or younger) are susceptible to severe disease symptoms caused by *B. pertussis*; 5-week-old and older pigs are completely resistant to infection and clinical disease presentation[19]. Protection against infection with *B. pertussis* was associated with the developmentally regulated expression of porcine beta-defensin 1 (*pBD-1*), a porcine defensin homologue of human beta-defensin 2[20]. As mentioned earlier, *B. pertussis* exhibits a reduced growth rate and produces significantly lower levels of adenylate cyclase toxin at 39°C compared to at 37°C. Negative impacts resulting from the slightly higher temperature were hypothesized to have contributed to the reduced disease observed in macaques (core body temperature of 39°C) compared to the disease severity observed in baboons (core body temperature of 37°C)[6]. Although never tested, it is possible that the 39°C core body temperature of pigs could contribute to the need for young animals, high challenge doses, and an intrapulmonary route of inoculation for *B. pertussis* to cause serious disease in pigs[19]. It has been demonstrated that older pigs, between 4 and 5 weeks old, exhibit mild respiratory clinical symptoms, such as non-paroxysmal cough and breathing difficulties, after intrapulmonary challenge with relatively high doses of human and ovine strains of *B. parapertussis*[21]. In contrast to pigs challenged with *B. pertussis*, pigs challenged with *B. parapertussis* transmitted *B. parapertussis* to non-infected pigs by a direct route[21]. Using the newborn piglet model of pertussis, it has been demonstrated that maternal immunization provides protection against *B. pertussis* in newborn piglets[22,23].

6.3.2 *Bordetella bronchiseptica* pig model

Bordetella bronchiseptica is highly infectious among many companion, poultry, wildlife, and livestock species, including swine. Globally, *B. bronchiseptica* is widely prevalent and is a significant contributor to respiratory disease in pigs. In young pigs, it causes severe bronchopneumonia with high morbidity and, if untreated, mortality. It is a primary aetiological agent of atrophic rhinitis, causing a moderate to mild reversible form, and promotes colonization by toxigenic strains of *Pasteurella multocida*, usually leading to severe, progressive atrophic rhinitis[24–28]. Similarly, *B. bronchiseptica* infections enhance colonization and increase the severity of respiratory disease associated with other bacterial and viral pathogens and is thus a main contributing agent in porcine respiratory disease complex, a multifactorial disease state that is consistently listed as a top animal health research priority[29–40].

Bordetella bronchiseptica is frequently isolated from nasal turbinates and lung lesions of fattening pigs who may not exhibit clinical signs of respiratory disease. Nonetheless, field surveys document that subclinical pneumonia can result in substantial economic losses due to slower weight gain, increased days to market, and reduced feed efficiency[41,42]. A common trait of experimental *B. bronchiseptica* infections is they result in long-term to lifelong carriage[43–48]. This holds true despite the use of vaccines; *B. bronchiseptica* is frequently isolated from the nasal cavities of vaccinated animals suggesting that vaccines fail to protect animals from colonization[49,50]. More importantly, vaccinated animals then serve as asymptomatic carriers that continue to shed and transmit *B. bronchiseptica* to cohorts[45,51–55].

Clinical signs vary greatly among pigs infected with *B. bronchiseptica* depending on age, immune status, and co-infection with other pathogens. In uncomplicated disease, clinical signs typically appear 2–3 days after infection and are associated with rhinitis and bronchitis, including sneezing, nasal discharge, ocular discharge, and a dry, repeated cough[56]. More severe signs can occur in neonatal pigs when bronchopneumonia develops with associated dyspnoea and lethargy. Clinical presentation may also include progressive atrophic rhinitis, bronchopneumonia, or systemic disease upon co-infection with other bacterial or viral pathogens such as *Haemophilus parasuis* or *S. suis*[56]. Antibody passively acquired by piglets from the colostrum of infected or vaccinated sows does not protect against colonization, but does provide protection against disease symptoms such as turbinate lesions and pneumonia[57–60]. Vaccination of sows has been shown to delay clinical signs for several weeks[60], and when clinical signs are exhibited, lesions typically fail to develop or are significantly reduced in severity[26,61,62].

Given that pigs are natural hosts for *B. bronchiseptica* and clinical disease is easily reproducible in appropriate containment facilities, swine serve as a valuable model to experimentally test key aspects of *Bordetella* disease and transmission without the use of modified or immune-deficient animals. Virulence gene expression for both *B. pertussis* and *B. bronchiseptica* is regulated by a two-component sensory transduction system encoded by the *bvgAS* locus. BvgAS controls expression of a spectrum of phenotypic phases transitioning between a virulent (Bvg+) phase and a non-virulent (Bvg−) phase, a process referred to as phenotypic modulation. A proposed role for phenotypic modulation by the BvgAS signal transduction system is to coordinate the regulation of genes that are required for survival in the different environments encountered by *B. bronchiseptica*, including those encountered during transmission between hosts. In this proposed role, the Bvgi phase, or other phase intermediates between Bvgi and Bvg+, was proposed to be required for aerosol or indirect transmission[63–65]. However, when this hypothesis was experimentally tested in swine, both the wild-type isolate and a Bvg+ phase-locked mutant transmitted to naïve piglets housed in the same pen as well as to naïve piglets in pens across the room[48].

The results demonstrated that in a natural host, phenotypic modulation from the fully virulent Bvg+ phase is not required for *B. bronchiseptica* respiratory infection and host-to-host transmission by either direct or indirect/airborne routes[48]. The requirement for Bvg-repressed genes during the infectious cycle in swine has not been tested because a Bvg− phase-locked strain could not be recovered from directly challenged piglets[48]. *Bordetella bronchiseptica* has been observed to survive at least 45 days in soil[66] and for several weeks in lake water[67,68] so it is possible that Bvg-repressed genes could play a role in the transmission of *B. bronchiseptica* through environmental sources[69]. Furthermore, the genetic signatures of several *Bordetella* species were identified in meta-genomic analyses of many soil and water sources, indicating that these are important environmental niches for various members of this genus[70].

Among the Bvg-activated genes are those encoding the type three secretion system (T3SS), which mediates persistent colonization of the lower respiratory tract in mice[71]. When the contribution of the T3SS to the pathogenesis of *B. bronchiseptica* in swine was experimentally tested, the T3SS was found to be required for maximal disease severity and persistence in the lower respiratory tract[47]. However, this study demonstrated that the T3SS mutant was not required for transmission among swine by both direct and indirect routes, demonstrating that transmission can occur even with attenuated disease[47].

Important insights into key aspects of *Bordetella* disease and immune dynamics have been gained from using pigs as a model. Despite these benefits there are significant disadvantages to utilizing pigs as a model of *Bordetella* disease. As with baboons, a major disadvantage is a requirement for large and costly biocontainment facilities and highly trained veterinary staff suitable for the care of pigs. Various other requirements related to the housing and care issues elevate the cost of using pigs compared to rodents, albeit not to the cost level associated with baboons. Since *B. bronchiseptica* is pervasive in swine herds, a significant challenge to utilizing pigs to study *B. bronchiseptica* pathogenesis is obtaining pigs that are not colonized by *B. bronchiseptica* prior to challenge. To circumvent this issue, it is possible to use piglets that are referred to as naturally farrowed early-weaned (NFEW) for *B. bronchiseptica* studies. Unlike humans, maternal antibodies are not transferred through the placenta from sow to piglets before birth. Instead, maternal antibodies are passively acquired by newborn piglets through the ingestion of colostrum[72]. NFEW piglets are removed from the sow shortly after receiving colostrum to prevent vertical transmission of *B. bronchiseptica* from sow to piglets. Despite the presence of maternal antibodies, disease pathogenesis, transmission, and immune response to *B. bronchiseptica* infection can be successfully and, more importantly, consistently evaluated using NFEW piglets. Another challenge associated with utilizing pigs to study *Bordetella* pathogenesis is the lack of reagents, such as purified antigens and other immune-related reagents, immortalized cells lines, and the inability to genetically modify the host. Similar to the baboon model, while the outbred nature of pigs better replicates human vaccination and infection processes, it presents challenges such as variation among or between experimental groups as well as difficulty in obtaining statistical significance. While pigs are valuable animal model for investigating key aspects of *Bordetella* disease, immune dynamics, and transmission, these disadvantages need to be fully considered before beginning studies utilizing pigs as a model for *Bordetella* disease.

6.4 Rats

Other than primates, rats are the only animal model that have been reported to exhibit both leucocytosis and paroxysmal coughing, hallmarks of pertussis human infections. Hornibrook and Ashburn first reported rats coughing following infection with *B. pertussis*[73]. This original report was later confirmed by Hall et al. in which the authors report Sprague-Dawley rats exhibiting a paroxysmal cough and leucocytosis following intrabronchial inoculation with *B. pertussis*[74]. The coughing-rat model was further developed in subsequent studies and used to investigate *B. pertussis* pathogenesis and immunity[75–79]. Using the coughing-rat model, a transposon-insertion mutant of *B. pertussis* lacking pertussis toxin failed to induce a paroxysmal cough compared to a wild-type *B. pertussis* strain[75]. Additionally, vaccination using a whole-cell pertussis vaccine was demon-

strated to reduce the incidence of paroxysmal coughing and leucocytosis in a coughing-rat model of pertussis[77]. In contrast to humans and non-human primates, rats do not secret mucus following infection with *B. pertussis* and do not transmit *B. pertussis* infection to other cage cohorts[80].

Rats have served as a valuable model to experimentally test key aspects of *B. bronchiseptica* disease dynamics, immune responses to infection, and vaccination. The knowledge gained from these key aspects has significantly contributed and positively impacted our general knowledge regarding *B. pertussis* pathogenesis and therapeutics. For example, using a rat model, it was demonstrated that expression and subsequent production of the genes encoding flagella, a classical Bvg⁻ phase phenotype, during respiratory infection (Bvg⁺ conditions) led to a decrease in tracheal colonization[81]. This was a significant advancement for the field because these findings demonstrated the significance of BvgAS regulation on pathogenesis. Other highlights include demonstrating that dermonecrotizing toxin damages bone formation[82], the *bvgAS* loci of *B. pertussis* and *B. bronchiseptica* are interchangeable during a respiratory infection[83], full activation (from Bvg-intermediate to Bvg⁺ or wild type) is required for optimal colonization of the upper respiratory tract of rats (nasal cavity, larynx, and trachea)[64], filamentous haemagglutinin is required for tracheal colonization by aiding in mediating clearance mechanism of the mucociliary escalator[84], and the requirement of a functional T3SS and fimbriae for persistent tracheal colonization[71,85].

While rats are efficient for experiments such as those previously described, there are limitations to their use relative to other animals. They lack some of the size and physiological attributes that pigs share with humans and other primates. They also lack some of the strengths of the mouse model, such as the remarkable reproducibility allowed with inbred strains and the many genetic mutations that have been made in mice that allow the relationship to be probed on the host side.

6.5 Mice

Over the last few decades, there has been an increasing focus on the use of laboratory-raised mice for the study of the complex interactions between pathogens and host. There are several advantages to the use of mice, including their size, ease of handling and low-cost housing, as well as their fecundity. Reaching sexual maturity within 2 months and a gestational period of less than a month allow for three or four generation times per year, and litter sizes of six to ten pups allow for rapid expansion of colonies and provide the animals necessary for detailed and reproducible experiments to be statistically robust. Consistency was bred into these animals by virtue of repeated and systematic inbreeding. After many generations, inbred mouse strains lack virtually all genetic diversity; all offspring of the same sex are assumed to be genetically identical.

The advantages described above have led to increasing momentum in the use of laboratory mice, which leads to more advantages. Since the great majority of prior work has been done in mice, there is the expectation that some explanation may accompany and justify publications using other animals. That momentum comes with its own self-reinforcing advantages. Data generated in the same inbred mouse strain should be readily comparable between experiments performed years apart in laboratories in various countries, allowing experiments to be reproduced in different laboratories. In general, *B. pertussis* seems to interact with the murine and human immune systems in a similar manner, inducing an analogous inflammatory response and being controlled by a complex combination of antibodies and T-cell responses. Since mice are the main experimental animal of immunology, a long and growing list of immunological tools have been developed in mice that are not available in any other animal, as will be discussed below (see also Chapter 3).

Despite these advantages, there are several important limitations to the use of mice in the study of *B. pertussis*, most notably the dramatic cough for which the disease is named. There are aspects of the cough that are clinically important and scientifically interesting, appearing to involve a neurological component in their extraordinary compulsion that appears able to overcome even the instinct to breath. But the paroxysmal cough itself cannot currently be studied in mouse experimental infections because mice do not cough in a way that resembles the human cough. Another weakness of the mouse infection model is the relatively poor efficiency of *B. pertussis*

in colonizing mice, requiring tens of thousands of organisms to initiate an infection. This precludes the study of important aspects of the transmission and initial colonization by *B. pertussis*, which is highly infectious among humans.

Fortunately, the weaknesses of the mouse model are effectively being overcome. In some cases, that involves the use of more expensive and difficult animal systems; both pigs and baboons are susceptible to the cough induced by *B. pertussis*. But in other cases, the observed weaknesses are being overcome by the noted strengths of the mouse model itself, especially its broad use in other areas of science including immunology and, more recently, the study of the microbiome. For example, Weyrich et al. showed that *B. pertussis* colonization of mice is orders of magnitude more efficient when the resident nasal microbiota are first disrupted with antibiotics[86]. The possibility of significant immunological differences between humans and mice can be minimized by the generation of mice with various manipulations that 'humanize' aspect of their immune system and/or microbiota.

6.6 Pathogenesis

The study of bacterial mechanisms involved in various aspects of infection and disease were first advanced by the generation of transposon mutants by Weiss and Falkow[87]. Since then, tools have been developed that allow for engineered mutations that avoid polar effects and can definitively attribute observed effects to the specific gene targeted. Most of the study of *B. pertussis* pathogenesis, reviewed in detail in Chapter 2, has been performed in the mouse model for the reasons described earlier in section 6.5. Most importantly, the consistency of inbred mice allows the results of experiments performed in many laboratories over the course of decades to be readily compared.

6.7 Immune modulation

Bordetella pertussis makes several molecules recognized by host pathogen recognition receptors that contribute to stimulating a robust immune response. But *B. pertussis* is not a passive player in the competition with the host and it generates a number of factors known to modulate the host response in a variety of ways. In fact, its remarkable success in inducing a robust response that serves its needs but does not efficiently mediate clearance suggests the bacterium has evolved a winning strategy involving multiple immunomodulators. The coughing illness induced is distinctively paroxysmal and severe, effectively propelling the agent into the environment to facilitate its spread. Although the pathology leads to a strong antibody and T-cell response, these are often not sufficient to end the disease for months. Together, these observations support the model that *B. pertussis* manipulates the immune response in a variety of ways, only some of which are yet discovered. The continued circulation of *B. pertussis* and its re-emergence in vaccinated populations further support the expectation that additional mechanisms will be revealed in the coming years.

6.8 Immune mice

Bordetella pertussis appears to cause the most severe and distinctive disease in naïve individuals, contributing to whooping cough being considered a childhood disease. However, there is a growing recognition that many infections in countries with high vaccine coverage occur in hosts previously exposed to some or all of its antigens. The ability to colonize and spread among partially-immune convalescent and/or vaccinated individuals appears to be relatively well accepted, but to have poorly permeated the study of its pathogenesis. The vast majority of *B. pertussis* factors studied have been examined in naïve mice only. Very few have been examined for their impact in immune animals. Kirimanjeswara et al.[88] demonstrated that the effects of pertussis toxin are even more dramatic in partially immune mice than in naïve mice, demonstrating an important role in allowing *B. pertussis* to avoid rapid antibody-mediated clearance. The re-emergence of *B. pertussis* within populations that have been previously vaccinated and/or infected highlights the need to assess the mechanisms this bacterium uses to successfully colonize and transmit among immune hosts.

6.9 Vaccine development

Current vaccines were tested and developed using the mouse model, and ongoing efforts to improve vaccines primarily involve mice. Without a full understanding of the immunological mechanisms of protection, vaccine development was guided by efforts to optimize antibody titres generated in mice, for example. The assays of protection involved decreasing the symptoms of disease in the somewhat artificial cerebral challenge model. Importantly, these vaccines were developed with the goal of decreasing disease in mice challenged with extraordinary doses and/or via unnatural (e.g. intracerebral) routes. They were not tested for their ability to block colonization, since there were not robust experimental systems to study efficient colonization with natural, very low doses of bacteria. It should not be overly surprising, therefore, that current vaccines might be more effective in preventing disease than they are in preventing colonization and transmission of *B. pertussis*.

6.10 Experimental models of transmission

The greatest challenge in controlling *B. pertussis* is its remarkably high rate of transmission between hosts. Although a critical aspect of its biology, there has been very little progress in the study of transmission because of the lack of an effective experimental system. Recent work in three areas provides hope for advances: (1) other animals—*B. pertussis* has been observed to transmit among baboons and pigs; (2) modified microbiota—Weyrich et al.[86] recently showed that altering the microbiota of mice can greatly increase their susceptibility to *B. pertussis* colonization, raising the possibility that transmission might be observed and studied in mice; and (3) other *Bordetella* species—the classical *Bordetella* species are considered subspecies and include both *B. pertussis* and *B. bronchiseptica*, which naturally transmits efficiently among various animals including mice. Since these subspecies share most of the known factors involved in infection, understanding how they contribute to transmission of *B. bronchiseptica* among mice should provide a mechanistic understanding that can guide confirmatory experiments with *B. pertussis* in pigs and baboons. The combination of these three recent advances provides hope that there could be substantially improved understanding of the mechanistic basis for transmission between animals in the near future.

6.11 Microbiota

Bordetella pertussis is known to be highly efficient in transmitting between humans, but only poorly colonizes mice, requiring large numbers, generally over 10,000 colony-forming units, to initiate a vigorously growing and persistent infection. This is commonly believed to be due to host specificity, perhaps resembling the receptor–ligand type specificity observed with other bacterial pathogens. However, Weyrich et al.[86] observed that *B. bronchiseptica*, which colonizes mice much more efficiently than *B. pertussis*, also depletes the resident respiratory microbiota. Based on these observations, they hypothesize that *B. pertussis* failed to efficiently colonize mice because it fails to compete with resident microbiota in the mouse nose. To test this hypothesis, they inoculated mice that had been depleted of nasal microbiota and observed them to be highly susceptible to colonization by very low numbers of *B. pertussis*. This revealed not only that there are respiratory commensals that can block *B. pertussis* infection, but that a simple manipulation can allow mice to serve as a much more efficient host, opening the door to efficient and detailed examination of the mechanisms involved in the colonization and transmission processes.

6.12 The future of animal models

The various developments described in this chapter reveal an active and exciting research area poised to make significant advances in our understanding of the disease and the organisms that cause it. Unfortunately, there have been very limited investments made in these studies over recent years, and there remains little commitment to improve this. In addition, there is increasing pressure from the public to justify animal experimentation on the one hand, and increasing burdensome regulatory hurdles on the other, that add to the difficulty. Despite these hurdles, the tools and approaches are improving, and animal models are becoming more and more useful and valuable, and are the most likely approach to contribute to molecular mechanistic

understanding of the diseases caused by the Bordetellae and the only viable path to improved treatments and vaccines.

References

1. Murthy KK, Salas MT, Carey KD, et al. Baboon as a nonhuman primate model for vaccine studies. *Vaccine* 2006;**24**:4622–4.
2. Shearer MH, Dark RD, Chodosh J, et al. Comparison and characterization of immunoglobulin G subclasses among primate species. *Clin Diagn Lab Immunol* 1999;**6**:953–8.
3. Williamson ED, Packer PJ, Waters EL, et al. Recombinant (F1+V) vaccine protects cynomolgus macaques against pneumonic plague. *Vaccine* 2011;**29**:4771–7.
4. Trainor EA, Nicholson TL, Merkel TJ. Bordetella pertussis transmission. *Pathog Dis* 2015;**73**:ftv068.
5. Warfel JM, Merkel TJ. The baboon model of pertussis: effective use and lessons for pertussis vaccines. *Expert Rev Vaccines* 2014;**13**:1241–52.
6. Warfel JM, Beren J, Kelly VK, et al. Nonhuman primate model of pertussis. *Infect Immun* 2012;**80**:1530–6.
7. Warfel JM, Zimmerman LI, Merkel TJ. Acellular pertussis vaccines protect against disease but fail to prevent infection and transmission in a nonhuman primate model. *Proc Natl Acad Sci U S A* 2014;**111**:787–92.
8. Warfel JM, Zimmerman LI, Merkel TJ. Comparison of three whole-cell pertussis vaccines in the baboon model of pertussis. *Clin Vaccine Immunol* 2015;**23**:47–54.
9. Warfel JM, Papin JF, Wolf RF, et al. Maternal and neonatal vaccination protects newborn baboons from pertussis infection. *J Infect Dis* 2014;**210**:604–10.
10. Warfel JM, Merkel TJ. Bordetella pertussis infection induces a mucosal IL-17 response and long-lived Th17 and Th1 immune memory cells in nonhuman primates. *Mucosal Immunol* 2013;**6**:787–96.
11. Eby JC, Gray MC, Warfel JM, et al. Use of a toxin neutralization assay to characterize the serologic response to adenylate cyclase toxin after infection with Bordetella pertussis. *Clin Vaccine Immunol* 2017;**24**: e00370-16.
12. Eby JC, Gray MC, Warfel JM, et al. Quantification of the adenylate cyclase toxin of Bordetella pertussis in vitro and during respiratory infection. *Infect Immun* 2013;**81**:1390–8.
13. Perry ME, Mustafa Y, Licence ST, et al. Pig palatine tonsil as a functional model for the human. *Clin Anat* 1997;**10**:358.
14. Pracy JP, White A, Mustafa Y, et al. The comparative anatomy of the pig middle ear cavity: a model for middle ear inflammation in the human? *J Anat* 1998;**192**:359–68.

15. Kuper CF, Koornstra PJ, Hameleers DM, et al. The role of nasopharyngeal lymphoid tissue. *Immunol Today* 1992;**13**:219–24.

16. Wu HY, Nguyen HH, Russell MW. Nasal lymphoid tissue (NALT) as a mucosal immune inductive site. *Scand J Immunol* 1997;**46**:506–13.

17. Lugton I. Mucosa-associated lymphoid tissues as sites for uptake, carriage and excretion of tubercle bacilli and other pathogenic mycobacteria. *Immunol Cell Biol* 1999;**77**:364–72.

18. Brandtzaeg P. Function of mucosa-associated lymphoid tissue in antibody formation. *Immunol Invest* 2010;**39**: 303–55.

19. Elahi S, Brownlie R, Korzeniowski J, et al. Infection of newborn piglets with Bordetella pertussis: a new model for pertussis. *Infect Immun* 2005;**73**:3636–45.

20. Elahi S, Buchanan RM, Attah-Poku S, et al. The host defense peptide beta-defensin 1 confers protection against Bordetella pertussis in newborn piglets. *Infect Immun* 2006;**74**:2338–52.

21. Elahi S, Thompson DR, Strom S, et al. Infection with Bordetella parapertussis but not Bordetella pertussis causes pertussis-like disease in older pigs. *J Infect Dis* 2008;**198**:384–92.

22. Elahi S, Buchanan RM, Babiuk LA, et al. Maternal immunity provides protection against pertussis in newborn piglets. *Infect Immun* 2006;**74**:2619–27.

23. Elahi S, Thompson DR, Van Kessel J, et al. Protective role of passively transferred maternal cytokines against Bordetella pertussis infection in newborn piglets. *Infect Immun* 2017;**85**:e01063-16.

24. Cross RF. Bordetella bronchiseptica-induced porcine atrophic rhinitis. *J Am Vet Med Assoc* 1962;**141**:1467–8.

25. de Jong MF, Nielsen JP. Definition of progressive atrophic rhinitis. *Vet Rec* 1990;**126**:93.

26. Duncan JR, Ross RF, Switzer WP, et al. Pathology of experimental Bordetella bronchiseptica infection in swine: atrophic rhinitis. *Am J Vet Res* 1966;**27**:457–66.

27. Pedersen KB, Barfod K. The aetiological significance of Bordetella bronchiseptica and Pasteurella multocida in atrophic rhinitis of swine. *Nord Vet Med* 1981;**33**:513–22.

28. Rutter JM. Virulence of Pasteurella multocida in atrophic rhinitis of gnotobiotic pigs infected with Bordetella bronchiseptica. *Res Vet Sci* 1983;**34**.:287–95.

29. Duncan JR, Ramsey FK, Switzer WP. Pathology of experimental Bordetella bronchiseptica infection in swine: pneumonia. *Am J Vet Res* 1966;**27**:467–72.

30. Palzer A, Ritzmann M, Wolf G, et al. Associations between pathogens in healthy pigs and pigs with pneumonia. *Vet Rec* 2008;**162**:267–71.

31. Dunne HW, Kradel DC, Doty RB. Bordetella bronchiseptica (Brucella bronchiseptica) in pneumonia in young pigs. *J Am Vet Med Assoc* 1961;**139**:897–9.

32. Brockmeier SL, Halbur PG, Thacker EL. Porcine respiratory disease complex. In: Brogden KA, Guthmiller JM (eds), *Polymicrobial Diseases*. Washington, DC: ASM Press; 2002, pp 231–58.

33. Brockmeier SL, Palmer MV, Bolin SR, et al. Effects of intranasal inoculation with Bordetella bronchiseptica, porcine reproductive and respiratory syndrome virus, or a combination of both organisms on subsequent infection with Pasteurella multocida in pigs. *Am J Vet Res* 2001;**62**:521–5.

34. Brockmeier SL. Prior infection with Bordetella bronchiseptica increases nasal colonization by Haemophilus parasuis in swine. *Vet Microbiol* 2004;**99**:75–8.

35. Brockmeier SL, Palmer MV, Bolin SR. Effects of intranasal inoculation of porcine reproductive and respiratory syndrome virus, Bordetella bronchiseptica, or a combination of both organisms in pigs. *Am J Vet Res* 2000;**61**:892–9.

36. Brockmeier SL, Loving CL, Nicholson TL, et al. Coinfection of pigs with porcine respiratory coronavirus and Bordetella bronchiseptica. *Vet Microbiol* 2008;**128**: 36–47.

37. Brockmeier SL, Lager KM. Experimental airborne transmission of porcine reproductive and respiratory syndrome virus and Bordetella bronchiseptica. *Vet Microbiol* 2002;**89**:267–75.

38. Loving CL, Brockmeier SL, Vincent AL, et al. Influenza virus coinfection with Bordetella bronchiseptica enhances bacterial colonization and host responses exacerbating pulmonary lesions. *Microb Pathog* 2010;**49**: 237–45.

39. Vecht U, Arends JP, van der Molen EJ, et al. Differences in virulence between two strains of Streptococcus suis type II after experimentally induced infection of newborn germ-free pigs. *Am J Vet Res* 1989;**50**:1037–43.

40. Vecht U, Wisselink HJ, van Dijk JE, et al. Virulence of Streptococcus suis type 2 strains in newborn germfree pigs depends on phenotype. *Infect Immun* 1992;**60**:550–6.

41. Boessen CR, Kliebenstein JB, Cowart RP, et al. Effective use of slaughter checks to determine economic losses from morbidity in swine. *Acta Vet Scand* 1988;**84**:366 8.

42. Guerrero RJ (ed). *Respiratory Disease: An Important Global Problem in the Swine Industry*. Lausanne: IPVS; 1990.

43. Goodnow RA. Biology of Bordetella bronchiseptica. *Microbiol Rev* 1980;**44**:722–38.

44. Mattoo S, Cherry JD. Molecular pathogenesis, epidemiology, and clinical manifestations of respiratory infections due to Bordetella pertussis and other Bordetella subspecies. *Clin Microbiol Rev* 2005;**18**:326–82.

45. Zhao Z, Wang C, Xue Y, et al. The occurrence of Bordetella bronchiseptica in pigs with clinical respiratory disease. *Vet J* 2011;**188**:337–40.

46. Nicholson TL, Brockmeier SL, Loving CL. Contribution of Bordetella bronchiseptica filamentous hemagglutinin and pertactin to respiratory disease in swine. *Infect Immun* 2009;**77**:2136–46.

47. Nicholson TL, Brockmeier SL, Loving CL, et al. The Bordetella bronchiseptica type III secretion system is required for persistence and disease severity but not transmission in swine. *Infect Immun* 2014;**82**:1092–103.

48. Nicholson TL, Brockmeier SL, Loving CL, et al. Phenotypic modulation of the virulent Bvg phase is not required for pathogenesis and transmission of Bordetella bronchiseptica in swine. *Infect Immun* 2012;**80**:1025–36.

49. Shin EK, Jung R, Hahn TW. Polymorphism of pertactin gene repeat regions in Bordetella bronchiseptica isolates from pigs. *J Vet Med Sci* 2007;**69**:771–4.

50. Shin EK, Seo YS, Han JH, et al. Diversity of swine Bordetella bronchiseptica isolates evaluated by RAPD analysis and PFGE. *J Vet Sci* 2007;**8**:65–73.

51. Ellis JA. How well do vaccines for Bordetella bronchiseptica work in dogs? A critical review of the literature 1977–2014. *Vet J* 2015;**204**:5–16.

52. Bemis DA, Carmichael LE, Appel MJ. Naturally occurring respiratory disease in a kennel caused by Bordetella bronchiseptica. *Cornell Vet* 1977;**67**:282–93.

53. Bemis DA. Bordetella and Mycoplasma respiratory infections in dogs and cats. *Vet Clin North Am Small Anim Pract* 1992;**22**:1173–86.

54. Coutts AJ, Dawson S, Binns S, et al. Studies on natural transmission of Bordetella bronchiseptica in cats. *Vet Microbiol* 1996;**48**:19–27.

55. Schulz BS, Kurz S, Weber K, et al. Detection of respiratory viruses and Bordetella bronchiseptica in dogs with acute respiratory tract infections. *Vet J* 2014;**201**:365–9.

56. Brockmeier SL, Register KB, Nicholson TL, et al. Bordetellosis. In: Zimmerman JJ, Karriker LA, Ramirez A, et al. (eds), *Diseases of Swine*. Oxford: Wiley-Blackwell; 2012. pp 670–708.

57. Kobisch M, Pennings A. An evaluation in pigs of Nobi-Vac AR and an experimental atrophic rhinitis vaccine containing P multocida DNT-toxoid and B bronchiseptica. *Vet Rec* 1989;**124**.:57–61.

58. Magyar T, King VL, Kovacs F. Evaluation of vaccines for atrophic rhinitis—a comparison of three challenge models. *Vaccine* 2002;**20**:1797–802.

59. Riising HJ, van Empel P, Witvliet M. Protection of piglets against atrophic rhinitis by vaccinating the sow with a vaccine against Pasteurella multocida and Bordetella bronchiseptica. *Vet Rec* 2002;**150**:569–71.

60. Rutter JM, Taylor RJ, Crighton WG, et al. Epidemiological study of Pasteurella multocida and Bordetella bronchiseptica in atrophic rhinitis. *Vet Rec* 1984;**115**:615–9.

61. de Jong MF, Akkermans JP. Investigation into the pathogenesis of atrophic rhinitis in pigs. I. Atrophic rhinitis caused by Bordetella bronchiseptica and Pasteurella multocida and the meaning of a thermolabile toxin of P. multocida. *Vet Q* 1986;**8**:204–14.

62. Giles CJ, Smith IM, Baskerville AJ, et al. Clinical bacteriological and epidemiological observations on infectious atrophic rhinitis of pigs in southern England. *Vet Rec* 1980;**106**:25–8.

63. Cotter PA, Jones AM. Phosphorelay control of virulence gene expression in Bordetella. *Trends Microbiol* 2003;**11**:367–73.

64. Cotter PA, Miller JF. A mutation in the Bordetella bronchiseptica bvgS gene results in reduced virulence and increased resistance to starvation, and identifies a new class of Bvg-regulated antigens. *Mol Microbiol* 1997;**24**:671–85.

65. Stockbauer KE, Fuchslocher B, Miller JF, et al. Identification and characterization of BipA, a Bordetella Bvg-intermediate phase protein. *Mol Microbiol* 2001;**39**:65–78.

66. Mitscherlich E, Marth EH. Microbial survival in the environment. In: *Bacteria and Rickettsiae Important in Human and Animal Health*: Springer-Verlag; 1984, pp 45–7.

67. Porter JF, Parton R, Wardlaw AC. Growth and survival of Bordetella bronchiseptica in natural waters and in buffered saline without added nutrients. *Appl Environ Microbiol* 1991;**57**:1202–6.

68. Porter JF, Wardlaw AC. Long-term survival of Bordetella bronchiseptica in lakewater and in buffered saline without added nutrients. *FEMS Microbiol Lett* 1993;**110**:33–6.

69. Coote JG. Environmental sensing mechanisms in Bordetella. *Adv Microb Physiol* 2001;**44**:141–81.

70. Hamidou Soumana I, Linz B, Harvill ET. Environmental origin of the genus Bordetella. *Front Microbiol* 2017;**8**:28.

71. Yuk MH, Harvill ET, Miller JF. The BvgAS virulence control system regulates type III secretion in Bordetella bronchiseptica. *Mol Microbiol* 1998;**28**:945–59.

72. Butler JE, Kehrli ME Jr. Immunoglobulins and immunocytes in the mammary gland and its secretions. In: Mestecky J, Lamm M, Strober W, et al. (eds), *Mucosal Immunology*, 3rd ed. Burlington, MA: Elsevier Academic Press; 2005, pp 1776–84.

73. Hornibrook JW, Ashburn LL. A study of experimental pertussis in the young rat. *Public Health Rep* 1939;**54**:439–44.

74. Hall E, Parton R, Wardlaw AC. Cough production, leucocytosis and serology of rats infected intrabronchially with Bordetella pertussis. *J Med Microbiol* 1994;**40**:205–13.

75. Parton R, Hall E, Wardlaw AC. Responses to Bordetella pertussis mutant strains and to vaccination in the coughing rat model of pertussis. *J Med Microbiol* 1994;**40**:307–12.

76. Hall E, Parton R, Wardlaw AC. Differences in coughing and other responses to intrabronchial infection with Bordetella pertussis among strains of rats. *Infect Immun* 1997;**65**:4711–7.

77. Hall E, Parton R, Wardlaw AC. Responses to acellular pertussis vaccines and component antigens in a coughing-rat model of pertussis. *Vaccine* 1998;**16**:1595–603.

78. Hall E, Parton R, Wardlaw AC. Time-course of infection and responses in a coughing rat model of pertussis. *J Med Microbiol* 1999;**48**:95–8.

79. Wardlaw AC, Hall E, Parton R. Coughing rat model of pertussis. *Biologicals* 1993;**21**:27–9.

80. Woods DE, Franklin R, Cryz SJ Jr, et al. Development of a rat model for respiratory infection with Bordetella pertussis. *Infect Immun* 1989;**57**:1018–24.

81. Akerley BJ, Cotter PA, Miller JF. Ectopic expression of the flagellar regulon alters development of the Bordetella-host interaction. *Cell* 1995;**80**:611–20.

82. Horiguchi Y, Okada T, Sugimoto N, et al. Effects of Bordetella bronchiseptica dermonecrotizing toxin on bone formation in calvaria of neonatal rats. *FEMS Immunol Med Microbiol* 1995;**12**:29–32.

83. Martinez de Tejada G, Miller JF, et al. Comparative analysis of the virulence control systems of Bordetella pertussis and Bordetella bronchiseptica. *Mol Microbiol* 1996;**22**:895–908.

84. Cotter PA, Yuk MH, Mattoo S, et al. Filamentous hemagglutinin of Bordetella bronchiseptica is required for efficient establishment of tracheal colonization. *Infect Immun* 1998;**66**:5921–9.

85. Mattoo S, Miller JF, Cotter PA. Role of Bordetella bronchiseptica fimbriae in tracheal colonization and development of a humoral immune response. *Infect Immun* 2000;**68**:2024–33.

86. Weyrich LS, Feaga HA, Park J, et al. Resident microbiota affect Bordetella pertussis infectious dose and host specificity. *J Infect Dis* 2014;**209**:913–21.

87. Weiss AA, Hewlett EL, Myers GA, et al. Tn5-induced mutations affecting virulence factors of Bordetella pertussis. *Infect Immun* 1983;**42**:33–41.

88. Kirimanjeswara GS, Agosto LM, Kennett MJ, et al. Pertussis toxin inhibits neutrophil recruitment to delay antibody-mediated clearance of Bordetella pertussis. *J Clin Invest* 2005;**115**:3594–601.

CHAPTER 7

The human immune responses to pertussis and pertussis vaccines

Françoise Mascart, Violette Dirix, and Camille Locht

Abstract

Two types of pertussis vaccines are currently available: the first-generation, whole-cell (wP) and more recent, acellular (aP) vaccines. The aP vaccine has replaced the wP vaccine in most industrialized countries, based on an improved safety profile and comparable efficacy of the former compared to the latter. Both the aP and wP vaccines, along with prior infection, protect well against disease from future exposure to *Bordetella pertussis*, albeit by different mechanisms, and the human immune response to both natural infection and vaccination has been extensively studies over the past few decades. Shortly after the discovery of the causative agent *B. pertussis*, both agglutinating antibodies and complement-binding antibodies have been identified in the serum of convalescent patients. However, how much they contribute to protection against disease or infection is still not known. Nevertheless, passive transfer of convalescent serum can significantly attenuate the disease and placental transfer of maternal antibodies induced by vaccination during pregnancy has recently been shown to provide strong protection against severe disease in the offspring. Natural infection and wP vaccination have both been shown to induce a strong Th-1-oriented T-cell response, whereas administration of aP vaccine shifts the response to a Th-2 profile, which may be a reason for the fast waning of immunity upon aP vaccination, compared to wP vaccination and natural infection. Neither the wP nor aP vaccines induce life-long, sterilizing immunity and, thus, even in highly vaccinated populations continued circulation of *B. pertussis* remains common. Therefore, new vaccines are needed that protect both against disease and infection. One such candidate, live attenuated BPZE1, designed to prevent *B. pertussis* infection, is currently in clinical development.

7.1 Introduction

It was the specific immune response naturally induced in humans by infection with the whooping cough agent that helped Jules Bordet and Octave Gengou to establish the identity of the causative agent[1] as what was initially referred to as *Bacillus pertussis* and now known as *Bordetella pertussis*. When complement was mixed with heat-inactivated antiserum from convalescent children in the presence of Gram-negative bacilli isolated very early at the onset of the disease, the complement was adsorbed via the antiserum to the bacilli and was no longer available to lyse red blood cells, a phenomenon termed complement deviation. This occurred only with serum from children recovering from whooping cough, and not with serum from children

Mascart, F., Dirix, V., and Locht, C., *The human immune responses to pertussis and pertussis vaccines*. In: *Pertussis: epidemiology, immunology, and evolution*. Edited by Pejman Rohani and Samuel V. Scarpino: Oxford University Press (2019). © Oxford University Press. DOI: 10.1093/oso/9780198811879.003.0007 recovering from other respiratory infections. More-over, serum from convalescent children could agglutinate the whooping cough bacilli, whereas serum from uninfected children or from children suffering from other respiratory infections did not. Besides conclusively establishing the identity of the whooping cough agent, this study also provided the first evidence of a specific immune response induced during pertussis disease. However, early on it was already recognized that the agglutination activities of convalescent antisera varied between children and could differentiate various *B. pertussis* strains[2]. Furthermore, initial attempts to treat per-tussis patients by serotherapy had met with limited success[2]. Nevertheless, these early studies identified two major immune factors in human *B. pertussis*-specific antisera: complement-binding antibodies and agglutinating antibodies. The former may be relevant as effector molecules required for comple-ment-mediated bacterial killing, although the role of this mechanism in protection against pertussis is still not conclusively established. Also, the nature of the complement deviation factor is not yet fully identified, although initial studies suggested that it may be composed of lipids[3]. The *B. pertussis* lipooli-gosaccharide (LOS) is thus a likely candidate for the induction of complement-binding serum antibodies.

Today, we know that the *B. pertussis* factors respon-sible for the induction of agglutinating antibodies, initially called agglutinogens, are essentially fimbriae (Fim), and to a lesser extent pertactin and LOS. *Bordetella pertussis* produces two serotypically distinct Fim, called Fim2 and Fim3[4]. Some isolates produce both, some only one, and some no Fim at all, pro-viding an explanation for the heterogeneity in agglutination assays depending on the isolates used and the antiserum tested.

7.2 Human B-cell responses to *Bordetella pertussis* infection and vaccination

In contrast to the initial unsuccessful attempts[2], subsequent studies have shown that prophylactic treatments with serum from convalescent patients may have some effectiveness in the prevention of the disease in highly exposed children[5]. Similarly, human hyperimmune serum, obtained by repeated injections of a whole-cell pertussis (wP) vaccine preparation, administered to highly exposed infants

provided significant protection, by either prevent-ing whooping cough or markedly attenuating the disease[6]. Shortly thereafter, the active component in the hyperimmune serum was identified as gamma globulin with agglutinating properties. The purified gamma-globulin preparations from human hyper-immune serum were found to provide total or par-tial protection in highly exposed infants, provided they were given before the paroxysmal stage[7]. Some improvement of the symptoms was also observed when the gamma globulins were given to patients with full-blown pertussis disease. These studies clearly established that infection by *B. pertussis* and vaccination with wP induce serum antibodies in humans that have the potential to protect against whooping cough or modify its course to a milder form. Serum antibodies were thus the first immune effectors studied in the context of *B. pertussis* infec-tion. Initially this was done by functional antibody testing, such as bacterial agglutination, complement deviation, and opsonization assays[3]. Agglutinating antibodies and opsonic antibodies induced by infec-tion or vaccination have thus been judged to be important for protection against *B. pertussis*[8], although Waddell and L'Engle warned early on that agglutin-ation responses cannot be used as a measure of immunity to pertussis[9].

The quest for *B. pertussis* antigens that induce protective immunity in humans has continued over the last 70 years and is still ongoing. Over the years, various serological tests were developed using enzyme-linked-immunosorbent assays to detect anti-bodies against various *B. pertussis* antigens, the most frequent being filamentous haemagglutinin (FHA), pertussis toxin (PT), pertactin, and Fim. These tests have helped to detect specific immuno-globulin (Ig)-G in the sera of both infected and vac-cinated subjects, and IgG levels against PT are now currently used for pertussis serodiagnosis[10], as PT is the only *B. pertussis*-specific antigen. Anti-PT IgG is believed to play a major role in protection against whooping cough, as children vaccinated with a monovalent acellular pertussis (aP) vaccine con-taining only PT were better protected against a severe *B. pertussis* infection when their sera con-tained high anti-PT IgG levels[11]. It has, however, still not been possible to define a level of antibodies that would predict protection, and reported antibody levels are highly variable between different studies.

In addition to their titres, antibodies induced by different vaccine types differ by their antigenic specificities, as well as by their affinities. This is well illustrated by results obtained in premature babies vaccinated at 2–3 and 4 months of age either with wP or aP vaccine[12] (Figure 7.1). Whereas in contrast to full-term infants, 6-month-old preterm infants primed with wP vaccine had no detectable serum anti-FHA and anti-PT IgG, significant IgG against a whole-cell *B. pertussis* lysate were detected at 6 months after vaccination (Figure 7.1a, b). In contrast, aP vaccine induced similar anti-FHA and anti-PT IgG titres in preterm and full-term infants at 6 months of age, although their IgG responses to the whole-cell lysate were low. However, even though the antibody levels in both groups of infants were comparable, the avidity indexes of anti-FHA IgG were lower in the preterm than in the full-term infants (Figure 7.1c). Merely measuring specific IgG concentrations in response to a restricted number of antigens without functional evaluation may therefore not be sufficient to assess the quality of the immune responses to vaccines.

Furthermore, compared to IgG, relatively little attention has been paid so far to other immunoglobulin isotypes. As *B. pertussis* is essentially a mucosal pathogen, it is expected to activate mucosal B cells that circulate as IgA-secreting plasmablasts before their final maturation in the mucosa. This was initially shown by Nagel et al.[13], who reported that IgA was induced by *B. pertussis* infection and not by vaccination, and the measurement of serum IgA has been used in Australia for the diagnosis of recent *B. pertussis* infection. However, more recently, anti-PT IgG assays were found to be superior to anti-PT IgA assays for the diagnosis of pertussis from a single sample[14].

Both after wP and aP vaccinations, as well as after *B. pertussis* infection, specific IgG levels decrease quite rapidly. Decay curves of antibody titres are biphasic with a change from a rapid to a slower decay beginning around 3–5 months after an antigen encounter[15]. The antibody decay depends on the type of pertussis vaccine given and on the antigen specificity analysed. The most rapid rate of antibody decrease was observed for anti-PT IgG, whereas longer persistence of antibodies was noted for anti-FHA and anti-pertactin antibodies[16]. Nevertheless, anti-PT levels achieved 1 month after booster vaccination seem to be predictive of the persistence of immunity. In vaccinated children, antibody levels are often low before the preschool booster, approximately 5 years after the previous vaccine injection[17,18,] and detectable antibody levels persist for at least 5 years after the adolescent booster[16]. However, the decline of antibody levels does not necessarily translate into an increase in susceptibility to infection. As some of the early activated B cells evolve into long-term antigen-specific memory B (B_{mem}) cells that continuously circulate in the blood, the enumeration of antigen-specific B_{mem} cells in the blood may therefore perhaps better reflect persistent humoral immune responses than antibody concentrations. These cells can be detected by enzyme-linked immunosorbent spot assay after their *in vitro* differentiation into antibody-secreting cells. They were shown to be present both after infection and in vaccinated children whose antibody levels had already waned[19–21]. However, the number of these B_{mem} cells also progressively decreases after an antigen encounter[20,21], initially quite rapidly, but may remain still detectable 5 years after priming with aP vaccine[22] and even longer after priming with wP vaccine[23]. Furthermore, due to regular boosting, their number progressively increases with age[19]. This was also shown in a cross-sectional study on children and adults with a history of past *B. pertussis* infection, in whom the frequencies of specific B_{mem} cells detected during the acute phase of the infection increased with age[21].

7.3 Human T-cell responses to *Bordetella pertussis* infection and vaccination

Several vaccine efficacy trials have indicated that protection against *B. pertussis* is not always correlated to serum antibody titres[24], suggesting that additional immune mechanisms are involved. As Th1-type immune responses were shown to play a role in the clearance of *B. pertussis* in a mouse model[25], the characterization of *B. pertussis*-induced cellular immune responses in humans has also been the focus of several studies. Convalescent patients were shown to selectively induce antigen-specific Th1 cells[26], which are already present in the acute phase of the infection (i.e. 4–35 days after the onset

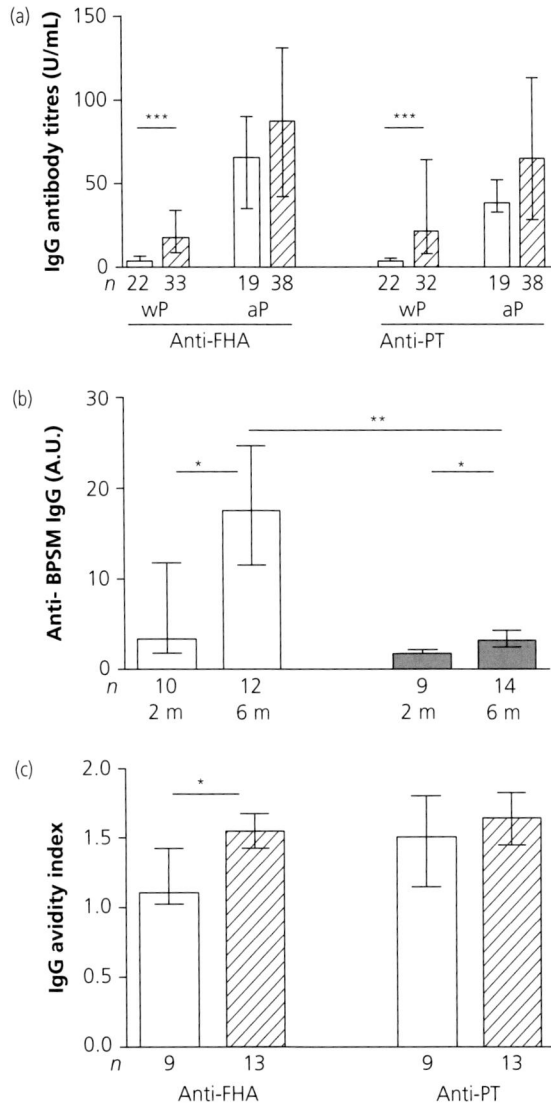

Figure 7.1 Serum anti-*B. pertussis* IgG. IgG titres were measured in sera from highly preterm infants vaccinated with wP or aP, as described in[12]. (a) Anti-FHA and anti-PT IgG titres were compared between 6-month-old preterm infants (open columns) and 6-month-old full-term infants (hatched columns). The cohort, described in[29], was primed at 2, 3, and 4 months of age either with wP or aP vaccines, as indicated. Preterm and full-term infants were compared two by two in each vaccine groups using the Mann–Whitney U test. (b) IgG titres against whole-cell *B. pertussis* lysate (BPSM) were measured in 2-month-old unvaccinated and 6-month-old primed preterm infants, who were vaccinated with wP (white columns) or aP (grey columns). Results were compared two by two by the Mann–Whitney U test for unpaired data and by a Wilcoxon test for paired data. (c) The IgG avidity indexes of anti-FHA and anti-PT IgG were measured as described[12] in 6-month-old infants primed with aP and compared between preterm (open columns, cohort of[12]) and full-term (hatched columns, cohort of[29]) infants. Results were compared two by two by the Mann–Whitney U test. * $p < 0.05$; ** $p < 0.001$; *** $p < 0.0001$. m, months; n = number of infants included in each category. Columns represent medians and 25–75th interquartiles of the results.

of the symptoms)[26,27]. Notably, even very young infants (1–4 months of age) infected with *B. pertussis* are able to mount strong Th1 immune responses to *B. pertussis* antigens[27]. Similarly, a high number of interferon (IFN)-γ-secreting cells are induced upon FHA or PT stimulation of peripheral blood mononuclear cells (PBMCs) from 3-month-old infants vaccinated 2 weeks earlier with wP[27]. In contrast, no secretion of Th2 type cytokines was detected in *B. pertussis*-infected or in wP-vaccinated infants. Even highly premature infants born before 31 weeks of gestational age are able to mount a Th1 response to FHA and PT already at 3–6 months of age, after a first injection (3 months) or first full course of wP vaccination (6 months), contrasting with the absence of detectable anti-FHA and anti-PT IgG in these infants[12].

In contrast to wP, which induces essentially a Th1 pattern of cytokine secretions, priming infants with aP results in a mixed Th1-Th2 profile[28,29,30]. High FHA- and PT-induced interleukin (IL)-5 concentrations are secreted by PBMCs from 6-month-old infants in addition to IFN-γ (Figure 7.2), resulting in a Th2 skewing. This Th2 skewing is not restricted to *B. pertussis* antigens, but extends to other antigens, either administrated within the same vaccine, such as tetanus toxoid, or independent of the vaccine, such as ß-lactoglobulin[29]. It is accompanied by a delayed maturation in aP-vaccinated infants of the IFN-γ-secretion capacity in response to a polyclonal stimulation of the lymphocytes with phytohemagglutinin (PHA), resulting in lower PHA-induced IFN-γ/IL-5 ratios in aP- compared to wP-vaccinated infants[29,30]. These differences in the Th1/Th2 responses to PHA induced

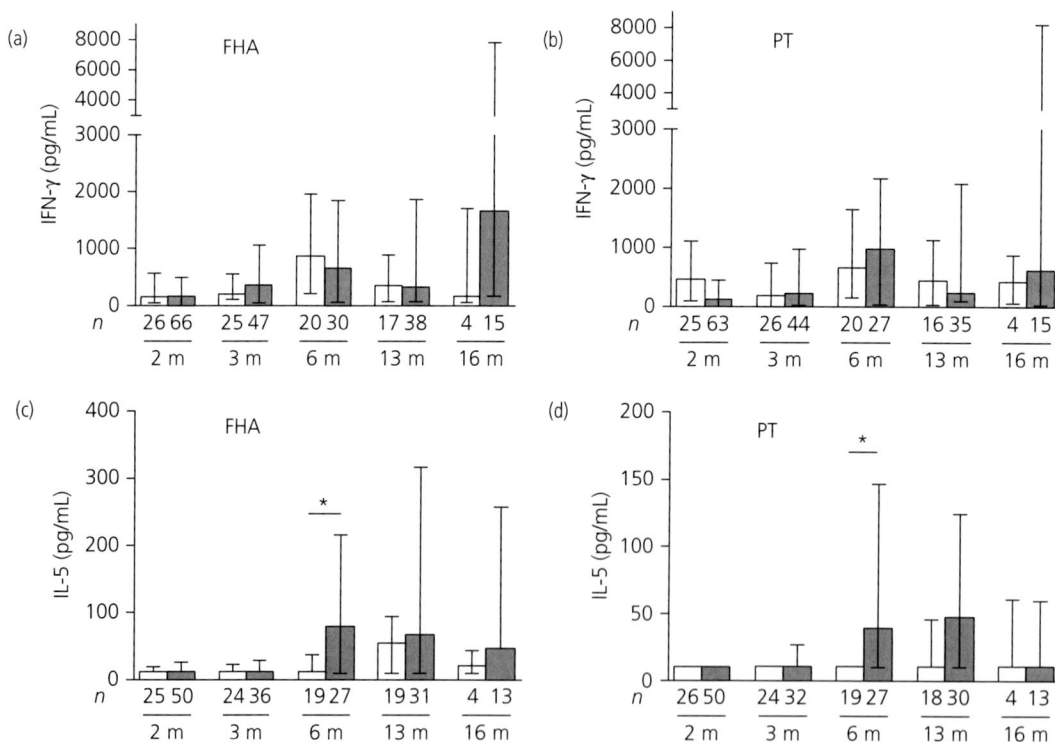

Figure 7.2 Longitudinal study of FHA- and PT-induced cytokines in children primed with wP or aP vaccines. Peripheral blood mononuclear cells (PBMCs) were collected from infants at 2 months, before vaccination, and at 3, 6, 13, and 16 months in two cohorts of infants primed at 2, 3, and 4 months of age with wP (Tetracoq®, Sanofi Pasteur) (white columns) or aP vaccines (Tetravac®, Sanofi Pasteur) (grey columns), and boosted at 15 months with the Tetravac vaccine[31]. PBMCs were *in vitro* stimulated as described[31] with FHA (a, c) or PT (b, d). IFN-γ (a, b) and IL-5 (c, d) concentrations were measured in cell culture supernatants. Columns represent the medians and 25–75th interquartiles of the results. m, months; n = number of infants included in each columns. Results were compared two by two by the Mann–Whitney U test. * $p < 0.05$.

by different pertussis vaccines are no longer observed at 13 months of age, before the first vaccine booster, at least in full-term born infants[31]. A similarly general Th2 imprinting by aP vaccine was observed in highly premature infants, but it persists in this case at least until 16–17 months of age, and the degree of Th2 imprinting is related to the type of aP vaccine administered[12,32]. Furthermore, a neonatal aP vaccine dose given before the routine vaccine administrations is associated with an even stronger Th2 bias in the *B. pertussis*-specific cytokine responses when analysed at 8 months of age[33].

Despite the Th2 imprinting induced by aP vaccine during priming, both wP and aP vaccines induce specific IFN-γ secretions in most infants[29] (Figure 7.2). CD4$^+$ T lymphocytes are the major source of PT- and FHA-specific IFN-γ both after infection and vaccination, but CD8$^+$ T lymphocytes also contribute to this IFN-γ synthesis[27,34,35]. Surprisingly, CD8$^+$ T lymphocytes may also sometimes inhibit the IFN-γ production by the CD4$^+$ T lymphocytes, suggesting a possible role of CD8$^+$ regulatory T cells[35].

Not all 6-month-old vaccinated infants do produce *B. pertussis*-specific IFN-γ. For aP-vaccinated infants, this has been shown to be at least partially the consequence of inhibition by monocyte-derived IL-10[36]. PT induces the secretion of IL-10 by antigen-presenting cells[37], and its pivotal role in the Th1/Th17 balance was demonstrated in a human *in vitro* model[38]. Through a fine-tuning control mechanism of regulatory cytokine expression by antigen-presenting cells, IL-10 not only blocks IFN-γ secretion but also concurrently enhances IL-17 production[38]. It can therefore be anticipated that vaccinated children who fail to induce *B. pertussis*-specific IFN-γ may in fact produce IL-17, a cytokine, which has not yet been investigated thoroughly in *B. pertussis*-infected or vaccinated humans. Only very low secretions of IL-17 were reported to be induced by FHA and pertactin in 4-year-old Dutch children primed with aP vaccine[39].

The duration of *B. pertussis*-specific cellular immune responses has been analysed in several cohorts of children, with the aim to investigate the effect of age on specific T-cell immunity to *B. pertussis*. These studies are difficult to compare as they are essentially cross sectional and influenced by the schedules and types of vaccine administrated, as well as

by the persistent *B. pertussis* circulation, three factors that vary between the different countries where the studies were performed. Moreover, head-to-head comparisons between wP and aP vaccines became progressively impossible, as aP vaccines have almost universally replaced wP vaccines in the countries where most studies are performed.

In an Italian study performed in aP-vaccinated children, memory T-cell responses, as assessed by their lymphoproliferative responses to *B. pertussis* antigens, remained detectable only in a small proportion of children before the preschool booster dose, and IFN-γ concentrations were extremely low[18]. Similar results were more recently obtained in another Italian study reporting very low specific IFN-γ concentrations secreted by PT-stimulated PBMCs from aP-primed children before the preschool boost[40]. This study compared two groups of children primed with two different aP vaccines and found that significant PT-induced lymphoproliferation was still detectable in children primed with the Hexavac®, whereas this was not the case for those primed with Infanrix hexa®. These results stress again the influence of even minor differences in the composition of aP vaccines on the T-cell responses induced. In contrast, a Dutch study reported persistence in aP-primed 4-year-old children (Infanrix®) of significant *B. pertussis*-induced cytokine secretions (tumour necrosis factor (TNF)-α, IFN-γ, IL-5, IL-17, IL-10) by PBMCs, responses that were, however, significantly higher than those seen in wP-primed children[39]. Another cross-sectional study performed in Belgium, analysed the specific T-cell responses at the cellular level by flow cytometry. Similarly to the Dutch study, the percentages of IFN-γ-containing CD4$^+$ T lymphocytes detected after *in vitro* stimulation with FHA or PT were still significant in 5–7-year-old children primed with aP vaccine (Infanrix hexa®) and who had not yet received a booster dose.

Similarly to the booster vaccination at 15 months[31], preschool boosters with aP in aP-primed children were reported to have only a marginal effect on the *B. pertussis*-induced IFN-γ responses[39,40]. The analysis of IFN-γ production at the cellular level confirmed this marginal effect, which reaches statistical significance only for the percentages of FHA-induced IFN-γ-containing CD4$^+$ T lymphocytes

($p < 0.05$) of aP-primed children boosted between 5.5 and 8.2 years (Tetravac® or Infanrix®-IPV) (Figure 7.3). In contrast, the booster dose may have a major effect on the Th2-type cytokine production, which significantly increases after an aP boost in aP-primed children, and thereby augments the Th2 skewing, with, in turn, was reported to be associated with an increased risk for injection site reactions[34,41,42]. No such increase in Th2 cytokine production early after the preschool booster was

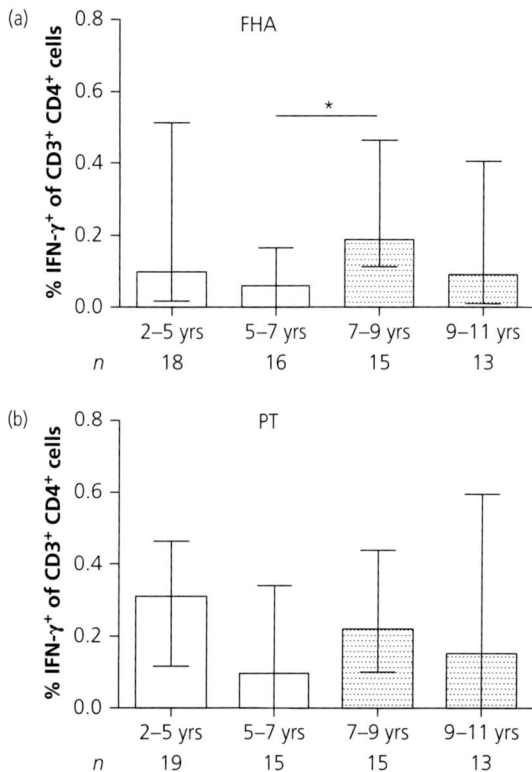

Figure 7.3 Cross-sectional study of FHA- and PT-induced IFN-γ-containing CD4+ T lymphocytes in children primed with aP vaccine. Blood was collected from two groups of children (2–5 years and 5–7 years) primed with Infanrix-hexa® at 2, 3, and 4 months of age and boosted at 15 months (white columns), and from two groups of children (7–9 years and 9–11 years) primed with Tetravac®, boosted around 15 months with Tetravac®, and around 7 years with a Tetravac® or with an Infanrix®-IPV (dotted columns). The proportion of (a) FHA- and (b) PT-induced IFN-γ-containing CD4+ T lymphocytes was analysed by flow cytometry after *in vitro* stimulation with the antigens, as described[44]. Columns represent the medians and 25–75th percentiles of the results. n = number of infants included in each columns. These unpublished results were obtained within a study partially reported[44].

noted in wP-primed children[34], suggesting that the priming course of immunization may have an impact on the immune response generated after booster immunization.

Only a few studies have characterized the long-term effects of the preschool booster dose on the *B. pertussis*-induced T-cell responses. Two years post booster, PT- and FHA-induced IFN-γ secretions were reported to be higher than pre-boost values both for wP- and aP-primed Dutch children, whereas this was not the case early after the booster dose, at least for aP-primed children[23]. This study also reported similar concentrations of *B. pertussis* antigen-induced Th2 cytokines in wP- and aP-primed children 2 years after the booster. Another Dutch study indicated that almost all T-cell responses increase with age in Dutch wP-primed children, which are higher in 9-year-old than in 4-year-old children just after the preschool booster[43]. The persistence of specific T-cell responses in most aP- or wP-primed preadolescents (9–11 years old) was also reported in a Belgian study, in which, however, wP-primed children had more frequently persistent antigen-specific cytokine responses than aP-primed children[44]. The time elapsed since the last vaccine booster was significantly longer for wP- compared to aP-vaccinated children[44], indicating that, even if the *B. pertussis*-specific cellular immune responses of preadolescents may be influenced by high circulation of *B. pertussis*, wP-primed children display longer-lasting cellular immune responses than aP-primed children. Specific cellular immune responses analysed by lymphoproliferation were also reported to be persistent for at least up to 5 years after an adolescent booster dose[16].

Very little is known about cell-mediated immune responses to *B. pertussis* antigens in adults. In a limited study, most healthy Italian adults were reported to have *B. pertussis*-specific T-cell responses characterized by typical production of Th1 cytokines[45], and this was attributed to exposure-induced immunity, which was presumably high as the country had low vaccine coverage of infants. Repeated exposure could lead to progressively higher T-cell responses to *B. pertussis* antigens in adults compared to children, as suggested by higher responses detected in 9-year-old compared to 6- and 4-year-old Dutch children[43]. However, during an acute infection with

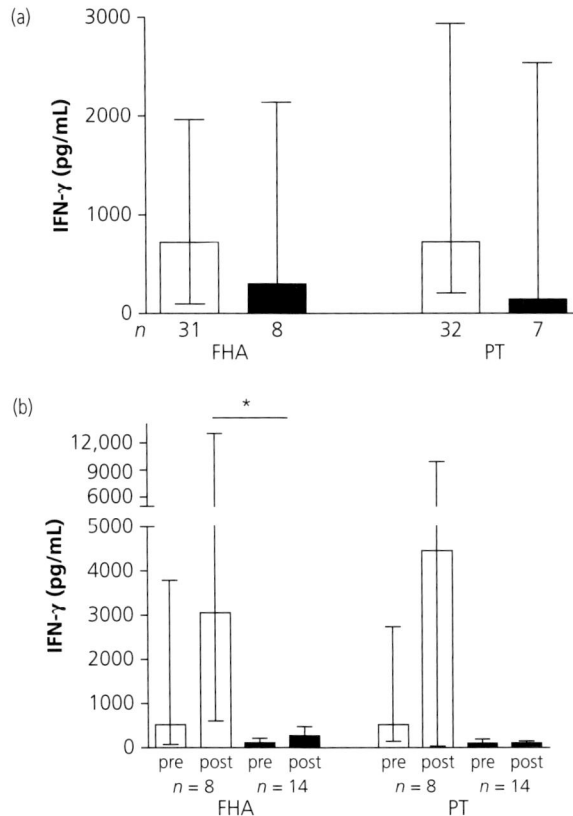

Figure 7.4 Comparison between FHA- and PT-induced IFN-γ secretions in children and adults. Peripheral blood mononuclear cells (PBMCs) were collected from infants[27] (white columns) and from adults (black columns) during the acute phase of whooping cough (a), as well as from 1-year-old children before and after the Tetravac® 15 month-booster[31] (white columns) compared to adults before and after a Boostrix® (black columns) (b). The PBMCs were stimulated *in vitro* with FHA or PT and IFN-γ concentrations released in the cell culture supernatants were measured[27,31]. Columns represent the medians and 25–75th percentiles of the results. n = number of infants included in each columns. * $p < 0.01$.

B. pertussis, the FHA- and PT-induced IFN-γ secretions are not higher in adults compared to infants (Figure 7.4). Similarly, the FHA- or PT-induced IFN-γ secretions by PBMCs isolated 1 month after a booster dose are lower in adults than in 16-month-old children. Even though the adults were boosted with a Boostrix® and the children with a Tetravac®, having a higher antigenic content, these results suggest that IFN-γ responses from children are not dramatically lower than those of adults.

7.4 Current pertussis vaccines

7.4.1 Whole-cell pertussis vaccines

The first attempts to develop a pertussis vaccine were undertaken shortly after the description of

B. pertussis by Bordet and Gengou[1], with, however, variable success, which led the American Medical Association to withdraw its approval of pertussis vaccines already on the market in the 1930s[46]. The first convincing evidence for a protective efficacy of a home-made pertussis vaccine came from the studies reported in the 1930s by Madsen[47] and Sauer[48], and later by Kendrick and Eldering[49]. In these successful studies, wP vaccines were prepared from freshly grown *B. pertussis* organisms on blood agar plates suspended in solutions containing phenol or merthiolate. It was recommended that vaccines should be prepared only from recently isolated, strongly haemolytic bacilli grown on blood-containing media, and that commercial vaccines made from old laboratory strains should not be used[48].

However, even when vaccines were carefully prepared from freshly grown, recent phase I isolates, they were not always effective in preventing whooping cough[50], illustrating the difficulties in manufacturing consistently potent pertussis vaccine lots. In addition to the importance of the bacterial growth conditions, vaccine producers also have to consider the impact of different detoxification methods. Early non-commercial vaccine preparations that induced significant protection also induced leucocytosis in the vaccinated children, at a level comparable to that seen during whooping cough[48]. Leucocytosis has long been recognized as a hallmark of pertussis[51] and is caused by PT, one of the major virulence factors of *B. pertussis* (reviewed in[52]).

During the 1940s, the pertussis vaccine manufacturing processes became more and more standardized and manufacturing control protocols were established. Animal models were developed to assess potency and safety. A mouse intracerebral challenge test was established to evaluate potency of the vaccines, in which vaccinated mice were injected with 40,000 colony-forming units (CFU) of the *B. pertussis* 18323 strain, and death was monitored over 2 weeks[53]. A mouse weight gain test was used to assess residual toxicity of vaccine preparations (reviewed in[54]). This test is based on the fact that intraperitoneal injection of pertussis vaccine results in initial weight loss, which should be recovered 3 days later. The World Health Organization (WHO) gathered expert groups to draft recommendations for vaccine producers[55]. Large controlled efficacy trials were conducted by the British Medical Research Council, which concluded that the five vaccines tested were efficacious against pertussis, especially against the most severe forms of the disease[56]. Furthermore, these studies found a significant correlation between protection in children and the potency in the intracerebral challenge model[57], which has subsequently been used as a lot-release criterion. Interestingly, the protective effects of wP in children also correlated with agglutinin production in mice, pointing to a protective role of Fim.

Since then, wP vaccines have been widely used in all parts of the world, and thanks to the extended programme on immunization of the WHO, pertussis vaccine coverage, combined with diphtheria and tetanus vaccines, referred to as DTwP, has now reached more than 85 per cent of infants[58]. This has led to a spectacular decrease in pertussis incidence in all countries in which routine pertussis vaccination was implemented[59], demonstrating the effectiveness of wP vaccination in reducing pertussis incidence. Conversely, in countries, such as Sweden, where pertussis vaccination was interrupted, the incidence of the disease immediately increased again[60].

Despite the undisputable effectiveness of the wP vaccine, and despite rigorous control procedures for vaccine lots, variations in wP vaccine efficacy can still occur. In efficacy trials conducted in the 1990s, where wP vaccine was used as a comparator of the newer aP vaccine, efficacy of the wP vaccine varied from greater than 95 per cent against severe pertussis in one study[60] to less than 50 per cent in other studies[62,63], although in both cases the vaccines had successfully passed the intracerebral challenge test. Moreover, in parallel with the decline of pertussis in vaccinated populations, adverse publicity was voiced about alleged potentially severe side effects of wP vaccine, which has led to a mistrust of the vaccines and consequently a drop in compliance in several countries[59]. The adverse events ranged from usually mild local reactions, such as local redness, swelling, and pain, to systemic reactions, mostly fever and drowsiness, and in rarer cases, vomiting and persistent crying[64]. All of these were more frequent in children vaccinated with DTwP vaccines than with DT vaccines lacking the pertussis component.

7.4.2 Acellular pertussis vaccines

In addition to decreasing vaccine uptake, these allegations of wP-induced side effects, whether justified or not, also resulted in efforts to develop safer, less reactogenic vaccines, which ultimately led to the generation and use in countries that can afford them, of defined aP vaccines. Better-defined aP vaccines were first developed in Japan in the 1970s and then used for mass vaccination in that country since 1981. The Japanese vaccine contained essentially FHA and PT detoxified with formaldehyde, had a low lipopolysaccharide content, and was formulated with aluminium hydroxide[65]. This vaccine showed potency comparable to that of wP vaccine in the mouse intracerebral challenge model, but was less reactogenic, both in animals and in children. Its efficacy in protection against pertussis disease in children was also comparable to that of wP vaccine[66].

However, in Japan, aP vaccine was initially only used in 2-year-old children, and the first safety and efficacy trials in younger children were carried out in Sweden in the second part of the 1980s[24]. In the first Swedish trial, two aP vaccines were compared, one containing only detoxified PT, and the other containing PT and FHA. The study showed that both aP vaccines caused fewer adverse reactions than wP vaccine and showed significant protection, but the two-component vaccine showed slightly better protection (roughly 70 per cent) than the mono-component vaccine (roughly 55 per cent). Protection was better against more severe disease and, for both vaccines, reached approximately 80 per cent against culture-confirmed pertussis with greater than 30 days of cough. The study failed to show any correlation between post-vaccination serum antibody levels and protection in children. A 3-year, unblinded passive surveillance study of the trial cohort confirmed that the two-component vaccine provided better protection against culture-confirmed pertussis than the mono-component vaccine[67]. A subsequent trial with a mono-component aP vaccine combined with diphtheria and tetanus toxoids (DTaP) confirmed a vaccine efficacy against laboratory-confirmed pertussis of 55 per cent after two doses but showed efficacy of over 70 per cent after a third dose[68]. The vaccine failures in this study had less severe disease than the patients in the control group, and 4 weeks after the third dose the anti-PT IgG levels were higher in the children that did not develop pertussis compared to those that later had pertussis.

In the early 1990s, several other placebo-controlled, double-blind efficacy trials were conducted in Sweden[62,69], Italy[63], Germany[70], and Senegal[61], in which different aP vaccines were compared head to head, sometimes also to the wP vaccine. A study carried out in Senegal[61], so far, the only efficacy trial conducted in an African country, indicated a significantly lower protective efficacy after three doses of a two-component aP compared to wP vaccine, with efficacy rates against laboratory-confirmed pertussis with 21 days or more of paroxysmal cough of 96 per cent and 85 per cent for wP and aP vaccines, respectively. Gustafsson et al.[62] compared two DTaP vaccines (a two-component and a five-component

vaccine) with a DTwP vaccine routinely used in the United States. The five-component vaccine contained, in addition to PT and FHA, pertactin and both serotypes of FIM. The trial confirmed that rates of adverse reactions were significantly higher for DTwP than for DTaP, and vaccine efficacy of the two-component DTaP was close to 60 per cent after three doses and 85 per cent for the five-component vaccine. DTwP performed very poorly in this study, with an efficacy of less than 50 per cent. A similarly low efficacy of DTwP was also found in the Italian trial[63], which included the same wP vaccine as a comparator to two other DTaP. The main difference between the two DTaP was the mode of PT detoxification, a chemical detoxification for one of them and a genetic detoxification for the other. Both vaccines offered close to 85 percent protection after three doses. There was no statistically significant difference between the two DTaP vaccines, although the genetically detoxified vaccine induced higher levels of PT-neutralizing antibodies than the chemically inactivated vaccine. A subsequent trial, again performed in Sweden, directly compared three doses of a two-component, a three-component, and a five-component aP vaccine with a wP vaccine from a different manufacturer[69]. In this study, the five-component and the three-component vaccines showed equally high efficacy as wP vaccine against culture-confirmed pertussis with 21 days or more of cough, but the three-component vaccine was less efficacious than the wP vaccine comparator when all culture-positive cases were considered, irrespective of cough.

The general consensus emerging from these and other trials[70–72] was that a two-component might be more efficacious than a mono-component pertussis vaccine, and that the addition of pertactin and/or Fim might further increase vaccine efficacy[73]. Protection, especially against mild disease, seemed to increase with the number of vaccine components[69], and the best aP vaccine appeared to provide protection at levels similar to good wP vaccines with considerably less severe reactions. Therefore, many industrialized countries progressively switched from wP to aP vaccines, and by the early 2000s, almost all European countries, the United States, Canada, Australia, and Japan had switched to aP vaccines for the primary immunization series.

7.4.3 Serological correlates of protection

Given the high global vaccination coverage with efficacious DTwP or DTaP, efficacy trials using a placebo control seem no longer possible. Unfortunately, efforts to identify serological correlates of protection in these clinical trials have largely remained unsuccessful, although the earlier studies showed some level of correlation between clinical protection induced by wP vaccine and post-vaccination serum agglutinin titres[56,57]. Intuitively, serological correlates of protection should be easier to establish with molecularly defined aP vaccines but the initial efficacy trials failed to do so. Among several explanations, one may be that protective immunity to pertussis not only relies on serum antibodies, but also on other effector mechanisms, including cell-mediated immunity[25]. Nevertheless, follow-up logistic regression analyses using the data of some of these trials with five-component aP vaccine identified a statistically significant correlation between clinical protection and post-vaccination serum IgG to pertactin, Fim, and PT, but not to FHA[74]. The strongest regression coefficient was found for the anti-pertactin antibodies, and the weakest for the anti-PT IgG. These observations are in line with the Italian study, in which two vaccines that induced significantly different anti-PT serum IgG showed similar protective efficacy[63]. Thus, the major serological correlate of aP vaccine-induced protection may be serum antibodies to Fim and pertactin, consistent with the findings that three- and five-component aP vaccines offer better protection than mono- or two-component vaccines. However, using a multivariate analysis, Cherry et al.[75] reported that only antibodies to pertactin and PT were significantly correlated with protection. They also reported that subjects with low anti-pertactin serum antibodies, regardless of the presence of anti-PT IgG, had a higher chance of *B. pertussis* infection than subjects with high anti-pertactin antibodies. In this household study, no pertussis occurred in subjects with both high anti-pertactin and high anti-PT antibodies. From these studies, it was concluded that the presence of high anti-pertactin and anti-Fim serum antibodies correlates with protection against typical and mild pertussis, whereas anti-PT antibodies correlate only with protection against typical

pertussis[76]. Anti-FHA antibodies did not correlate with protection, which is in contrast to the conclusion of the efficacy trials showing that a two-component vaccine provided superior protection over a mono-component vaccine[24], suggesting that FHA contributes to protection.

7.4.4 Waning of vaccine-induced immunity

In spite of all the efforts to develop safe and effective vaccines and despite a wide global coverage of greater than 85 per cent of pertussis vaccination, the disease is not under control and has not been eliminated in any country. In fact, in recent years, it has made a spectacular come-back in several countries, especially in the industrialized world, where exclusively aP vaccines are used[77,78]. In particular, the incidence of pertussis among fully vaccinated preadolescents and adolescents has substantially increased in some of those countries, although globally the highest rates remain in infant populations. Faster waning of aP-induced compared to wP-induced immunity became evident in the early 2000s[79]. A case–control study undertaken in California on 4–12-year-old children, all vaccinated with five doses of DTaP, showed a significant increase in the odds of acquiring laboratory-confirmed pertussis each year after the fifth dose[80], demonstrating unexpectedly fast waning of aP vaccine-induced protection. A parallel study, also carried out in California, confirmed this rapid waning and showed a reduction in vaccine effectiveness after the fifth dose from 98 per cent during the first year to barely 70 per cent after 5 years[81]. In an Australian case–control study, three doses of DTaP were found to result in roughly 80 per cent effectiveness in 1-year-old infants, but less than 60 per cent in 3-year-old children, which was substantially lower than the effectiveness of DTwP in 3-year-olds[82]. A convincing illustration of the differential longevity of protective immunity by the two types of vaccines was reported by Sheridan et al.[83] who showed, after observing pertussis cases in vaccinated children over more than 10 years, that those who had received three doses of DTaP in the primary course had substantially higher rates of pertussis than those who had received three doses of DTwP. Individuals who had received a mixed course, with DTwP as the initial dose, had intermediate incidence rates. These differences persisted for more than a decade.

7.4.5 Repeated vaccination, cocooning, and maternal immunization

Considering the short-lived immunity induced by aP vaccination, repeated vaccinations with DTaP have been proposed. However, it has been observed for more than 20 years that, although DTaP induces less adverse reactions than DTwP, the frequencies of both local and systemic adverse reactions increase with the number of doses administered[84]. Therefore, aP vaccines with reduced antigen contents (Tdap) were developed. They were shown to be safe in adolescents and adults, yet able to induce strong serum antibody responses to *B. pertussis* antigens[85]. A large multicentre, randomized, double-blind efficacy trial provided a protective efficacy of over 90 per cent over a period of 2.5 years after a single booster dose of a stand-alone reduced antigen content aP vaccine[86].

Since the recommendation to use Tdap for teenagers and adults, several effectiveness studies have been undertaken. Wei et al.[87] reported effectiveness of Tdap vaccination of approximately 68 percent during a school outbreak in children 11 years of age or more. All children had received a full course of DTaP vaccination in infancy. An Australian field study on DTwP-primed teenagers indicated a Tdap booster effectiveness of more than 85 per cent within the first 2 years after vaccination[88], suggesting that the type of vaccine used for the primary course may influence the protective immunity boosted by Tdap. A population-based effectiveness study gave a Tdap effectiveness of 65 per cent among the 11–12-year-olds, of 47 per cent among the 13–16-year-olds, and of 66 per cent along the 17–19-year-olds[89]. The former two groups most likely had received DTaP for the first series in their infancy, whereas the latter group was likely to have been primed by DTwP, confirming a greater Tdap booster effect in DTwP-primed than in DTaP-primed subjects.

Although Tdap booster vaccination is effective in reducing pertussis rates in adolescents and adults, the duration of this effectiveness is not known.

Furthermore, Tdap vaccination of adolescents and adults has limited effects on pertussis incidence in infants. Since 75 percent of infants infected with *B. pertussis* acquire their infection from a household contact, mostly from their mothers[90], it was thought that post-partum vaccination with Tdap of the mother, but also vaccination of all household contacts of a newborn, a so-called cocoon strategy, would significantly reduce the incidence of pertussis in infants less than 3 months of age[91]. However, cocooning has been notoriously difficult to implement and, so far, has failed to show any effectiveness[92].

A more successful strategy to limit infant pertussis by using Tdap appears to be maternal immunization. This strategy is based on the concept that protective antibodies generated by Tdap vaccination during pregnancy can be actively transported through the placenta to the yet unborn child and thereby protect the infant from birth on until the first course of active vaccination has initiated. The first evidence of the effectiveness of maternal Tdap immunization was obtained after an important pertussis outbreak throughout England, where vaccine effectiveness against pertussis in infants less than 3 months old reached greater than 90 per cent[93]. Other studies have since confirmed the high effectiveness of maternal vaccination to protect neonates against pertussis, but also pointed to the importance of timing of maternal immunization[94,95] and suggest that the ideal time window for maternal immunization would be 27–31 weeks of gestation. Several countries have now recommended maternal immunization in the third trimester of pregnancy.

As promising as these studies are, they also show that universal maternal immunization is difficult to implement[96]. In addition, many issues still need to be addressed. Given the short half-life of anti-PT IgG[97], maternal immunization at first pregnancy is likely to be insufficient to protect infants born after subsequent pregnancies. Therefore, maternal immunization at each pregnancy may be required. However, it is currently unknown whether immunogenicity and safety of maternal Tdap vaccination at multiple pregnancies remains similar to those at the first pregnancy. Tdap vaccination during pregnancy is considered safe for both mother and infant, but reactions to the vaccination do occur relatively frequently, which may lead some mothers to refuse repeat vaccination in a future pregnancy[98]. It remains to be seen whether these reactions increase in severity and/or frequency at each pregnancy. Furthermore, all infant–mother pairs so far investigated included DTwP-primed mothers. No information is currently available for infant–mother pairs with DTaP-primed mothers. This open question is particularly relevant considering the fact that the Tdap booster effect is greater in DTwP-primed than in DTaP-primed subjects[89]. Finally, the effect of maternal immunization on the primary course of infant vaccination also needs careful attention. Earlier studies have shown an inverse relationship between maternal antibodies and antibodies induced in children after DTwP vaccination[99]. Although this seems to be less the case for the antibody responses to aP vaccination, a certain degree of blunting by maternal antibodies has also been reported for the response to DTaP[100]. The clinical relevance of this blunting is currently unknown, but if maternal immunization reduces vaccine-induced immunity in the infants after the primary course, it may ultimately give rise to an increase in pertussis incidence in older children[101]. Finally, whether maternal immunization influences T-cell responses induced by the primary course of infant vaccination has also not yet been studied.

7.5 Improved pertussis vaccines

Although maternal Tdap immunization in the beginning of the third trimester seems to be currently the best option to limit pertussis in the most vulnerable age group, it is unlikely to solve the global occurrence of pertussis in the long run. Tdap is unlikely to stop the transmission cycle of *B. pertussis* because of its inability to prevent infection and asymptomatic transmission of the organism, as recently shown in a non-human primate model[102] and discussed in detail elsewhere[103]. Moreover, a mathematical modelling study concluded that asymptomatic transmission is the most parsimonious explanation for the resurgence of pertussis in countries with high-vaccination coverage[104]. Therefore, for the ultimate control of pertussis, vaccines are needed which not only protect against pertussis symptoms, but also effectively limit infection and thereby induce a sufficient level of herd immunity to prevent transmission.

7.5.1 New formulations, new antigens

Since the introduction of DTaP and Tdap, no novel pertussis vaccine has reached clinical use. Several new avenues have been explored, but generally, they have not proceeded beyond initial studies in mice, as recently discussed[105]. wP vaccines with reduced endotoxin contents have been shown to be less reactogenic in mice, without affecting potency[106]. However, in a head-to-head trial in children, comparing conventional wP vaccine with reduced endotoxin wP vaccine, both vaccines showed similar reactogenicity and immunogenicity[107]. *Bordetella pertussis* outer membrane vesicles containing many of the major antigens, including LOS, have been shown to be less reactogenic than wP[108] and, combined with aluminium hydroxide, to provide strong protection in mice[109], but have not yet undergone clinical evaluations in humans.

Attempts to improve aP vaccines have also been undertaken. Based on the differences in the type of immune responses elicited between wP and aP vaccines, various formulations and adjuvants have been tested in an effort to bring the immune responses induced by aP vaccine close to those elicited by wP vaccine. Formulations in different micro- or nanoparticles and the addition of Toll-like receptor (TLR) ligands were shown to enhance and accelerate antibody responses in mice, such as IgG2a, indicative of Th1 induction, as well as Th1 and Th17 cytokines[110–114]. An interesting, recently discovered TLR-2 agonist with adjuvant properties is the *B. pertussis* lipoprotein BP1569[115], and a synthetic lipopeptide corresponding to the N-terminal part of BP1569 was found to enhance the protective effect of aP vaccines in mice.

The effect of additional antigens to aP vaccine has also been investigated. The adenylate cyclase toxin, a strong *B. pertussis* immunogen[116], combined with aP vaccine increased the potency of aP vaccine in mice[117]. The outer membrane protein BrkA, although not protective on its own, improved the protective potential of a two-component, but not of a three-component aP vaccine in mice[118]. The addition of the iron-regulated proteins IRP1-3[119] and AfuA[120] were also reported to provide protection in mice and, when added to aP, augmented its potency. Finally, LOS has been considered based on the fact it is a major *B. pertussis* surface antigen and able to induce bactericidal antibodies. The core oligosaccharide portion of the *Bordetella* LOS conjugated to a carrier protein was indeed found to induce bactericidal antibodies in mice[121], but so far, its protective effect has not yet been assessed.

7.5.2 Live attenuated pertussis vaccines

With the exception of a reduced endotoxin wP vaccine, none of the above-mentioned formulations has yet undergone clinical evaluation in humans. It is also not known whether any of them will prolong vaccine-induced immunity over the current vaccines or whether they will significantly reduce carriage and prevent transmission, two of the major objectives for new pertussis vaccines[104,122].

Studies with baboons have shown that, in contrast to vaccination with aP or wP, prior infection leads to sterilizing immunity[102]. Furthermore, many studies have indicated that infection-induced protection is longer lasting than vaccine-induced protection[123,124], although none of them result in lifelong protection. Therefore, the ultimate protection against whooping cough may perhaps be best achieved by live *B. pertussis* inoculation, as already proposed by Huang et al. in 1962[125]. This notion provided the basis for the development of live attenuated nasal vaccines[126].

The first attempt to develop a live pertussis vaccine was based on the observation that auxotrophic mutations may strongly attenuate pathogens. Therefore, a *B. pertussis aroA* mutant has been constructed and tested in mice[127]. Compared to the parental strain, the *aroA* mutant was unable to efficiently colonize the mouse lung, but induced a certain level of protection after multiple nasal applications. Cornford-Nairns et al.[128] targeted *aroQ*, another gene in the prechorismate biosynthetic pathway, and found that an *aroQ* mutant persisted longer in the lungs than the *aroA* mutant and that it induced protection in mice after a single nasal administration.

An alternative way to genetically attenuate *B. pertussis* is to target bona fide virulence factors. This approach was taken to develop BPZE1[129], currently the most advanced new pertussis vaccine candidate. BPZE1 was constructed by genetically inactivating or removing three major *B. pertussis* toxins: PT,

dermonecrotic toxin, and tracheal cytotoxin. These genetic alterations did not affect the ability of the strain to colonize the mouse respiratory tract, but strongly attenuated the organism, and, given as a single nasal dose, protected mice against virulent *B. pertussis* as well as against infection by *B. parapertussis*[129] and *B. bronchiseptica*[130]. Importantly, nasal administration of BPZE1 also induced specific IgA in the bronchoalveolar lavage fluids, whereas they were not induced by DTaP[131]. This observation may be relevant, as *B. pertussis* is a strictly mucosal pathogen, and disseminated disease has only rarely been observed[132]. Therefore, the induction of mucosal immunity may play an important role in protection against *B. pertussis* infection. BPZE1 has undergone extensive preclinical safety and efficacy evaluation in mice (reviewed in[105,133]), indicating that BPZE1-induced immunity is long lived[134], comparable to that induced by prior infection with virulent *B. pertussis*[135], and that it induces bactericidal antibodies[136]. Recently, BPZE1 has also been tested in the baboon model and was found to induce strong protection after a single administration, both against pertussis disease and nasopharyngeal colonization by a highly virulent *B. pertussis* isolate[137].

Furthermore, BPZE1 was found to be safe in both animal models, as well as in severely immunocompromised mice, including IFN-γ receptor-deficient mice[138]. Interestingly, instead of exacerbating them, as seen with virulent *B. pertussis*, it also protected mice against heterologous, inflammatory diseases, such as pneumonia caused by influenza A virus[139], respiratory syncytial virus-induced bronchiolitis[140], asthma[141,142], and contact hypersensitivity of the skin[142]. The mechanism of this anti-inflammatory activity of BPZE1 is not known but is possibly related to the ability of the vaccine strain to induce unconventional suppressor T cells that express ecto-enzymes, which use ATP and NAD$^+$ as substrates to generate adenosine[143], a molecule with potent anti-inflammatory properties[144]. This may explain why BPZE1 displays anti-inflammatory properties without being immunosuppressive. Alternatively, BPZE1-treated human monocyte-derived dendritic cells have been shown to induce *in vitro* a population of T lymphocytes that are able to suppress the proliferation of syngeneic T cells by a cell contact-dependent mechanism[145]. Together with the well-documented genetic stability of BPZE1[146], these safety data have allowed BPZE1 to be downgraded from a biosafety level 2 organism to a biosafety level 1 organism in several countries, a prerequisite for its clinical development.

7.5.3 Clinical development of BPZE1

A first-in-man, double-blind, placebo-controlled, dose-escalating trial has been successfully completed and shown that BPZE1 given as 100 µL nasal drops containing up to 10^3, 10^5, or 10^7 CFU is safe in healthy adult volunteers and able to transiently colonize the human nasopharynx and to induce serum antibody responses to PT, FHA, pertactin, and Fim[147]. No statistically significant differences in general safety and local tolerability in the respiratory tract were noted between any of the vaccine dosage groups and the placebo group. Nasopharyngeal colonization was detected in 1 out of 12 subjects in the low-dose, 1 out of 12 in the medium-dose, and 5 out of 12 in the high-dose group. Colonization was transient and usually lasted for approximately 2 weeks.

All colonized subjects significantly increased their serum antibody titres to the *B. pertussis* antigens tested, and the antibody titres at 6 months were at least as high as at 1 month after vaccination. Most vaccinees seroconverted to all four antigens. In contrast, no increase in serum antibodies to any of these antigens was seen in the non-colonized subjects, neither in the placebo group, nor in the vaccine groups. Interestingly, the antibody titres in the colonized low-dose vaccinees were as high as those of the high-dose subjects, suggesting that, similar to what was observed in baboons[137], serum antibody titres are not strictly dose dependent. Interestingly, the proportion of colonized vaccinees who produced increased levels of anti-Fim serum IgG was substantially higher than that seen after natural infection with *B. pertussis*[148]. Since anti-Fim antibodies have been linked to protection[4], this may be an encouraging observation for the vaccine potential of BPZE1.

Although this initial clinical trial showed promise, even with the highest dose tested, only roughly 40 per cent of the study subjects were colonized. This may be due to several reasons. It is possible that the optimal BPZE1 dose for humans is higher than 10^7 CFU, and/or that the volume used was suboptimal.

It is also possible that pre-existing immunity may have hampered vaccine take in some of the subjects, as suggested by the fact that prior to BPZE1 administration anti-FHA, anti-pertactin, and anti-Fim IgG levels were significantly higher in the non-colonized than in the colonized subjects. As this study was conducted in Sweden on volunteers who were born at the time this country had completely stopped pertussis vaccination, elevated antibody titres to the *B. pertussis* antigens could not be due to vaccination, but were most likely the result of prior *B. pertussis* exposure.

To address these hypotheses, a second clinical trial is currently under way, in which higher volumes (400 µL) are being used, and the doses are increased from 10^7 CFU to 10^8 to 10^9 CFU. Furthermore, subjects with high pre-existing serum antibody levels to PT or pertactin have been excluded. However, a fourth group has been added in this trial with subjects having high pre-existing anti-pertactin antibody levels, in order to determine whether an increase in dose and/or volume can overcome the blockade by prior *B. pertussis* exposure. Hopefully this second trial will provide guidance for further clinical development of BPZE1.

T-cell responses in BPZE1-vaccinated volunteers have not yet been analysed. However, B-cell responses have been studied in these volunteers, and a significant increase in antigen-specific plasmablasts was seen in blood 7 and 14 days after BPZE1 administration in the colonized, but not in the non-colonized subjects[149]. Furthermore, in all colonized volunteers a significant increase in antigen-specific B_{mem} cells was detected at day 28 post-vaccination, which then declined in the blood, as they may have homed to secondary lymphoid organs. Both activated B_{mem} cells and tissue-like B_{mem} cells were found in these subjects.

The effect of BPZE1 on human innate immune cells has been studied in *ex vivo*-generated human monocyte-derived dendritic cells. Upon *in vitro* stimulation with BPZE1, they upregulated the surface expression of maturation markers and produced IL-10, IL-1ß, IL-23, IL-6, and IL-12p70 in a dose-dependent manner[145]. They are also able to drive the expansion of IFN-γ- and IL-17-producing effector T cells, including IFN-γ/IL-17 double-positive cells. This may be of particular interest, as efficient protection against extracellular bacteria requires synergy between Th1 and Th17 lineages[150]. In contrast, the

Th2 responses were decreased by the BPZE1 treatment. BPZE1-treated dendritic cells could also sense CCL21 gradients necessary for lymphatic migration, whereas dendritic cells treated with the virulent parent strain were blocked in their ability to migrate. Therefore, BPZE1 is expected to more rapidly and efficiently activate the adaptive immune responses than virulent *B. pertussis*.

7.5.4 Future outlook

Ideally, the ultimate target for a vaccine like BPZE1 would be neonates, who could be nasally vaccinated at birth to protect them before any encounter with virulent *B. pertussis*. However, this may be hampered by the relative immaturity of the neonatal immune system, especially with respect to its ability to induce T-cell responses[151]. Nevertheless, previous studies have shown that less than 2-month-old children naturally infected with *B. pertussis* are able to mount strong IFN-γ responses to PT and FHA[27]. In the veterinary field, a live attenuated *B. bronchiseptica* vaccine can provide protection against atrophic rhinitis to 2-day-old piglets when given nasally[152], showing that live *Bordetella* vaccines can induce strong protection in neonates. It remains of course to be seen whether BPZE1 can protect equally well 2-day-old infants against pertussis.

If BPZE1 could be used as a priming vaccine for neonates, the standard DTaP vaccination course may then serve as a booster to further strengthen immunity. This would thus not disturb well-established standard vaccination schedules and would simply require the addition of a nasal dose early in life. This has been successfully tested in mice[153]. Although potentially feasible, it will be a long and winding road before neonatal vaccination with BPZE1 can become clinical practice. A faster way to clinical use of BPZE1 may be its application as a pertussis stand-alone vaccine for adolescents and adults. In contrast to aP vaccines BPZE1 is expected to provide strong protection against *B. pertussis* infection and prevent the spread of the virulent bacilli. This should lead to a higher cocooning effectiveness than currently observed with aP vaccination and may therefore have an important impact on the reduction of infant pertussis in addition to protecting the adolescents/adults against whooping cough.

Furthermore, effective protection against nasopharyngeal carriage by virulent *B. pertussis* may have an important public health impact that by far exceeds that of protection against pertussis. Recent studies suggest that subclinical *B. pertussis* infection may be an important, largely overlooked cause of inflammatory and autoimmune diseases, such as coeliac disease[154], multiple sclerosis[155], and Alzheimer's disease[156]. The role of *B. pertussis* infection in the pathogenesis of the latter disease has been experimentally addressed in a mouse model of Alzheimer's disease[157] showing that *B. pertussis* infection causes significant infiltration of inflammatory T cells and NKT cells in the brain of APP/PS1 mice, which was accompanied by increased glial activation and amyloid-ß deposition. Thus, as also shown for other live vaccines[158], BPZE1 may have beneficial non-specific effects to protect against various off-target diseases, whether infections or non-infectious.

Acknowledgements

We thank K. Smits and F. Vermeulen for sharing unpublished results.

References

1. Bordet J, Gengou O. Le microbe de la coqueluche. *Ann Inst Pasteur* 1906;**20**:731–41.
2. Bordet J, Gengou O. Note complémentaire sur le microbe de la coqueluche. *Ann Inst Pasteur* 1907:**21**:721–6.
3. Wollstein M. The Bordet-Gengou bacillus of pertussis. *J Exp Med* 1909:**11**:41–54.
4. Poolman JT, Hallander HO. Acellular pertussis vaccines and the role of pertactin and fimbriae. *Expert Rev Vaccines* 2007;**6**:47–56.
5. McGuinness AC, Stokes J Jr, Mudd S. The clinical use of human serums preserved by the lyophile process. *J Clin Invest* 1937;**16**:185–96.
6. McGuinness AC, Bradford WL, Armstrong JG. The production and use of hyperimmune human whooping cough serum. *J Pediatr* 1940;**16**:21–9.
7. Lapin JH. Serum in the prophylaxis of contacts and the treatment of whooping cough. *J Pediatr* 1945;**26**:555–9.
8. Cravitz L, Williams JW. A comparative study of the 'immune response' to various pertussis antigens and the disease. *J Pediatr* 1946;**28**:172–86.
9. Waddell WW Jr, L'Engle CS Jr. Immune response to early administration of pertussis vaccine. *J Pediatr* 1946;**29**:487–92.

10. Guiso N, Berbers G, Fry NK, et al. What to do and what not to do in serological diagnosis of pertussis: recommendations from EU reference laboratories. *Eur J Clin Microbiol Infect Dis* 2011;**30**:307–12.

11. Taranger J, Trolifors B, Lagergard T, et al. Correlation between pertussis toxin IgG antibodies in postvaccination sera and subsequent protection against pertussis. *J Infect Dis* 2000;**181**:1010–3.

12. Vermeulen F, Verscheure V, Damis E, et al. Cellular immune responses of preterm infants after vaccination with whole-cell or acellular pertussis vaccines. *Clin Vaccine Immunol* 2010;**17**:258–62.

13. Nagel J, Poot-Scholtens EJ. Serum IgA antibody to Bordetella pertussis as an indicator of infection. *J Med Microbiol* 1983;**16**:417–26.

14. May ML, Doi SA, King D, et al. Prospective evaluation of an Australian pertussis toxin IgG and IgA enzyme immunoassay. *Clin Vaccine Immunol* 2012:**19**:190–7.

15. Hallander HO, Ljungman M, Storsaeter J, et al. Kinetics and sensitivity of ELISA IgG pertussis antitoxin after infection and vaccination with Bordetella pertussis in young children. *APMIS* 2009;**117**:797–807.

16. Edelman K, He Q, Mäkinen J, et al. Immunity to pertussis 5 years after booster immunization during adolescence. *Clin Infect Dis* 2007;**44**:1271–7.

17. Le T, Cherry JD, Chang SJ, et al. Immune responses and antibody decay after immunization of adolescents and adults with an acellular pertussis vaccine: the APERT Study. *J Infect Dis* 2004;**190**:535–44.

18. Esposito S, Agliardi T, Giammanco A, et al. Long-term pertussis-specific immunity after primary vaccination with a combined diphtheria, tetanus, tricomponent acellular pertussis, and hepatitis B vaccine in comparison with that after natural infection. *Infection Immun* 2001;**69**:4516–20.

19. Hendrikx LH, Öztürk K, de Rond LGH, et al. Identifying long-term memory B-cells in vaccinated children despite waning antibody levels specific for Bordetella pertussis proteins. *Vaccine* 2011;**29**:1431–7.

20. Hendrikx LH, Felderhof MK, Öztürk K, et al. Enhanced memory B-cell immune responses after a second acellular pertussis booster vaccination in children 9 years of age. *Vaccine* 2011;**30**:51–8.

21. van Twillert I, van Gaans-van den Brink JAM, Poelen MCM, et al. Age related differences in dynamics of specific memory B cell populations after clinical pertussis infection. *PLoS One* 2014;**9**:e85227.

22. Carollo M, Pandolfi E, Tozzi AE, et al. Humoral and B-cell memory responses in children five years after pertussis acellular vaccine priming. *Vaccine* 2014;**32**:2093–9.

23. Schure RM, Hendrikx LH, de Rond LG, et al. Differential T- and B-cell responses to pertussis in acellular vaccine-primed versus whole-cell vaccine-primed children 2

years after preschool acellular vaccination. *Clin Vaccine Immunol* 2013;**20**:1388–95.

24. Ad hoc group for the study of pertussis vaccines. Placebo-controlled trial of two acellular pertussis vaccines in Sweden—protective efficacy and adverse events. *Lancet* 1988;**1**:955–60.

25. Mills KHG, Barnard A, Watkins J, et al. Cell-mediated immunity to Bordetella pertussis: role of Th1 cells in bacterial clearance in a murine respiratory infection model. *Infection Immun* 1993;**61**:399–410.

26. Ryan M, Murphy G, Gothefors L, et al. Bordetella pertussis respiratory infection in children is associated with preferential activation of type 1 T helper cells. *J Infect Dis* 1997;**175**:1246–50.

27. Mascart F, Verscheure V, Malfroot A, et al. Bordetella pertussis infection in 2-month-old infants promotes type 1 T cell responses. *J Immunol* 2003;**170**:1504–9.

28. Ausiello CM, Urbani F, La Sala A, et al. Vaccine- and antigen-dependent type 1 and type 2 cytokine induction after primary vaccination of infants with whole-cell or acellular pertussis vaccines. *Infection Immun* 1997;**65**:2168–74.

29. Mascart F, Hainaut M, Peltier A, et al. Modulation of the infant immune responses by the first pertussis vaccine administrations. *Vaccine* 2007;**25**:391–8.

30. Rowe J, Macaubas C, Monger TM, et al. Antigen-specific responses to diphtheria-tetanus-acellular pertussis vaccine in human infants are initially Th2 polarized. *Infection Immun* 2000;**68**:3873–7.

31. Dirix V, Verscheure V, Goetghebuer T, et al. Cytokine and antibody profiles in 1-year-old children vaccinated with either acellular or whole-cell pertussis vaccine during infancy. *Vaccine* 2009;**27**:6042–7.

32. Vermeulen F, Dirix V, Verscheure V, et al. Persistence at one year of age of antigen-induced cellular immune responses in preterm infants vaccinated against whooping cough: comparison of three different vaccines and effect of a booster dose. *Vaccine* 2013;**31**:1981–6.

33. White OJ, Rowe J, Richmond P, et al. Th2-polarisation of cellular immune memory to neonatal pertussis vaccination. *Vaccine* 2010;**28**:2648–52.

34. Ryan M, Murphy G, Nilsson L, et al. Distinct T-cell subtypes induced with whole cell and acellular pertussis vaccines in children. *Immunology* 1998;**93**:1–10.

35. Dirix V, Verscheure V, Vermeulen F, et al. Both CD4+ and CD8+ lymphocytes participate in the IFN-γ response to filamentous hemagglutinin from Bordetella pertussis in infants, children, and adults. *Clin Dev Immunol* 2012;2012:795958.

36. Dirix V, Verscheure V, Goetghebuer T, et al. Monocyte-derived interleukin-10 depresses the Bordetella pertussis-specific gamma interferon response in vaccinated infants. *Clin Vaccine Immunol* 2009;**16**:1816–21.

37. Ausiello CM, Fedele G, Urbani F, et al. Native and genetically inactivated pertussis toxins induce human dendritic cell maturation and synergize with lipopolysaccharide in promoting T helper type 1 responses. *J Infect Dis* 2002;**186**:351–60.

38. Nasso M, Fedele G, Spensieri F, et al. Genetically detoxified pertussis toxin induced Th1/Th17 immune response through MAPKs and IL-10-depedent mechanisms. *J Immunol* 2009;**183**:1892–9.

39. Schure RM, Hendrickx LH, de Rond LG, et al. T-cell responses before and after the fifth consecutive acellular pertussis vaccination in 4-year-old Dutch children. *Clin Vaccine Immunol* 2012;**19**:1879–86.

40. Palazzo R, Carollo M, Bianco M, et al. Persistence of T-cell immune response induced by two acellular pertussis vaccines in children five years after primary vaccination. *New Microbiol* 2016;**39**:35–47.

41. Ryan EJ, Nilsson L, Kjellman N, et al. Booster immunization of children with an acellular pertussis vaccine enhances Th2 cytokine production and serum IgE concentrations against pertussis toxin but not against common allergens. *Clin Exp Immunol* 2000;**121**: 193–200.

42. Rowe J, Yerkovich ST, Richmond P, et al. Th2-associated local reactions to the acellular diphtheria-tetanus-pertussis vaccine in 4 to 6 year-old children. *Infection Immun* 2005;**73**:8130–5.

43. Schure RM, de Rond L, Öztürk K, et al. Pertussis circulation has increased T-cell immunity during childhood more than a second acellular booster vaccination in Dutch children 9 years of age. *PLoS One* 2012;**7**:e41928.

44. Smits K, Pottier G, Smet J, et al. Different T cell memory in preadolescents after whole-cell or acellular pertussis vaccination. *Vaccine* 2013;**32**:111–8.

45. Ausiello CM, Lande R, la Sala A, et al. Cell-mediated immune response of healthy adults to Bordetella pertussis vaccine antigens. *J Infect Dis* 1998;**178**:466–70.

46. Lawson G. Immunity studies in pertussis. *Am J Hyg* 1939;**29**:119–31.

47. Madsen T. Vaccination against whooping cough. *J Am Med Assoc* 1933;**101**:187–8.

48. Sauer LW. Whooping cough: résumé of a seven year study. *J Pediatr* 1933;**2**:740–9.

49. Kendrick PL, Eldering G. Progress report on pertussis immunization. *Am J Public Health* 1936;**26**:8–12.

50. Doull JA, Shibley GS, McClelland JE. Active immunization against whooping cough. Interim report of the Cleveland experience. *Am J Public Health* 1936;**26**:1097–105.

51. Fröhlich J. Beitrag sur Pathologie des Keuchhustens. *Jahrbuch für Kinderhusten* 1897;**44**:53–8.

52. Locht C, Coutte L, Mielcarek N. The ins and outs of pertussis toxin. *FEBS J* 2011;**278**:4668–82.

53. Kenrick PL, Eldering G, Dixon MK, et al. Mouse protection tests in the study of pertussis vaccine. *Am J Public Health* 1947;**37**:803–10.

54. Pittman M, Cox CB. Pertussis vaccine testing for freedom-from-toxicity. *Appl Microbiol* 1965;**13**:447–56.

55. WHO. Report of expert committee on biological standardization. *WHO Technical Report Series* 1979;**638**:60–80.

56. MRC. The prevention of whooping cough by vaccination: a Medical Research Council investigation. *Br Med J* 1951;**1**:1463–71.

57. MRC. Vaccination against whooping cough: relation between protection in children and results of laboratory tests. *Br Med J* 1956;**2**:454–62.

58. WHO. Global routine vaccination coverage, 2011. *Wkly Epidemiol Rec* 2012;**44**:432–5.

59. Storsaeter J, Wolter J, Locht C. Pertussis vaccines. In Locht C (ed), *Bordetella Molecular Microbiology.* Norfolk: Horizon Press; 2007, pp 245–88.

60. Romanus V, Jonsell R, Bergquist SO.Pertussis in Sweden after the cessation of general immunization in 1979. *Pediatr Infect Dis J* 1987;**6**:364–71.

61. Simondon F, Preziosi MP, Yam A, et al. A randomized double-blind trial comparing a two-component acellular and a whole-cell pertussis vaccine in Senegal. *Vaccine* 1997;**15**:1606–12.

62. Gustafsson L, Hallander HO, Olin P, et al. A controlled trial of a two-component acellular, a five-component acellular, and a whole-cell pertussis vaccine. *N Engl J Med* 1996;**334**:349–55.

63. Greco D, Salmaso S, Mastrantonio P, et al. A controlled trial of two acellular vaccines and one whole-cell vaccine against pertussis. *N Engl J Med* 1996;**334**:341–8.

64. Cody CL, Baraff LJ, Cherry JD, et al. Nature and rates of adverse reactions associated with DTP and DT immunizations in infants and children. *Pediatrics* 1981;**68**:650–60.

65. Sato Y, Kimura M, Fukumi H. Development of a pertussis component vaccine in Japan. *Lancet* 1984;**1**:122–6.

66. Aoyama T, Murase Y, Kato T, et al. Efficacy of an acellular pertussis vaccine in Japan. *J Pediatr* 1985;**107**. 180–2.

67. Storsaeter J, Olin P. Relative efficacy of two acellular pertussis vaccines during three years of passive surveillance. *Vaccine* 1992;**10**:142–4.

68. Trollfors B, Taranger J, Lagergard T, et al. A placebo-controlled trial of a pertussis-toxoid vaccine. *N Engl J Med* 1995;**333**:1045–50.

69. Olin P, Rasmussen F, Gustafsson L, et al. Randomised controlled trial of two-component, three-component, and five-component acellular pertussis vaccines compared with whole-cell pertussis vaccine. *Lancet* 1997;**350**:1569–77.

70. Schmitt HJ, von König CH, Neiss A, et al. Efficacy of acellular pertussis vaccine in early childhood after household exposure. *JAMA* 1996;**275**:37–41.

71. Liese JG, Meschievitz CK, Harzer E, et al. Efficacy of a two-component acellular pertussis vaccine in infants. *Pediatr Infect Dis J* 1997;**16**:1038–44.

72. Stehr K, Cherry JD, Heininger U, et al. A comparative efficacy trial in Germany in infants who received either the Lederle/Takeda acellular pertussis component DTP (DTaP) vaccine, the Lederle whole-cell component DTP vaccine, or DT vaccine. *Pediatrics* 1998;**101**:1–11.

73. Olin P. The best acellular pertussis vaccines are multi-component. *Pediatr Infect Dis J* 1997;**16**:517–9.

74. Storsaeter J, Hallander HO, Gustafsson L, et al. Levels of anti-pertussis antibodies related to protection after household exposure to Bordetella pertussis. *Vaccine* 1998;**20**:1907–16.

75. Cherry JD, Gornbein J, Heininger U, et al. A search for serologic correlates of immunity to Bordetella pertussis cough illnesses. *Vaccine* 1998;**16**:1901–6.

76. Olin P, Hallander HO, Gustafsson L, et al. How to make sense of pertussis immunogenicity data. *Clin Infect Dis* 2001;**33**:S288–91.

77. Chiappini E, Stival A, Galli L, et al. Pertussis re-emergence in the post-vaccination era. *BMC Infect Dis* 2013;**13**:151.

78. Cherry JD. Epidemic pertussis in 2012—the resurgence of a vaccine-preventable disease. *N Engl J Med* 2012;**367**:785–7.

79. Lacombe K, Yam A, Simondon F, et al. Risk factors for acellular and whole-cell pertussis vaccine failure in Senegalese children. *Vaccine* 2004;**23**:623–8.

80. Klein NP, Bartlett J, Rowhani-Rahbar A, et al. Waning protection after fifth dose of acellular pertussis vaccine in children. *N Engl J Med* 2012;**367**:1012–9.

81. Misegardes LK, Winter K, Harriman K, et al. Association of childhood pertussis with receipt of 5 doses of pertussis vaccine by time since the last vaccine dose, California, 2010. *JAMA* 2012;**308**:2126–32.

82. Quinn HE, Snelling TL, Macartney KK, et al. Duration of protection after first dose of acellular pertussis vaccine in infants. *Pediatrics* 2014;**133**:e513–9.

83. Sheridan SL, Ware RS, Grimwood K, et al. Number and order of whole cell pertussis vaccines in infancy and disease protection. *JAMA* 2012;**308**:454–6.

84. Halperin SA, Eastwood BJ, Barreto L, et al. Adverse reactions and antibody response to four doses of acellular or whole cell pertussis vaccine combined with diphtheria and tetanus toxoids in the first 19 months of life. *Vaccine* 1996;**14**:767–72.

85. Pichichero ME, Rennels MB, Edwards KM, et al. Combined tetanus, diphtheria, and 5-component pertussis vaccine for use in adolescents and adults. *JAMA* 2005;**293**:3003–12.

86. Ward JI, Cherry JD, Chang SJ, et al. Efficacy of an acellular pertussis vaccine among adolescents and adults. *N Engl J Med* 2005;**353**:1555–63.

87. Wei SC, Tatti K, Cushing K, et al. Effectiveness of adolescent and adult tetanus, reduced-dose diphtheria, and acellular pertussis vaccine against pertussis. *Clin Infect Dis* 2010;**51**:315–21.

88. Rank C, Quinn HE, McIntyre PB. Pertussis vaccine effectiveness after mass immunization of high school students in Australia. *Pediatr Infect Dis J* 2009;**28**:152–3.

89. Liko J, Robinson SG, Cieslak PR. Pertussis vaccine performance in an epidemic year—Oregon, 2012. *Clin Infect Dis* 2014;**59**:261–3.

90. Bisgard KM, Pascual FB, Ehresmann KR, et al. Infant pertussis: who was the source? *Pediatr Infect Dis J* 2004;**23**:985–9.

91. Van Rie A, Hethcote HW. Adolescent and adult pertussis vaccination: computer simulations of five new strategies. *Vaccine* 2004;**22**:3154–65.

92. Healy CM, Rench MA, Wootton SH, et al. Evaluation of the impact of a pertussis cocooning program on infant pertussis infection. *Pediatr Infect Dis J* 2015;**34**:22–6.

93. Amirthalingam G, Andrews N, Campbell H, et al. Effectiveness of maternal pertussis vaccination in England: and observational study. *Lancet* 2014;**384**:1521–8.

94. Winter K, Cherry JD, Harriman K. Effectiveness of prenatal tetanus, diphtheria, and acellular pertussis vaccination on pertussis severity in infants. *Clin Infect Dis* 2017;**64**:9–14.

95. Winter K, Nickell S, Powell M, et al. Effectiveness of prenatal versus postpartum tetanus, diphtheria, and acellular pertussis vaccination in preventing infant pertussis. *Clin Infect Dis* 2017;**64**:3–8.

96. Healy CM. Pertussis vaccination in pregnancy. *Hum Vaccin Immunother* 2016;**12**:1972–81.

97. Van Savage J, Decker MD, Edwards KM, et al. Natural history of pertussis antibody in the infant and effect on vaccine response. *J Infect Dis* 1990;**161**:487–92.

98. Perry J, Towers CV, Weitz B, et al. Patient reaction to Tdap vaccination in pregnancy. *Vaccine* 2017;**35**:3064–6.

99. Englund JA, Anderson EL, Reed GF, et al. The effect of maternal antibody on the serologic response and the incidence of adverse reactions after primary immunization with acellular and whole-cell pertussis vaccines combined with diphtheria and tetanus toxoids. *Pediatrics* 1995;**96**:580–4.

100. Maertens K, Caboré RN, Huygen K, et al. Pertussis vaccination during pregnancy in Belgium: results of a prospective controlled cohort study. *Vaccine* 2016;**34**:142–50.

101. Bento AI, Rohani P. Forecasting epidemiological consequences of maternal immunization. *Clin Infect Dis* 2016;**63**:S205–12.

102. Warfel JM, Zimmerman LI, Merkel TJ. Acellular pertussis vaccines protect against disease but fail to prevent infection and transmission in a nonhuman primate model. *Proc Natl Acad Sci U S A* 2014;**111**:787–92.

103. Locht C. Live pertussis vaccines: will they protect against carriage and spread of pertussis. *Clin Microbiol Infect* 2016;**22**:S96–102.

104. Althouse BM, Scarpino SV. Asymptomatic transmission and the resurgence of Bordetella pertussis. *BMC Med* 2015;**13**:146.

105. Locht C. Will we have new pertussis vaccines? *Vaccine* 2018;**36**:5460–9.

106. Dias WO, van der Ark A, Sakauchi MA, et al. An improved whole cell pertussis vaccine with reduced content of endotoxin. *Hum Vaccin Immunother* 2012;**9**:339–48.

107. Zorzeto TQ, Higashi HG, da Silva MT, et al. Immunogenicity of a whole-cell pertussis vaccine with low lipopolysaccharide content in infants. *Clin Vaccine Immunol* 2009;**16**:544–50.

108. Rumbo M, Hozbor D. Development of improved pertussis vaccine. *Hum Vaccin Immunother* 2014;**10**:2450–3.

109. Roberts R, Moreni G, Bottero D, et al. Outer membrane vesicles as acellular vaccine against pertussis. *Vaccine* 2008;**26**:4639–46.

110. Polewicz M, Garcia A, Garlapati S, et al. Novel vaccine formulations against pertussis offer earlier onset of immunity and provide protection in the presence of maternal antibodies. *Vaccine* 2013;**31**:3148–55.

111. Bruno C, Agnolon V, Berti F, et al. The preparation and characterization of PLG nanoparticles with an entrapped synthetic TLR7 agonist and their preclinical evaluation as adjuvant for an adsorbed DTaP vaccine. *Eur J Pharm Biopharm* 2016;**105**:1–8.

112. Li P, Asokanathan C, Liu F, et al. PGLA nano/micro particles encapsulated with pertussis toxoid (PTd) enhances Th1/Th17 immune responses in a murine model. *Int J Pharm* 2016;**513**:183–90.

113. Elahi S, Van Kessel J, Kiros TG, et al. C-di-GMP enhances protective innate immunity in a murine model of pertussis. *PLoS One* 2014;**9**:e109778.

114. Agnolon V, Bruno C, Leuzzi R, et al. The potential of adjuvants to improve immune responses against TdaP vaccines: a preclinical evaluation of MF59 and monophosphoryl lipid A. *Int J Pharm* 2015;**492**:169–76.

115. Dunne A, Mielke LA, Allen AC, et al. A novel TLR2 agonist from Bordetella pertussis is a potent adjuvant that promotes protective immunity with an acellular pertussis vaccine. *Mucosal Immunol* 2015;**8**:607–17.

116. Arciniega JL, Hewlett EL, Johnson FD, et al. Human serologic response to envelope-associated proteins and adenylate cyclase toxin of Bordetella pertussis. *J Infect Dis* 1991;**163**:135–42.

117. Cheung GY, Xing D, Prior S, et al. Effect of different forms of adenylate cyclase toxin of Bordetella pertussis on protection afforded by an acellular pertussis vaccine in a murine model. *Infection Immun* 2006;**74**:6797–805.

118. Marr N, Oliver DC, Laurent V, et al. Protective activity of the Bordetella pertussis BrkA autotransporter in the murine lung colonization model. *Vaccine* 2008;**26**:4306–11.

119. Alvarez Hayes, J, Erben, E, Lamberti, Y, et al. Identification of a new protective antigen of Bordetella pertussis. *Vaccine* 2011;**29**:8731–9.

120. Alvarez Hayes J, Erben E, Lamberti Y, et al. Bordetella pertussis iron regulated proteins as potential vaccine components. *Vaccine* 2013;**31**:3543–8.

121. Kubler-Kielb J, Vinogradov E, Lagergard T, et al. Oligosaccharide conjugates of Bordetella pertussis and bronchiseptica induce bactericidal antibodies, an addition to pertussis vaccine. *Proc Natl Acad Sci U S A* 2011;**108**:4087–92.

122. Gambhir M, Clark TA, Cauchemez S, et al. A change in vaccine efficacy and duration of protection explains recent rises in pertussis incidence in the United State. *PLoS Comput Biol* 2015;**11**:e1004138.

123. Wendelboe AM, Van Rie A, Salmaso S, et al. Duration of immunity against pertussis after natural infection or vaccination. *Pediatr Infect Dis J* 2005;**24**:S58–61.

124. Wearing HJ, Rohani P. Estimating the duration of pertussis immunity using epidemiological signatures. *PLoS Pathogens* 2009;**10**:e1000647.

125. Huang CC, Chen PM, Kuo JK, et al. Experimental whooping cough. *N Engl J Med* 1962;**266**:105–11.

126. Locht C, Mielcarek N. Live attenuated vaccines against pertussis. *Expert Rev Vaccines* 2014;**13**:1147–58.

127. Roberts M, Maskell D, Novotny P, et al. Construction and characterization in vivo of Bordetella pertussis aroA mutants. *Infection Immun* 1990;**58**:732–9.

128. Cornford-Nairns R, Daggard G, Mukkur T. Construction and preliminary immunobiological characterization of a novel, non-reverting, intranasal live attenuated whooping cough vaccine candidate. *J Microbiol Biotechnol* 2012;**22**:856–65.

129. Mielcarek N, Debrie AS, Raze D, et al. Live attenuated B. pertussis as a single-dose nasal vaccine against whooping cough. *PLoS Pathog* 2006;**2**:e65.

130. Kammoun H, Feunou PF, Foligne B, et al. Dual mechanism of protection by live attenuated Bordetella pertussis BPZE1 against Bordetella bronchiseptica in mice. *Vaccine* 2012;**30**:5864–70.

131. Mielcarek N, Debrie AS, Mahieux S, et al. Dose response of attenuated Bordetella pertussis BPZE1-induced protection in mice. *Clin Vaccine Immunol* 2010;**17**:317–24.

132. Troseid M, Jonassen TO, Steinbakk M. Isolation of Bordetella pertussis in blood culture from a patient with multiple myeloma. *J Infect* 2006;**52**:e11–3.

133. Locht C, Mielcarek N. Live attenuated vaccines against pertussis. *Expert Rev Vaccines* 2014;**13**:1147–58.

134. Feunou PF, Kammoun H, Debrie AS, et al. Long-term immunity against pertussis induced by a single nasal administration of live attenuated B. pertussis BPZE1. *Vaccine* 2010;**28**:7047–53.

135. Skerry CM, Mahon BP. A live, attenuated Bordetella pertussis vaccine provides long-term protection against virulent challenge in a murine model. *Clin Vaccine Immunol* 2011;**18**:187–93.

136. Feunou PF, Bertout J, Locht C. T- and B-cell-mediated protection induced by novel, live attenuated pertussis vaccine in mice. Cross protection against parapertussis. *PLoS One* 2010;**5**:e10178.

137. Locht C, Papin JF, Lecher S, et al. Live attenuated pertussis vaccine BPZE1 protects baboons against B. pertussis disease and infection. *J Infect Dis* 2017;**216**:117–24.

138. Skerry CM, Cassidy JP, English K, et al. A live attenuated Bordetella pertussis candidate vaccine does not cause disseminating infection in gamma interferon receptor knockout mice. *Clin Vaccine Immunol* 2009;**16**:1344–51.

139. Li R, Lim A, Phoon MC, et al. Attenuated Bordetella pertussis protects against highly pathogenic influenza A viruses by dampening the cytokine storm. *J Virol* 2010;**84**:7105–13.

140. Schnöller C, Roux X, Sawant D, et al. Attenuated Bordetella pertussis vaccine protects against respiratory syncytial virus disease via an IL-17-dependent mechanism. *Am J Respir Crit Care Med* 2014;**189**:194–202.

141. Kavanagh H, Noone C, Cahill E, et al. Attenuated Bordetella pertussis vaccine strain BPZE1 modulates allergen-induced immunity and prevents allergic pulmonary pathology in a murine model. *Clin Exp Allergy* 2010;**40**:933–41.

142. Li R, Cheng C, Chong SZ, et al. Attenuated Bordetella pertussis BPZE1 protects against allergic airway inflammation and contact dermatitis in mouse models. *Allergy* 2012;**67**:1250–8.

143. Fedele G, Sanseverino I, D'Agostino K, et al. Unconventional, adenosine-producing suppressor T cells induced by dendritic cells exposed to BPZE1 pertussis vaccine. *J Leukoc Biol* 2015;**98**:631–9.

144. Ohta A, Sitkovsk M. Extracellular adenosine-mediated modulation of regulatory T cells. *Frontiers Immunol* 2014;**5**:304.

145. Fedele G, Bianco M, Debrie AS, et al. Attenuated Bordetella pertussis vaccine candidate BPZE1 promotes human dendritic cell CCL21-induced migration and drives a Th1/Th17 response. *J Immunol* 2011;**186**:5388–96.

146. Feunou PF, Ismaili J, Debrie AS, et al. Genetic stability of the live attenuated Bordetella pertussis vaccine candidate BPZE1. *Vaccine* 2008;**26**:5722–7.

147. Thorstensson R, Trollfors B, Al-Tawil N, et al. A phase I clinical study of a live attenuated Bordetella pertussis vaccine—BPZE1; a single centre, double-blind, placebo-controlled, dose-escalating study of BPZE1 given intranasally to healthy adult male volunteers. *PLoS One* 2014;**9**:e83449.

148. Hallander HO, Ljungman M, Jahnmatz M, et al. Should fimbriae be included in pertussis vaccines? Studies on ELISA IgG anti-Fim2/3 antibodies after vaccination and infection. *APMIS* 2009;**117**:660–71.

149. Jahnmatz M, Amu S, Ljungman M, et al. B-cell responses after intranasal vaccination with the novel attenuated Bordetella pertussis vaccine strain BPZE1 in a randomized phase I clinical trial. *Vaccine* 2014;**32**:3350–6.

150. Lin L, Ibrahim AS, Xu X, et al. Th1-Th17 cells mediate protective adaptive immunity against Staphylococcus aureus and Candida albicans infection in mice. *PLoS Pathog* 2009;**5**:e1000703.

151. Lewis DB, Yu CC, Meyer J, et al. Cellular and molecular mechanisms for reduced interleukin-4 and interferon-gamma production by neonatal T cells. *J Clin Invest* 1991;**87**:194–202.

152. De Jong MF. Prevention of atrophic rhinitis in piglets by means of intranasal administration of a live non-AR pathogenic Bordetella bronchiseptica vaccine. *Vet Q* 1987;**9**:123–33.

153. Feunou PF, Kammoun H, Debrie AS, et al. Heterologous prime-boost immunization with live attenuated B. pertussis BPZE1 followed by acellular pertussis vaccine in mice. *Vaccine* 2014;**32**:4281–8.

154. Rubin K, Glazer S. Potential role of Bordetella pertussis in celiac disease. *Int J Celiac Dis* 2015;**3**:75–6.

155. Rubin K, Glazer S. The potential role of subclinical Bordetella pertussis colonization in the etiology of multiple sclerosis. *Immunobiology* 2016;**221**:512–5.

156. Rubin K, Glazer S. The pertussis hypothesis: Bordetella pertussis colonization in the pathogenesis of Alzheimer's disease. *Immunobiology* 2017;**222**:228–40.

157. McManus RM, Higgins SC, Mills KH, et al. Respiratory infection promotes T cell infiltration and amyloid-ß deposition in APP/PS1 mice. *Neurobiol Aging* 2014;**35**:109–21.

158. Bardenheier BH, McNeill MM, Wodi AP, et al. Risk of non-targeted infectious disease hospitalizations among U.S. children following inactivated and live vaccines, 2005–2014. *Clin Infect Dis* 2017;**65**:729–37.

CHAPTER 8

Evolutionary epidemiology theory of vaccination

Sylvain Gandon

Abstract

The aim of vaccination is to prevent or limit the risk of pathogen infections for individual hosts but large vaccination coverage often has dramatic epidemiological consequences at the scale of the whole host population. This massive perturbation of the ecology and transmission of the pathogen can also have important evolutionary effects. In particular, vaccine-driven evolution may lead to the spread of new pathogen variants that may erode the benefits of vaccination. This chapter presents a theoretical framework for modelling the short- and long-term epidemiological and evolutionary consequences of vaccination. This framework can be used to make quantitative predictions about the speed of such evolutionary processes. This work helps identify the relevant phenotypic traits that need to be measured in specific parasite populations in order to evaluate the potential evolutionary consequences of vaccination. In particular, this may help in the debate regarding the involvement of evolution in the re-emergence of pertussis in spite of the high coverage of vaccination.

8.1 Introduction

The epidemiological consequences of vaccination have received a considerable amount of attention, both from an empirical and a theoretical standpoint[1-5]. One of the most significant conceptual developments from this research is the finding that there is a critical vaccination coverage above which a parasite can be driven to extinction because of herd immunity[1]. But the dramatic ecological perturbation induced by vaccination may also have major evolutionary consequences on pathogen populations. The objective of this chapter is to present a general theoretical framework that allows a broad diversity of epidemiological and evolutionary scenarios

following vaccination to be considered[6]. More specifically, this framework can be used to investigate the contribution of different factors in the resurgence of pertussis in spite of high vaccination coverage. In principle, we may distinguish between different non-mutually exclusive explanations for the re-emergence of pertussis: (1) insufficient coverage of vaccination[4,7-9,] (2) waning immunity of vaccination[8], (3) imperfect immunity of vaccination[8,10,11,] and (4) pathogen evolution following vaccination[12,13]. The theoretical framework presented below can be used to disentangle the interplay between all these factors. This is a necessary first step to understand the consequences of vaccination on the epidemiology

Gandon, S., *Evolutionary epidemiology theory of vaccination*. In: *Pertussis: epidemiology, immunology, and evolution*. Edited by Pejman Rohani and Samuel V. Scarpino: Oxford University Press (2019). © Oxford University Press.
DOI: 10.1093/oso/9780198811879.003.0008

and the evolution of pathogens. This framework can also be tailored to specific pathogens and may thus help understand the driving forces leading to pertussis re-emergence.

First, I use this model to derive vaccination thresholds under different epidemiological scenarios. Second, I present how adaptive dynamics can be used to predict the long-term evolution of pathogens after vaccination. Third, I show that this theoretical framework can also be used to study the short-term evolution of pathogens following vaccination. Fourth, I discuss briefly the potential impact of spatial structure on vaccine-driven evolution. At the end of each section, I discuss how this model can be used to give some insights into the dynamics of pertussis.

8.2 Epidemiology

The epidemiological consequences of vaccination can be analysed using a classical SIR model, modified to include imperfect vaccination[14]. The host can be either susceptible (S), infectious (I), or recovered (R). Vaccination adds another level of heterogeneity among the hosts because I explicitly assume that vaccination may be leaky, so that vaccinated hosts may also become infected. In other words, both susceptible and infectious hosts can be either naïve or vaccinated (indicated with a subscript N or V, respectively). All recovered hosts (vaccinated or not) are assumed to be fully immune to reinfection, and therefore the different types of recovered hosts (vaccinated or not) are pooled in a single host class. In the first step of this analysis, it is useful to assume the pathogen population to be monomorphic and to neglect its evolution. This yields the following set of differential equations:

$$\dot{S}_N = \lambda(1-p) + \omega_V S_V + \omega_R R - (\delta + \Lambda_N)S_N$$
$$\dot{S}_V = \lambda p - (\delta + \Lambda_V + \omega_V)S_V$$
$$\dot{I}_N = \Lambda_N S_N - (\delta + \alpha_N + \gamma_N)I_N$$
$$\dot{I}_V = \Lambda_V S_V - (\delta + \alpha_V + \gamma_V)I_V$$
$$\dot{R} = \gamma_N I_N + \gamma_V I_V - (\delta + \omega_R)R. \tag{1}$$

The per capita rate of arrival of new susceptible hosts in the population (immigrants and newborns) is λ. Among those individuals, a proportion p is vaccinated. Susceptible hosts become infected

with rates $\Lambda_N = \beta_{NN}I_N + \beta_{VN}I_V$ or $\Lambda_V = \beta_{NV}I_N + \beta_{VV}I_V$ when they are naïve or vaccinated, respectively. The rates of infection depend on the densities of infected hosts (I_N and I_V) and on the parasite transmission rates β_{ij} from host i to host j (where i and j can be either naïve, N, or vaccinated, V). Hosts have a natural mortality rate δ, and infected hosts suffer extra mortality due to the presence of the parasite (i.e. parasite virulence). Parasite virulence may differ between naïve and vaccinated hosts (α_N and α_V, respectively). Recovery rates may also differ between these two hosts (γ_N and γ_N, respectively). I also assume that immunity induced by vaccination or natural infections may wane at rates ω_V and ω_R, respectively.

The above-described dynamical system has two equilibria. In the first, the pathogen is absent. The stability of this equilibrium depends on the ability of a parasite to invade a fully susceptible population, which is given by the basic reproduction ratio of the parasite[2,15–18]. When the transmission rate from host i to host j can be written as the product $\beta_{ij} = \pi_i\phi_j$ of the production π_i of propagules by host i and the susceptibility ϕ_j of host j the basic reproduction ratio is[16,17]:

$$R_0 = \left(1 - \frac{p\delta}{\delta + \omega_V}\right)R_0^N + R_0^V p \tag{2}$$

where
$$R_0^N = \frac{\beta_{NN}}{(\delta + \alpha_N + \gamma_N)}\frac{\lambda}{\delta} \text{ and } R_0^V = \frac{\beta_{VV}}{(\delta + \alpha_V + \gamma_V)}\frac{\lambda}{\delta + \omega_V}$$
are the basic reproduction ratios in unvaccinated and 100 per cent vaccinated host populations, respectively. Parasite invasion of a fully susceptible host population will occur whenever $R_0 > 1$. In this case, the spread of the pathogen will yield a new endemic equilibrium characterized by the densities $\bar{S}_N, \bar{S}_V, \bar{I}_N, \bar{I}_V, \bar{R}$ (the 'overbar' refers to the endemic equilibrium).

Equation (2) can be used to derive the critical vaccination coverage leading to disease eradication:

$$p_c = \frac{(R_0^N - 1)(\delta + \omega_V)}{R_0^N \delta - R_0^V(\delta + \omega_V)}. \tag{3}$$

When the vaccination coverage is above this threshold, $p > p_c$, the parasite cannot survive in the host population and is driven to extinction. In contrast, when

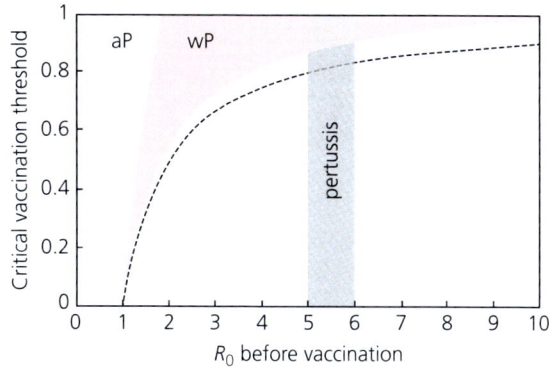

Figure 8.1 Vaccination threshold for perfect vaccines (dashed line), for the imperfect whole-cell pertussis (wP) vaccine (red shaded area), and for the imperfect acellular pertussis (aP) vaccine (brown shaded area). The light grey area refers to an estimate of the basic reproduction ratio of pertussis. The data used for these curves is retrieved from a 2015 World Health Organization report (see main text[55]). I model the efficacy of the vaccine with the parameter r where $\beta_{VV} = \beta_{NV} = (1-r)\beta_{NN}$. Perfect vaccine: $r = 1$; wP vaccine: where $r \in [0.46, 0.92]$; aP: $r \in [0.80, 0.89]$ with $\omega_V = 0.01$ and $\delta = 0.01$.

the pathogen has a large basic reproduction ratio R_0^N and when available vaccines are not very effective (i.e. when R_0^V is also relatively large), eradication is not feasible (Figure 8.1). In this case, vaccination leads to a new endemic equilibrium with a reduced incidence of the disease. Before reaching the new endemic equilibrium, the disease incidence can oscillate during a transient phase. In particular, when coverage is close to the vaccination threshold and waning immunity is low, the system may rapidly reach a very low incidence without leading the parasite towards extinction. After this 'honeymoon period', several epidemics will occur (damped oscillations) before the system settles to the new endemic equilibrium (Figure 8.2).

More specific predictions for pertussis thus require estimates of the basic reproduction ratio in naïve and vaccinated populations. A recent report from the World Health Organization provides an estimation of the basic reproduction ratio of pertussis, $R_0^N \sim 5.5$, and indicates that the protection conferred by both whole-cell and acellular vaccination is imperfect[19]. The effectivity of the protection is estimated to be between 46 and 92 per cent for the whole-cell vaccine and between 80 and 89 per cent for the acellular vaccines[19]. In addition, the immunity induced by the acellular vaccine is suspected to wane over time[8,20,21]. I used these estimations and equation (3) to plot the critical vaccination coverage for both types of vaccines (see Figure 8.1).

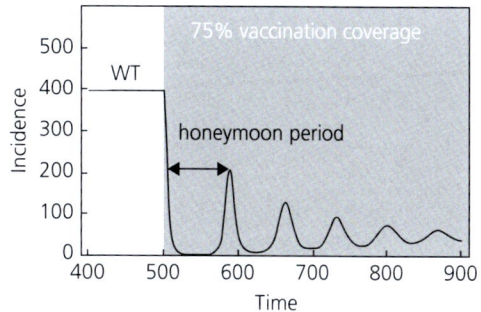

Figure 8.2 Epidemiological dynamics (incidence through time) of the model (equation 1) for a monomorphic pathogen population where only the wild type (WT) circulates. Vaccination is assumed to start at $t = 500$ at a coverage $p = 0.75$. Parameter values: $\lambda = 1000$, $\delta = 0.01$, $\omega_R = \omega_V = 0$, $\alpha_N = \alpha_V = 1$, $\beta_{NN} = 10^{-4}$, $\beta_{NV} = \beta_{NN}/10$, $\beta_{VV} = \beta_{NN}/20$, $\beta_{VN} = \beta_{NN}/2$, $\gamma_N = \gamma_V = 1$. For these parameter values: $R_0^N \approx 4.97$, $R_0^V \approx 0.25$ and $p_c \approx 0.84$.

Figure 8.1 should be used with caution because it is based on rough estimates of the parameters of the model. Yet the figure indicates that only very high vaccination coverage with the whole-cell vaccine may allow pertussis eradication. Eradication with the acellular vaccine seems unfeasible.

8.3 Evolution

In this section, I will examine the evolutionary dynamics taking place in pathogen populations after vaccination. There are two largely separate bodies of research on this question (e.g.[22,23]), and each addresses different aspects of how vaccination affects pathogen evolution. The first line of enquiry focuses on the so-called escape mutants and is directed towards understanding how vaccination selects for parasite strains that are able to evade the protective effects of the vaccine[14,22,24–27]. The second line of enquiry focuses on the so-called virulence or life-history mutants and is directed towards understanding how vaccination causes evolutionary changes in the extent to which a parasite harms its host (i.e. evolutionary changes in its virulence[6,13–16]). The central premise behind this research is that virulence evolves as a result of constraints among parasite life-history characteristics, and that vaccination can alter the form of these constraints, thereby causing evolutionary changes in virulence. The evolutionary model I develop below accounts for a broad range of pathogen mutations. In practice, different strains are allowed to circulate in the host population. For the sake of simplicity, however, I will restrict the analysis to cases where the infection of an already infected host is impossible (but see how to relax this assumption in[16,28,31]). Each parasite strain is characterized by three critical life-history characteristics that are displayed during an infection of either a naïve or a vaccinated host: its transmission rate, β; its recovery rate, γ; and its virulence, α. These three traits are assumed to be constant throughout the infectious period (but see how to relax this assumption in[32]). We refer to the predominant strain present prior to vaccination as the wild-type strain and to strains that are selectively favoured in vaccinated hosts as vaccine-favoured variants (see Figure 8.3). The selective advantage of a vaccine-favoured variant arises from the differences in one or more of its three life-history parameters compared with the wild type, when measured in a vaccinated host. For example, in the dichotomy mentioned previously, escape mutants might be viewed as variants that have reduced recovery rates in the vaccinated hosts (relative to the wild type) due to the changes in epitope. Similarly, life-history or virulence mutants might be viewed as vaccine-favoured variants that have an increased transmission in vaccinated hosts due to an increased rate of replication. More generally, however, vaccine-favoured variants can differ from the wild type in all three

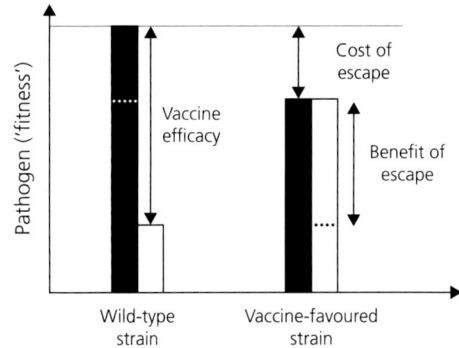

Figure 8.3 Schematic of the comparison between a wild-type strain and a vaccine-favoured variant. The lifetime reproductive success (a relevant measure of fitness at endemic equilibrium) is plotted in naïve hosts (black) and in vaccinated hosts (white) for both strains. The efficacy of the vaccine can be evaluated by the reduced performance of wild-type strain in vaccinated hosts. The cost of adaptation to vaccination is evaluated in naïve hosts, while its benefit is evaluated in vaccinated hosts (in both cases, relative to the performance of the wild-type strain). When the population is away from the endemic equilibrium, other measures of pathogen fitness may be more relevant (see main text and Figure 8.5). Day T, Gandon S. The evolutionary epidemiology of vaccination. *Journal of the Royal Society Interface* 2007; 4: 803–817. Copyright © 2007, © 2007 The Royal Society.

life-history parameters. What matters is that the combined effect of these changes results in higher fitness in vaccinated hosts than the wild type (I will define fitness in the following sections). I will also typically suppose that vaccine-favoured variants suffer some fitness cost in naïve hosts (see Figure 8.3), since otherwise they would probably already have reached appreciable frequencies in the absence of vaccination. As with the benefit enjoyed by a vaccine-favoured variant, the cost paid in naïve hosts also arises from differences in one or more of its three life-history components compared with the wild type, when measured in naïve hosts.

8.3.1 Long-term evolution

Most studies of the spread of escape mutants or the evolution of virulence in response to vaccination have assumed that epidemiological dynamics are very fast relative to evolution. This separation of timescales derives from explicit models of population dynamics and population genetics[33,34] where the phenotypic variation among the different genotypes is small (i.e. small mutation steps). It is also at the heart of game-theoretic analyses and adaptive dynamics[35,36]. The parasite population is assumed to reach an endemic epidemiological equilibrium before mutation introduces genetic variation (new mutant strains with potentially different phenotypic traits). At this endemic equilibrium, under the simplifying assumption that $\beta_{ij} = \pi_i \phi_j$, the expected number of new infections produced by a randomly chosen infected host by the resident strain (i.e. wild type) during its total infectious period (its lifetime reproductive success) is:

$$R = \frac{\beta_{NN}}{\left(\delta + \alpha_N + \gamma_N\right)}\bar{S}_N + \frac{\beta_{VV}}{\left(\delta + \alpha_V + \gamma_V\right)}\bar{S}_V. \quad (4)$$

Since the number of infected hosts remains constant at endemic equilibrium, this necessarily yields $R = 1$. At this endemic equilibrium, one may formalize evolution by focusing on the invasion of a rare mutant strain with different phenotypic traits α_N^m, α_V^m, β_{NN}^m, β_{VV}^m, γ_N^m, γ_V^m. The mutant will initially spread if its lifetime reproductive success, R^m, is higher than the lifetime reproductive success of the resident. Therefore, a condition for the mutant to invade the population is[17]:

$$R^m = \frac{\beta_{NN}^m}{\left(\delta + \alpha_N^m + \gamma_N^m\right)}\bar{S}_N + \frac{\beta_{VV}^m}{\left(\delta + \alpha_V^m + \gamma_V^m\right)}\bar{S}_V > R = 1. \quad (5)$$

The lifetime reproductive success of the mutant is a measure of the fitness of the mutant and the above description of evolutionary dynamics has been used to describe the spread of escape mutants after a vaccination campaign[14,24,25]. The same rationale can be used to characterize virulence evolution as a result of vaccination[16,28–30]. The adaptive dynamics approach is based on the assumption that the introduction of the vaccine leads to a new endemic equilibrium of the resident strain and that a mutant pathogen arises with altered life-history components. Typically, it is supposed that these life-history traits satisfy some constraint, for instance, a positive relationship between transmission rate and virulence. In this case, the transmission rate β_{NN}^m can be modelled as a function of the virulence α_N^m of the mutant. Again, one then asks whether or not such a mutant can spread. The ultimate goal of this type of analysis, however, is to establish whether or not there is a vaccine-favoured strain that, once present in the population,

can resist invasion by all other possible mutant strains. In the model I considered, a strain that maximizes R^m is evolutionarily stable because no other strategy can invade it. This approach can be used to explore the effect of different types of imperfect (i.e. leaky) vaccines on the evolution of pathogens where the specificity of the mode of action of each vaccine affects the life-history traits of the pathogens in vaccinated hosts. For example, Gandon et al.[16,28] used this approach to show that vaccines limiting infection or transmission have no effect on virulence evolution (in the absence of multiple infections), while vaccines reducing the within-host growth rate of the parasites may favour more virulent strains. Indeed, unlike the other two vaccines, anti-growth rate vaccines reduce the fitness cost associated with disease-induced mortality and consequently modify the balance between the cost (host death) and the benefit (transmission) of pathogen virulence.

Interestingly, the above model can also be tailored to incorporate the effects of other types of vaccines that induce antitoxin immunity against pertussis[37]. A major component of the disease is caused by the production of toxins that can suppress the host immune and inflammatory responses[38]. The production of these toxins is costly but it has positive effects on the within-host growth rate of the bacteria because it reduces host defence. Under these assumptions, the lifetime reproductive success of the mutant at the endemic equilibrium set by the resident becomes[37]:

$$R^m = \frac{\beta\left[\alpha^m\right]e^{-c\alpha^m}}{\left(\delta + \alpha^m + \gamma\right)}\bar{S}_N + \frac{\beta\left[(1-r)\alpha^m\right]e^{-c\alpha^m}}{\left(\delta + (1-r)\alpha^m + \gamma\right)}\bar{S}_V \quad (6)$$

where pathogen virulence α^m is governing the production of the toxin, pathogen transmission β is an increasing function of toxin production, the function $e^{-c\alpha^m}$ measures the fitness cost associated with toxin production, and r measures the efficacy of the antitoxin vaccine. Because they reduce the efficacy of these toxins, antitoxin vaccines may select for new pathogen strategies. If the cost of producing the toxin is high, vaccination selects for lower production of this useless and costly toxin. In contrast, when the cost of producing the toxin is low, the pathogen may overproduce the toxin in order to

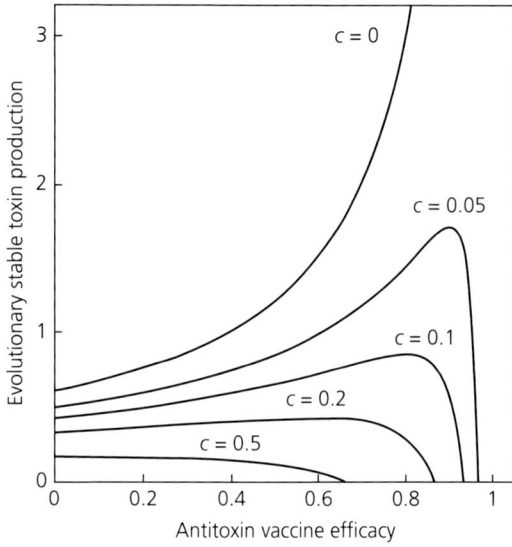

Figure 8.4 Evolutionary stable toxin production for different levels of antitoxin vaccine efficacy, r, and different toxin production costs, c, derived using equation (6). In this figure, all the hosts are assumed to be vaccinated (i.e. $p = 1$) and the following transmission function was used: $\beta\left[\alpha^m\right] = 0.1\left(0.2 + \alpha^m\right)^{0.5}$. Other parameter values: $\delta = 0.01$, $\omega_R = \omega_V = 0$ and $\gamma_V^m = 1$. Reproduced with permission from the authors Gandon S et al. *Nature* 2002;417: 609–10.

compensate for its lower efficacy in vaccinated hosts. Figure 8.4 shows that maximal levels of evolutionary stable toxin production can be reached for intermediate efficacy of antitoxin vaccines. Note that ω_V, the waning immunity of the vaccine, is expected to act indirectly on pathogen evolution via the effect of ω_V on the 'effective coverage' of vaccination (i.e. $\bar{S}_V / (\bar{S}_N + \bar{S}_V)$ at this equilibrium.

8.3.2 Transitory evolution

The analysis in section 8.3.1 is based on the assumption that after the start of a vaccination campaign the system reaches a new epidemiological equilibrium before any new mutation arises in the pathogen population. This is clearly unlikely when the transient dynamics following the start of the vaccination is long and/or when there is some standing genetic variance in the pathogen population because, for instance, of large pathogen mutation rates. To handle these situations we need a more general theoretical framework that allows evolutionary change to occur on any timescale relative to that of the epidemiological

dynamics. More specifically we need a model where the pathogen population may be polymorphic at any given time. In the following, I assume that n different asexually reproducing pathogen strains are competing within a heterogeneous host population composed of naïve and vaccinated individuals. We use q_N^i and q_V^i to denote the frequencies of parasite strain i infecting naïve and vaccinated hosts, respectively (where the subscripts N and V refer to the different types of hosts). We also assume that mutation occurs with rate μ in all strains. A strain j mutates to a strain i with probability m_{ji}. These strains may differ in one or several life-history traits (e.g. transmission, virulence, or recovery) and, consequently, may have different per capita growth rates in the population. Under these assumptions, the epidemiological dynamics depends on the mean parasite trait values in the two types of hosts:

$$\dot{S}_N = \lambda\left(1-p\right) + \omega_V S_V + \omega_R R - \left(\delta + \overline{\Lambda_N}\right)S_N$$

$$\dot{S}_V = \lambda p - \left(\delta + \overline{\Lambda_V} + \omega_V\right)S_V$$

$$\dot{I}_N = \overline{\Lambda_N}S_N - \left(\delta + \overline{\alpha_N} + \overline{\gamma_N}\right)I_N \qquad (7)$$

$$\dot{I}_V = \overline{\Lambda_V}S_V - \left(\delta + \overline{\alpha_V} + \overline{\gamma_V}\right)I_V$$

$$\dot{R} = \overline{\gamma_N}I_N + \overline{\gamma_V}I_V - \left(\delta + \omega_R\right)R$$

where, for instance, $\overline{\alpha_N} = \sum_{i=1}^n q_N^i \alpha_N^i$, $\overline{\Lambda_N} = \overline{\beta_{NN}}I_N + \overline{\beta_{VN}}I_V$, and the average of $\overline{\beta_{AB}}$ is calculated over the distribution of q_A^i. Note that the equation (1) is a special case of equations (7) when there is no variance among strains in their transmission, virulence, or recovery. Following Day and Gandon[39], it is also possible to derive the rate of change in frequency of a given strain i in both types of hosts:

$$\dot{q}_N^i = q_N^i\left(r_{NN}^i - \overline{r_{NN}}\right) + \mu\left(\sum_{j=1}^n m_{ji}q_N^j - q_N^i\right)$$

$$+ \frac{I_V}{I_N}\overline{r_{VN}}\left(q_V^i - q_N^i\right) + \frac{I_V}{I_N}q_V^i\left(r_{VN}^i - \overline{r_{VN}}\right) \qquad (8a)$$

$$\dot{q}_V^i = q_V^i\left(r_{VV}^i - \overline{r_{VV}}\right) + \mu\left(\sum_{j=1}^n m_{ji}q_V^j - q_V^i\right)$$

$$+ \frac{I_N}{I_V}\overline{r_{NV}}\left(q_N^i - q_V^i\right) + \frac{I_N}{I_V}q_N^i\left(r_{NV}^i - \overline{r_{NV}}\right).$$

The quantities r_{AB}^i refer to the instantaneous per capita growth rate of strain i in host type B originating from host type A:

$$r^i_{NN} = \beta^i_{NN}S_N - \left(\delta + \alpha^i_N + \gamma^i_N\right)$$

$$r^i_{VN} = \beta^i_{VN}S_N. \tag{8b}$$

$$r^i_{VV} = \beta^i_{VV}S_V - \left(\delta + \alpha^i_V + \gamma^i_V\right).$$

$$r^i_{NV} = \beta^i_{NV}S_V.$$

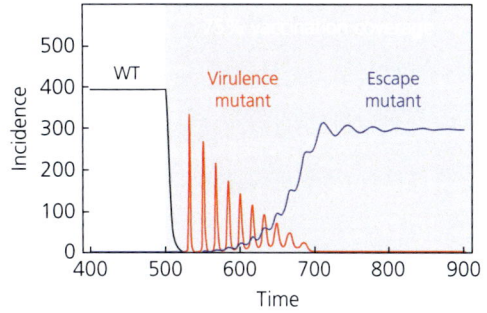

Figure 8.5 Transient evolution after the start of vaccination for the same parameter values as in Figure 8.2 but with some mutation ($\mu = 10^{-3}$) allowing the emergence of two alternative pathogen strains. The virulence mutant (in red): $\alpha^1_V = 11$, $\beta^1_{NN} = 5 \times 10^{-4}$, $\beta^1_{NV} = \beta^1_{NN}/10$, $\beta^1_{VN} = \beta^1_{NN}/2$, $\beta^1_{VV} = \beta^1_{NN}/20$, $R^{N,1}_0 \approx 4.16$, $R^{V,1}_0 \approx 2.08$. The escape mutant (in blue): $\alpha^2_V = 1$, $\beta^2_{NN} = 0.85$ 10^{-4}, $\beta^2_{NV} = \beta^2_{NN}/10$, $\beta^2_{VN} = \beta^2_{NN}/2$, $\beta^2_{VV} = \beta^2_{NN}/20$, $R^{N,2}_0 \approx 4.22$, $R^{V,2}_0 \approx 2.11$.

Equations (8a and 8b) link evolutionary dynamics with the epidemiological equations (7) and can be used to better understand the three different forces acting on the evolution of parasite strain frequencies in both types of hosts (selection, mutation, and migration). For example, let us focus on the evolution of the parasite population infecting the subpopulation of naïve hosts. The first term in (8a) refers to the action of natural selection in naïve hosts. For example, strain i will increase in frequency in naïve hosts if it has a higher per capita growth rate (new infections minus the death and clearance of previous infections) than average in the naïve host subpopulation. The second term refers to the action of mutation. Whether this will result in an increase or a decrease of strain i frequency depends on all strain frequencies, as well as on the mutation model. The third term refers to the effect of immigration (i.e. transmission) of parasites from the subpopulation of vaccinated hosts into naïve hosts. This effect is weighted by the relative size of the two subpopulations of hosts (because the effect of immigration on strain frequency depends on the relative size of the two subpopulations) and depends on the difference between the frequency of strain i in the two subpopulations. For example, if the frequency of the focal strain is higher in parasites infecting vaccinated hosts, then immigration (transmission) from this subpopulation to naïve hosts will increase the frequency of this strain in naïve hosts. The fourth term is also due to immigration from the other subpopulation and illustrates that such transmission can have evolutionary consequences even if the frequency of strain i does not differ between the two subpopulations. Indeed, this fourth term expresses the fact that the frequency of strain i in naïve hosts may increase if this strain is over-represented among immigrants from the population of parasites infecting the vaccinated hosts[6]. For example, if strain i is more transmissible than other strains, the immigration (i.e. transmission) from vaccinated to naïve hosts

will increase the frequency of strain i in naïve hosts even if the two subpopulations do not initially differ in strain frequencies.

This model clarifies the interplay between selection, mutation, and migration on pathogen evolution and can be used to understand the short-term consequences of vaccination. For instance, in Figure 8.5, I examine the consequences of 75 per cent coverage with an imperfect vaccine on pathogen dynamics.

I used exactly the same parameters as in Figure 8.2 except that I allow the wild type to mutate towards two different vaccine-favoured variants. The virulence mutant has a lower reproduction ratio than the escape mutant. Yet, the virulence mutant outcompetes the other pathogen variants during the transient phase that immediately follows the start of the vaccination campaign. This is because the virulence is assumed to be associated with a higher transmission rate. During the 'honeymoon' period of the vaccination campaign (see Figure 8.2), many hosts are uninfected and the availability of this new ecological niche selects for the ability to spread and to recolonize the vaccinated host population. The virulence mutant emerges soon after the start of vaccination and produces several epidemics. But as soon as the oscillatory dynamic damps down, the escape mutant takes over and this yields a new endemic equilibrium.

Note that in the long-term, the maximization of the basic reproduction ratio does predict the evolutionary outcome. Yet, this example illustrates that these predictions may be altered by transitory perturbations of the epidemiology after the initiation of a vaccination campaign. In other words, to predict the evolutionary trajectories of the pathogen, it is also important to accurately predict its epidemiological dynamics[40].

8.3.3 Spatial structure

The theoretical framework described in section 8.3.2 does not take into account the fact that the host population may be spatially structured. Yet, spatial structure and isolation by distance can dramatically affect the epidemiology of pathogens. In particular, spatial structure is known to slow down the speed of spreading epidemics[41–43]. Spatial structure may also affect life-history evolution and, in general, limited dispersal tends to promote less virulent pathogens[44]. Interestingly, however, spatial structure has been shown to interact with vaccine-driven evolution. Zurita-Gutiérrez and Lion[45] showed that imperfect anti-growth rate vaccines could select for higher levels of pathogen virulence in very viscous environments (see figure 3a in[45]). Most pertussis vaccines act through an anti-growth rate effect via the neutralization of the toxin that affects both virulence and transmission (as in equation (6) and Figure 8.4). One may thus expect that taking into account spatial structure may magnify the effects of imperfect vaccines on the evolution of pertussis vaccination.

Another important effect of spatial structure is the potential clustering of vaccination coverage. This clustering has been found to affect the re-emergence of many infectious diseases including pertussis[9,46–48]. The evolutionary consequences of vaccination clustering, however, are less clear. In the earlier-described theoretical framework, vaccination clustering is likely to reduce the rate of transmission between types of hosts (β_{NV} and β_{VN}) relative to the transmission taking place within the same type of hosts (β_{NN} and β_{VV}). This pattern of transmission is likely to promote the coexistence between different types of pathogens that may be specialized on either vaccinated or unvaccinated hosts[17]. But a proper analysis of the effects of vaccination on the epidemiology and evolution of such a spatially structured host population remains to be carried out.

8.4 Concluding remarks

I briefly reviewed previous attempts to model parasite evolution after a vaccination campaign and this theoretical framework can be used to analyse very different epidemiological scenarios. In particular, this model can be used to understand the dynamics of vaccine-favoured mutants. In the long term, the model shows that the evolutionary outcome is mainly governed by the basic reproduction ratio of the different pathogens. In other words, the classical dichotomy between escape and virulence mutants is a bit artificial because, in the end, all the life-history parameters of these different variants (in both naïve and vaccinated hosts) govern the evolutionary outcome. It is likely that there is a continuum of phenotypic variants between escape and virulence mutants. Whether or not a particular mutant is favoured depends on its life-history parameters as well as the type of vaccine and the coverage of vaccination. For instance, imperfect vaccines that act through a reduction of the within-host replication of the pathogen can select for higher virulence and transmission (see Figure 8.4).

The model is also useful to clarify the distinction between short-term and long-term evolution. If the basic reproduction ratio is a relevant measure of fitness for long-term evolutionary predictions, it can be a very poor predictor of short-term evolution during the transient phase following the initiation of vaccination. I illustrate this with an example where a very transmissible (and highly virulent) mutant may be very successful during this transient phase, before being outcompeted by a less transmissible and virulent mutant (but with a larger basic reproduction ratio). To generate both short-term and long-term predictions on the evolution of pathogens following a vaccination campaign we need (1) a quantification of the life-history parameters of all the different pathogen variants (in both naïve and vaccinated hosts) as well as (2) good predictions of the epidemiological dynamics because the availability of susceptible hosts has a major impact on selection (see equation (8b)). In addition, because pathogen evolution is likely to feed back on epidemiological dynamics, it is also important to allow evolution and epidemiology to interact to generate more accurate predictions[40].

In principle, the above models could help us understand the consequences of vaccination in a broad range of infectious diseases. In practice, however, several important obstacles have to be recognized. First of all, it is very difficult to obtain good estimations of the most basic epidemiological traits (e.g. transmission, virulence, and recovery) in both naïve and vaccinated hosts of the different pathogens circulating at a given point in time[49]. The work of Read et al.[50] on Marek's disease provides a notable exception. A careful examination of the phenotypic effects of different virus strains in naïve and vaccinated chickens led to a convincing demonstration of the implication of vaccination on the evolution of virulence in this pathogen. It is obviously impossible to carry out similar studies with human pathogens but without a genotype-to-phenotype map it is impossible to predict evolutionary trajectories of pathogens[40]. Furthermore, the described theoretical framework is based on several simplifying assumptions. I have already discussed the lack of spatial structure but the above models rely also on the assumption that pathogen reproduction is asexual. Most pathogens, however, can recombine their genomes when they co-infect the same host. If key life-history traits of the pathogens (such as transmission and virulence) are governed by multiple loci, recombination may break coadapted gene associations (positive epistasis) and may slow down pathogen adaptation[51]. The impact of recombination on the evolutionary consequences of vaccination remain, however, poorly studied (but see[52]).

There is no doubt that *B. pertussis* has evolved since the start of vaccination in the 1940s[53,54]. However the causal relationship between vaccination and these genotypic changes remains controversial. Since *B. pertussis* has no non-human reservoir, any changes in human immunity, and in particular those induced by vaccination, are likely to have major consequences on the fitness landscape of this pathogen. Yet, there is a lack of information on the phenotypes associated with specific mutations[54]. In the absence of these functional relationships, it is difficult to use the described theoretical framework to predict changes in genotype frequencies. Besides, it is important to notice that several successive shifts in allele frequencies occurred after the implementation of a high-coverage vaccination[53]. Three shifts resulted in antigenic divergence and a fourth shift involved the increase of toxin production. At a given point in time, different vaccine-favoured mutations may coexist and circulate in the same host population. Although *B. pertussis* is generally assumed to reproduce clonally, recombination within the same host may allow the bacteria to acquire several of these mutations. To better understand pertussis evolution, it is important to characterize the epistasis between these mutations[51]. We see that, again, the main factor that limits the predictive power of these models is the lack of information about the fitness costs and the fitness benefits associated with these mutations in naïve and vaccinated hosts. The present book is a very timely contribution because it is an attempt to provide a comprehensive synthesis on pertussis. The theoretical framework discussed in this chapter provides a heuristic tool to understand the multiple consequences of vaccination on epidemiology and evolution. It may inspire more experimental work on the estimation of epidemiological parameters associated with different vaccine-favoured mutations. Ultimately, the integration of this data into the modelling framework described in this chapter may provide a way to quantify the relative impact of vaccine-driven evolution in the recent resurgence of pertussis.

Acknowledgements

I would like to thank Pej Rohani and Samuel Scarpino for very useful comments on an earlier version of this manuscript. I also want to thank Troy Day, Sébastien Lion, Margaret Mackinnon, Sean Nee, and Andrew Read for many inspiring discussions.

References

1. Anderson RM, May RM, Anderson B. *Infectious Diseases of Humans: Dynamics and Control*. Oxford: Oxford University Press; 1992.
2. McLean AR, Blower SM. Modelling HIV vaccination. *Trends Microbiol* 1995;**3**:458–63.
3. Earn DJD, Rohani P, Bolker BM, et al. A simple model for complex dynamical transitions in epidemics. *Science* 2000;**287**:667–70.

4. Rohani P, Earn DJ, Grenfell BT. Impact of immunisation on pertussis transmission in England and Wales. *Lancet* 2000;**355**:285–6.

5. Tildesley MJ, Savill NJ, Shaw DJ, et al. Optimal reactive vaccination strategies for a foot-and-mouth outbreak in the UK. *Nature* 2006;**440**:83–6.

6. Gandon S, Day T. The evolutionary epidemiology of vaccination. *J R Soc Interface* 2007;**4**:803–17.

7. Gangarosa EJ, Galazka AM, Wolfe CR, et al. Impact of anti-vaccine movements on pertussis control: the untold story. *Lancet* 1998;**351**:356–61.

8. de Cellès DM, Magpantay FMG, King AA, et al. The pertussis enigma: reconciling epidemiology, immunology and evolution. *Proc Biol Sci* 2016;**283**:20152309.

9. Atwell JE, Otterloo JV, Zipprich J, et al. Nonmedical vaccine exemptions and pertussis in California, 2010. *Pediatrics* 2013;**132**:624–30.

10. Warfel JM, Zimmerman LI, Merkel TJ. Acellular pertussis vaccines protect against disease but fail to prevent infection and transmission in a nonhuman primate model. *Proc Natl Acad Sci U S A* 2014;**111**:787–92.

11. Althouse BM, Scarpino SV. Asymptomatic transmission and the resurgence of Bordetella pertussis. *BMC Med* 2015;**13**:146.

12. Lam C, Octavia S, Ricafort L, et al. Rapid increase in pertactin-deficient Bordetella pertussis isolates, Australia. *Emerg Infect Dis* 2014;**20**:626–33.

13. Bart MJ, Harris SR, Advani A, et al. Global population structure and evolution of Bordetella pertussis and their relationship with vaccination. *mBio* 2014;**5**:e01074.

14. Scherer A, McLean A. Mathematical models of vaccination. *Br Med Bull* 2002;**62**:187–99.

15. Dushoff J. Incorporating immunological ideas in epidemiological models. *J Theor Biol* 1996;**180**:181–7.

16. Gandon S, Mackinnon M, Nee S, et al. Imperfect vaccination: some epidemiological and evolutionary consequences. *Proc R Soc B Biol Sci* 2003;**270**:1129–36.

17. Gandon S. Evolution of multihost parasites. *Evolution* 2004;**58**:455–469.

18. McLean AR, Anderson RM. Measles in developing countries. Part II. The predicted impact of mass vaccination. *Epidemiol Infect* 1988;**100**:419–42.

19. World Health Organization. Pertussis vaccines: WHO position paper—September 2015. *Releve Epidemiol Hebd* 2015;**90**:433–58.

20. Gustafsson L, Hessel L, Storsaeter J, et al. Long-term follow-up of Swedish children vaccinated with acellular pertussis vaccines at 3, 5, and 12 months of age indicates the need for a booster dose at 5 to 7 years of age. *Pediatrics* 2006;**118**:978–84.

21. Witt MA, Katz PH, Witt DJ. Unexpectedly limited durability of immunity following acellular pertussis vaccination in preadolescents in a North American outbreak. *Clin Infect Dis Off Publ Infect Dis Soc Am* 2012;**54**:1730–5.

22. van Boven M, Mooi FR, Schellekens JFP, et al. Pathogen adaptation under imperfect vaccination: implications for pertussis. *Proc Biol Sci* 2005;**272**:1617–24.

23. Restif O, Grenfell BT. Integrating life history and cross-immunity into the evolutionary dynamics of pathogens. *Proc Biol Sci* 2006;**273**:409–16.

24. Mclean AR. Vaccination, evolution and changes in the efficacy of vaccines: a theoretical framework. *Proc R Soc Lond B Biol Sci* 1995;**261**:389–93.

25. McLean AR. Vaccines and their impact on the control of disease. *Br Med Bull* 1998;**54**:545–56.

26. Wilson JN, Nokes DJ, Carman WF. Current status of HBV vaccine escape variants—a mathematical model of their epidemiology. *J Viral Hepat* 1998;**5**:25–30.

27. Wilson JN, Nokes DJ, Carman WF. The predicted pattern of emergence of vaccine-resistant hepatitis B: a cause for concern? *Vaccine* 1999;**17**:973–8.

28. Gandon S, Mackinnon MJ, Nee S, et al. Imperfect vaccines and the evolution of pathogen virulence. *Nature* 2001;**414**:751–6.

29. André J-B, Gandon S. Vaccination, within-host dynamics, and virulence evolution. *Evol Int J Org Evol* 2006;**60**:13–23.

30. Ganusov VV, Antia R. Imperfect vaccines and the evolution of pathogens causing acute infections in vertebrates. *Evol Int J Org Evol* 2006;**60**:957–69.

31. Day T, Proulx SR. A general theory for the evolutionary dynamics of virulence. *Am Nat* 2004;**163**:E40–63.

32. Day T. Virulence evolution and the timing of disease life-history events. *Trends Ecol Evol* 2003;**18**:113–8.

33. Beck K. Coevolution: mathematical analysis of host-parasite interactions. *J Math Biol* 1984;**19**:63–77.

34. Andreasen V, Christiansen FB. Slow coevolution of a viral pathogen and its diploid host. *Philos Trans R Soc Lond B Biol Sci* 1995;**348**:341–54.

35. Geritz SAH, Kisdi E, Meszé NA, et al. Evolutionarily singular strategies and the adaptive growth and branching of the evolutionary tree. *Evol Ecol* 1998;**12**:35–57.

36. Waxman D, Gavrilets S. 20 questions on adaptive dynamics. *J Evol Biol* 2005;**18**:1139–54.

37. Gandon S, Mackinnon MJ, Nee S, et al. Microbial evolution (communication arising): antitoxin vaccines and pathogen virulence. *Nature* 2002;**417**:610.

38. do Vale A, Cabanes D, Sousa S. Bacterial toxins as pathogen weapons against phagocytes. *Front Microbiol* 2016;**7**:42.

39. Day T, Gandon S. Applying population-genetic models in theoretical evolutionary epidemiology. *Ecol Lett* 2007;**10**:876–88.

40. Gandon S, Day T, Metcalf CJE, et al. Forecasting epidemiological and evolutionary dynamics of infectious diseases. *Trends Ecol Evol* 2016;**31**:776–88.

41. Griette Q, Raoul G, Gandon S. Virulence evolution at the front line of spreading epidemics. *Evolution* 2015;**69**:2810–9.

42. Lion S, Gandon S. Evolution of spatially structured host–parasite interactions. *J Evol Biol* 2015;**28**:10–28.

43. Lion S, Gandon S. Spatial evolutionary epidemiology of spreading epidemics. *Proc Biol Sci* 2016;**283**:20161170.

44. Lion S, Boots M. Are parasites 'prudent' in space? *Ecol Lett* 2010;**13**:1245–55.

45. Zurita-Gutiérrez YH, Lion S. Spatial structure, host heterogeneity and parasite virulence: implications for vaccine-driven evolution. *Ecol Lett* 2015;**18**:779–89.

46. Omer SB, Enger KS, Moulton LH, et al. Geographic clustering of nonmedical exemptions to school immunization requirements and associations with geographic clustering of pertussis. *Am J Epidemiol* 2008;**168**:1389–96.

47. Ruijs WL, Hautvast JL, van der Velden K, et al. Religious subgroups influencing vaccination coverage in the Dutch Bible belt: an ecological study. *BMC Public Health* 2011;**11**:102.

48. Aloe C, Kulldorff M, Bloom BR. Geospatial analysis of nonmedical vaccine exemptions and pertussis outbreaks in the United States. *Proc Natl Acad Sci* 2017;**114**: 7101–5.

49. Gandon S, Day T. Evidences of parasite evolution after vaccination. *Vaccine* 2008;**26** Suppl 3:C4–7.

50. Read AF, Baigent SJ, Powers C, et al. Imperfect vaccination can enhance the transmission of highly virulent pathogens. *PLOS Biol* 2015;**13**:e1002198.

51. Day T, Gandon S. The evolutionary epidemiology of multilocus drug resistance. *Evol Int J Org Evol* 2012; **66**:1582–97.

52. Watkins ER, Penman BS, Lourenço J, et al. Vaccination drives changes in metabolic and virulence profiles of Streptococcus pneumoniae. *PLoS Pathog* 2015;**11**: e1005034.

53. Mooi FR. Bordetella pertussis and vaccination: the persistence of a genetically monomorphic pathogen. *Infect Genet Evol J Mol Epidemiol Evol Genet Infect Dis* 2010;**10**:36–49.

54. Schmidtke AJ, Boney KO, Martin SW, et al. Population diversity among Bordetella pertussis Isolates, United States, 1935–2009. *Emerg Infect Dis* 2012;**18**:1248–55.

55. World Health Organization. *Fact Sheet: World Malaria Report 2016.* 2016. http://www.who.int/malaria/media/world-malaria-report-2016/en/.

CHAPTER 9

Temporal patterns of *Bordetella pertussis* genome sequence and structural evolution

Michael R. Weigand, Margaret M. Williams, and Glen Otero

Abstract

Population genetic studies of *Bordetella pertussis* have long wrestled with a conflicting dichotomy of low gene sequence variation and high restriction fragment profile diversity. Recent applications of high-throughput sequencing and bioinformatics have deepened understanding of *Bordetella* evolution confirming that frequent chromosome rearrangement is a significant source of diversity in this species, in addition to gradual modification of protein-coding gene sequences. This chapter summarizes recent progress in the study of *B. pertussis* genomics to characterize temporal genetic shifts in the circulating population. Much of the presented work reinforces the dichotomy of *B. pertussis* genome evolution, concurrent mutation both of nucleotide sequences and gene arrangements, which still presents challenges in the genomic era. Disentangling the specific contributions of these processes to disease resurgence, as well as exploring their potential utility for vaccine development and novel therapeutics, provides a wealth of future research opportunities.

9.1 Introduction

9.1.1 Background

Taxonomically, *Bordetella pertussis* falls within the family Alcaligenaceae of the beta-proteobacteria. The genus *Bordetella* presently consists of 16 recognized species, including human and animal pathogens as well as recently described environmental species[1–3]. Although some *Bordetella* species have been recovered from the environment, such as *Bordetella petrii* from an anaerobic bioreactor[4,5] and *Bordetella muralis*, *Bordetella tumulicola*, and *Bordetella tumbae* from painted plaster surfaces[2], most Bordetellae are human

or animal pathogens. *Bordetella bronchiseptica* causes infection in several mammals, causing kennel cough in dogs, and was recently shown to survive multiple life cycles of the soil amoeba *Dictyostelium discoideum* while maintaining the ability to infect mice[6]. *Bordetella avium* is a frequent respiratory pathogen of turkeys and other birds, and *Bordetella parapertussis* has been isolated from sheep[7–11]. While *B. pertussis* has long been recognized as an obligate human pathogen and the primary agent of whooping cough[12], it is not the only *Bordetella* species capable of human respiratory infection. Indeed, there are increasing reports of pertussis-like cough

Weigand, M. R., Williams, M. M., and Otero, G., *Temporal patterns of* Bordetella pertussis *genome sequence and structural evolution*. In: *Pertussis: epidemiology, immunology, and evolution.* Edited by Pejman Rohani and Samuel V. Scarpino: Oxford University Press (2019). DOI: 10.1093/oso/9780198811879.003.0009 illness resulting from *Bordetella holmesii*, *B. parapertussis*, and *B.bronchiseptica*, as well as the potentially zoonotic *Bordetella hinzii*. Such infections were once thought to be rare, or perhaps even opportunistic[13–16]. However, recent improvements in molecular diagnostics have begun to reveal more widespread infection by these species, which can frequently be misdiagnosed as *B. pertussis*, including state-wide outbreaks[17–21]. Only in *B. parapertussis* have human-derived isolates been shown by RAPD (random amplified polymorphic DNA) to be monophyletic and genetically distinct from isolates recovered from other hosts (sheep)[11]. Many clinical laboratories continue to identify pertussis infection by polymerase chain reaction (PCR) assays that only target insertion sequence (IS)-*481*, a mobile element present in *B. pertussis*, but also *B. holmesii* and, occasionally, *B. bronchiseptica*[14]. Although retrospective analysis of nasopharyngeal specimens considered pertussis positive by PCR demonstrated 0 to less than 1 per cent contained *B. holmesii* misidentified as *B. pertussis*, it remains unclear the extent to which human respiratory infection by these congenerics contributes to pertussis, or pertussis-like, disease incidence, a topic explored in Chapter 11 of this book[14,22].

Phylogenetically, the closely related 'classic Bordetellae', *B. pertussis*, *B. parapertussis*, and *B. bronchiseptica*, form a discrete clade of pathogenic species within a genus of diverse ecologies[23,24]. In the pre-genomics study of *Bordetella*, the classic Bordetellae were often regarded as subspecies, differentiated by host adaptation[24]. The first release of complete genomes of *B. pertussis*, *B. parapertussis*, and *B. bronchiseptica* by Parkhill et al.[25] confirmed the similarity of genome sequence, but also demonstrated diversity in several characteristics, including genome size, IS element content, and gene order arrangement. Gene complements of the three species suggest that *B. pertussis* and *B. parapertussis* are independent derivatives of a *B. bronchiseptica*-like ancestral organism. With little net gene gain by the derivative species, the overall trend is that *B. pertussis*, and to a lesser extent *B. parapertussis*, have experienced genome reduction during the process of speciation[23–26].

9.1.2 Pre-genomic understanding of *Bordetella pertussis* population genetics

Before whole-genome sequencing (WGS) platforms made bacterial genomes easily attainable, several molecular typing techniques were employed to characterize *B. pertussis* strains and analyse them in the context of epidemiological data[27–32]. As data accumulated from multiple typing methods, two pictures of *B. pertussis* diversity emerged. Nucleotide- or protein-level molecular characterization methods such as multiple core enzyme electrophoresis (MLEE), multiple-locus variable number tandem repeat analysis (MLVA), and multilocus sequence typing (MLST) indicated that little variability exists among circulating *B. pertussis* strains[28,33–35]. By contrast, global typing techniques that examined large fragments across the entire genome, including restriction fragment length polymorphism (RFLP)[27] and pulsed-field gel electrophoresis (PFGE)[30,36,37], painted a picture of higher *B. pertussis* strain diversity.

At a locus level, early molecular typing focused on virulence-related genes, in particular those that encode the immunogenic components of most acellular vaccines: pertactin (*prn*), pertussis toxin (*ptx*), filamentous haemagglutinin (*fhaB*), and fimbriae[38–40]. Even with this limited set of targets, it became clear that *B. pertussis* strains from around the world have diverged from reference strains used in the manufacture of whole and acellular vaccines[35,41,42]. Similarly, comparisons of *B. pertussis* isolates before and after acellular vaccine introduction demonstrated shifts in molecular type (e.g. in pertussis toxin subunit A (ptxA) and the repeat region of *prn*[41]), but on the whole, *B. pertussis* was considered a monomorphic species. Minimal nucleotide sequence variability also held true between the three classic Bordetellae, concluded from DNA–DNA hybridization analysis, MLEE, and comparative sequence analysis of 16S and 23S rRNA genes, *rpoB*, and *gyrB*[24,43].

More recently, *B. pertussis* strain diversity measured by a combination of MLVA and four-locus MLST was lower than that observed through PFGE of the same isolates[32]. Comparison of diversity indices of the three typing methods demonstrated that PFGE had the highest discriminatory power, followed by MLVA, then MLST[29]. PFGE has revealed several profiles in circulating *B. pertussis*, sometimes within a single epidemic[30,32,36,37]. In surveillance within the United States and nine European countries, dynamic shifts in predominate PFGE profiles have been observed across years[30,36,37,44]. In an epidemic in Washington state, United States, 2011–2013, 31 PFGE profiles were observed in isolates across the state, though subgroups of isolates from small communities each shared one PFGE profile[32]. Similarly, in a related genome-wide typing method, RFLP found that small outbreak clusters of isolates, for instance from a convent or single family, shared the same fragment size pattern but *B. pertussis* isolates obtained from the same Dutch town over 6 years without an epidemiological link were variable[27].

Observed diversity in PFGE profiles was linked to chromosomal diversity in a series of elegant genetic experiments which provided the first evidence of genome structural fluidity in *B. pertussis*. In 1997, Stibitz and Yang described a novel approach to chromosomal mapping by directly measuring the distance between genes and illustrated its utility with laboratory strains of *B. pertussis*[45]. Through this 'chromosomal surveying' they observed considerable variation in genome organization among the five strains tested, noting that most could be explained by inversion of large chromosomal segments. Moreover, they noted that the map of *B. parapertussis* was also quite divergent, despite the close phylogenetic relatedness of these two human pathogens inferred by DNA–DNA hybridization[46]. Stibitz and Yang later applied their method to characterize a group of 14 *B. pertussis* clinical isolates with diverse PFGE profiles recovered during an epidemic in Alberta, Canada[47]. Comparing these maps revealed that large-scale chromosomal rearrangements were present in natural, circulating populations as well as laboratory references strains, suggesting a high rate of genomic fluidity. The authors astutely noted that the high copy number of IS481 may catalyse homologous recombination and observed inversions included key virulence factors such as the *fha* and *bvg* operons, speculating on both the cause and consequence of such profound genome rearrangements.

The apparent dichotomy of *B. pertussis* population dynamics, simultaneously appearing as a monomorphic species despite dynamic genome structural fluidity, has developed in parallel as technologies and methodologies progressed. Recent advances in whole-genome mapping and sequencing platforms have overcome many previous limitations for genomic analysis and provided researchers new insight through comparison of hundreds of draft and complete genome assemblies[48–54]. Such studies have greatly deepened the understanding of *Bordetella* species genome evolution, as outlined in

recent excellent reviews[55,56]. Yet this dichotomy continues to confound the study of *B. pertussis* evolution, presenting unique challenges to comparative genomics not experienced in many other bacterial species. In the present chapter, we summarize recent progress in the study of *B. pertussis* genomics with emphasis on characterizing temporal genetic shifts in the circulating population. Much of the presented work confirms the complexity of *B. pertussis* genome evolution, which appears to evolve through both mutation of nucleotide sequence and gene arrangement, concurrently. While these processes have contributed to speciation and adaptation in response to vaccine pressure, disentangling their contributions to each remains difficult.

9.2 Nucleotide sequence variation

Although *B. pertussis* has earned a reputation as a monomorphic pathogen[55,57], it has changed significantly between 1920 and 2010[48]. Previous genomic studies have identified a small number of mutations within a few genes that have spread widely throughout the circulating population with little evidence of geographic restriction[35,48,58–62]. Allelic variants in genes of the acellular vaccine immunogens, such as pertussis toxin subunit A (*ptxA*), pertussis toxin promoter (*ptxP*), pertactin (*prn*), serotype 3 fimbriae (*fim3*), and serotype 2 fimbriae (*fim2*), have emerged such that currently circulating strains no longer match references used to synthesize the acellular vaccine components[49,51], leading many to conclude that they have arisen in response to vaccine-driven selection[53,57,60,63–65]. It is important to understand why, how, and where such genetic changes in the acellular vaccine components have emerged so that improved vaccine development and disease surveillance can be accomplished. While other single nucleotide polymorphisms (SNPs) linked to specific vaccine immunogen alleles have been discovered, they have not become fixed in the population[48,62]. These SNPs reside in genes whose proteins have various functions in different cell compartments. It is possible that many different combinations of these SNPs may collectively provide a small selective advantage, making it unlikely that any one SNP becomes fixed in the population and making it difficult to attribute a distinct advantage to any single SNP. We will briefly describe the major trends that have been reported in *B. pertussis* allelic variation as it pertains to acellular vaccine immunogens, virulence-related factors, and other functional gene categories.

9.2.1 Pertactin

Pertactin (Prn) is an important component of current acellular pertussis vaccines[66]. While at least 17 functional *prn* alleles in *B. pertussis* have been reported, currently alleles *prn1–3* dominate the global population with *prn2* being the most common[48,67]. Several studies have shown that prior to 1960, the *prn1* allele was predominant, but since then *prn2* and *prn3* have risen in frequency to where *prn2* is now the most prevalent allele and *prn1* and *prn3* are similar in frequency[48,68,69,70] (see Figure 9.1). Allelic variation in *prn* is mainly caused by the insertion or

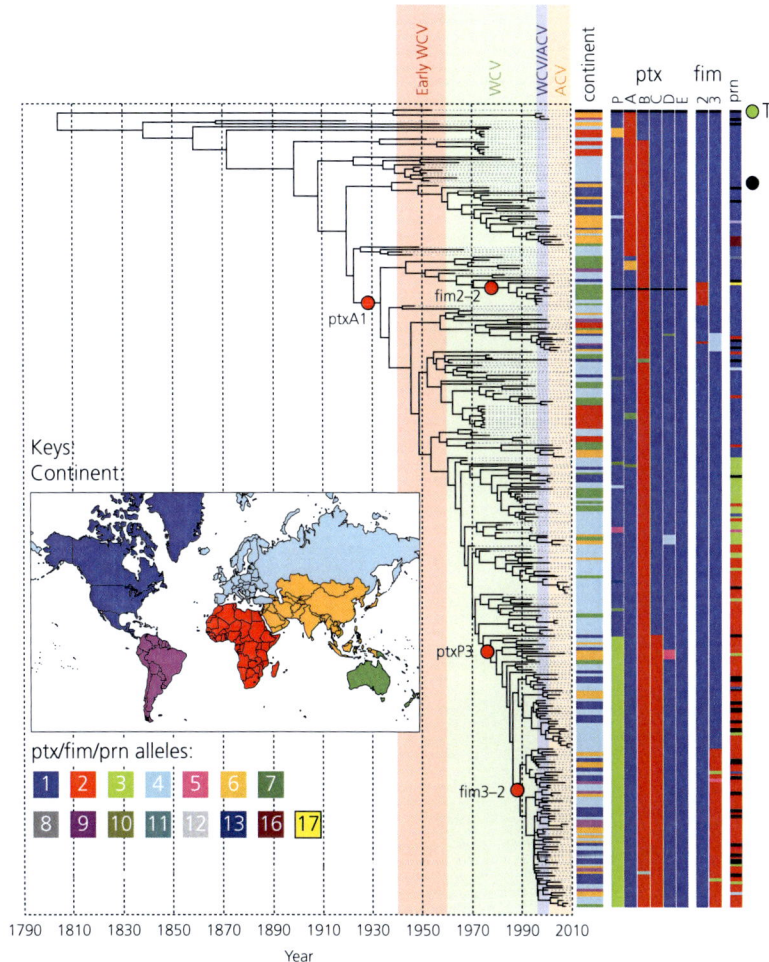

Figure 9.1 Global phylogeny and distribution of vaccine immunogen alleles. Bayesian phylogeny of samples for which date information was available within the most common clade of *B. pertussis*. The position of a node along the x-axis of the tree represents the median date reconstructed for that node across all sampled trees. Dates of whole-cell vaccine (WCV) and acellular vaccine (ACV) periods are shown as background colours behind the tree. To the right of the tree, the continent of origin of isolates is indicated by the first column of horizontal bars, coloured according to the inset key. The remaining nine columns represent loci within the *ptx* operon, the *fim2* and *fim3* loci, and the *prn* locus, with assigned numerical alleles coloured according to the key. The position of reference strain Tohama I (T) is indicated with a green filled circle and a black filled circle represents the American vaccine strain B310 (B). Red circles indicate the major changes in antigen gene alleles in proteins used in current aPs (from *ptxA2* to *ptxA1*, *fim2-1* to *fim2-2*, *ptxP1* to *ptxP3*, and *fim3-1* to *fim3-2*). Reproduced with permission from the authors Bart MJ et al *mBio* 5(2):e01074-14. doi:10.1128/mBio.01074-14. This work is licensed under a Creative Commons Attribution 3.0 Unported license (CC BY 3.0) https://creativecommons.org/licenses/by/3.0/.

deletion of repeats in two regions, R1 and R2, with most variations occurring in R1[71].

In addition to global *prn* allele frequency fluctuations, isolates recovered in the United States have become predominantly Prn deficient in recent years by one of at least 16 different mutations in the *prn* gene[48,64]. Although first detected in the United States in a 1994 *B. pertussis* isolate, the Prn-deficient phenotype did not begin steadily increasing until 2010[32,64]. Lack of Prn does not appear to inhibit *B. pertussis* transmission as emergence of Prn deficiency in France, Italy, Japan, Finland, Australia, and the United States is well documented, with several different genotypes attributed to the Prn-deficient phenotype, including IS*481* insertions in forward or reverse orientation and in multiple locations, 5′ signal sequence deletion, promoter and 5′ coding region deletion, and premature stop codons (see Table 9.1)[32,64,72–79]. While Prn deficiency appears more prevalent in isolates recovered from fully vaccinated patients, it does not appear to have an impact on acellular pertussis vaccine efficacy[75,80,81]. Prn deficiency has also been reported

not to impact the virulence of *B. pertussis* in the murine model of respiratory infection or in human infants less than 6 months of age[82]. As a result of these observations, Prn deficiency has been posited to confer a fitness advantage during infection[81,84], thereby lending support to the premise that B. pertussis is adapting to evade acellular vaccine-mediated immunity[32,35,48,64,75,85]. More recently, Prn-deficient strains have emerged that also do not express filamentous haemagglutinin (FHA)[83,86] or pertussis toxin (PT)[87].

9.2.2 Pertussis toxin

PT is a multi-subunit complex and shifts in allele frequency within specific subunits have been documented in several countries[68–70]. The worldwide emergence of eight *ptxA* alleles that give rise to six protein variants (PtxA1, PtxA3, PtxA4, PtxA5, PtxA9, and PtxA10) differing by only one or two amino acids has been reported[48]. Three alleles have been predominant, *ptxA1*, *ptxA2*, and *ptxA4*, with *ptxA2* and *ptxA4* predominant prior to 1960. The current predominant allele, *ptxA1*, likely arose

Table 9.1 List of *prn* alleles and observed disrupting mutations. Wild-type and mutant pertactin (*prn*) genotypes that result in pertactin production or deficiency, respectively

prn	Allele	GenBank accession no. (genome position)	Reference
Wild type	1	NC_002929 (BP1054)	25,52,70
	2	CP011687 (ABB80_14075)	54,70
	3	AJ011093	70
	4	AJ011015	69
	5	AJ011016	161
	6	HE965805 (2714973–2717696)	162
	7	AJ133784	70
	8	AJ133245	70
	9	AF456356	35
	10	AJ430832	48
	11	AJ507642	163
	12	AB278117	164
	13	EF486277	GenBank
	14	HQ165753	35
	15	JX100834	GenBank
	16	KC981248	48
	17	KC981249	48
	18	KJ433480	165

prn	Allele	GenBank accession no. (genome position)	Reference
Deficiency mutation	*prn1*::26del109	AB670735	64,74
	prn1::IS*481*-1599fwd	AB670737	74
	prn1::IS*481*-1599rev	AB670736	74
	prn2::32del80	Not available	78
	prn2::IS*481*-240fwd	Not available	52,64,79
	prn2::IS*481*-240rev	CP010839 (2997657–2993861)	52,54
	prn2::IS*481*-1613fwd	CP010254 (1079800–1083596)	52,54,64,79,85
	prn2::IS*481*-1613rev	CP010253 (2977857–2974061)	52,54,85
	prn2::IS*1002*-1613fwd	Not available	73
	prn2::IS*481*-2735fwd	KF804023	54,64
	prn2::IS*481*-2735rev	CP013899 (1079795–1083571)	54
	prn2::Stop-C223T	CP011752 (1079787–1082534)	54,78
	prn2::del 666 2 bp	CP021403 (2979800–2977055)	166
	prn2::C638T (A213V)	CP013881 (2975764–2972945)	54
	prn2::Stop-C739T	CP011185 (2973049–2970257)	54
	prn2::Stop-C760T	KF804024	64
	prn2::Stop-C1273T	CP010251 (1080858–1083605)	64,79
	prn2::InsG-1185	KF804025	64
	prn2::−1513del145	CP021402 (1008517–1011120)	166
	prn2::del520 and upstream IS1663	FJ480200	85
	prn2::−2090del478	KF804026	64
	prn promoter, −283del−40	CP021401 (1079512–1082303)	166
	prn 5′ and promoter deletion	CP022362 (1085097–1086504)	166
	prn9::IS*481*-1613fwd	CP012087 (1080853–1084664)	54
	prn promoter disruption, bp −74	KF804027	52, 64
	prn gene deletion	Not available	73

between 1921 and 1932 and increased in frequency from only 5 per cent prior to 1960 to 90 per cent since 2000[48] (see Figure 9.1).

Since PT is understood to act as a critical component of pertussis pathogenicity and virulence, it is somewhat difficult to imagine the emergence of *B. pertussis* strains deficient in the eponymous toxin. Yet strains deficient in both Prn and PT have been reported, resulting from deletion of the neighbouring toxin biosynthetic (*ptx*) and transport (*ptl*) operons[85,87]. Whether PT-deficient strains proliferate in the population in a manner similar to the spread of Prn-deficient strains remains to be seen as such types are currently rare in the circulating population[88].

9.2.3 Pertussis toxin promoter

Several studies suggest that variation in the pertussis toxin promoter (*ptxP*) is linked to clonal sweeps[39,61,62] as the *ptxP3* allele has swiftly replaced the resident *ptxP1* strains in many European countries, the United States, and Australia[39,61,62,89,90]. Strains with *ptxP1* were most common until 1995 but strains with *ptxP3* have since increased in frequency and *ptxP3* is currently the predominant allele. Bayesian analysis suggested that the mutation resulting in the *ptxP3* allele arose between 1974 and 1977[48]. Isolates collected during the pre-vaccine era were mainly harbouring *prn1*, *ptxA2*, and *ptxP1* alleles but since the introduction of whole cell vaccination, isolates harbouring *prn2*, *ptxA1*, and *ptxP3* alleles have become predominant[35,67].

Strains carrying the *ptxP3* allele have been shown to produce more PT *in vitro*[90] and have been associated with increases in pertussis notifications in at least two countries[90,91]. *In vivo* studies have shown that variation in *ptxA*, *ptxP*, and *prn* affect bacterial colonization of naïve and vaccinated mice[92–97], underlining the biological significance of these changes. However, in one study, the effects were not observed[98]. Moreover, comparison of isogenic mutants suggested that both the *ptxP3* allele and its genetic background independently enhanced colonization in mice[99]. While the exact cause(s) of elevated PT and its effects is a topic of ongoing investigation, an improved assay for PT production should allow some light to be shed on the topic[88].

9.2.4 Fimbriae

Fimbriae are important in pertussis immunity and trials with whole-cell pertussis vaccines have revealed that the agglutinogen response in mice and children correlates with protection[100]. *Bordetella pertussis* can produce two closely related but antigenically distinct fimbriae. Designated serotype 2 and 3 fimbriae, they are composed of the major subunits Fim2 and Fim3, respectively, and bind to sulphated sugars[100,101]. Recently, a change to the nomenclature of the *fim2* and *fim3* genes has been proposed to better conform to naming standards and avoid confusion in allele typing. The suggested changes are *fimW* (*fim2*) and *fimH* (*fim3*)[52].

In a recent global population study, only two *fimW* alleles, *fimW1* and *fimW2*, were observed[48], the products of which differ by a single amino acid. The allele *fimW1* predominated in all periods investigated while *fimW2* was found at low frequencies (2–23 per cent) in the same periods[48]. Phylogenetic analysis indicated that mutations in the *fimW2* allele have likely arisen multiple times between 1970 and 2002[48] (see Figure 9.1). Five alleles were identified in *fimH* giving rise to four distinct proteins: FimH-1, FimH-2, FimH-3, and FimH-6, with alleles *fimH1* and *fimH2* predominant[48]. While the *fimH1* allele has been historically predominant, the *fimH2* allele has increased in frequency from 1 to 37 per cent since 2000. The mutation resulting in the *fimH2* allele

is predicted to have occurred between 1986 and 1989[48] (see Figure 9.1) and the polymorphic amino acid residue is located in a surface epitope that has been shown to interact with human serum[102].

9.2.5 Filamentous haemagglutinin

Numerous studies have demonstrated that FHA contributes to colonization and persistent infection by acting as an adhesin that mediates attachment to the respiratory epithelium, promoting biofilm formation and modulating immune responses[103–105]. Mutants lacking FHA exhibit reduced colonization[103,106] and elicit a more robust inflammatory response, leading to faster clearance compared to the wild type[107]. FHA is both a surface-associated and secreted immunogenic protein which undergoes a series of processing and translocation steps during maturation[104]. Both the full-length protein and processed 'mature' protein appear to have roles in virulence[108].

Interestingly, unlike the other acellular vaccine components, there is no reported allelic variation of the structural gene *fhaB* in *B. pertussis* and the FHA proteins found in *B. pertussis*, *B. parapertussis*, and *B. bronchiseptica* are very similar[109]. This lack of allelic diversity underscores the central role FHA plays in host colonization and disease. However, although very rare, FHA-deficient *B. pertussis* strains have been identified. Two strains included homopolymeric frame-shift mutation within *fhaB* that resulted in premature translational termination, another was disrupted by IS*481* insertion, while others had no discernible mutations in the *fhaB* gene, its promoter, or other genes known to be required for its expression[49,86]. Locating a mutation responsible for FHA deficiency outside of the *fhaB* gene itself may be difficult as FHA is processed and secreted via the complex two-partner secretion pathway. Perhaps not surprisingly, some deficient strains have mutations in genes not previously known to participate in FHA biosynthesis[86]. The *fhaB* gene has three IS*481* insertion target sites, like *prn*, but IS*481* disruption at different positions in *fhaB* may confer different phenotypes[54,86].

Like PT, FHA appears fundamental to persistent infection and the emergence of FHA-deficient isolates would represent a remarkable shift in *B. pertussis* host–pathogen interaction. If FHA truly is a requisite for infection, then it is possible that

FHA-deficient strains might only be isolated from patients that are co-infected with an FHA-producing strain. This scenario may also hold true in the case of an FHA-deficient and Prn-deficient isolate[83]. The robustness of FHA-deficient strains and their ability to spread in the population will need to be understood in order to design effective vaccines.

9.2.6 Virulence-associated and non-virulence-associated genes

Multiple studies have reported SNPs in virulence-associated genes, or their promoter regions, that may confer a fitness advantage to *B. pertussis*[48,50,52]. These studies have also detected SNPs in other functional gene categories such as sulphur metabolism, energy metabolism, inorganic ion transport, and amino acid transport, among others[48,50,52]. Specific SNPs have also been linked to the emergence of the now predominant lineage bearing allele *ptxP3* and others appear to discriminate *B. pertussis* isolates collected before and after vaccine introduction[50].

In addition to detection of individual SNPs, SNP density has been utilized to infer which loci may be undergoing selective pressure. Statistically significant increases in SNP density (when compared to that of the whole genome) have been reported in virulence-associated genes and genes encoding transport/binding proteins[48,53]. When genes are categorized by subcellular localization, increased SNP densities have been found in genes encoding outer membrane proteins[48]. The group of genes activated by the BvgAS two-component system has also been found to contain elevated SNP densities compared to those genes repressed or not activated by BvgAS[48]. Elevated SNP densities have also been found in the *ptx* operon promoter region, the intergenic region between the *bvg* operon and *fhaB*, a putative methylase, and two hypothetical proteins[48]. Two genes, *ptxA* and *cysB*, which play a role in sulphur metabolism, were the only individual gene loci reported to have elevated SNP densities[48].

SNPs that arise independently in multiple branches of a phylogenetic tree are also indicative of adaptation. These homoplasic SNPs have been identified in BvgAS-activated genes (*fimW* and *fimH*), a type III secretion protein (*bscI*), a PT transport protein (*ptlB*), a periplasmic solute-binding protein (*smoM*) involved in transport of mannitol, and *cysM* which is involved in cysteine biosynthesis and sulphate assimilation[48].

A recent study of closed genome assemblies from *B. pertussis* epidemic isolates reported unique SNPs in 14 of 44 virulence-related genes[52]. Some SNPs were found in all epidemic isolates and in two cases, also in a vaccine strain. Analysis of all non-synonymous SNPs revealed that 34 unique protein-coding genes were mutated in at least 78 per cent of the strains and a significant proportion of them (46 per cent) were involved in energy production, inorganic ion transport, and amino acid transport, corroborating results observed by others[48]. However, no significant SNP density differences were observed between virulence genes and the rest of the genome, contradictory to studies conducted with larger sample sizes[48,53]. Interestingly, Prn-deficient and Prn-producing strains co-mingled throughout the phylogenetic tree constructed from the variants. Further, no variants outside of *prn* were found to be exclusive to the Prn-deficient strains. The lack of any compensatory mutations associated with Prn deficiency is an important observation that perhaps necessitates a re-evaluation of the role Prn plays in pathogenesis[52].

9.2.7 Genome-level sequence and strain diversity

Studies of *B. pertussis* populations have identified two lineages that have been globally predominant the last 60 years. Isolates collected during the pre-vaccine era mainly harboured the *prn1*, *ptxA2*, and *ptxP1* alleles, but since the introduction of whole-cell vaccination, isolates containing the *prn2*, *ptxA1*, and *ptxP3* alleles have been predominant[67]. While current circulating strains are monomorphic with respect to *ptx*, *ptxP*, and *prn* (when expressed), the rest of the genome remains remarkably fluid, as evidenced by the existence of strains deficient in one or more virulence factors, for example, Prn, PT, and FHA[32,48,49,64,85–87]. Furthermore, lack of PT does not appear to be caused by a multitude of gene disruptions like those observed underlying Prn and, more rarely, FHA deficiency[49,86]. This genomic fluidity has produced circulating strains that no longer match the commonly used reference strain, Tohama I, and other strains used to synthesize acellular vaccine components[49,51,54,110]. As will be discussed below, genomic rearrangement plays a significant role in this observed genomic polymorphism.

The search for SNPs linked to virulence, the various *prn* disruptions, and the discovery of strains lacking expression of other acellular vaccine immunogens underscore the need for improved strain diversity classification. Despite leveraging gene sequence information for a handful of virulence genes, MLST and MLVA are still not as discriminatory as PFGE for classifying *B. pertussis* diversity[29,32]. Unfortunately, PFGE lacks significant genotype information and standardization between laboratories and so is often paired with MLST to provide a more complete picture of *B. pertussis* variability. WGS-based approaches have the potential to provide SNP-level typing resolution of the whole genome, but still face barriers to adoption, such as cost and lack of bioinformatic standardization. A possible solution to some of these challenges may be whole-genome MLST, which offers allele-based typing for thousands of gene targets based on SNP variation detected directly from raw sequencing data[111]. Importantly, such gene-by-gene approaches are more readily standardized across laboratories, making them well suited to disease surveillance and molecular epidemiology. Despite these and other barriers to adoption it is likely only a matter of time until WGS is leveraged for routine SNP-based classification of *B. pertussis*. As will be explained in section 9.3, structural variability will likely need to be included in future developments of strain typing.

9.3 Chromosome structure variation

9.3.1 Speciation through rearrangement

When complete reference assemblies of the closely-related respiratory pathogens *B. bronchiseptica*, *B. parapertussis*, and *B. pertussis* were first compared, the extent to which rearrangement shaped speciation within the genus became apparent[25]. In what has become a foundational study to the field of *B. pertussis* genomics, Parkhill et al. concluded that *B. pertussis* and *B. parapertussis* had diverged from a *B. bronchiseptica*-like ancestor, not through the gradual accumulation of nucleotide sequence polymorphisms, but rather through considerable rearrangement and gene

deletion. Much of these structural differences were bordered by identical copies of IS elements, in both species, providing genomic support for their suspected involvement in rearrangement. Moreover, host-range and gene content variation among the three species was attributed to gene inactivation or loss in *B. pertussis* and *B. parapertussis* through such IS-mediated deletion, not the acquisition of host-specific factors. Only 23 genes detected in *B. pertussis* and *B. parapertussis* were absent in *B. bronchiseptica*. More recent comparative genomics has shown this prevalence towards deletion extends more broadly across the genus *Bordetella*[23].

In that initial comparison of the classic Bordetellae, rearrangements and large-scale deletions appeared most prolific in *B. pertussis*, almost exclusively bound by copies of IS*481*. The element IS*481* has expanded widely throughout the *B. pertussis* genome to more than 240 copies in recent clinical isolates[52]. Its absence in *B. parapertussis* and rare observation in *B. bronchiseptica*[33,112] suggests recent acquisition and, therefore, subsequent rapid expansion has been a major driving force in *B. pertussis* speciation. Most IS element insertions are expected to be deleterious, disrupting protein-coding genes, and thus out-competed through purifying selection. However, when effective population size is small, such as due to population bottlenecks, non-lethal insertions can accumulate through random drift[113]. As a result, IS element expansion has been commonly observed in bacteria with host-restricted ecology, likely resulting from a combination of factors such as reduced strength of purifying selection and fewer essential genes[113,114]. Theoretical modelling also suggests that IS element accumulation may provide a source of adaptive mutation in changing environments[115], similar to varied selection pressure on pathogenic bacteria during cycles of infection and transmission. Genome reduction has been observed in diverse species of bacteria, particularly obligate pathogens or symbionts, and frequently results from gene inactivation and rearrangement following IS element proliferation[116]. For example, analyses of *Yersinia pestis*, *Mycobacterium leprae*, and *Serratia symbiotica* have contributed to the understanding of how such processes lead to divergence of a single species through reduction within a genus of primarily free-living organisms[117–120]. Therefore, it appears that expansion of IS*481* in *B. pertussis* results from both a recent evolutionary bottleneck and subsequent ecological restriction experienced during niche transition to the human host[121]. The resulting high copy number of IS*481* has supplied an abundance of opportunities for structural mutations that have characterized *B. pertussis* species evolution.

9.3.2 Regions of difference

Such processes of structural evolution are unlikely to cease following species divergence and would continue to shape the *B. pertussis* genome, consistent with the original reports of frequent inversion observed through chromosomal mapping[45,47]. Before additional closed assemblies were available for direct comparative analyses, Parkhill's reference genome of Tohama I facilitated further investigation of genome structural variation by higher-throughput methods. Using microarray-based technologies such as comparative genome hybridization, many studies queried the genetic content of circulating *B. pertussis* isolates and frequently identified 'regions of difference' (RDs) compared to Tohama I[26,59,60,122–126]. Most of these reported RDs appeared to be deletions of contiguous protein-coding genes that encoded varied functions. Comparison of isolates from different countries revealed little impact from geography or vaccination history[125,126], but temporal analyses did suggest that *B. pertussis* was continuing to lose genetic material[59]. Some RDs appeared to associate with specific phylogenetic lineages defined by sequence type, suggesting they had become fixed through recent evolutionary history as *B. pertussis* continued adaptation through genome reduction[125–127]. More recent studies employing WGS further solidify these conclusions[48,53,62,63,128]. Two studies attempted to also document gene acquisition but were unsuccessful[50,122]. Microarray hybridization mapping by Brinig et al.[122] also indicated that recent isolates differed from Tohama I, and each other, by multiple rearrangements. In many of these studies, reported RDs and rearrangements were flanked by insertions of IS*481*, again implicating its central role in *B. pertussis* genome evolution. However, identification of RDs in *B. parapertussis* through similar comparative analyses

suggested that such processes are not unique to *B. pertussis*[123].

Although much attention has been given to conserved genome reduction, RDs resulting from large duplications have also been reported in isolates of *B. pertussis*, albeit rarely. Observed duplications have been restricted to a few discrete regions but some isolates have contained more than one duplication. The highly haemolytic strain APV23ai exhibited unstable duplication of the *cya* locus, which encodes adenylate cyclase toxin, by both Southern blot[129] and DNA microarray hybridization[26]. Duplication of flagellar biosynthesis genes has been observed in French vaccine strain IM1416[26], Finnish clinical isolate KKK1330[124], and vaccine reference strains 6229 and 25525[130]. Lastly, Senegalese clinical isolate IS8235 included duplication of iron metabolism genes according to DNA microarray hybridization[131]. At least the *cya* and flagellar duplications were flanked by insertions of IS*481*[26,124] and have been analysed in other isolates using WGS[86]. Together these reports indicate that *B. pertussis* genome content variation is more plastic than reduction.

9.3.3 Long-read sequencing technology facilitates assembly

The development and economization of high-throughput, shotgun sequencing technology has revolutionized all aspects of microbiology. However, popular short-read sequencing platforms are poorly suited to evaluate rearrangements due to the inherent challenges of assembling repeats present in bacterial genomes[132]. Because there are greater than 240 encoded IS elements, draft assemblies of *B. pertussis* comprise hundreds of contigs, most with identical IS elements at their ends, with no indication of their organizational structure. These same IS elements often flank rearrangements, making the structural differences between isolates impossible to infer. Genome sequence-level characterization of rearrangement, like that first described by Parkhill et al.[25], requires complete assemblies and traditional approaches to genome 'finishing' are time-consuming and costly.

Genome sequence-level characterization of structural variation has recently become possible through advances in long-read sequencing technology, specifically the Pacific Biosciences RS II platform, capable of spanning IS elements to permit closed, circular assembly. Many genomes of *B. pertussis* isolates have already been completely assembled using this technology[49,51,52,54,86,87,130,133–135]. Newer sequencing platforms, such as from Oxford Nanopore Technologies, promise even longer read lengths with a dramatically reduced instrument footprint[136] for greater access to closed genome assembly for bacterial pathogens[137]. Important to the study of pertussis disease resurgence, such complete genomic data broadens the scope of measurable variation in a pathogen with little nucleotide sequence diversity[54]. However, additional validation will be critical if such *de novo* assemblies become routine inputs for comparative analyses of chromosome structural rearrangement. Various whole-genome mapping technologies based on restriction enzyme digestion, nicking, or tagging offer valuable, independent confirmation of assembly architecture[138,139].

9.3.4 Whole-genome alignment confirms rearrangement

Multiple-alignment of complete assemblies immediately confirmed previous observations of dynamic gene organization in circulating *B. pertussis*, primarily observable as large inversions and almost always flanked by copies of IS481[49,51] (see Figure 9.2). Akin to the chromosomal mapping of Stibitz and Yang[47], Bowden et al.[52] showed through whole-genome alignment that isolates within defined, state-wide epidemics differed in genome structure, bringing rearrangement analysis fully into the genomic era. Such initial comparisons of complete assemblies revealed considerable rearrangement plasticity among *B. pertussis* isolates[49,51,52,55], but were limited to small sample sizes due to high sequencing costs, leaving their analyses of rearrangement largely observational.

The availability of additional closed assemblies permitted Weigand et al.[54] to perform the first quantitative analysis of rearrangements in circulating *B. pertussis* through alignment of more than 250 genomes, most of which were derived from clinical isolates recovered in the United States during 2010–2014. That study identified 62 discrete genome structures, 53 of which were unique to only

a single isolate. Again, IS481 was frequently found at predicted rearrangement breakpoints, but so too were rRNA operons, of which *B. pertussis* has three copies. The identified structures correlated with PFGE profiles such that genomes from isolates with the same PFGE profile were frequently aligned without gaps or rearrangement ('collinear'). However, minor variants of abundant structures were also detected within smaller groups of collinear isolates and among unique structures, revealing that genome structure continues to diversify.

Using the methods described in Weigand et al.[54], the 287 complete *B. pertussis* genome assemblies available from the National Center for Biotechnology Information GenBank (downloaded on 1 March 2017) can be classified into 22 collinear groups of two or more genomes (see Figure 9.3). Seventy-three of these genomes have unique structures not shared with any others in the database, suggesting that rearrangements remain under-sampled. This pattern is not surprising as many initial complete genome sequencing projects have employed biased selection strategies to prioritize isolated diversity measured by various molecular typing methods, including PFGE. The distribution of common structures in Figure 9.3 reflects the relative abundance of predominant PFGE profiles in the United States during 2010–2014 as described in Weigand et al.[54], the study from which most of the available data are derived.

9.3.5 Symmetric inversion

Bacterial chromosomes exhibit organizational patterns such that core gene order is often conserved, at least within species or lineages[140]. Such organization results, in part, from strand biases linked to bidirectional replication[140,141], which coordinates gene expression[142–144] and ultimately imposes constraints on rearrangement[141,145]. Much of the genomic rearrangements observed in *B. pertussis*, initially by chromosomal mapping[45,47], later by DNA hybridization[122], and eventually whole-genome multiple alignment[52,54], can be explained by large inversions. More specifically, these inversions frequently appear centred around the replication origin or terminus and can thus be described as 'symmetric'. Broader comparative analyses across many species indicates that selection favours replichore balance and,

(a)

(b)

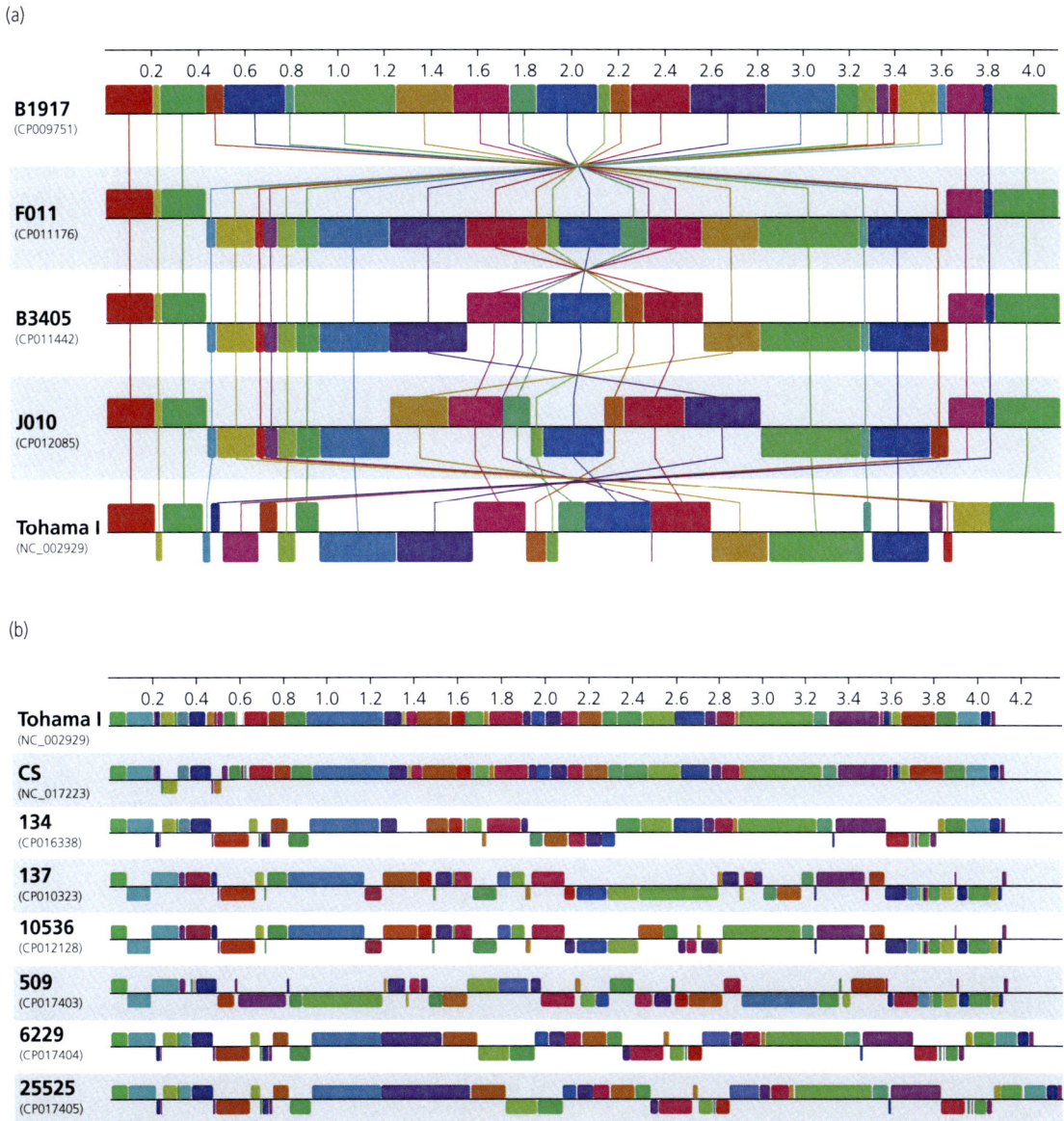

Figure 9.2 Chromosome structural fluidity among isolates of *B. pertussis*. (a) The order and orientation of genetic content varies between clinical isolates (B1917, F011, B3405, and J010) and compared to reference strain Tohama I due to numerous rearrangements. (b) Alignment of available complete genome assemblies from eight vaccine reference strains revealed that their chromosome structures are also heterogeneous. Connecting lines omitted for clarity.

therefore, such symmetric inversions are likely a common feature of bacterial genome evolution[141,145,146].

Genes affected by symmetric inversion remain the same relative distance from the replication origin and do not switch between the leading and lagging strands. Such events presumably have a less negative impact on expression, but may still disrupt regulatory networks by physically separating genes of coordinated function that are frequently encoded nearby. Conversely, inversions which occur within a single replichore, between the replication origin and

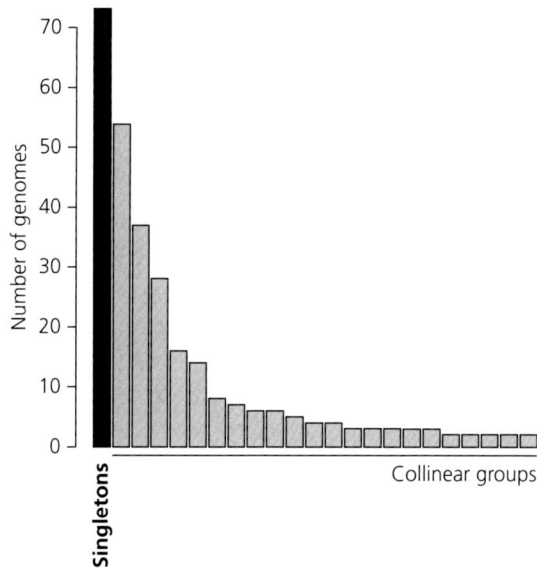

Figure 9.3 Chromosome structure variability among 287 complete genome assemblies of B. *pertussis*. The abundance distribution of unique structures present among all available complete assemblies (downloaded 1 March 2017) included 22 clustered groups and 73 structures present in only one genome (singletons).

terminus ('asymmetric'), likely come at a greater cost as they disrupt both gene distance from the origin and strand placement. The fixation of rare, asymmetric inversions implies that their resulting impact on fitness must be sufficiently beneficial to overcome organizational constraints. Speciation and niche specialization of B. *pertussis* appears to have included such asymmetric inversion during divergence from B. *bronchiseptica* and B. *parapertussis*[25]. Furthermore, conserved asymmetric inversion was also shown to follow deep branching within the B. *pertussis* phylogeny, distinguishing vaccine reference strains and clinical isolates with varied *ptxP* alleles, suggesting their fixation contributed to further adaptation[54]. Minor variations of common genome structures have also been observed, which occasionally differ by small, asymmetric inversions[54].

9.3.6 Phylogenetic concordance of rearrangement

Early studies of gene content variation among circulating B. *pertussis* compared to the Tohama I reference genome revealed that RDs, namely deletions, were often associated with specific phylogenetic lineages defined by sequence type[26,59,60,122–126]. These results were interpreted as evidence for continued genetic reduction of B. *pertussis* as it adapted to the human respiratory niche. Chromosome structural analyses of complete genome assemblies have further shown that some structures, as well as specific rearrangements, are also conserved within discrete phylogenetic lineages[54]. For example, phylogenetic divergence of the *ptxP1* and *ptxP3* allele lineages is well documented[48] and the two groups are distinguishable by a conserved gene content[125–127] and rearrangement[54]. Likewise, alleles *fimH1* and *fimH2* differ by a single base substitution and genomes with allele profiles *ptxP3-fimH1* and *ptxP3-fimH2* also differ by a conserved symmetric inversion despite the observation of multiple discrete structures associated with each sequence type[54]. More broadly, the various genome structures observed in clinical isolates of the *ptxP3* allele lineages are more similar to each other than to Tohama I[54], consistent with reported patterns of gene content[125–127].

Concurrent analyses of sequence and structural variation can shed light on the evolutionary history of rearrangements, highlighting specific changes in local gene order linked to the emergence of predominant types. PFGE profile CDC237 is currently the most abundant profile observed among clinical isolates collected from across the United States[37] and genomes from these isolates are frequently collinear[54]. The phylogenetic placement of CDC237 suggested to Weigand et al. a potential path of sequential inversion from other common types and producing novel gene order conformations, perhaps conferring an advantage to this type[54]. Phylogenetic placement of structures has also indicated that some rearrangements have occurred repeatedly, or even reversibly, leading to dynamic shifts in the frequency of predominant types[54]. Such observations of homoplasy may provide further evidence to suggest that specific, local gene arrangements may be under selection within a background of genome-wide structural heterogeneity. Whether reversible rearrangements result from phase variation, or neutral polymorphism, remains to be tested.

9.3.7 Insertion sequence element variation

The main driver of *B. pertussis* genome evolution appears to be IS*481*, which continues to shape the genome through rearrangement and reduction. Within the genus *Bordetella*, this process seems unique to *B. pertussis* as no other species harbours so many copies of a single genetic element. Genomes of *B. holmesii* include approximately 50 copies each of IS*481* and IS*1001*-like elements[19,23,147] but not enough complete assemblies are yet available to determine if these facilitate rearrangement plasticity similar to that observed in *B. pertussis*. Therefore, initial IS*481* infection may have been a critical, potentiating event in the speciation of *B. pertussis*. Comparative genomics of many bacterial species harbouring transposable elements suggests that long-term coexistence between elements and their host genomes are often stable, balanced by duplication and deletion, following a neutral model[114]. However, outliers exist due to unexpectedly high IS element copy numbers, which are interpreted as resulting from transient imbalances. In the case of *B. pertussis*, this suggests that the species has only recently become infected by IS*481*, whose subsequent proliferation coincides with recent host restriction and genetic isolation that allow fixation of many weakly deleterious insertions through population bottlenecks[113,114].

Early pyrosequencing[59] and more recent complete genome assembly[52,54] of circulating isolates indicated that IS*481* copy number has increased compared to the older Tohama I reference. However, alignment of complete assemblies has also shown that many IS*481* copies consist of neighbouring, adjacent insertions[54]. Such 'tandem' insertions were also reported in an evaluation of IS*481*-targeted PCR for *B. pertussis* diagnosis[148] and together these data have indicated that simple IS*481* copy number does not equal the number of unique loci disrupted by insertion. In fact, whole-genome multiple alignment suggests that copy number variation between isolates largely results from differences in neighbouring, tandem insertions at conserved sites rather than insertion at unique sites[54]. The two other IS elements found in *B. pertussis*, IS*1002* and IS*1663*, have far fewer copies and do not appear to exhibit the same level of variation[54].

Many unoccupied insertion target sites have been predicted in the genome and nearly equal the number of observed IS*481* insertions in circulating strains[54]. Approximately 55 per cent of these predicted, unoccupied target sites fall within coding regions suggesting that further gene loss is possible through IS*481*-mediated disruption but that many such genes are also likely essential for growth. Many target sites have also been predicted in intergenic regions where insertion is assumed to be less deleterious. However, the distribution of intergenic IS elements in many bacterial taxa indicate that insertion between convergent neighbours is more common as insertion upstream of coding regions likely impact regulation[149]. At least two cases of upstream IS*481* insertion influencing gene expression have been reported in *B. pertussis* through the action of outward-facing regulatory features[150,151]. Perhaps fluctuations in neighbouring IS*481* insertion function to further modulate expression of downstream genes.

There remains no clear mechanistic understanding of IS*481* replication, beyond identification of a target site motif, which becomes duplicated upon insertion[152]. Classification remains challenging but IS elements are characterized according to such features as the sequence homology of their transposases, their mechanism of transfer, and the nature of their terminal repeats[153]. In *B. pertussis*, simple differences in IS*481* content between strains has been attributed to recent accumulation, both at genome-scale comparisons of total copy number[52,59] and inactivation of specific genes, such as *prn*[54,64]. Conversely, copy number reduction has been observed only through recombination events between pairs of IS*481* insertions, which result in deletion of intervening protein-coding genes (RDs), a process responsible for much of the reductive evolution underlying *B. pertussis* speciation[26,59,60,123–127]. Initial comparison of IS*481* content variation between complete genomes has provided early evidence of phylogenetic concordance, suggesting that IS element patterns may provide information for strain typing[54]. However, the utility of such information for molecular epidemiology requires accurate models of IS element excision and replication mechanisms.

9.4 Evolutionary implications

Initial comparative genomics of the classic Bordetellae indicated that *B. pertussis* speciation included genome rearrangement and reduction mediated by IS*481*[25]. As discussed throughout this chapter, there is clear evidence to suggest that such processes have continued as part of the ongoing adaptation of *B. pertussis* to its niche in the human respiratory tract. Moreover, such ongoing changes to chromosome structure may be the primary source of mutational diversity within the circulating population, rather than simple allelic diversity of protein-coding genes. Experiments in other species indicate that selection preserves bacterial chromosome organization and imposes constraints on rearrangement[141,145]. However, some gene order changes could confer adaptive gains in fitness and rearrangement-mediated adaptation has been observed in laboratory evolution experiments with diverse bacterial species[154,155]. In *B. pertussis*, the impact of IS element-mediated rearrangement has long been suspected to be manifest at the transcriptional level, as affected genes may be put under control of alternative regulatory elements[122]. The transcriptional effects, and thus their adaptive implications, of rearrangement in *B. pertussis* have not yet been explored. However, some supporting evidence of selection for specific gene order may already exist in temporal fluctuations of PFGE profiles, which can be taken as a proxy for genome structure and generally change in the absence of significant allelic diversity in *B. pertussis*. Shifting distribution of specific chromosome structures, observed as PFGE profiles, suggest that some are more successful, perhaps due to beneficial changes in gene order. The availability of large genomic datasets of complete assemblies are currently novel and evolutionary analyses of rearrangement requires further development of appropriate models of rate heterogeneity.

Phylogenetic concordance of chromosome structure will likely confound analyses of the adaptive nature of rearrangement as it will be difficult to determine which is under selection and which is hitchhiking. For example, one comparison of isogenic mutants reported that both the *ptxP3* allele and its genetic background enhanced colonization in mice[156]. Such challenges appear common to the study of genome association in microbes with high linkage disequilibrium, as highlighted in recent reviews[157–159]. The potential for epistatic interactions between mutations in nucleotide sequences and chromosome structures also need to be explored. Advances in bioinformatic algorithms, larger collections of sequenced *B. pertussis* isolates, and isogenic laboratory mutants are most certainly needed to approach such questions. However, contextualizing genome rearrangement within a SNP phylogeny, Weigand et al. observed not only association of chromosome structures with specific sequence types, but also evidence of identical inversions in divergent backgrounds[54]. These structures did not appear monophyletic in either background, suggesting the inversion had occurred multiple times, or even reversibly, in each. Perhaps the effects of such homoplastic rearrangements, either as mechanisms of phase variation or neutral polymorphism, may be easier to evaluate due to fewer linked SNPs. Of course, such association analyses require linked phenotypes, such as epidemiologically derived data or laboratory assays, to categorize isolates, which is beyond the scope of this chapter.

9.5 Future questions and challenges

Complete assembly of genomes from circulating *B. pertussis* has decisively confirmed decades-old evidence from low-resolution approaches that genome evolution in this species is marked by dramatic rearrangement fluidity. The availability of such data permits characterization of chromosome structure with nucleotide-level resolution, facilitating new opportunities to address questions surrounding pertussis disease resurgence, which are not without their own hurdles. Many such questions, and challenges, have been encountered throughout this chapter and are summarized as follows:

1. What are the phenotypic, or adaptive, effects of chromosome rearrangements? Although conservation of discrete structures may be inferred from comparative genomics, discerning their impact on *B. pertussis* physiology, pathogenicity, or host interaction will require controlled laboratory analyses with isogenic mutants.

2. Can chromosome structural information, and perhaps specific IS*481* content, be incorporated into a more comprehensive molecular typing strategy for *B. pertussis*? Existing methods of strain typing, including PFGE, MLST, MLVA, and SNPs, fail to adequately inform molecular epidemiology. However, such detailed characterization is not currently possible using short-read sequencing platforms available in public health laboratories, but may be overcome with further development of ultraportable, nanopore sequencers.

3. How does chromosome rearrangement contribute to disease resurgence or geographically defined epidemics? A more comprehensive typing strategy, incorporating structural rearrangement, and linked epidemiological data will be necessary to elucidate possible associations.

4. Can analyses of chromosome rearrangement and IS*481* content aid new acellular vaccine design? The rapid spread of Prn deficiency and recent emergence of multi-immunogen deficient mutants raises concerns about the future of vaccine efficacy. Perhaps a deeper understanding of how structural variation contributes to the ongoing evolution of *B. pertussis* will inform development of improved vaccine formulations.

5. How does tandem IS*481* insertion shape genome evolution? Further study is needed to clearly elucidate the mechanism of IS*481* transposition to determine if target site duplication promotes future, neighbouring insertion and its effects on expression of neighbouring protein-coding genes.

6. Does horizontal gene transfer occur in *B. pertussis* populations? Although some reports suggest that horizontal gene transfer has taken place among the classic Bordetellae[160,161] an exhaustive study in *B. pertussis* is lacking. Homologous recombination between IS elements likely mediates chromosome rearrangement and may also facilitate genetic exchange within a population. However, the unusually low SNP density observed among current circulating isolates may impose limits on their detection.

Disclaimer

The findings and conclusions in this chapter are those of the authors and do not necessarily represent the official position of the Centers for Disease Control and Prevention.

References

1. Vandamme PA, Peeters C, Cnockaert M, et al. *Bordetella bronchialis* sp. nov., *Bordetella flabilis* sp. nov. and *Bordetella sputigena* sp. nov., isolated from human respiratory specimens, and reclassification of Achromobacter sediminum Zhang et al. 2014 as Verticia sediminum gen. nov., comb. nov. *Int J Syst Evol Microbiol* 2015;**65**:3674–82.

2. Tazato N, Handa Y, Nishijima M, et al. Novel environmental species isolated from the plaster wall surface of mural paintings in the Takamatsuzuka tumulus: *Bordetella muralis* sp. nov., *Bordetella tumulicola* sp. nov. and *Bordetella tumbae* sp. nov. *Int J Syst Evol Microbiol* 2015;**65**:4830–8.

3. Ivanov YV, Linz B, Register KB, et al. Identification and taxonomic characterization of *Bordetella pseudohinzii sp.* nov. isolated from laboratory-raised mice. *Int J Syst Evol Microbiol* 2016;**66**:5452–9.

4. Gross R, Guzman CA, Sebaihia M, et al. The missing link: *Bordetella petrii* is endowed with both the metabolic versatility of environmental bacteria and virulence traits of pathogenic Bordetellae. *BMC Genomics* 2008;**9**:449.

5. von Wintzingerode F, Schattke A, Siddiqui RA, et al. *Bordetella petrii* sp. nov., isolated from an anaerobic bioreactor, and emended description of the genus Bordetella. *Int J Syst Evol Microbiol* 2001;**51**:1257–65.

6. Taylor-Mulneix DL, Bendor L, Linz B, et al. *Bordetella bronchiseptica* exploits the complex life cycle of *Dictyostelium discoideum* as an amplifying transmission vector. *PLOS Biol* 2017;**15**:e2000420.

7. Bemis DA, Greisen HA, Appel MJG. Pathogenesis of canine Bordetellosis. *J Infect Dis* 1977;**135**:753–62.

8. Magyar T, Chanter N, Lax AJ, et al. The pathogenesis of turbinate atrophy in pigs caused by *Bordetella bronchiseptica*. *Vet Microbiol* 1988;**18**:135–46.

9. Raffel TR, Register KB, Marks SA, et al. Prevalence of *Bordetella avium* infection in selected wild and domesticated birds in the eastern USA. *J Wildl Dis* 2002;**38**:40–6.

10. Welsh R. *Bordetella bronchiseptica* infections in cats. *J Am Anim Hosp Assoc* 1996;**32**:153–8.

11. Yuk MH, Heininger U, Martinez de Tejada G, et al. Human but not ovine isolates of *Bordetella parapertussis* are highly clonal as determined by PCR-based RAPD fingerprinting. *Infection* 1998;**26**:270–3.

12. Cotter PA, Miller JF. Bordetella. In: Groisman EA (ed), *Principles of Bacterial Pathogenesis*. New York: Academic Press; 2001, pp 619–74.

13. Brady C, Ackerman P, Johnson M, et al. *Bordetella bronchiseptica* in a pediatric cystic fibrosis center. *J Cyst Fibros* 2014;**13**:43–8.

14. Burgos-Rivera B, Lee AD, Bowden KE, et al. Evaluation of level of agreement in Bordetella species identification in three U.S. laboratories during a period of increased pertussis. *J Clin Microbiol* 2015;**53**:1842–7.

15. Dworkin MS, Sullivan PS, Buskin SE, et al. *Bordetella bronchiseptica* infection in human immunodeficiency virus-infected patients. *Clin Infect Dis* 1999;**28**:1095–9.

16. Norman KF, John D, Henry M, et al. Bordetella petrii clinical isolate. *Emerg Infect Dis J* 2005;**11**:1131.

17. Koepke R, Bartholomew ML, Eickhoff JC, et al. Widespread *Bordetella parapertussis* infections-Wisconsin, 2011–2012: clinical and epidemiologic features and antibiotic use for treatment and prevention. *Clin Infect Dis* 2015;**61**:1421–31.

18. Bouchez V, Guiso N. *Bordetella pertussis, B. parapertussis*, vaccines and cycles of whooping cough. *Pathog Dis* 2015;**73**:ftv055.

19. Pittet LF, Emonet S, Schrenzel J, et al. *Bordetella holmesii*: an under-recognised Bordetella species. *Lancet Infect Dis* 2014;**14**:510–9.

20. Qin X, Zerr DM, Kronman MP, et al. Comparison of molecular detection methods for pertussis in children during a state-wide outbreak. *Ann Clin Microbiol Antimicrob* 2016;**15**:28.

21. Rodgers L, Martin SW, Cohn A, et al. Epidemiologic and laboratory features of a large outbreak of pertussis-like illnesses associated with cocirculating *Bordetella holmesii* and *Bordetella pertussis*—Ohio, 2010–2011. *Clin Infect Dis* 2013;**56**:322–31.

22. Pittet LF, Emonet S, Francois P, et al. Diagnosis of whooping cough in Switzerland: differentiating *Bordetella pertussis* from *Bordetella holmesii* by polymerase chain reaction. *PLoS One* 2014;**9**:e88936.

23. Linz B, Ivanov YV, Preston A, et al. Acquisition and loss of virulence-associated factors during genome evolution and speciation in three clades of Bordetella species. *BMC Genomics* 2016;**17**:767.

24. von Wintzingerode F, Gerlach G, Schneider B, et al. Phylogenetic relationships and virulence evolution in the genus Bordetella. *Curr Top Microbiol Immunol* 2002;**264**:177–99.

25. Parkhill J, Sebaihia M, Preston A, et al. Comparative analysis of the genome sequences of *Bordetella pertussis, Bordetella parapertussis* and *Bordetella bronchiseptica*. *Nat Genet* 2003;**35**:32–40.

26. Caro V, Hot D, Guigon G, et al. Temporal analysis of French Bordetella pertussis isolates by comparative whole-genome hybridization. *Microbes Infect* 2006;**8**:2228–35.

27. van der Zee A, Vernooij S, Peeters M, et al. Dynamics of the population structure of *Bordetella pertussis* as measured by IS1002-associated RFLP: comparison of pre- and post-vaccination strains and global distribution. *Microbiology* 1996;**142**:3479–85.

28. van Amersfoorth SC, Schouls LM, van der Heide HG, et al. Analysis of *Bordetella pertussis* populations in European countries with different vaccination policies. *J Clin Microbiol* 2005;**43**:2837–43.

29. Advani A, Van der Heide HG, Hallander HO, et al. Analysis of Swedish *Bordetella pertussis* isolates with three typing methods: characterization of an epidemic lineage. *J Microbiol Methods* 2009;**78**:297–301.

30. Advani A, Hallander HO, Dalby T, et al. Pulsed-field gel electrophoresis analysis of *Bordetella pertussis* isolates circulating in Europe from 1998 to 2009. *J Clin Microbiol* 2013;**51**:422–8.

31. Advani A, Gustafsson L, Carlsson RM, et al. Clinical outcome of pertussis in Sweden: association with pulsed-field gel electrophoresis profiles and serotype. *APMIS* 2007;**115**:736–42.

32. Bowden KE, Williams MM, Cassiday PK, et al. Molecular epidemiology of the pertussis epidemic in Washington State in 2012. *J Clin Microbiol* 2014;**52**:3549–57.

33. van der Zee A, Mooi F, Van Embden J, et al. Molecular evolution and host adaptation of Bordetella spp.: phylogenetic analysis using multilocus enzyme electrophoresis and typing with three insertion sequences. *J Bacteriol* 1997;**179**:6609–17.

34. Schouls LM, van der Heide HG, Vauterin L, et al. Multiple-locus variable-number tandem repeat analysis of Dutch *Bordetella pertussis* strains reveals rapid genetic changes with clonal expansion during the late 1990s. *J Bacteriol* 2004;**186**:5496–505.

35. Schmidtke AJ, Boney KO, Martin SW, et al. Population diversity among *Bordetella pertussis* isolates, United States, 1935–2009. *Emerg Infect Dis* 2012;**18**:1248–55.

36. Hardwick TH, Cassiday P, Weyant RS, et al. Changes in predominance and diversity of genomic subtypes of *Bordetella pertussis* isolated in the United States, 1935 to 1999. *Emerg Infect Dis* 2002;**8**:44–9.

37. Cassiday PK, Skoff TH, Jawahir S, et al. Changes in predominance of pulsed-field gel electrophoresis profiles of *Bordetella pertussis* isolates, United States, 2000–2012. *Emerg Infect Dis* 2016;**22**:442–8.

38. Mooi FR, Hallander H, Wirsing von Konig CH, et al. Epidemiological typing of *Bordetella pertussis* isolates: recommendations for a standard methodology. *Eur J Clin Microbiol Infect Dis* 2000;**19**:174–81.

39. Van Loo IH, Mooi FR. Changes in the Dutch *Bordetella pertussis* population in the first 20 years after the introduction of whole-cell vaccines. *Microbiology* 2002;**148**:2011–8.

40. Tsang RS, Lau AK, Sill ML, et al. Polymorphisms of the fimbria fim3 gene of *Bordetella pertussis* strains isolated in Canada. *J Clin Microbiol* 2004;**42**:5364–7.

41. Guiso N, Boursaux-Eude C, Weber C, et al. Analysis of *Bordetella pertussis* isolates collected in Japan before and after introduction of acellular pertussis vaccines. *Vaccine* 2001;**19**:3248–52.

42. Litt DJ, Neal SE, Fry NK. Changes in genetic diversity of the *Bordetella pertussis* population in the United Kingdom between 1920 and 2006 reflect vaccination coverage and emergence of a single dominant clonal type. *J Clin Microbiol* 2009;**47**:680–8.

43. Musser JM, Hewlett EL, Peppler MS, et al. Genetic diversity and relationships in populations of Bordetella spp. *J Bacteriol* 1986;**166**:230–7.

44. Weber C, Boursaux-Eude C, Coralie G, et al. Polymorphism of *Bordetella pertussis* isolates circulating for the last 10 years in France, where a single effective whole-cell vaccine has been used for more than 30 years. *J Clin Microbiol* 2001;**39**:4396–403.

45. Stibitz S, Yang MS. Genomic fluidity of *Bordetella pertussis* assessed by a new method for chromosomal mapping. *J Bacteriol* 1997;**179**:5820–6.

46. Kloos WE, Mohapatra N, Dobrogosz WJ, et al. Deoxyribonucleotide sequence relationships among *Bordetella species*. *Int J Syst Evol Microbiol* 1981;**31**:173–6.

47. Stibitz S, Yang MS. Genomic plasticity in natural populations of *Bordetella pertussis*. *J Bacteriol* 1999;**181**:5512–5.

48. Bart MJ, Harris SR, Advani A, et al. Global population structure and evolution of *Bordetella pertussis*. and their relationship with vaccination. *MBio* 2014;**5**:e01074.

49. Bart MJ, van der Heide HG, Zeddeman A, et al. Complete genome sequences of 11 *Bordetella pertussis* strains representing the pandemic ptxP3 lineage. *Genome Announc* 2015;**3**.

50. Bart MJ, van Gent M, van der Heide HG, et al. Comparative genomics of prevaccination and modern *Bordetella pertussis* strains. *BMC Genomics* 2010;**11**:627.

51. Bart MJ, Zeddeman A, van der Heide HG, et al. Complete genome sequences of *Bordetella pertussis* isolates B1917 and B1920, representing two predominant global lineages. *Genome Announc* 2014;**2**.

52. Bowden KE, Weigand MR, Peng Y, et al. Genome structural diversity among 31 *Bordetella pertussis* isolates from two recent U.S. whooping cough statewide epidemics. *mSphere* 2016;**1**:e00036-16.

53. Sealey KL, Harris SR, Fry NK, et al. Genomic analysis of isolates from the United Kingdom 2012 pertussis outbreak reveals that vaccine antigen genes are unusually fast evolving. *J Infect Dis* 2015;**212**:294–301.

54. Weigand MR, Peng Y, Loparev V, et al. The history of *Bordetella pertussis* genome evolution includes structural rearrangement. *J Bacteriol* 2017;**199**:e00806-16.

55. Belcher T, Preston A. *Bordetella pertussis* evolution in the (functional) genomics era. *Pathog Dis* 2015;**73**:ftv064.

56. Sealey KL, Belcher T, Preston A. *Bordetella pertussis* epidemiology and evolution in the light of pertussis resurgence. *Infect Genet Evol* 2016;**40**:136–43.

57. Mooi FR. *Bordetella pertussis* and vaccination: the persistence of a genetically monomorphic pathogen. *Infect Genet Evol* 2010;**10**:36–49.

58. Bottero D, Gaillard ME, Basile LA, et al. Genotypic and phenotypic characterization of *Bordetella pertussis* strains used in different vaccine formulations in Latin America. *J Appl Microbiol* 2012;**112**:1266–76.

59. Bouchez V, Caro V, Levillain E, et al. Genomic content of *Bordetella pertussis* clinical isolates circulating in areas of intensive children vaccination. *PLoS One* 2008;**3**:e2437.

60. Kallonen T, Grondahl-Yli-Hannuksela K, Elomaa A, et al. Differences in the genomic content of *Bordetella pertussis* isolates before and after introduction of pertussis vaccines in four European countries. *Infect Genet Evol* 2011;**11**:2034–42.

61. Petersen RF, Dalby T, Dragsted DM, et al. Temporal trends in *Bordetella pertussis* populations, Denmark, 1949–2010. *Emerg Infect Dis* 2012;**18**:767–74.

62. van Gent M, Bart MJ, van der Heide HG, et al. Small mutations in *Bordetella pertussis* are associated with selective sweeps. *PLoS One* 2012;**7**:e46407.

63. Octavia S, Maharjan RP, Sintchenko V, et al. Insight into evolution of *Bordetella pertussis* from comparative genomic analysis: evidence of vaccine-driven selection. *Mol Biol Evol* 2011;**28**:707–15.

64. Pawloski LC, Queenan AM, Cassiday PK, et al. Prevalence and molecular characterization of pertactin-deficient *Bordetella pertussis* in the United States. *Clin Vaccine Immunol* 2014;**21**:119–25.

65. Xu Y, Liu B, Grondahl-Yli-Hannuksila K, et al. Whole-genome sequencing reveals the effect of vaccination on the evolution of *Bordetella pertussis. Sci Rep* 2015;**5**:12888.

66. Cherry JD. Why do pertussis vaccines fail? *Pediatrics* 2012;**129**:968–70.

67. Bouchez V, Hegerle N, Strati F, et al. New data on vaccine antigen deficient *Bordetella pertussis* isolates. *Vaccines (Basel)* 2015;**3**:751–70.

68. Cassiday P, Sanden G, Heuvelman K, et al. Polymorphism in *Bordetella pertussis* pertactin and pertussis toxin virulence factors in the United States, 1935–1999. *J Infect Dis* 2000;**182**:1402–8.

69. Mooi FR, He Q, van Oirschot H, et al. Variation in the *Bordetella pertussis* virulence factors pertussis toxin and pertactin in vaccine strains and clinical isolates in Finland. *Infect Immun* 1999;**67**:3133–4.

70. Mooi FR, van Oirschot H, Heuvelman K, et al. Polymorphism in the *Bordetella pertussis* virulence factors P.69/pertactin and pertussis toxin in The Netherlands: temporal trends and evidence for vaccine-driven evolution. *Infect Immun* 1998;**66**:670–5.

71. Boursaux-Eude C, Guiso N. Polymorphism of repeated regions of pertactin in *Bordetella pertussis, Bordetella parapertussis,* and *Bordetella bronchiseptica. Infect Immun* 2000;**68**:4815–7.

72. Barkoff AM, Mertsola J, Guillot S, et al. Appearance of *Bordetella pertussis* strains not expressing the vaccine antigen pertactin in Finland. *Clin Vaccine Immunol* 2012;**19**:1703–4.

73. Lam C, Octavia S, Ricafort L, et al. Rapid increase in pertactin-deficient *Bordetella pertussis* isolates, Australia. *Emerg Infect Dis* 2014;**20**:626–33.

74. Otsuka N, Han HJ, Toyoizumi-Ajisaka H, et al. Prevalence and genetic characterization of pertactin-deficient *Bordetella pertussis* in Japan. *PLoS One* 2012;**7**:e31985.

75. Martin SW, Pawloski L, Williams M, et al. Pertactin-negative *Bordetella pertussis* strains: evidence for a possible selective advantage. *Clin Infect Dis* 2015;**60**:223–7.

76. Quinlan T, Musser KA, Currenti SA, et al. Pertactin-negative variants of *Bordetella pertussis* in New York State: a retrospective analysis, 2004–2013. *Mol Cell Probes* 2014;**28**:138–40.

77. Tsang RS, Shuel M, Jamieson FB, et al. Pertactin-negative *Bordetella pertussis* strains in Canada: characterization of a dozen isolates based on a sur-

vey of 224 samples collected in different parts of the country over the last 20 years. *Int J Infect Dis* 2014;**28**:65–9.

78. Zeddeman A, van Gent M, Heuvelman CJ, et al. Investigations into the emergence of pertactin-deficient *Bordetella pertussis* isolates in six European countries, 1996 to 2012. *Euro Surveill* 2014;**19**:20881.

79. Queenan AM, Cassiday PK, Evangelista A. Pertactin-negative variants of *Bordetella pertussis* in the United States. *N Engl J Med* 2013;**368**:583–4.

80. Breakwell L, Kelso P, Finley C, et al. Pertussis vaccine effectiveness in the setting of pertactin-deficient pertussis. *Pediatrics* 2016;**137**:e20153973.

81. Hegerle N, Dore G, Guiso N. Pertactin deficient *Bordetella pertussis* present a better fitness in mice immunized with an acellular pertussis vaccine. *Vaccine* 2014;**32**:6597–600.

82. Bodilis H, Guiso N. Virulence of pertactin-negative Bordetella pertussis isolates from infants, France. *Emerg Infect Dis* 2013;**19**:471–4.

83. Hegerle N, Paris AS, Brun D, et al. Evolution of French *Bordetella pertussis* and *Bordetella parapertussis* isolates: increase of Bordetellae not expressing pertactin. *Clin Microbiol Infect* 2012;**18**:E340–6.

84. Safarchi A, Octavia S, Luu LD, et al. Pertactin negative Bordetella pertussis demonstrates higher fitness under vaccine selection pressure in a mixed infection model. *Vaccine* 2015;**33**:6277–81.

85. Bouchez V, Brun D, Cantinelli T, et al. First report and detailed characterization of *B. pertussis* isolates not expressing pertussis toxin or pertactin. *Vaccine* 2009;**27**:6034–41.

86. Weigand MR, Pawloski LC, Peng Y, et al. Screening and genomic characterization of filamentous hemagglutinin-deficient *Bordetella pertussis. Infect Immun* 2018;**86**:e00869-17.

87. Williams MM, Sen K, Weigand MR, et al. *Bordetella pertussis* strain lacking pertactin and pertussis toxin. *Emerg Infect Dis* 2016;**22**:319–22.

88. Gates I, DuVall M, Ju H, et al. Development of a qualitative assay for screening of *Bordetella pertussis* isolates for pertussis toxin production. *PLoS One* 2017;**12**:e0175326.

89. Kallonen T, Mertsola J, Mooi FR, et al. Rapid detection of the recently emerged *Bordetella pertussis* strains with the ptxP3 pertussis toxin promoter allele by real-time PCR. *Clin Microbiol Infect* 2012;**18**:E377–9.

90. Mooi FR, van Loo IH, van Gent M, et al. *Bordetella pertussis* strains with increased toxin production associated with pertussis resurgence. *Emerg Infect Dis* 2009;**15**:1206–13.

91. Octavia S, Sintchenko V, Gilbert GL, et al. Newly emerging clones of *Bordetella pertussis* carrying prn2 and ptxP3 alleles implicated in Australian pertussis epidemic in 2008–2010. *J Infect Dis* 2012;**205**:1220–4.

92. Bottero D, Gaillard ME, Fingermann M, et al. Pulsed-field gel electrophoresis, pertactin, pertussis toxin S1 subunit polymorphisms, and surfaceome analysis of vaccine and clinical *Bordetella pertussis* strains. *Clin Vaccine Immunol* 2007;**14**:1490–8.

93. Gzyl A, Augustynowicz E, Gniadek G, et al. Sequence variation in pertussis S1 subunit toxin and pertussis genes in *Bordetella pertussis* strains used for the whole-cell pertussis vaccine produced in Poland since 1960: efficiency of the DTwP vaccine-induced immunity against currently circulating *B. pertussis* isolates. *Vaccine* 2004;**22**:2122–8.

94. King AJ, Berbers G, van Oirschot HF, et al. Role of the polymorphic region 1 of the *Bordetella pertussis* protein pertactin in immunity. *Microbiology* 2001; **147**:2885–95.

95. Komatsu E, Yamaguchi F, Abe A, et al. Synergic effect of genotype changes in pertussis toxin and pertactin on adaptation to an acellular pertussis vaccine in the murine intranasal challenge model. *Clin Vaccine Immunol* 2010;**17**:807–12.

96. van Gent M, van Loo IH, Heuvelman KJ, et al. Studies on Prn variation in the mouse model and comparison with epidemiological data. *PLoS One* 2011;**6**:e18014.

97. Watanabe M, Nagai M. Effect of acellular pertussis vaccine against various strains of *Bordetella pertussis* in a murine model of respiratory infection. *J Health Sci* 2002;**48**:560–4.

98. Denoel P, Godfroid F, Guiso N, et al. Comparison of acellular pertussis vaccines-induced immunity against infection due to *Bordetella pertussis* variant isolates in a mouse model. *Vaccine* 2005;**23**:5333–41.

99. King AJ, van der Lee S, Mohangoo A, et al. Genome-wide gene expression analysis of *Bordetella pertussis* isolates associated with a resurgence in pertussis: elucidation of factors involved in the increased fitness of epidemic strains. *PLoS One* 2013;**8**:e66150.

100. Geuijen CA, Willems RJ, Mooi FR. The major fimbrial subunit of *Bordetella pertussis* binds to sulfated sugars. *Infect Immun* 1996;**64**:2657–65.

101. Geuijen CA, Willems RJ, Bongaerts M, et al. Role of the *Bordetella pertussis* minor fimbrial subunit, FimD, in colonization of the mouse respiratory tract. *Infect Immun* 1997;**65**:4222–8.

102. Williamson P, Matthews R. Epitope mapping the Fim2 and Fim3 proteins of *Bordetella pertussis* with sera from patients infected with or vaccinated against whooping cough. *FEMS Immunol Med Microbiol* 1996;**13**:169–78.

103. Cattelan N, Dubey P, Arnal L, et al. Bordetella biofilms: a lifestyle leading to persistent infections. *Pathog Dis* 2016;**74**:ftv108.

104. Scheller EV, Cotter PA. Bordetella filamentous hemagglutinin and fimbriae: critical adhesins with unrealized vaccine potential. *Pathog Dis* 2015;**73**: ftv079.

105. Villarino Romero R, Osicka R, Sebo P. Filamentous hemagglutinin of *Bordetella pertussis*: a key adhesin with immunomodulatory properties? *Future Microbiol* 2014;**9**:1339–60.

106. Henderson MW, Inatsuka CS, Sheets AJ, et al. Contribution of Bordetella filamentous hemagglutinin and adenylate cyclase toxin to suppression and evasion of interleukin-17-mediated inflammation. *Infect Immun* 2012;**80**:2061–75.

107. Serra DO, Conover MS, Arnal L, et al. FHA-mediated cell-substrate and cell-cell adhesions are critical for *Bordetella pertussis* biofilm formation on abiotic surfaces and in the mouse nose and the trachea. *PLoS One* 2011;**6**:e28811.

108. Melvin JA, Scheller EV, Noel CR, et al. New insight into filamentous hemagglutinin secretion reveals a role for full-length FhaB in Bordetella virulence. *MBio* 2015;**6**:e01189-15.

109. Inatsuka CS, Julio SM, Cotter PA. Bordetella filamentous hemagglutinin plays a critical role in immunomodulation, suggesting a mechanism for host specificity. *Proc Natl Acad Sci U S A* 2005;**102**:18578–83.

110. Caro V, Bouchez V, Guiso N. Is the sequenced *Bordetella pertussis* strain Tohama I representative of the species? *J Clin Microbiol* 2008;**46**:2125–8.

111. Maiden MC, Jansen van Rensburg MJ, Bray JE, et al. MLST revisited: the gene-by-gene approach to bacterial genomics. *Nat Rev Microbiol* 2013;**11**:728–36.

112. Diavatopoulos DA, Cummings CA, Schouls LM, et al. *Bordetella pertussis*, the causative agent of whooping cough, evolved from a distinct, human-associated lineage of *B. bronchiseptica*. *PLoS Pathog* 2005;**1**:e45.

113. Siguier P, Gourbeyre E, Chandler M. Bacterial insertion sequences: their genomic impact and diversity. *FEMS Microbiol Rev* 2014;**38**:865–91.

114. Iranzo J, Gomez MJ, Lopez de Saro FJ, et al. Large-scale genomic analysis suggests a neutral punctuated dynamics of transposable elements in bacterial genomes. *PLoS Comput Biol* 2014;**10**:e1003680.

115. Startek M, Le Rouzic A, Capy P, et al. Genomic parasites or symbionts? Modeling the effects of environmental pressure on transposition activity in asexual populations. *Theor Popul Biol* 2013;**90**:145–51.

116. McCutcheon JP, Moran NA. Extreme genome reduction in symbiotic bacteria. *Nat Rev Microbiol* 2011;**10**:13–26.

117. Burke GR, Moran NA. Massive genomic decay in Serratia symbiotica, a recently evolved symbiont of aphids. *Genome Biol Evol* 2011;**3**:195–208.

118. Chain PS, Hu P, Malfatti SA, et al. Complete genome sequence of *Yersinia pestis* strains Antiqua and

Nepal516: evidence of gene reduction in an emerging pathogen. *J Bacteriol* 2006;**188**:4453–63.

119. Parkhill J, Wren BW, Thomson NR, et al. Genome sequence of *Yersinia pestis,* the causative agent of plague. *Nature* 2001;**413**:523–7.

120. Cole ST, Eiglmeier K, Parkhill J, et al. Massive gene decay in the leprosy bacillus. *Nature* 2001;**409**:1007–11.

121. Preston A, Parkhill J, Maskell DJ. The bordetellae: lessons from genomics. *Nat Rev Microbiol* 2004;**2**:379–90.

122. Brinig MM, Cummings CA, Sanden GN, et al. Significant gene order and expression differences in *Bordetella pertussis* despite limited gene content variation. *J Bacteriol* 2006;**188**:2375–82.

123. Cummings CA, Brinig MM, Lepp PW, et al. *Bordetella* species are distinguished by patterns of substantial gene loss and host adaptation. *J Bacteriol* 2004;**186**:1484–92.

124. Heikkinen E, Kallonen T, Saarinen L, et al. Comparative genomics of *Bordetella pertussis* reveals progressive gene loss in Finnish strains. *PLoS One* 2007;**2**:e904.

125. King AJ, van Gorkom T, Pennings JL, et al. Comparative genomic profiling of Dutch clinical *Bordetella pertussis* isolates using DNA microarrays: identification of genes absent from epidemic strains. *BMC Genomics* 2008;**9**:311.

126. King AJ, van Gorkom T, van der Heide HG, et al. Changes in the genomic content of circulating *Bordetella pertussis* strains isolated from the Netherlands, Sweden, Japan and Australia: adaptive evolution or drift? *BMC Genomics* 2010;**11**:64.

127. Lam C, Octavia S, Sintchenko V, et al. Investigating genome reduction of *Bordetella pertussis* using a multiplex PCR-based reverse line blot assay (mPCR/RLB). *BMC Res Notes* 2014;**7**:727.

128. Safarchi A, Octavia S, Wu SZ, et al. Genomic dissection of Australian *Bordetella pertussis* isolates from the 2008–2012 epidemic. *J Infect* 2016;**72**:468–77.

129. Dalet K, Weber C, Guillemot L, et al. Characterization of adenylate cyclase-hemolysin gene duplication in a *Bordetella pertussis* isolate. *Infect Immun* 2004;**72**:4874–7.

130. Weigand MR, Peng Y, Loparev V, et al. Complete genome sequences of Four *Bordetella pertussis* vaccine reference strains from Serum Institute of India. *Genome Announc* 2016;**4**:e01404-16.

131. Njamkepo E, Cantinelli T, Guigon G, et al. Genomic analysis and comparison of *Bordetella pertussis* isolates circulating in low and high vaccine coverage areas. *Microbes Infect* 2008;**10**:1582–6.

132. Koren S, Harhay GP, Smith TP, et al. Reducing assembly complexity of microbial genomes with single-molecule sequencing. *Genome Biol* 2013;**14**:R101.

133. Weigand MR, Peng Y, Loparev V, et al. Complete genome sequences of *Bordetella pertussis* vaccine reference strains 134 and 10536. *Genome Announc* 2016;**5**:e00979-16.

134. Boinett CJ, Harris SR, Langridge GC, et al. Complete genome sequence of *Bordetella pertussis* D420. *Genome Announc* 2015;**3**:e00657-15.

135. Eby JC, Turner L, Nguyen B, et al. Complete genome sequence of *Bordetella pertussis* strain VA-190 isolated from a vaccinated 10-year-old patient with whooping cough. *Genome Announc* 2016;**4**:e00972-16.

136. Jain M, Olsen HE, Paten B, et al. The Oxford Nanopore MinION: delivery of nanopore sequencing to the genomics community. *Genome Biol* 2016;**17**:239.

137. Quick J, Quinlan AR, Loman NJ. A reference bacterial genome dataset generated on the MinION portable single-molecule nanopore sequencer. *Gigascience* 2014;**3**:22.

138. Wu CW, Schramm TM, Zhou S, et al. Optical mapping of the *Mycobacterium avium* subspecies paratuberculosis genome. *BMC Genomics* 2009;**10**:25.

139. Zhou S, Kile A, Bechner M, et al. Single-molecule approach to bacterial genomic comparisons via optical mapping. *J Bacteriol* 2004;**186**:7773–82.

140. Kang Y, Gu C, Yuan L, et al. Flexibility and symmetry of prokaryotic genome rearrangement reveal lineage-associated core-gene-defined genome organizational frameworks. *MBio* 2014;**5**:e01867.

141. Esnault E, Valens M, Espeli O, et al. Chromosome structuring limits genome plasticity in *Escherichia coli. PLoS Genet* 2007;**3**:e226.

142. Montero Llopis P, Jackson AF, Sliusarenko O, et al. Spatial organization of the flow of genetic information in bacteria. *Nature* 2010;**466**:77–81.

143. Sobetzko P, Travers A, Muskhelishvili G. Gene order and chromosome dynamics coordinate spatiotemporal gene expression during the bacterial growth cycle. *Proc Natl Acad Sci U S A* 2012;**109**:E42–50.

144. Valens M, Penaud S, Rossignol M, et al. Macrodomain organization of the *Escherichia coli* chromosome. *EMBO J* 2004;**23**:4330–41.

145. Campo N, Dias MJ, Daveran-Mingot ML, et al. Chromosomal constraints in Gram-positive bacteria revealed by artificial inversions. *Mol Microbiol* 2004;**51**:511–22.

146. Eisen JA, Heidelberg JF, White O, et al. Evidence for symmetric chromosomal inversions around the replication origin in bacteria. *Genome Biol* 2000;**1**: RESEARCH0011.

147. Tatti KM, Sparks KN, Boney KO, et al. Novel multitarget real-time PCR assay for rapid detection of Bordetella species in clinical specimens. *J Clin Microbiol* 2011;**49**:4059–66.

148. Glare EM, Paton JC, Premier RR, et al. Analysis of a repetitive DNA sequence from *Bordetella pertussis* and its application to the diagnosis of pertussis using

the polymerase chain reaction. *J Clin Microbiol* 1990;**28**:1982–7.

149. Plague GR. Intergenic transposable elements are not randomly distributed in bacteria. *Genome Biol Evol* 2010;**2**:584–90.

150. DeShazer D, Wood GE, Friedman RL. Molecular characterization of catalase from *Bordetella pertussis*: identification of the katA promoter in an upstream insertion sequence. *Mol Microbiol* 1994;**14**:123–30.

151. Han HJ, Kuwae A, Abe A, et al. Differential expression of type III effector BteA protein due to IS481 insertion in *Bordetella pertussis*. *PLoS One* 2011;**6**:e17797.

152. Stibitz S. IS481 and IS1002 of *Bordetella pertussis* create a 6-base-pair duplication upon insertion at a consensus target site. *J Bacteriol* 1998;**180**:4963–6.

153. Siguier P, Gourbeyre E, Varani A, et al. Everyman's guide to bacterial insertion sequences. *Microbiol Spectr* 2015;**3**:MDNA3-0030-2014.

154. Douillard FP, Ribbera A, Xiao K, et al. Polymorphisms, chromosomal rearrangements, and mutator phenotype development during experimental evolution of *Lactobacillus rhamnosus* GG. *Appl Environ Microbiol* 2016;**82**:3783–92.

155. Raeside C, Gaffe J, Deatherage DE, et al. Large chromosomal rearrangements during a long-term evolution experiment with *Escherichia coli*. *MBio* 2014;**5**:e01377-14.

156. de Gouw D, Hermans PW, Bootsma HJ, et al. Differentially expressed genes in *Bordetella pertussis* strains belonging to a lineage which recently spread globally. *PLoS One* 2014;**9**:e84523.

157. Chen PE, Shapiro BJ. The advent of genome-wide association studies for bacteria. *Curr Opin Microbiol* 2015;**25**:17–24.

158. Power RA, Parkhill J, de Oliveira T. Microbial genome-wide association studies: lessons from human GWAS. *Nat Rev Genet* 2017;**18**:41–50.

159. Read TD, Massey RC. Characterizing the genetic basis of bacterial phenotypes using genome-wide association studies: a new direction for bacteriology. *Genome Med* 2014;**6**:109.

160. Buboltz AM, Nicholson TL, Karanikas AT, et al. Evidence for horizontal gene transfer of two antigenically distinct O antigens in *Bordetella bronchiseptica*. *Infect Immun* 2009;**77**:3249–57.

161. Park J, Zhang Y, Chen C, et al. Diversity of secretion systems associated with virulence characteristics of the classical bordetellae. *Microbiology* 2015;**161**:2328–40.

162. Mastrantonio P, Spigaglia P, van Oirschot H, et al. Antigenic variants in *Bordetella pertussis* strains isolated from vaccinated and unvaccinated children. *Microbiology* 1999;**145**:2069–75.

163. Park J, Zhang Y, Buboltz AM, et al. Comparative genomics of the classical *Bordetella* subspecies: the evolution and exchange of virulence-associated diversity amongst closely related pathogens. *BMC Genomics* 2012;**13**:545.

164. Poynten M, McIntyre PB, Mooi FR, et al. Temporal trends in circulating *Bordetella pertussis* strains in Australia. *Epidemiol Infect* 2004;**132**:185–93.

165. Han HJ, Kamachi K, Okada K, et al. Antigenic variation in *Bordetella pertussis* isolates recovered from adults and children in Japan. *Vaccine* 2008;**26**:1530–4.

166. Weigand MR, Peng Y, Cassiday PK, Loparev VN, Johnson T, Juieng P, Nazarian EJ, Weening K, Tondella ML, Williams MM. 2017. Complete genome sequences of *Bordetella pertussis* isolates with novel pertactin-deficient deletions. Genome Announc 5:e00973-17.

CHAPTER 10

Vaccine-driven selection and the changing molecular epidemiology of *Bordetella pertussis*

Ruiting Lan and Sophie Octavia

Abstract

Pertussis remains one of the least controlled vaccine-preventable diseases despite high vaccine coverage in many countries. There are ongoing debates about the causes of its resurgence. Major changes have occurred in the *Bordetella pertussis* population since the introduction of vaccination. Currently circulating strains in Australia and many other high income countries are grouped in single nucleotide polymorphism (SNP) cluster I (also known as *ptxP3* strains). The emergence and expansion of SNP cluster I has been associated with two major genetic changes in *B. pertussis*: a change in its pertussis toxin promoter (to *ptxP3*) which leads to increased pertussis toxin production and the change of the acellular vaccine pertactin gene allele to *prn2*. More recently, strains that lack pertactin have emerged independently in different lineages. The resurgence of pertussis in highly vaccinated populations can be, at least in part, explained by genetic changes that increase the fitness of circulating *B. pertussis* strains.

10.1 Introduction

Pertussis, or whooping cough, is an acute respiratory disease caused by *Bordetella pertussis*. It is estimated that 20–40 million cases and 400,000 deaths occur annually worldwide[1]. Pertussis is most severe in unimmunized infants with milder disease in older children and adults. However, the latter group, with partial immunity from immunization and/or pertussis infection, is an important reservoir of infection. The burden of pertussis is underestimated, because symptoms are often non-specific and available diagnostic tests are not ideal[2]. In countries with established pertussis immunization and surveillance programmes, such as the United States, Canada, and Australia, the incidence of pertussis fell by 90 per cent in the two decades after the introduction of routine vaccination in the 1950s. However, concerns about adverse effects of the original whole-cell vaccine (WCV) led to its replacement in many high income countries in the 1990s with acellular vaccines (ACVs) which are associated with fewer adverse effects such as fever and local reactions[3]. In Australia, ACVs were licensed for use in 1997 and were funded, initially, for fourth and fifth (booster) doses and then for the primary course in 1999 when it almost completely replaced WCV[4]. The most commonly used

Lan, R. and Octavia, S., *Vaccine-driven selection and the changing molecular epidemiology of* Bordetella pertussis. In: *Pertussis: epidemiology, immunology, and evolution.* Edited by Pejman Rohani and Samuel V. Scarpino: Oxford University Press (2019). © Oxford University Press. DOI: 10.1093/oso/9780198811879.003.0010

ACV in Australia contains three components, pertussis toxin (PT), pertactin (Prn), and filamentous haemagglutinin (FHA). Pertussis vaccine uptake in Australia is high (92 per cent for three doses at 12 months of age, 95 per cent at 24 months, and 90 per cent for four doses at 5 years of age)[5], but the disease incidence is the highest of all the vaccine-preventable diseases targeted by the Australian standard vaccination schedule.

In many countries, there have been substantial increases in pertussis notification rates since the 1990s[6]. Factors postulated to explain this include more sensitive diagnostic methods and, therefore, better recognition of atypical disease, more rapidly waning immunity after ACV than after WCV and, potentially, changes in *B. pertussis* itself. The reduction in the number of antigens from, presumably, hundreds (which were poorly defined) in the WCV to as few as one (PT alone) to five (PT, Prn, FHA, and fimbrial serotypes 2 and 3, in various combinations) purified antigens used in ACVs has led to the vehemently contested hypothesis that vaccine-driven selection is responsible for the resurgence of pertussis. In this chapter, we review the recent molecular epidemiology of pertussis with an emphasis on Australian data, currently circulating *B. pertussis* clones worldwide, evidence of vaccine selection pressure, and their implications for strategies to control the disease.

10.2 Pertussis epidemics in Australia

In common with many other countries[7,8], Australia has had a resurgence of pertussis notifications over the past 20 years despite progressive increases in vaccine coverage among children. Over 310,000 cases, from a population of just over 20 million, have been notified since 1991. Epidemic cycles occur every 3–5 years with six peak years: 1994, 1997, 2001, 2005, 2011, and 2015.

However, the 2008–2012 prolonged epidemic lasted 5 years (see Figure 10.1a). The typical seasonal variation of pertussis also disappeared during the prolonged epidemic. At its peak in 2011, 39,000 cases were reported, and the notification rate was 173 per 100,000 population, compared with the pre-epidemic notification rate of 23 per 100,000 population in 2007[9]. Hospitalization rates also increased threefold, compared with the previous 5-year average[9,10]. Children

aged 5–9 years have featured prominently in the 2008–2012 epidemic and most cases (approximately 70 per cent) were in immunized children[10]. The most recent epidemic in 2015 showed a similar trend where children aged 5–9 years were the most commonly affected age group (Figure 10.1b).

Initial increases in notification rates in the early 1990s coincided with the widespread use of locally developed diagnostic serological test for diagnosis of pertussis in older children and adults[11]. The serological test uses a single serum immunoglobulin (Ig)-A level measured by enzyme immunoassay using a whole *B. pertussis* antigen for diagnosis, with the cut-off set to give high specificity but low sensitivity for recent infection. Although this was instrumental in drawing attention to atypical disease in older children and adults, it cannot explain recent epidemics or those in countries where serological tests were not widely used. Other factors, including changes to the vaccine schedule (addition and removal of 18-month booster) and more widespread availability of diagnostic polymerase chain reaction (PCR) tests for pertussis may also have affected notification rates[12]. Although the relative influence of these factors is unknown, most major changes in notification rates appear to reflect real changes in disease incidence.

10.3 Major clones circulating in Australia and globally

Bordetella pertussis is a homogeneous species with little genetic diversity, much of which lies in single base substitutions, infrequent gene deletions, and genome rearrangement[13–22]. *Bordetella pertussis* evolution must be understood within this framework. The diversity has been studied using various methods including multiple locus variable number tandem repeat analysis (MLVA), single nucleotide polymorphism (SNP) typing, pulsed-field gel electrophoresis (PFGE), antigenic gene variation of selected genes encoding ACV antigens and/or virulence, and genome sequencing.

To assess the genetic variation of *B. pertussis* in Australia and potential vaccine-driven selection, Kurniawan et al.[23] analysed 208 Australian isolates collected from the 1970s using MLVA. Of the 37 MLVA types (MTs) identified, four (MT27, MT29, MT64, and MT70) were the most prevalent; they represented 66

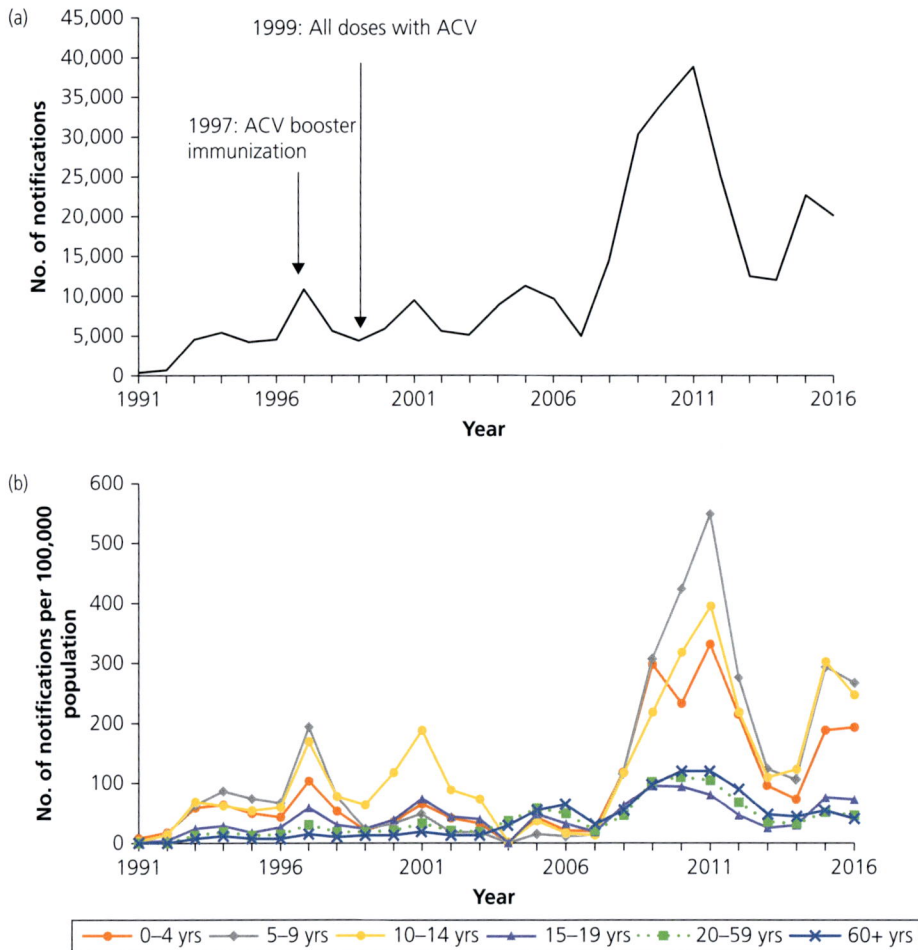

Figure 10.1 Pertussis trends in Australia. (a) Number of pertussis notifications for the period of 1991–2016 in Australia. (b) Age-specific pertussis notification rates per 100,000 population for the period 1991–2016. Mandatory pertussis reporting was instituted in 1991. Whole-cell vaccine was used up to 1999. Acellular vaccine (ACV) was introduced as booster immunizations in 1997 and by 1999–2000, ACVs were used for all pertussis vaccinations.

per cent of strains overall, but occurred with varying frequency during different time periods. MT27 (13 per cent) included one isolate from 1973 and the others were from the 1990s to 2008: MT29 (21 per cent) from 1972; MT70 (21 per cent) from 1996 to 2005; and MT64 (9 per cent) from 1989 and 2002. The same set of isolates was analysed by Octavia et al.[24] using a set of 65 SNPs, identified by comparative genome sequencing of 10 representative isolates. SNP typing divided these isolates into 42 SNP profiles (SPs), most of which grouped into four SNP clusters—I, II, III, and IV—consisting, predominantly, of MT27, MT29, MT64, and MT70, respectively.

Analysis of 194 Australian isolates collected from four states (up to 2010 at the time of the study), during the 2008–2012 pertussis epidemic showed that the majority (86 per cent) belonged to three SNP profiles from cluster I (SPs 13, 14, and 16, in order of decreasing frequency), with similar distributions between states. MLVA typing differentiated these isolates into 27 MTs, of which MT27 (46.3 per cent) and MT214 (13.4 per cent) were the most common. Within each predominant SP, the number of MTs varied between states. Overall, MLVA typing showed a greater diversity amongst the isolates.

The studies of Kurniawan et al.[23] and Octavia et al.[24], also included representative isolates from Canada, the United States, France, and Japan, which were variably distributed among the SNP clusters: isolates from Canada, the United States, and France were predominantly SNP cluster I/MT27 whereas the Japanese isolates were predominantly cluster IV/MT186, with some MT27[25]. Others have also reported a predominance of MT27 in the United Kingdom, United States, and the Netherlands[26–28]. Advani et al.[26] identified 22 MTs among 126 Swedish isolates from pertussis patients from 1970 to 2004, of which 41 per cent were MT27. In the Netherlands, MT27 was not detected in 1995 but became predominant from 1999 to 2004[18,28]. Similarly, in the United Kingdom, MT27 was first observed in 1982 and thereafter steadily increased to become the predominant MLVA type (70 per cent) from 2002 to 2006[27].

A different 87-SNP typing scheme[29] was used to differentiate 125 isolates into 14 SNP types (STs), mostly from European countries and Africa[30]. ST11 and ST12 were predominant in Europe and also present in North America. Based on the four Australian isolates included in that study, ST11 and ST12 can be inferred to belong to SNP cluster I of Octavia et al.[24]. However, as the SNP data from these studies are not directly comparable, they cannot be used collectively to assess the trends globally. An internationally standardized SNP typing scheme is much needed. The recently sequenced genomes of two Dutch isolates (B1917 and B1831) were placed in SNP cluster I. Nevertheless, it is clear that SNP cluster I has largely replaced other *B. pertussis* strains in Australia and other countries since the 1990s.

Bart et al.[20] sequenced 343 global strains isolated from 1920 to 2010. There were two divergent lineages before the WCV era and one of the early divergent lineages seems to have gone extinct. All currently circulating strains were derived from one lineage. The study included 110 *ptxP3* (SNP cluster I) strains and confirmed the stepwise evolution and global dissemination of *ptxP3* strains which will be discussed in more detail later in this chapter. Genome sequence data showed *B. pertussis* spread across the globe rather rapidly without geographical separation[20,21,31].

PFGE has been used for cross-country comparison in European countries and elsewhere[26,32–37]. The proportion of five predominant PFGE profiles, BpSR11, BpSR10, BpSR3, BpSR5, and BpSR12[37], have increased in the last 10 years in Europe, from 44 per cent to 70 per cent of isolates. PFGE group IVβ which contains BpSR11 was also a predominant group in Poland[38]. These PFGE profiles probably belong to SNP cluster I since they contain *ptxP3* and/or *prn2* alleles.

10.4 Epidemiology data on correlation of the changing over of pertussis clones and vaccines

The cause of the increased incidence of pertussis in the last 20 years remains controversial. Some studies suggest that *B. pertussis* were 'escaping' from vaccine-induced immunity, based on temporal correlation between the emergence of new genotypes and changes from the use of WCV to ACV[39,40] but others disagree[41]. Australian data from both MLVA and SNP typing showed a significant correlation. Only the four prevalent MLVA types have sufficient isolates for comparison between three periods, based on vaccine(s) in use: WCV prior to 1997; the transition period (1997 to 1999) when both were used; and ACV, from 2000 onwards. Between these periods, MT64 (cluster III) remained steady, MT29 (cluster II) decreased, while MT27 (cluster I) and MT70 (cluster IV) increased[23]. The same set of isolates used in the MLVA study was also analysed using SNP typing as described in section 10.3[24]. Division by SNP clusters allowed the inclusion of the minor MLVA types in the analysis and reaffirmed the trend observed through MLVA data[24]. It is possible that the correlation is coincidental. However, as discussed in section 10.5, the corresponding changes in antigen-related genes suggest that the changeover to ACV played a role at least in driving the increase of the new clones.

10.5 Changeover of clones is associated with changes in antigenic genes

Allelic variation in antigen-related genes has been quite extensively studied to monitor changes in *B. pertussis*[20,23,24,27,35,36,39,42,43]. The major changes involve genes encoding two major antigens included in ACVs—PT subunit A (*ptxA*) and Prn (*prn*) and the *ptx* promoter (*ptxP*). The corresponding alleles in the *B. pertussis* strain Tohama I which is used to produce the ACV antigens in a number of ACVs are *prn1* and *ptxA2*. There have been at least three key changes in the recent history of *B. pertussis*: from *ptxA2* to *ptxA1* (see Figure 10.2), *prn1* to *prn2*, and a change from *ptxP1* (previously present in most strains) to *ptxP3* at the most recent expansion[24]. Early studies showing changes in allelic prevalence generated debate about vaccine selection[41], but the significance of these changes, in relation to vaccine efficacy, was uncertain. Analysis of allelic changes in relation to changes in clonal distribution, which has been feasible only with the use of MLVA and SNP typing, provides more insight into bacterial

population changes over time, since different alleles are generally associated with different clones. With a few exceptions (such as changes to *prn2* from other *prn* alleles[24]), they generally have not arisen independently.

The possible role of vaccine-driven selection on changes in clone distribution can be seen in the results of MLVA and SNP typing of Australian isolates. Allelic profiles of genes encoding all five antigens used in ACVs (*prn*, *ptxA*, *fim2*, *fim3*, and *fhaB*) and *ptxP* were determined for more than 200 isolates from 1970s to the most recent and compared with those encoding the ACV antigens (*prn1*, *ptxA2*, *fhaB1*, *fim3A*, and *fim2-1*). Isolates from the four predominant Australian SNP clusters share *ptxA1* and *fhaB1* alleles, all clusters have *ptxP1* except cluster I,

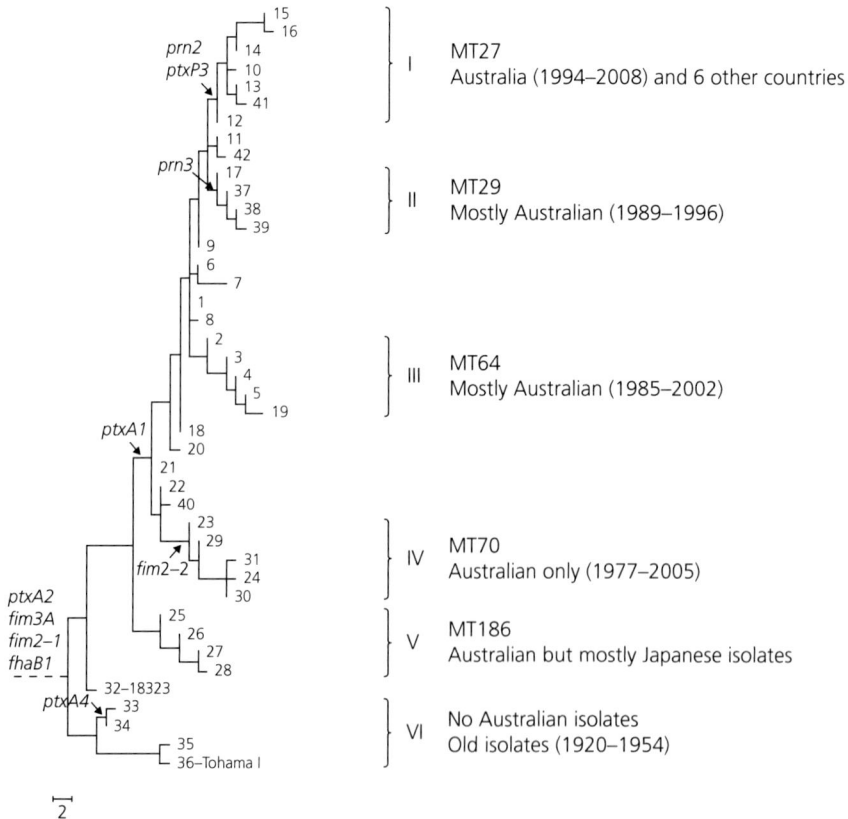

Figure 10.2 Phylogenetic relationships of Australian and global isolates. The SNP clusters were marked with roman numerals. Main MLVA types and isolates distribution are marked on the right. Numbers at the terminal branches are SNP profiles. Changes of ACV antigen gene alleles are marked on the branches. Inferred ancestral alleles were marked at the root of the tree. The bar represents unit of branch length in the number of SNPs. Reproduced with permission from the authors Octavia et al *MBE* 2010 Jan; 28(1) 707–715 Copyright © 2010, Oxford University Press. All Rights Reserved.

Table 10.1   Antigenic gene alleles of major single nucleotide polymorphisms clusters in *B. pertussis* isolates in Australia

	prn	*fha*	*fim2*	*fim3*	*ptx(A)*	*ptxP*
ACV	1	1	2-1	3-A	A2	–
Cluster I	2	1	2-1	3-A/B	A1	*ptxP3*
Cluster II	3	1	2-1	3-A	A1	*ptxP1*
Cluster III	1	1	2-1	3-A	A1	*ptxP1*
Cluster IV	1	1	2-2	3-A	A1	*ptxP1*

Summary from data in Octavia et al.[24]

ACV, acellular vaccine, alleles were based on Tohama I which was used for producing the ACV. Clusters refer to those identified by SNP typing (see text for details).
Genes studied and products: *fha*—filamentous haemagglutinin; *fim*—fimbriae; *prn*—pertactin; *ptx*—pertussis toxin; *ptxP*—pertussis toxin promoter.

which has *ptxP3*. Allelic profiles are shown in Table 10.1[24] and Figure 10.2. Differences compared with five-component ACV are *ptxA1* in all clusters plus additional changes for different clusters: cluster I, *prn2* (and also *fim3-B* in a subset of cluster I); cluster II, *prn3*; cluster III, none; and cluster IV, *fim2-2*. Strains with non-ACV antigen types belong to clusters I and IV, which have emerged and increased since the introduction of ACV. Cluster I is also associated with another important event in *B. pertussis* evolution: the change of *ptx* promoter from *ptxP1* to *ptxP3*, which is discussed in section 10.6[44].

An increase in MT70 (SNP cluster IV, *fim2-2* allele) correlated with the introduction of ACV, suggesting that the change in the *fim2* alleles may be driven by selection pressure[23]. However, five-component ACV (with fimbrial antigens) represents a very small proportion of the ACV used in Australia, and our recent data showed that MT70 has not been identified in Australia since 2010 among more than 200 isolates tested. Therefore, it is unlikely that the increase of MT70 in first few years after ACV introduction was driven by selection on *fim2* from *fim2-1* to *fim2-2*.

There have also been changes in clone distribution in the *B. pertussis* population in the Netherlands. SNP typing[30] of 704 isolates from 1949 to 2010 showed replacement of the local *B. pertussis* population occurred at least on four occasions at 6–19-year intervals, which was referred to as 'selective sweeps'. Each sweep was associated with changes in antigenic gene alleles of *ptxA*, *ptxP*, *prn*, and/or *fim3*[30]. Australian data are also consistent with the occurrence of selective sweeps in the Netherlands and

thus this is a global phenomenon. Taken together, these studies suggest that, when an advantageous allele arises in one population, it spreads to other regions where the selection pressure is similar.

Non-*ptxP3* strains are still more prevalent in many countries where WCVs are in use or have just been replaced by ACVs, therefore supporting ACV-driven selection of SNP cluster I (*ptxP3*) strains. For example, China has only completely replaced the WCV with an ACV in 2012 and predominantly has a *ptxP1* genotype even though an isolate with *ptxP3* was first detected in 2000[45]. In Poland, where WCV had been (and is still) used, exclusively for 50 years[38], the predominant strains were *ptxP1* although the frequency of *ptxP3* has increased from 1.2 per cent in 1995–1999 to 17.5 per cent in 2000–2013[46]. However, in Argentina where WCVs continue to be used, the predominant strains are *ptxP3*[47]. Note that in the study by Hozbor et al.[47], the Argentinean isolates were typed by *prn* but not *ptxP*, with 79 per cent of the isolates carrying *prn2*. By inference, these isolates also harboured *ptxP3* since *prn2* is strongly associated with *ptxP3*[48]. These findings suggest that, once introduced, *ptxP3* strains can compete well and may eventually replace the old genotypes in the WCV-immunized populations.

10.6   Emergence and expansion of the *ptxP3* lineage

Strains carrying the novel *ptxP3* allele were first discovered by Mooi et al.[44] and are now seen as a pandemic lineage that has spread across the globe[20]. The *ptxP3* allele differs from the *ptxP1* allele by a single base mutation in the *ptx* promoter *ptxP*. It has been shown that *ptxP3* increases PT production by about 60 per cent, presumably by improving binding of the global virulence regulator, BvgA, to the PT promoter[44]. Increased PT production potentially provides a selective advantage since it is a key virulence factor and also the major immunogen in all ACVs. Increased hospitalization rates among children infected with *B. pertussis* containing the *ptxP3* allele have been reported in the Netherlands[44], suggesting that increased PT production causes more severe disease. An increased proportion of severe disease in children infected by *ptxP3* strains was also observed in Australia[49].

The prevalence of *ptxP3* varied between countries from less than 10 per cent to 100 per cent[36,43,47,50–55]. The earliest reported appearance of *ptxP3* strains was in 1988 in the Netherlands,[29] when the WCV was still in use. The ACV was introduced in 2001, as a booster, and only replaced the WCV in 2005. This could be interpreted to indicate that there is no role for the ACV in the emergence of *ptxP3* strains since they did not become the predominant strains until 2004, during the transitional period[56].

In other European countries, *ptxP3* has also reached a high frequency. In Sweden, *ptxP3* was first detected in 1997 and subsequently became predominant[36]. In the Gothenburg region, where a PT mono-component vaccine was exclusively used between 1996 and 1999, while the rest of Sweden used predominantly a three-component vaccine, *ptxP3* isolates were found with a particularly high frequency, suggesting vaccine-driven selection[36]. The use of mono-PT vaccine could have facilitated the emergence of *ptxP3* strains in the Gothenburg region[36]. In Denmark, where a mono-component PT ACV was used from 1997[43], *ptxP3* isolates were first found in 1995 and reached 63 per cent of isolates between 1997 and 2010. The rapid increase may be associated with the ACV use as well as imports of *ptxP3* strains from neighbouring countries where *ptxP3* strains are prevalent[43]. In Finland, the earliest detection of *ptxP3* was in 1994, and by 2003 *ptxP3* isolates were the majority and by 2007, almost universal[53]. However, the ACV only replaced the WCV in 2005, after *ptxP3* strains became predominant. Therefore, it was concluded that the increase in *ptxP3* could not be directly linked with the resurgence, although it

coincided with pertussis epidemics in 1999 and 2003–2004[53].

In Australia, *ptxP3* strains were first observed in 1997 and belonged to SNP cluster I[48]. The frequency of *ptxP3* strains reached 86 per cent in the 2008–2012 epidemic and the expansion of *ptxP3* strains coincided with the introduction of ACV in Australia[24].

In the United States, the earliest *ptxP3* (MT27) isolates were found in 1989, reached a high frequency in 1991–1996 within the early ACV period, and are now almost universal[52]. Similarly, *prn2* emerged during the period when the *B. pertussis* population was relatively diverse and subsequently the *ptxP3*–*prn2* genotype became predominant. However, again, these changes occurred about 10 years before the transition of WCV to ACV and so apparently do not support the selection of the *ptxP3*–*prn2* genotype by ACV. In contrast, the emergence and predominance of *fim3B* correlated with the changeover to ACV[52].

10.7 Multiple independent emergence and expansion of pertactin-deficient strains

In contrast to *ptxP3* strains, which are likely to have emerged once and spread across the globe[20], there are clearly multiple independent emergence and expansion events of Prn-deficient strains in different countries. Convergent evolution to the same Prn deficiency phenotype is perhaps the best evidence of vaccine selection from an evolutionary perspective. Martin et al.[57] found that patients who received at least one dose of ACV have a two- to fourfold higher risk of being infected by a Prn-deficient strain than unvaccinated patients. Prn-deficient strains also showed increased fitness in ACV-immunized hosts in mouse studies[58,59]. However, Breakwell[60] reported that ACV is similarly effective against both Prn-deficient and Prn-producing strains.

Infection by Prn-deficient strains does not cause more severe disease[49,61]. Some evidence indicates that Prn-deficient strains may be associated with a less severe disease, which may lead to increased transmission due to delayed or under-diagnosis of the infected cases[49].

Prn-deficient strains were first reported in France[62], then in Japan[25] and now in many countries including

United Kingdom, Canada, United States, Australia, and Israel[57,58,63–70] (see Figure 10.3). Prn-deficient isolates were first found in Japan in 1995 with MT186 and the closely related MT194 and MT226[25], which carry *prn1* and belonged to SNP cluster V[24].

The French isolates carried *prn2* and thus most likely belonged to SNP cluster I. Interestingly, none of the MT27 (SNP cluster I) isolates from Japan were Prn deficient[25]. The Prn-deficient strains found in Australia all belonged to SNP cluster

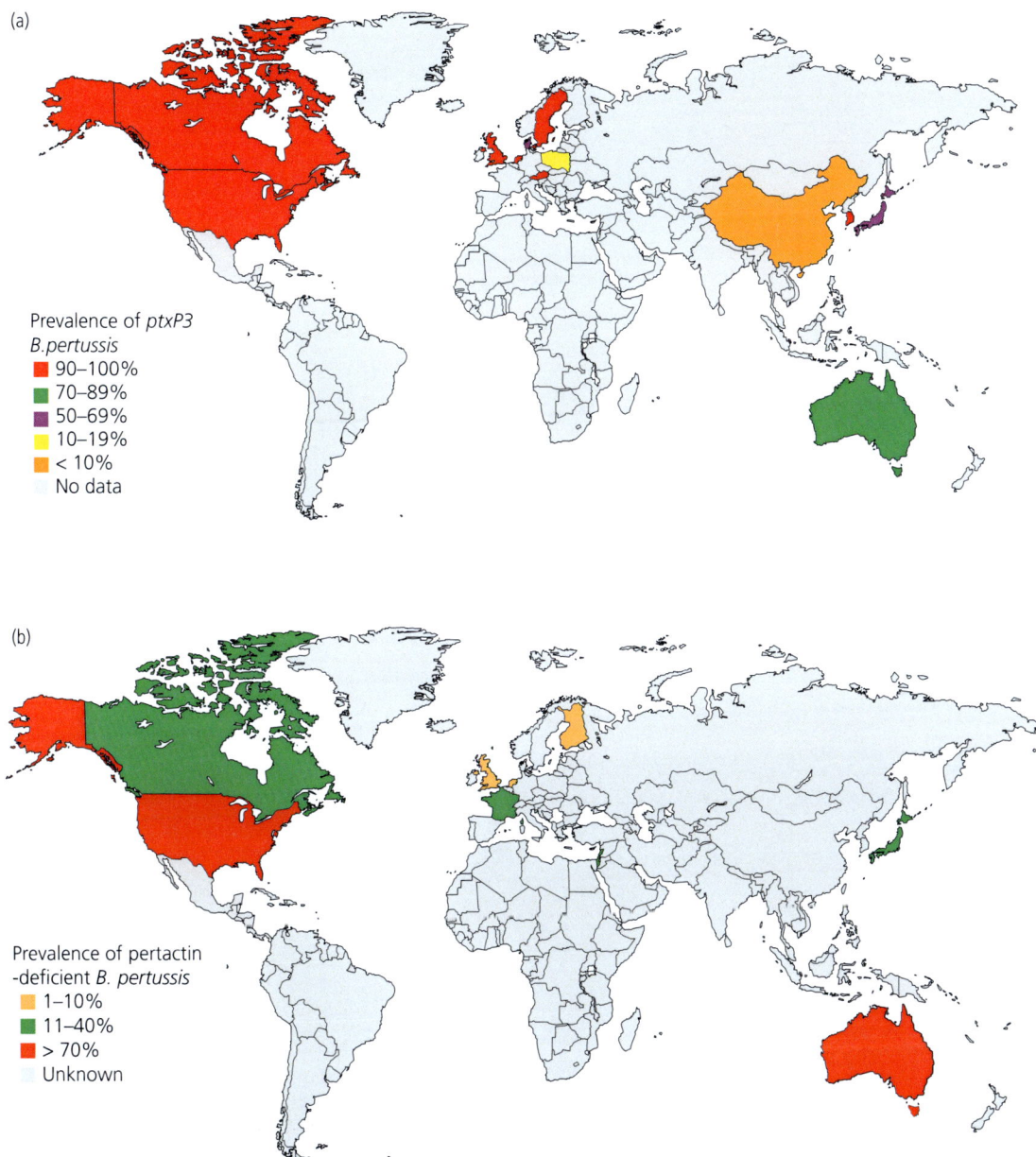

Figure 10.3 World map showing prevalence of (a) *ptxP3* and (b) Prn-deficient *B. pertussis* strains in countries where assessments have been made (coloured according to legend). The maps were created using mapchart.net (http://mapchart.net). Panel (b): Carbonetti et al *CVI* 2016 Nov 23 (11): 842–850 Copyright © 2016, American Society for Microbiology. All Rights Reserved.

I. Prn-deficient strains were at a very low frequency at the start of the 2008 to 2012 epidemic but by 2012 they became predominant with a prevalence of 78 per cent[64]. In the United States, the prevalence of Prn-deficient strains varied from 18 per cent to 85 per cent[57,66–68]. The prevalence in European countries is generally low, from 1 to 13 per cent[58,63].

Prn-deficient strains have been reported to be decreasing in Japan from 41 per cent in 2005–2007 to 8 per cent in 2014–2016[71]. The decrease was due to clonal replacement from MT186 to MT27, that is, from non-*ptxP3* (SNP cluster V) to *ptxP3* strains (SNP cluster I). It should be noted that Prn-deficient strains reported in all other countries belonged to *ptxP3* lineage.

Non-expression of Prn was mainly due to damage of the *prn* gene by insertion sequence (IS)-*481* which can insert at the same site in either the forward or reverse direction relative to the *prn* coding region[25,62,64,65]. It can also be disrupted by the insertion of IS*1002*, deletion of a single base in a poly(G) tract of the coding region, deletion of the entire *prn* gene, inversion of the promoter region, deletion of 49 bases in the signal sequence, or deletion of 28 amino acids in the signal peptide[22,25,62,64,65]. Some Prn-deficient strains had no damage in the coding region or the promoter of *prn*, suggesting that other regulatory defects led to the non-production of Prn[64]. The multiple mechanisms of inactivation of *prn* clearly show that Prn-deficient strains emerged in parallel multiple times, pointing to strong selection pressure from the vaccine. The independent origins of Prn-deficient strains have been confirmed by genome data from Australia[21] and the United States[22]. Even the same mechanism of inactivation of the *prn* gene by IS*481* insertion can occur independently[21].

10.8 Pertussis toxin and filamentous haemagglutinin non-expression strains

There were also reports of isolates lacking PT and FHA expression[62,72,73]. However, the frequency of PT- and FHA-deficient strains is low. The first PT-deficient isolate was collected in France in 2007 from a non-vaccinated patient suspected of pertussis[62], while the first FHA-deficient isolate was collected in 1954 in France[74]. Two FHA-deficient isolates were collected in Sweden in 2009[72]. FHA deficiency in one strain was caused by an insertion in a homopolymeric tract while in the other strain no damage was found in the gene or its promoter[72]. One isolate deficient in both PT and FHA was isolated from an unvaccinated infant in the United States in 2013[73]. Interestingly, both the United States and French isolates lacking PT expression shared the same 28 kb deletion containing the entire *ptx* operon[73]. Despite their low frequency, the data suggest that these vaccine antigen deficient strains can potentially escape vaccine-induced immunity targeted against these antigens.

10.9 Functional significance of variation in antigenic genes

Functional differences in *prn* variation are detectable. Prn is a colonization factor and also contributes to resistance to neutrophil-mediated clearance[75]. However, van Gent et al.[76] showed that strains carrying *prn2* or *prn3* alleles were less able to colonize unimmunized mice than strains carrying the *prn1* allele. Thus, the newly emerged alleles are not necessarily providing advantages in the colonization of the host with respect to the role of pertactin.

The major differences between the Prn variants are in repeat region 1 (R1). It has been shown that the difference induces type-specific antibodies, with little cross-reactivity between Prn1 and Prn2[77]. R1 pairs with other repeat regions of Prn to form a new conformational epitope, which is a potential mechanism whereby *prn2* strains can escape from allele-specific (anti-Prn1) immune responses[78]. However, a subsequent study by the same group could not detect type specific antibodies in human sera[79].

Although changeover of the *ptxB* allele from *ptxB1* to *ptxB2* has only been seen in Japan, the difference in a single amino acid between the two alleles showed functional differences in receptor recognition and toxicity[80]. PtxB1 protein binds better to the glycoprotein, fetuin, and Jurkat T cells *in vitro* while PtxB2 protein was more effective at promoting lymphocytosis in mice[80].

The effect of antigen gene changes may have an additive effect. Using isogenic mutants of Tohama I, Komatsu et al.[81] showed that the double mutant carrying both *ptxA1* and *prn2* is better than either a *ptxA1* or *prn2* single mutant in survival in the ACV-immunized environment. The ACV used in the study was derived from Tohama I, which carries *ptxA2* and *prn1* alleles.

10.10 Effect of strain variation on vaccine protection *in vivo*

The intranasal mouse challenge model[82] is one of the assays used for licensing vaccines as advocated by the World Health Organization. Pertussis vaccine efficacies found in mice are positively correlated with that in humans[83]. In this model, differences in the clearance of bacteria from the trachea and lungs are measured by bacteria counts after challenge with sublethal doses of different *B. pertussis* strains in immunized mice[82]. Several studies have shown that strain variation affects vaccine efficacy in the mouse model including both WCV and ACV[34,58,76,81].

Three studies used different WCVs against circulating strains—a Dutch WCV (made from a *prn1* strain)[84], a custom WCV prepared by the investigators[34] and a WCV used in Poland[85]. All three studies showed better protection against infection with the matched strains than with the genotypically mismatched strains. Vaccine-induced immunity cleared *B. pertussis* more efficiently after intranasal challenge with a strain carrying the same antigenic gene alleles than when challenged with a strain with different antigenic gene alleles. These studies showed good sensitivity of the mouse model as well as detectable differences in protection against strains with mismatches in key antigens.

There are few studies that tested whether the *ptxP3* strains have any advantages over non-*ptxP3* strains in the ACV-immunized host. Safarchi et al.[86] used a mixed-infection mouse model to compare the competitive fitness between a *ptxP3–prn2* strain and a *ptxP1–prn3* strain and found that the *ptxP3–prn2* strain outcompeted the *ptxP1–prn3* strain in both naïve mice and mice vaccinated with a three-component vaccine containing PT, Prn, and FHA. Using Dutch clinical strains, King et al.[87] compared the colonization difference between *pxtP3* and *ptxP1* strains and found better colonization of *ptxP3* strains. In that study, they also constructed and tested an isogenic *ptxP1* mutant in a *ptxP3* back-

ground and a *ptxP3* mutant in a *ptxP1* background and found that the genetic background also affected colonization efficiency. Therefore, the increased fitness of *ptxP3* strains must involve more than the *ptxP3* allele[87].

The effect of the ACV on the fitness of Prn-deficient strains has been tested in the mouse model. Hegerle et al.[58] showed that Prn-deficient strains are fitter in immunized mice using *B. pertussis* strains isolated in France. Using a mixed-infection assay and isolates from Australia, Safarchi et al.[59] also showed that the Prn-deficient strain had a higher competitive fitness in immunized mice. The latter study further showed that the Prn-deficient strain was outcompeted by the Prn-positive strain in naïve mice which is consistent with the study of van Gent et al.[76] which found that the *prn* knockout mutant was less capable of colonizing mice than the wild type. Therefore, the loss of Prn has an adaptive advantage in the vaccine-immunized host but a disadvantage in non-immunized hosts. The loss of Prn helps *B. pertussis* to evade vaccine-induced immunity and the selection is likely to be strong enough to outweigh any negative effect from the loss of function of the *prn* gene.

One should point out that the dosage of the vaccine used may affect the sensitivity of the mouse model to detect strain differences. In the study by Komatsu et al.[81] that tested *ptxA2* and *prn1* isogenic mutants, the efficacy against isogenic strains was dose dependent; three different dosages were compared: 0.0025, 0.025, and 0.25 single human doses[81]. A sixfold higher dose was required to provide an equal level of protection against the double mutant (*ptxA1–prn2*) than was required to protect against the wild type. Although antibody levels were not measured, it was assumed that protective antibody levels would correlate with the single human doses. If these results are extrapolated to humans, they suggest that the level of protection afforded from the ACV is lower against the circulating strains carrying *ptxA1–prn2* alleles than against previously circulating strains carrying the vaccine-type alleles. Thus, the duration of ACV-induced immunity is effectively shortened. However, there is no clear correlation between antibody levels and protection in humans (probably because appropriate markers of immunity are ill defined) and there is evidence that undetectable antibody levels may offer protection[88]. Nevertheless, a recent serosurvey[89], which showed that an increased proportion of young children had undetectable levels of PT IgG antibody just before the 2008–2012 epidemic in Australia, suggests that antibody levels are important.

Therefore, based on studies in mice, it is clear that the pertussis vaccines are less effective against strains carrying antigenic gene alleles that differ from those contained in the vaccine. This phenomenon was shown for both the WCV and ACV. There may be differences between vaccines—probably depending on the number and immunogenicity of antigens used—and the number of doses given. However, any changes to the bacterium resulting in escape of vaccine-induced immunity may be applicable towards both the WCV and ACV.

10.11 Other factors leading to an increased pertussis notification rate

Apart from strain variation, there are other factors that may have contributed to the resurgence of pertussis including waning immunity, improved diagnosis, and differences in immunity induced by the WCV and ACV. PCR has become more widely available for pertussis diagnosis over the past decade and is acknowledged as one factor contributing to an increase in laboratory-notified cases. It is faster, more sensitive, less dependent on specimen quality, and remains positive for longer than culture and is applicable to all age groups, using easily collectable samples. A 'positive feedback loop' has been reported—the more cases reported, means greater awareness and more tests ordered by clinicians, leading to a surge of cases. In a large urban–suburban region of Ontario, Canada[90], a fivefold increase of pertussis (between 2005 and 2007) was paralleled by a sixfold increase in clinical laboratory testing, coinciding with the introduction of a highly sensitive assay. This also distorted the usual seasonality of pertussis[90], which typically occurs mainly during autumn, but testing for pertussis increases during winter, presumably in patients with cough due to winter respiratory viral infections.

The recently reported outbreaks of *B. holmesii* infections may falsely increase the number of pertussis cases reported by PCR diagnosis, which most commonly uses IS*481* as the target as it is also present in *B. holmesii*[91–93]. However, only a very small fraction of cases were due to *B. holmesii*, mostly in adolescents and adults[91–97] and are unlikely to significantly affect notification rates.

Waning immunity is also a key contributor. There is no agreed antibody level that correlates with immune protection, making it more difficult to address this factor. It is widely accepted that vaccine protection wanes, as shown in studies. The duration of immunity after vaccination with WCV or ACV is uncertain. Protective immunity in humans after vaccination was estimated to be between 4 to 12 years[98, 99]; some studies have shown similar durations of immunity after vaccination with ACV or WCV. However, there is ample evidence that ACV immunity lasts for a shorter time than WCV immunity[100–102]. ACV effectiveness wanes as early as 2–3 years after boosters[101]. Klein et al.[103] estimated that the risk of pertussis infection increased by 42 per cent a year after the fifth dose. One study estimated that only 10 per cent of children would be protected 8.5 years after the last dose of ACV[102]. Another study estimated that the duration of protection for ACV is as low as 5 years, three times shorter than WCV-induced immunity which is about 15 years[104].

An intriguing contributing factor is the difference in the initial priming effect between the WCV and ACV. Using data from Queensland, Australia, Sheridan et al.[105] found that children who received a primary course of WCV were better protected than those given ACV initially, for at least a decade. This study also supports previous findings that ACV-induced immunity wanes more rapidly[106] and may have contributed to the increase in pertussis. A United States study confirmed that WCV augments ACV in the duration of protection[107]. These findings highlight the waning vaccine protection over time and shorter ACV-induced immunity.

Further, ACVs and WCVs stimulate differential immune responses[108]. WCV, like natural pertussis infections, induces a largely T-helper cell (Th) type 1 response, characterized by the production of cytokines such as interleukin (IL)-2, tumour necrosis factor-α, and interferon-γ and to a lesser extent a Th17 response. In contrast, ACV leads more to a Th2 response with the induction of IL-4 and IL-5.

It is assumed that, partly due to these differences in immune responses, ACV is less effective than WCV in protection as has been shown by various human studies[105,107]. ACV is known to protect infants from severe disease as has been confirmed in a baboon model[109]. However, ACV immunization cannot prevent colonization and infected ACV-immunized baboons readily transmit infection to naïve animals. In contrast, WCV-immunized animals cleared infection significantly faster than ACV-immunized animals[109].

10.12 Concluding comments

Pertussis is potentially vaccine preventable but remains one of the most prevalent diseases notified annually. There are ongoing debates about the causes of its resurgence. There have been major changes in the *B. pertussis* population, which coincide with changes in vaccination schedules. Currently circulating *B. pertussis* strains are grouped in SNP cluster I, the emergence and expansion of which has been associated with two major changes in *B. pertussis* strains: increased PT production because of a change in its promotor (to *ptxP3*) and the change to the non-ACV *prn2* allele. These findings suggest greater fitness of SNP cluster I *B. pertussis*, at least in highly immunized populations, presumably because strains carrying *ptxp3–prn2*, which are almost always linked, can resist immune pressure. More recently, strains that lack the vaccine antigen, Prn, due to deletion or inactivation of *prn* have emerged and expanded. The emergence of Prn-deficient strains independently provides the best evidence of vaccine-driven selection.

Despite these changes, pertussis vaccination remains effective in recently vaccinated populations but protection wanes with time. Maintenance of high vaccination coverage and, presumably, more frequent boosters will be needed to improve the control of pertussis. Changes in pertussis epidemiology coincided with the introduction of ACV in some countries although possible evidence to the contrary also exists in others.

The resurgence of pertussis in highly vaccinated populations can be, at least in part, explained by genetic changes that increase the fitness of circulating *B. pertussis* isolates. Therefore, the mounting evidence of changes in antigenic gene alleles in current circulating strains necessitates the consideration of updating the current ACV. Emergence of Prn-deficient strains indicates that inclusion of different or more antigen variants from current strains will not be adequate and a novel approach may be required.

Acknowledgements

Research in the authors' laboratory is supported by grants from the National Health and Medical Council of Australia. We thank Laurence Luu for comments on the manuscript.

References

1. Tan T, Trindade E, Skowronski D. Epidemiology of pertussis. *Pediatr Infect Dis J* 2005;**24**:S10–S8.
2. Wood N, McIntyre P. Pertussis: review of epidemiology, diagnosis, management and prevention. *Paediatr Respir Rev* 2008;**9**:201–11.
3. McIntyre P, Forrest J, Heath T, et al. Pertussis vaccines: past, present and future in Australia. *Commun Dis Intell* 1998;**22**:125–32.
4. Torvaldsen S, Hull BP, McIntyre PB. Using the Australian Childhood Immunisation Register to track the transition from whole-cell to acellular pertussis vaccines. *Commun Dis Intell Q Rep* 2002;**26**:581–3.
5. Quinn HE. Pertussis control in Australia—the current state of play. *Commun Dis Intell Q Rep* 2014;**38**:E177–8.
6. Galanis E, King AS, Varughese P, et al. Changing epidemiology and emerging risk groups for pertussis. *CMAJ* 2006;**174**:451–2.
7. Tan T, Dalby T, Forsyth K, et al. Pertussis across the globe: recent epidemiologic trends from 2000 to 2013. *Pediatr Infect Dis J* 2015;**34**:e222–32.
8. Cherry JD. Epidemic pertussis in 2012—the resurgence of a vaccine-preventable disease. *N Engl J Med* 2012;**367**:785–7.
9. Pillsbury A, Quinn HE, McIntyre PB. Australian vaccine preventable disease epidemiological review series: pertussis, 2006–2012. *Commun Dis Intell Q Rep* 2014;**38**:E179–94.
10. Spokes PJ, Quinn HE, McAnulty JM. Review of the 2008–2009 pertussis epidemic in NSW: notifications and hospitalisations. *N S W Public Health Bull* 2010;**21**:167–73.
11. Poynten M, Hanlon M, Irwig L, et al. Serological diagnosis of pertussis: evaluation of IgA against whole cell and specific Bordetella pertussis antigens as markers of recent infection. *Epidemiol Infect* 2002;**128**:161–7.

12. Hale S, Quinn HE, Kesson A, et al. Changing patterns of pertussis in a children's hospital in the polymerase chain reaction diagnostic era. *J Pediatr* 2016;**170**:161–5.e1.

13. Maharjan RP, Gu C, Reeves PR, et al. Genome-wide analysis of single nucleotide polymorphisms in Bordetella pertussis using comparative genomic sequencing. *Res Microbiol* 2008;**159**:602–8.

14. Cummings CA, Brinig MM, Lepp PW, et al. Bordetella species are distinguished by patterns of substantial gene loss and host adaptation. *J Bacteriol* 2004;**186**:1484–92.

15. Brinig MM, Cummings CA, Sanden GN, et al. Significant gene order and expression differences in Bordetella pertussis despite limited gene content variation. *J Bacteriol* 2006;**188**:2375–82.

16. Caro V, Hot D, Guigon G, et al. Temporal analysis of French Bordetella pertussis isolates by comparative whole-genome hybridization. *Microbes Infect* 2006;**8**:2228–35.

17. Heikkinen E, Kallonen T, Saarinen L, et al. Comparative genomics of Bordetella pertussis reveals progressive gene loss in Finnish strains. *PLoS One* 2007;**2**:e904.

18. King AJ, van Gorkom T, Pennings JL, et al. Comparative genomic profiling of Dutch clinical Bordetella pertussis isolates using DNA microarrays: identification of genes absent from epidemic strains. *BMC Genomics* 2008;**9**:311.

19. Caro V, Bouchez V, Guiso N. Is the sequenced Bordetella pertussis strain Tohama I representative of the species? *J Clin Microbiol* 2008;**46**:2125–8.

20. Bart MJ, Harris SR, Advani A, et al. Global population structure and evolution of Bordetella pertussis and their relationship with vaccination. *mBio* 2014;**5**:e01074.

21. Safarchi A, Octavia S, Wu SZ, et al. Genomic dissection of Australian Bordetella pertussis isolates from the 2008–2012 epidemic. *J Infect* 2016;**72**:468–77.

22. Weigand MR, Peng Y, Loparev V, et al. The history of Bordetella pertussis genome evolution includes structural rearrangement. *J Bacteriol* 2017;**199**:e00806-16.

23. Kurniawan J, Maharjan RP, Chan WF, et al. Worldwide distribution of two predominant clones of Bordetella pertussis identified by multiple-locus variable-number tandem repeat analysis. *Emerging Infect Dis* 2010;**16**:297–300.

24. Octavia S, Maharjan RP, Sintchenko V, et al. Insight into evolution of Bordetella pertussis from comparative genomic analysis: evidence of vaccine-driven selection. *Mol Biol Evol* 2011;**28**:707–15.

25. Otsuka N, Han HJ, Toyoizumi-Ajisaka H, et al. Prevalence and genetic characterization of pertactin-deficient Bordetella pertussis in Japan. *PLoS One* 2012;**7**:e31985.

26. Advani A, Van der Heide HG, Hallander HO, et al. Analysis of Swedish Bordetella pertussis isolates with three typing methods: characterization of an epidemic lineage. *J Microbiol Methods* 2009;**78**:297–301.

27. Litt DJ, Neal SE, Fry NK. Changes in genetic diversity of the Bordetella pertussis population in the United Kingdom between 1920 and 2006 reflect vaccination coverage and emergence of a single dominant clonal type. *J Clin Microbiol* 2009;**47**:680–8.

28. Schouls LM, van der Heide HG, Vauterin L, et al. Multiple-locus variable-number tandem repeat analysis of Dutch Bordetella pertussis strains reveals rapid genetic changes with clonal expansion during the late 1990s. *J Bacteriol* 2004;**186**:5496–505.

29. van Gent M, Bart MJ, van der Heide HG, et al. SNP-based typing: a useful tool to study Bordetella pertussis populations. *PLoS One* 2011;**6**:e20340.

30. van Gent M, Bart MJ, van der Heide HG, et al. Small mutations in Bordetella pertussis are associated with selective sweeps. *PLoS One* 2012;**7**:e46407.

31. Octavia S, Wu SZ, Kaur S, et al. Whole-genome sequencing and comparative genomic analysis of Bordetella pertussis isolates from the 2007–2008 epidemic in Israel. *J Infect* 2017;**74**:204–7.

32. Hallander H, Advani A, Riffelmann M, et al. Bordetella pertussis strains circulating in Europe in 1999 to 2004 as determined by pulsed-field gel electrophoresis. *J Clin Microbiol* 2007;**45**:3257–62.

33. Caro V, Bouchez V, Guiso N, et al. Pertussis in Argentina and France. *Vaccine* 2007;**25**:4335–9.

34. Bottero D, Gaillard ME, Fingermann M, et al. Pulsed-field gel electrophoresis, pertactin, pertussis toxin S1 subunit polymorphisms, and surfaceome analysis of vaccine and clinical Bordetella pertussis strains. *Clin Vaccine Immunol* 2007;**14**:1490–8.

35. Hallander HO, Advani A, Donnelly D, et al. Shifts of Bordetella pertussis variants in Sweden from 1970 to 2003, during three periods marked by different vaccination programs. *J Clin Microbiol* 2005;**43**:2856–65.

36. Advani A, Gustafsson L, Ahren C, et al. Appearance of Fim3 and ptxP3-Bordetella pertussis strains, in two regions of Sweden with different vaccination programs. *Vaccine* 2011;**29**:3438–42.

37. Advani A, Hallander HO, Dalby T, et al. Pulsed-field gel electrophoresis analysis of Bordetella pertussis isolates circulating in Europe in 1998 to 2009. *J Clin Microbiol* 2013;**51**:422–8.

38. Mosiej E, Augustynowicz E, Zawadka M, et al. Strain variation among Bordetella pertussis isolates circulating in Poland after 50 years of whole-cell pertussis vaccine use. *J Clin Microbiol* 2011;**49**:1452–7.

39. Mooi FR, van Loo IH, King AJ. Adaptation of Bordetella pertussis to vaccination: a cause for its reemergence? *Emerging Infect Dis* 2001;**7**:526–8.

40. Mooi FR. Bordetella pertussis and vaccination: the persistence of a genetically monomorphic pathogen. *Infect Genet Evol* 2010;**10**:36–49.

41. Godfroid F, Denoel P, Poolman J. Are vaccination programs and isolate polymorphism linked to pertussis re-emergence? *Expert Rev Vaccines* 2005;**4**: 757–79.

42. van Loo IH, Heuvelman KJ, King AJ, et al. Multilocus sequence typing of Bordetella pertussis based on surface protein genes. *J Clin Microbiol* 2002;**40**:1994–2001.

43. Petersen RF, Dalby T, Dragsted DM, et al. Temporal trends in Bordetella pertussis populations, Denmark, 1949–2010. *Emerg Infect Dis* 2012;**18**:767–74.

44. Mooi FR, van Loo IH, van Gent M, et al. Bordetella pertussis strains with increased toxin production associated with pertussis resurgence. *Emerging Infect Dis* 2009;**15**:1206–13.

45. Xu Y, Zhang L, Tan Y, et al. Genetic diversity and population dynamics of Bordetella pertussis in China between 1950–2007. *Vaccine* 2015;**33**:6327–31.

46. Mosiej E, Zawadka M, Krysztopa-Grzybowska K, et al. Sequence variation in virulence-related genes of Bordetella pertussis isolates from Poland in the period 1959–2013. *Eur J Clin Microbiol Infect Dis* 2015; **34**:147–52.

47. Hozbor D, Mooi F, Flores D, et al. Pertussis epidemiology in Argentina: trends over 2004–2007. *J Infect* 2009;**59**:225–31.

48. Lam C, Octavia S, Bahrame Z, et al. Selection and emergence of pertussis toxin promoter ptxP3 allele in the evolution of Bordetella pertussis. *Infect Genet Evol* 2012;**12**:492–5.

49. Clarke M, McIntyre PB, Blyth C, et al. The relationship between Bordetella pertussis genotype and clinical severity in Australian children with pertussis. *J Infect* 2015;**72**:171–8.

50. Kim SH, Lee J, Sung HY, et al. Recent trends of antigenic variation in Bordetella pertussis isolates in Korea. *J Korean Med Sci* 2014;**29**:328–33.

51. Shuel M, Jamieson FB, Tang P, et al. Genetic analysis of Bordetella pertussis in Ontario, Canada reveals one predominant clone. *Int J Infect Dis* 2013;**17**:e413–7.

52. Schmidtke AJ, Boney KO, Martin SW, et al. Population diversity among Bordetella pertussis isolates, United States, 1935–2009. *Emerg Infect Dis* 2012;**18**:1248–55.

53. Kallonen T, Mertsola J, Mooi FR, et al. Rapid detection of the recently emerged Bordetella pertussis strains with the ptxP3 pertussis toxin promoter allele by real-time PCR. *Clin Microbiol Infect* 2012;**18**: E377–9.

54. Wagner B, Melzer H, Freymuller G, et al. Genetic variation of Bordetella pertussis in Austria. *PLoS One* 2015;**10**:e0132623.

55. Miyaji Y, Otsuka N, Toyoizumi-Ajisaka H, et al. Genetic analysis of Bordetella pertussis isolates from the 2008–2010 pertussis epidemic in Japan. *PLoS One* 2013;**8**:e77165.

56. King AJ, van Gorkom T, van der Heide HG, et al. Changes in the genomic content of circulating Bordetella pertussis strains isolated from the Netherlands, Sweden, Japan and Australia: adaptive evolution or drift? *BMC Genomics* 2010;**11**:64.

57. Martin SW, Pawloski L, Williams M, et al. Pertactin-negative Bordetella pertussis strains: evidence for a possible selective advantage. *Clin Infect Dis* 2015;**60**:223–7.

58. Hegerle N, Dore G, Guiso N. Pertactin deficient Bordetella pertussis present a better fitness in mice immunized with an acellular pertussis vaccine. *Vaccine* 2014;**32**:6597–600.

59. Safarchi A, Octavia S, Luu LD, et al. Pertactin negative Bordetella pertussis demonstrates higher fitness under vaccine selection pressure in a mixed infection model. *Vaccine* 2015;**33**:6277–81.

60. Breakwell L, Kelso P, Finley C, et al. Pertussis vaccine effectiveness in the setting of pertactin-deficient pertussis. *Pediatrics* 2016;**137**:e20153973.

61. Bodilis H, Guiso N. Virulence of pertactin-negative Bordetella pertussis isolates from infants, France. *Emerg Infect Dis* 2013;**19**:471–4.

62. Bouchez V, Brun D, Cantinelli T, et al. First report and detailed characterization of B. pertussis isolates not expressing pertussis toxin or pertactin. *Vaccine* 2009;**27**:6034–41.

63. Sealey KL, Harris SR, Fry NK, et al. Genomic analysis of isolates from the United Kingdom 2012 pertussis outbreak reveals that vaccine antigen genes are unusually fast evolving. *J Infect Dis* 2015;**212**:294–301.

64. Lam C, Octavia S, Ricafort L, et al. Multiple independent emergence and rapid expansion of pertactin-deficient Bordetella pertussis in Australia. *Emerg Infect Dis* 2014;**20**:626–33.

65. Queenan AM, Cassiday PK, Evangelista A. Pertactin-negative variants of Bordetella pertussis in the United States. *N Engl J Med* 2013;**368**:583–4.

66. Pawloski LC, Queenan AM, Cassiday PK, et al. Prevalence and molecular characterization of pertactin-deficient Bordetella pertussis in the United States. *Clin Vaccine Immunol* 2014;**21**:119–25.

67. Bowden KE, Williams MM, Cassiday PK, et al. Molecular epidemiology of the pertussis epidemic in Washington State in 2012. *J Clin Microbiol* 2014; **52**:3549–57.

68. Quinlan T, Musser KA, Currenti SA, et al. Pertactin-negative variants of Bordetella pertussis in New York State: a retrospective analysis, 2004–2013. *Mol Cell Probes* 2014;**28**:138–40.

69. Bamberger E, Abu Raya B, Cohen L, et al. Pertussis resurgence associated with pertactin-deficient and genetically divergent Bordetella pertussis isolates in Israel. *Pediatr Infect Dis J* 2015;**34**:898–900.

70. Tsang RS, Shuel M, Jamieson FB, et al. Pertactin-negative Bordetella pertussis strains in Canada: characterization of a dozen isolates based on a survey of 224 samples collected in different parts of the country over the last 20 years. *Int J Infect Dis* 2014;**28**:65–9.

71. Hiramatsu Y, Miyaji Y, Otsuka N, et al. Significant decrease in pertactin-deficient Bordetella pertussis isolates, Japan. *Emerg Infect Dis* 2017;**23**:699–701.

72. Bart MJ, van der Heide HG, Zeddeman A, et al. Complete genome sequences of 11 Bordetella pertussis strains representing the pandemic ptxP3 lineage. *Genome Announc* 2015;**3**:e01394-15.

73. Williams MM, Sen K, Weigand MR, et al. Bordetella pertussis strain lacking pertactin and pertussis toxin. *Emerg Infect Dis* 2016;**22**:319–22.

74. Bouchez V, Hegerle N, Strati F, et al. New data on vaccine antigen deficient Bordetella pertussis isolates. *Vaccines* 2015;**3**:751–70.

75. Inatsuka CS, Xu Q, Vujkovic-Cvijin I, et al. Pertactin is required for Bordetella species to resist neutrophil-mediated clearance. *Infect Immun* 2010;**78**:2901–9.

76. van Gent M, van Loo IH, Heuvelman KJ, et al. Studies on Prn variation in the mouse model and comparison with epidemiological data. *PLoS One* 2011;**6**:e18014.

77. He Q, Makinen J, Berbers G, et al. Bordetella pertussis protein pertactin induces type-specific antibodies: one possible explanation for the emergence of antigenic variants? *J Infect Dis* 2003;**187**:1200–5.

78. Hijnen M, Mooi FR, van Gageldonk PG, et al. Epitope structure of the Bordetella pertussis protein P.69 pertactin, a major vaccine component and protective antigen. *Infect Immun* 2004;**72**:3716–23.

79. Hijnen M, He Q, Schepp R, et al. Antibody responses to defined regions of the Bordetella pertussis virulence factor pertactin. *Scand J Infect Dis* 2008;**40**:94–104.

80. Millen SH, Watanabe M, Komatsu E, et al. Single amino acid polymorphisms of pertussis toxin subunit S2 (PtxB) affect protein function. *PLoS One* 2015;**10**:e0137379.

81. Komatsu E, Yamaguchi F, Abe A, et al. Synergic effect of genotype changes in pertussis toxin and pertactin on adaptation to an acellular pertussis vaccine in the murine intranasal challenge model. *Clin Vaccine Immunol* 2010;**17**:807–12.

82. Guiso N, Capiau C, Carletti G, et al. Intranasal murine model of Bordetella pertussis infection. I. Prediction of protection in human infants by acellular vaccines. *Vaccine* 1999;**17**:2366–76.

83. Mills KH, Ryan M, Ryan E, et al. A murine model in which protection correlates with pertussis vaccine efficacy in children reveals complementary roles for humoral and cell-mediated immunity in protection against Bordetella pertussis. *Infect Immun* 1998;**66**:594–602.

84. King AJ, Berbers G, van Oirschot HF, et al. Role of the polymorphic region 1 of the Bordetella pertussis protein pertactin in immunity. *Microbiology* 2001;**147**:2885–95.

85. Gzyl A, Augustynowicz E, Gniadek G, et al. Sequence variation in pertussis S1 subunit toxin and pertussis genes in Bordetella pertussis strains used for the whole-cell pertussis vaccine produced in Poland since 1960: efficiency of the DTwP vaccine-induced immunity against currently circulating B. pertussis isolates. *Vaccine* 2004;**22**:2122–8.

86. Safarchi A, Octavia S, Luu LD, et al. Better colonisation of newly emerged Bordetella pertussis in the co-infection mouse model study. *Vaccine* 2016;**34**:3967–71.

87. King AJ, van der Lee S, Mohangoo A, et al. Genome-wide gene expression analysis of Bordetella pertussis isolates associated with a resurgence in pertussis: elucidation of factors involved in the increased fitness of epidemic strains. *PLoS One* 2013;**8**:e66150.

88. Watanabe M, Komatsu E, Sato T, et al. Evaluation of efficacy in terms of antibody levels and cell-mediated immunity of acellular pertussis vaccines in a murine model of respiratory infection. *FEMS Immunol Med Microbiol* 2002;**33**:219–25.

89. Campbell P, McIntyre P, Quinn H, et al. Increased population prevalence of low pertussis toxin antibody levels in young children preceding a record pertussis epidemic in Australia. *PLoS One* 2012;**7**:e35874.

90. Fisman DN, Tang P, Hauck T, et al. Pertussis resurgence in Toronto, Canada: a population-based study including test-incidence feedback modeling. *BMC Public Health* 2011;**11**:694.

91. Njamkepo E, Bonacorsi S, Debruyne M, et al. Significant finding of Bordetella holmesii DNA in nasopharyngeal samples from French patients with suspected pertussis. *J Clin Microbiol* 2011;**49**:4347–8.

92. Bottero D, Griffith MM, Lara C, et al. Bordetella holmesii in children suspected of pertussis in Argentina. *Epidemiol Infect* 2012;**141**:714–7.

93. Mooi FR, Bruisten S, Linde I, et al. Characterization of Bordetella holmesii isolates from patients with pertussis-like illness in the Netherlands. *FEMS Immunol Med Microbiol* 2012;**64**:289–91.

94. Rodgers L, Martin SW, Cohn A, et al. Epidemiologic and laboratory features of a large outbreak of pertussis-like illnesses associated with co-circulating Bordetella holmesii and Bordetella pertussis—Ohio, 2010–2011. *Clin Infect Dis* 2013;**56**:322–31.

95. Kamiya H, Otsuka N, Ando Y, et al. Transmission of Bordetella holmesii during pertussis outbreak, Japan. *Emerg Infect Dis* 2012;**18**:1166–9.

96. Le Coustumier A, Njamkepo E, Cattoir V, et al. Bordetella petrii infection with long-lasting persistence in human. *Emerg Infect Dis* 2011;**17**:612–8.

97. Miranda C, Porte L, Garcia P. Bordetella holmesii in nasopharyngeal samples from Chilean patients with suspected Bordetella pertussis infection. *J Clin Microbiol* 2012;**50**:1505.

98. Wendelboe AM, Van Rie A, Salmaso S, et al. Duration of immunity against pertussis after natural infection or vaccination. *Pediatr Infect Dis J* 2005;**24**:S58–61.

99. Hallander HO, Andersson M, Gustafsson L, et al. Seroprevalence of pertussis antitoxin (anti-PT) in Sweden before and 10 years after the introduction of a universal childhood pertussis vaccination program. *APMIS* 2009;**117**:912–22.

100. Witt MA, Katz PH, Witt DJ. Unexpectedly limited durability of immunity following acellular pertussis vaccination in preadolescents in a North American outbreak. *Clin Infect Dis* 2012;**54**:1730–5.

101. Burdin N, Handy LK, Plotkin SA. What is wrong with pertussis vaccine immunity? The problem of waning effectiveness of pertussis vaccines. *Cold Spring Harb Perspect Biol* 2017;**9**:a029454.

102. McGirr A, Fisman DN. Duration of pertussis immunity after DTaP immunization: a meta-analysis. *Pediatrics* 2015;**135**:331–43.

103. Klein NP, Bartlett J, Rowhani-Rahbar A, et al. Waning protection after fifth dose of acellular pertussis vaccine in children. *N Engl J Med* 2012;**367**:1012–9.

104. Choi YH, Campbell H, Amirthalingam G, et al. Investigating the pertussis resurgence in England and Wales, and options for future control. *BMC Med* 2016;**14**:121.

105. Sheridan SL, Ware RS, Grimwood K, et al. Number and order of whole cell pertussis vaccines in infancy and disease protection. *JAMA* 2012;**308**:454–6.

106. Rendi-Wagner P, Kundi M, Mikolasek A, et al. Hospital-based active surveillance of childhood pertussis in Austria from 1996 to 2003: estimates of incidence and vaccine effectiveness of whole-cell and acellular vaccine. *Vaccine* 2006;**24**:5960–5.

107. Witt MA, Arias L, Katz PH, et al. Reduced risk of pertussis among persons ever vaccinated with whole cell pertussis vaccine compared to recipients of acellular pertussis vaccines in a large US cohort. *Clin Infect Dis* 2013;**56**:1248–54.

108. Higgs R, Higgins SC, Ross PJ, et al. Immunity to the respiratory pathogen Bordetella pertussis. *Mucosal Immunol* 2012;**5**:485–500.

109. Warfel JM, Zimmerman LI, Merkel TJ. Acellular pertussis vaccines protect against disease but fail to prevent infection and transmission in a nonhuman primate model. *Proc Natl Acad Sci U S A* 2014;**111**: 787–92.

CHAPTER 11

Congenerics: what can be learned about pertussis from pertussis-like disease caused by other *Bordetella*?

Iain MacArthur and Andrew Preston

Abstract

The evolution of *Bordetella pertussis* from *Bordetella bronchiseptica* (or a *B. bronchiseptica*-like ancestor) occurred primarily through gene loss and genome rearrangement, mediated largely through the expansion in the copy number of insertion sequence element repeats in the *B. pertussis* genome. *Bordetella pertussis* is attributed as the main causative agent of whooping cough. However, *Bordetella parapertussis* and *Bordetella holmesii* also cause disease that is very similar to that caused by *B. pertussis* (here termed pertussis-like disease). The evolution of *B. parapertussis* and *B. holmesii* displays striking similarities to that of *B. pertussis* and thus this chapter explores what might be gained from comparative studies of *B. pertussis*, *B. parapertussis*, and *B. holmesii* with regard to the understanding of whooping cough.

11.1 Introduction

Bordetella pertussis is often described as *the* causative agent of whooping cough, or pertussis. However, *B. parapertussis* and *B. holmesii* also cause pertussis, or at least pertussis-like disease (PLD). Here, we use PLD to describe the spectrum of respiratory disease caused by these bacteria, recognizing that other *Bordetella* cause disease that is not readily distinguishable from that caused by *B. pertussis*. This includes cases that present with paroxysmal cough similar to that occurring during classically described pertussis disease, post-tussive vomiting, and/or chronic cough that in duration and severity resembles chronic cough caused by *B. pertussis*, particularly in adolescents and adults in which whoop may not be present. PLD due to *B. parapertussis* and

B. holmesii infection has been described variably as either milder than, similar to, or in many regards the same as that caused by *B. pertussis*. Current data suggest that *B. pertussis* is the predominant cause of PLD, but accurate determination of the burden of PLD due to the other two species is lacking. In this chapter, we ask what might be learned about pertussis from PLD due to *B. parapertussis* and *B. holmesii*, and suggest that, very probably, comparative studies of these three species will generate better understanding of classic pertussis disease.

11.2 *Bordetella parapertussis*

Bordetella parapertussis was identified in 1938 as bacteria isolated from cases of whooping cough which differed in colony and growth characteristics from

MacArthur, I. and Preston, A., *Congenerics: What can be learned about pertussis from pertussis-like disease caused by other Bordetella?* In: Pertussis: *epidemiology, immunology, and evolution.* Edited by Pejman Rohani and Samuel V. Scarpino: Oxford University Press (2019). © Oxford University Press. DOI:10.1093/oso/9780198811879.003.0011

B. pertussis[1]. In addition to causing PLD in humans, a distinct lineage of *B. parapertussis* has been isolated from the respiratory tract of both healthy and pneumonic lambs[2]. Only the human-adapted lineage is discussed here. Like *B. pertussis*, *B. parapertussis* is assumed to be restricted to its human host niche and it displays similar genome reduction to *B. pertussis*, often a hallmark of specialization of bacteria to restricted niches[3].

While it is clear that *B. parapertussis* causes PLD, there is uncertainty over its incidence and whether disease due to *B. parapertussis* is as severe as disease caused by *B. pertussis*. Uncertainty over incidence arises from difficulties in identifying *B. parapertussis* as the causative agent. Disease due to *B. parapertussis* is often regarded as less severe than that due to *B. pertussis*. This notion stems largely from an authoritative study conducted during one of the European trials of the acellular pertussis vaccines[4] in which 38 cases of disease due to *B. parapertussis* were compared to matched cases of disease due to *B. pertussis*. A number of clinical characteristics were measured including duration of cough, duration of whoop, temperature, the occurrence of paroxysms, vomiting, and both leucocytosis and lymphocytosis. Only the latter two parameters differed significantly between the two cohorts. Leucocytosis is perhaps the most serious manifestation of pertussis, contributing to the pulmonary hypertension that is thought to lead to infant deaths in cases of fatal pertussis[5]. Thus, in young children, *B. parapertussis* infection may be less dangerous than *B. pertussis* infection, but for many cases, symptoms appear to be very similar to those due to *B. pertussis* infection.

Recently a number of studies have analysed what appear to be outbreaks of *B. parapertussis*-disease and these have enabled some comparisons in some older children. For example, an outbreak of *B. parapertussis* was noted in Wisconsin, United States, between October 2011 and December 2012[6]. During this time (when the number of cases of *B. pertussis* disease was high), of 7022 diagnosed cases of *Bordetella* infections, 417 (5.9 per cent) were confirmed as caused by *B. parapertussis*. The age range of those affected was less than 1 year to 14 years old, with just nine cases (4 per cent) observed in patients older than this. All patients diagnosed with *B. parapertussis* infection reported cough (as expected, as cough was the symptom that led to the patients seeking medical care and thus being diagnosed). However, more severe symptoms, associated with classic pertussis disease, were also frequently reported: paroxysmal cough (60 per cent), whoop (15 per cent), and post-tussive vomiting (31 per cent). The median duration of cough was 15 days. Thus, *B. parapertussis* caused disease similar to that caused by *B. pertussis* in severity, although the duration of cough was shorter.

An outbreak in Minnesota, United States, in 2014 has been described[7], after retrospective analysis of cases from Mayo Medical Laboratories that provide diagnostic testing nationally. In Minnesota, 31 cases of *B. parapertussis* infection were diagnosed in 2014. Although whoop and paroxysms were not specifically noted, 40 per cent of patients reported post-tussive vomiting, suggesting severe cough was common among these cases. These and other, often anecdotal, reports demonstrate that *B. parapertussis* causes disease that in many cases is similar to that caused by *B. pertussis*.

Knowledge of the molecular basis for *B. parapertussis* infection is limited. Strikingly, *B. parapertussis* contains homologues of many of the factors regarded as important for infection by *B. pertussis*, including filamentous haemagglutinin (FHA), fimbriae, adenylate cyclase, type three secretion system, and the key regulatory system BvgAS[8]. A notable difference is the absence of expression of pertussis toxin (PT) in *B. parapertussis*. Although the PT genetic locus is present, the PT locus promoter is very similar to that of the *B. bronchiseptica* which is thought to be very weak, producing very low (if any) transcription of the locus[8,9]. Also, the toxin subunit gene *ptxB* is a pseudogene in *B. parapertussis*. However, insertion of a heterologous promoter upstream of the *B. parapertussis* PT locus led to the production of functional PT, albeit it at low levels, suggesting that the locus is capable of encoding toxin[9]. PT is regarded as essential to the induction of leucocytosis and the absence of this in cases of *B. parapertussis* disease suggests that during infection *B. parapertussis* does not express PT at levels sufficient to induce leucocytosis. Also, *B. parapertussis* differs from *B. pertussis* in the structure of its lipopolysaccharide (LPS). *B. parapertussis* LPS contains O-polysaccharide (O-PS), that is absent from

B. pertussis due to deletion of the *wbm* O-PS locus[10]. In addition, the terminal band A trisaccharide present in *B. pertussis* LPS is absent from that of *B. parapertussis* in which the O-PS links directly to the LPS core region[11]. *Bordetella parapertussis* O-PS is important for interactions with aspects of host immunity (see section 11.8). However, the high levels of conservation between many factors in *B. pertussis* and *B. parapertussis* suggest that the fundamentals of colonization are probably very similar between the two.

11.3 *Bordetella holmesii*

Bordetella holmesii was first reported in 1995 as the cause of bacteraemia in an asplenic patient[12]. The ability of *B. holmesii* to cause invasive disease, mainly bacteraemia but also infections of other systems/sites, is a major difference to *B. pertussis*. Many of these infections were in people with other underlying health issues suggesting that *B. holmesii* is an opportunistic, invasive pathogen in immunocompromised hosts. However, over the past decade *B. holmesii* has emerged as a cause of PLD, with numerous reports observing *B. holmesii* as a contributing cause in cases of whooping cough outbreaks worldwide[13]. Even less is known about *B. holmesii* than *B. parapertussis*, but the molecular basis for *B. holmesii* pathogenesis appears to be significantly different to that of *B. pertussis*. *Bordetella holmesii* lacks homologues of most of the well characterized virulence determinants of *B. pertussis* including PT, adenylate cyclase–haemolysin toxin, fimbriae, pertactin, and type 1 and 2 secretion systems[14]. *Bordetella holmesii* does contain a BvgAS homologue although BvgS appears to be more similar to that of *Bordetella avium* and *B. hinzii* than *B. pertussis*[15]. *B. holmesii* does produce an adhesin similar to FHA[16] but its role in adherence of *B. holmesii* is unknown. Thus, it is perhaps surprising that despite the lack of these key virulence determinants, *B. holmesii* has been implicated in causing PLD that in many cases appeared very similar to that caused by *B. pertussis*. It is commonly stated (similar to the case of *B. parapertussis*) that *B. holmesii* causes less severe disease than does *B. pertussis*. However, in one of the most notable studies of *B. holmesii* disease during a period of high levels of pertussis in Ohio, United States, during 2010–2011, 48 laboratory-confirmed cases of *B. holmesii* infection were identified, compared to 112 caused by *B. pertussis*[17]. Comparison of the clinical characteristics of these cases found there were no statistical differences in levels of paroxysms, posttussive vomiting, whoop, and cough for longer than 14 days in cases caused by the two species regardless of the age groups involved. Thus, in these respects disease severity was similar for the two species.

11.4 Diagnoses and incidence: the contribution of *Bordetella parapertussis* and *Bordetella holmesii* to pertussis-like disease

With increasing concern over increases in pertussis incidence in numerous countries, there is increasing interest in, but little certainty of, the role of *B. parapertussis* and *B. holmesii* in this. A major difficulty in establishing the incidence of *B. parapertussis* and *B. holmesii* disease is that the use of specific diagnoses for them is somewhat limited. Culture-based diagnoses are no longer widely used due to the difficulties of obtaining suitable swab samples and the relative lack of sensitivity of this approach. *B. parapertussis* and *B. pertussis* are very difficult to distinguish on the basis of colony morphology and it is not clear how many diagnostic laboratories take further steps to confirm species identity. Serology is being used more widely in which a level of anti-PtxA antibodies above a threshold level, in combination with clinical signs, is taken as an indication of infection. This is *B. pertussis* specific and thus incapable of detecting infection by either *B. parapertussis* or *B. holmesii*. Polymerase chain reaction-based diagnosis is now the most widely used method in many countries[18] in which insertion sequence (IS)-*481* is targeted to detect *B. pertussis*. This is not present in *B. parapertussis* for which IS*1001* is used as a specific target. IS*481* is present in *B. holmesii*. To distinguish between *B. pertussis* and *B. holmesii*, a *B. pertussis*- and/or *B. holmesii*-specific gene target can be included in a multiplex reaction, for example[19]. However, it is not clear what proportion of laboratories are using species-distinguishing assays and currently the reporting of incidence of *B. parapertussis* and *B. holmesii* infection is based on incomplete data. Several studies of the ability of laboratories to distinguish between *B. pertussis* and

B. holmesii revealed that less than 10 per cent of laboratories did this successfully (discussed in[13]), suggesting that misdiagnosis of *B. holmesii* infection could occur. Several studies suggest that both *B. parapertussis* and *B. holmesii* cause significant levels of PLD at times. For example, during 2008–2010, 9.5 per cent of samples from nine states in the United States submitted to a reference laboratory in California were positive for either IS*481* or IS*1001*[20]. Of these, 13.5 per cent were positive for IS*1001*, that is, were positive for *B. parapertussis* infection. *B. parapertussis* infection was almost entirely restricted to children less than 10 years of age, whereas 32 per cent of the *B. pertussis* infections recorded were in people older than this. There was some seasonality to the incidence of *B. pertussis* infection in this sample, whereas *B. parapertussis* was recorded at relatively consistent levels throughout the year. In an excellent review of *B. holmesii* biology, Pittet and colleagues[13] reviewed 29 publications detailing the analysis of *Bordetella*-positive clinical samples for which the incidence of *B. parapertussis* and/or *B. holmesii* was determined. In these reports, *B. parapertussis* contributed from 0 to 32 per cent of the *Bordetella*-positive samples, and for *B. holmesii* these values ranged from 0 to 29 per cent. A majority of these studies were focused on samples from a particular geographical region at a particular time and mostly represent analysis of pertussis outbreaks. However, they demonstrate clearly that both *B. parapertussis* and *B. holmesii* are recognizable causes of PLD in multiple countries over many years.

In addition, there is evidence of limited cases of co-infection. For example, during the Wisconsin outbreak discussed earlier, 26 samples tested positive for both *B. pertussis* and *B. parapertussis*[6]. This suggests that both of these species can be present in a single host simultaneously, but there are insufficient data to draw conclusions about disease severity during co-infection. Similarly, there are insufficient data to be able to describe any positive or negative associations between different *Bordetella* species within the same host.

11.5 The evolution of pertussis-like disease

These *Bordetella* species offer a fascinating insight into the evolution of pathogens and perhaps of PLD itself. On three separate occasions, a *Bordetella* species has evolved by genome reduction and rearrangement to produce an organism that causes pertussis, or PLD. In the case of *B. pertussis* and *B. parapertussis*, the resulting organisms are very closely related whereas *B. holmesii* is much more distinct.

Very little is known about the biology of *B. parapertussis* and *B. holmesii*. However, genome sequences have been generated for both species, although at present the genomes of just three isolates of *B. parapertussis* are available (along with an ovine-derived *B. parapertussis* isolate). These data provide a window into the biology of these species and enable hypotheses to be raised regarding traits involved in causing PLD.

Bordetella pertussis and *B. parapertussis* both evolved from *B. bronchiseptica*, or a *B. bronchiseptica*-like ancestor[8,21] (see Figure 11.1). Phylogenetic analysis suggests that *B. holmesii* is most closely related to *B. hinzii*[21] (see Figure 11.1). Both of these ancestral species are pathogens in certain animal species. Key features of *Bordetella* genomes are displayed in Table 11.1.

It is clear that their evolution has involved primarily genome reduction with all three species having considerably smaller genomes than their *B. bronchiseptica* and *B. hinzii* closest relatives. In addition, the genomes of all three species have suffered large-scale rearrangement. This has been driven by an expansion of the number of repeats of certain IS elements within the genomes of these species compared to the relatively few IS elements within the genomes of the other Bordetellae (see Table 11.1). The high copy number of IS element repeats in these genomes facilitates frequent recombination between repeats, resulting in deletion of intervening genome segments and inversions and transversions within the genome (see Figure 11.2). In addition, IS elements can cause deleterious mutations such as insertional inactivation of genes and all three species contain an unusually high number of pseudogenes. Ordinarily, the mutagenic effects of a large number of IS elements would be expected to decrease the fitness of the arising bacterium. In this event, the clones with high numbers of IS elements might be expected to be outcompeted by the higher-fitness progenitor clones and lost from the population. For a clone with high

(a)

(b)

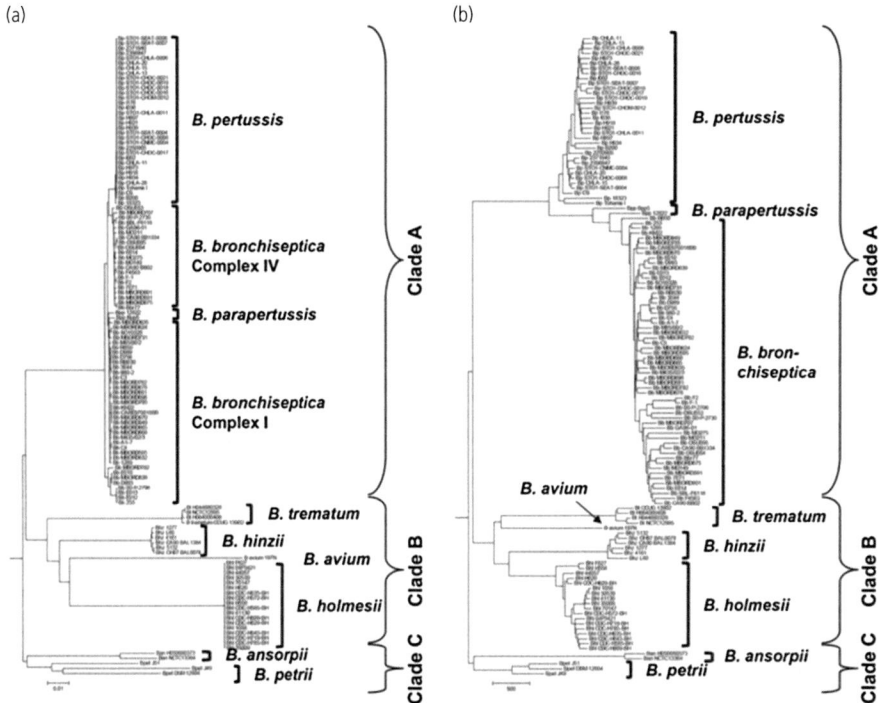

Figure 11.1 Phylogenetic trees depicting the phylogenetic relationships between *Bordetella* species based on genome-wide alignments (a) and the presence or absence of genes among species (b). The close genetic relationship between *B. pertussis/B. parapertussis* and *B. bronchiseptica* is evident. *Bordetella hinzii* is suggested as the most likely closest relative to *B. holmesii*. Reproduced with permission from the authors, Figure 1 Linz et al *BMC Genomics*. 2016 Sep 30;17(1):767. This work is licensed under a Creative Commons Attribution 4.0 http://creativecommons.org/licenses/by/4.0/.

numbers of IS elements to become fixed, it is thought that a reduction in competition is required, such as the arising clone occupying a new niche from which the progenitor bacteria are absent or present at low levels, or in which the new clone is fitter than the progenitor [3,22,23]. In addition, *B. pertussis* and *B. holmesii* (and quite possibly *B. parapertussis*) exhibit very low levels of genetic heterogeneity, suggesting a recent evolution from a small number of founding lineages [21]. Taken together, a genetically homogeneous bacterium with high numbers of IS element repeats is thought to have arisen by escape of the bacterium through a population bottleneck to a new niche with reduced competition, with subsequent expansion and evolution of these new clones. Thus, in the *Bordetella* genus this appears to have happened on three occasions, giving rise to *B. pertussis*, *B. parapertussis*, and *B. holmesii*. It is tempting to speculate that a niche change was the introduction of these species into humans. For *B. bronchiseptica*, there is evidence for a human-associated lineage

(complex IV[24]) suggesting that humans were hosts for *Bordetella* species before the emergence of modern day *B. pertussis* and *B. parapertussis*. Phylogenetic analyses suggest that *B. pertussis* is most closely related to complex IV *B. bronchiseptica* [21,24]. However, it is not clear whether *B. pertussis* evolved from an already human-associated ancestor or evolved under selection for traits that favoured them outcompeting *B. bronchiseptica* lineages within the human respiratory tract. Interestingly, *B. parapertussis* is more closely related to complex I *B. bronchiseptica* that are largely non-human-derived *B. bronchiseptica*, although some complex I strains have been isolated from human hosts[25].

Phylogenetic analyses strongly support each of these speciations being independent[21]. IS*481* was involved in this process in both *B. pertussis* and *B. holmesii*. It is the dominant IS element in *B. pertussis* whereas in *B. holmesii* an ISL3 family IS element (IS*1001Bhii*) and IS*bho1*, an IS element of the IS*407* subfamily of ISL3, have also expanded to

Table 11.1 Key features of the genomes of *Bordetella* species

Species[a]	Genome size (Mbp)[b]	GC %[c]	No. genes[d]	IS elements[e]
B. bronchiseptica (68)	5.210 (± 0.101)	68.07	4759	?
B. pertussis (613)	3.931 (± 0.198)	67.72	3576	IS*481*: 230–250 IS*1663*: 17 IS*1002*: 5
B. parapertussis (3)	4.782 (± 0.015)	68.1	4169	IS*1002*: 9 IS*1001*: 22
B. hinzii (10)	4.900 (± 0.039)	67.05	4456	?
B. holmesii (21)	3.631 (± 0.099)	62.7	3139	ISL*3*: 70–74 IS*3*: 95–96 IS*481*: 33–42
B. trematum (5)	4.391 (± 0.115)	65.7	3985	IS*3* family: 7
B. avium (2)	3.713 (± 0.027)	61.58	3279	–
B. petrii (3)	4.846 (± 0.566)	65.48	4718	IS*3* family: 92 IS*1663*: 6
B. ansorpii (2)	6.719 (± 0.037)	66.85	5357	31

[a] Number of genome sequences from which values were obtained. [b] Average size of the sequenced genomes, ± standard deviation.

[c] Median GC content of sequenced genomes. [d] Median number of genes per sequenced genome.

[e] Copy number of insertion sequence (IS) elements were determined from closed genome sequences using ISFinder[41] only due to the uncertainty regarding copy number of repeat elements in draft assemblies. For *B. bronchiseptica* and *B. hinzii*, no complete IS elements were detected in the few closed genome sequences available for these strains but many different IS elements are present in other strains, meaning the IS element repertoire of these species is undetermined. *B. ansorpii* contains low copy numbers of nine different IS elements as identified by ISFinder.

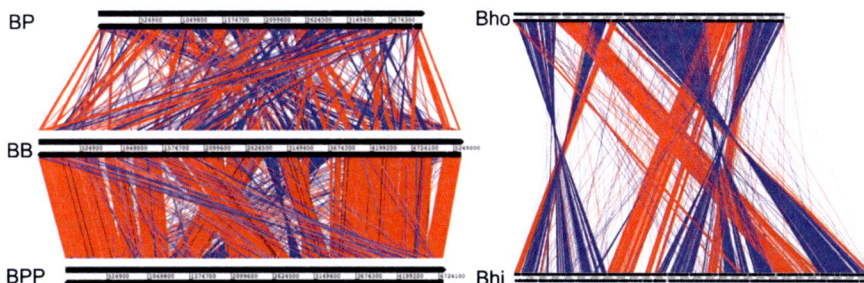

Figure 11.2 Genome reduction and rearrangement in *Bordetella pertussis*, *B. parapertussis*, and *B. holmesii*. Genome comparisons were made for representative genome sequences of *B. pertussis*, *B. parapertussis*, and *B. bronchiseptica* (BB) and for *B. holmesii* and *B. hinzii* using blastn (*B. pertussis*/*B. parapertussis* vs *B. bronchiseptica*) or tblastx comparison (*B. holmesii* vs *B. hinzii*) via the genome sequence alignment visualization tool, ACT[47]. Regions of homology are indicated by red (direct match) and blue (inverted match) lines. BB, *B. bronchiseptica*; Bhi, *B. hinzii*; Bho, *B. holmesii*; BP, *B. pertussis*; BPP, *B. parapertussis*.

similar copy numbers. Analysis of the five closed genome sequences available for *B. holmesii* reveals that each of the five strains contains a very similar repertoire of IS elements (see Table 11.2), suggesting a high level of genetic homogeneity among *B. holmesii*.

In *B. parapertussis*, IS*1001* is the dominant element. It is apparent that there is no single IS element that has been responsible for the genome evolution of the PLD-causing species.

The genetic homogeneity of both *B. pertussis* and *B. holmesii* is illustrated by analyses of the core gen-

Table 11.2   Insertion sequence (IS) elements identified in five strains of *B. holmesii* for which closed genome sequences are available, as identified by ISfinder[45]. Only closed genome sequences were analysed as accurate enumeration of repeat sequences is not possible using scaffold assemblies of genome sequences containing high numbers of repeats. Values indicate the number of complete IS elements identified in each strain with the number of partial elements in parentheses. ISL3 denotes the IS element also known as IS*1001Bhii*, and IS3 ssgr IS*407* is also known as IS*bho1*[46]

IS element	Strain				
	51541	F627	H558	44057	H903
ISL3	73 (10)	73 (12)	74 (11)	70 (20)	74 (11)
IS3 ssgr IS407					
(ISbho1)	96 (58)	96 (59)	95 (57)	95 (59)	95 (60)
IS481	33 (35)	42 (12)	41 (13)	34 (29)	40 (12)
IS66	3 (0)	3 (0)	3 (0)	3 (0)	3 (0)
IS30	0 (2)	0 (1)	0 (1)	0 (2)	0 (1)
IS3 ssgr IS150	0 (1)	0 (1)	0 (1)	0 (1)	0 (1)
IS21	8 (0)	8 (0)	8 (0)	8 (0)	8 (0)
IS3 ssgr IS51	8 (2)	8 (1)	8 (1)	8 (1)	8 (1)
Total	221 (108)	230 (86)	229 (84)	218 (112)	228 (86)

omes of the *Bordetella* genus and the individual species[21]. Among 34 *B. pertussis* isolates analysed, 2632 genes (out of approximately 3800 genes in total) were common to all of these isolates. Among 18 *B. holmesii* strains analysed, 2654 genes (out of a total of approximately 3700 genes) were common to all 18 strains. There are too few genome sequences available for *B. parapertussis* to fully assess its genetic homogeneity. Most of the differences in gene content within the species arise from the seemingly stochastic small-scale deletion of genes due to continuing recombination between IS elements, as opposed to gain of genes by any of the three species.

The genome reduction suffered by these three species is considered as genome streamlining. If the emergence from the population bottleneck that permitted fixation of high numbers of IS elements involved a niche change, then it is likely that the emerging species will contain genes that were involved in survival in the old niche, but that are not required in the new one. These genes can be lost from the new species without a decrease in fitness, or even with an increase in fitness if the streamlined genome carries a lower metabolic cost of replication than the larger ancestral genome, or some of the obsolete genes in fact inhibit survival in the new niche.

When comparing gene loss from *B. pertussis*, *B. parapertussis*, and *B. holmesii*, a similar pattern is observed in that the genes lost from these species encode mainly hypothetical proteins, transcriptional regulators, transport proteins, and a diverse array of proteins with metabolic functions[21]. The loss of metabolic and transport genes in particular has been interpreted as loss of metabolic diversity and adaptation to utilize a narrower range of resources. This fits with the idea that the speciation of *B. pertussis* and *B. parapertussis* from *B. bronchiseptica* involved adaptation to occupying solely the human niche compared to the seemingly wide host range and environmental niches of *B. bronchiseptica*[26]. The broadly similar gene loss from *B. holmesii* suggests that its genome reduction was also shaped by restriction to a narrower niche range although it is unclear if this might be restriction to a human host environment.

An obvious explanation for the common disease caused by these three species is that they have evolved to contain specific virulence factors involved in causing PLD. A number of *B. pertussis* factors involved in colonization, pathogenesis, and resisting host immune responses have been well characterized. However, many are also common to *B. bronchiseptica* for which they appear to perform the same or very similar functions as for *B. pertussis*.

Crucially, as mentioned previously, *B. holmesii* does not encode any of the key virulence factors identified as important for *B. pertussis* virulence, except for perhaps a homologue of FHA, although primary amino acid sequence homology to *B. pertussis* FHA is relatively low[16]. These observations very strongly suggest that no single well-characterized *B. pertussis* 'virulence factor' is a direct, unique cause of PLD. If specific genes do encode products that directly cause PLD then they are not classic virulence genes.

In fact, an extensive comparative analysis of the pan-genomes of the Bordetellae has not identified any trait specific to those species that can cause PLD, either in terms of genes present or absent[21]. In turn, this raises the possibility that PLD is a human-specific characteristic resulting from *Bordetella* colonization of the lower respiratory tract. In this scenario, it is the ability of *Bordetella* species to colonize and multiply in the human respiratory tract, resisting clearance by host defences, that are PLD-causing traits, and different gene products among the PLD-causing Bordetellae achieve this. There are some fundamental steps involved in this for which comparisons between *B. pertussis*, *B. parapertussis*, *B. holmesii*, and their ancestral species would be interesting.

11.6 Metabolism

The ability of these *Bordetella* species to grow and replicate within the human respiratory tract is probably essential for causing PLD and their metabolism is central to this. The *Bordetella* are asaccharolytic, and at least *in vitro* appear to utilize amino acids as carbon sources (for example, see[27]). The lack of a complete glycolysis pathway, due to the absence of glucokinase and phosphofructokinase, contributes to this[8]. This feature is shared by both the pathogenic *Bordetella* and the more distantly related environmental *B. petrii* suggesting this is an ancestral feature of *Bordetella* rather than an adaptation to infecting humans[28]. *Bordetella bronchiseptica* is regarded as metabolically more versatile than *B. pertussis* and this is thought to contribute to the ability of *B. bronchiseptica* to inhabit a wider range of niches than *B. pertussis*. In this regard, there is no evidence that *B. pertussis* has gained metabolic capabilities that are absent from *B. bronchiseptica* that allow *B. pertussis* to grow in humans, and this argues against a specific metabolic basis for the ability of the PLD-causing *Bordetella* to infect the human niche.

11.7 Adherence

Colonization of the human respiratory mucosa is critical for infection to cause PLD. For *B. bronchiseptica*, *B. pertussis*, and *B. parapertussis*, FHA and fimbriae appear to be key determinants of binding and the cilia of the upper airways appear to be the primary sites for initial colonization (reviewed in[29]). This interaction is thought to be an important determinant of host specificity for *Bordetella* infection. In a classic study, the adherence of *B. pertussis* to human ciliated respiratory cells was greater than that of *B. parapertussis* which, in turn, was greater than that of *B. bronchiseptica*[30]. In contrast, adherence to non-human ciliated cells was greatest for *B. bronchiseptica*, with low levels of adherence by *B. pertussis* and lower still by *B. parapertussis*. Detailed analysis of the adherence properties of *B. holmesii* and the other *Bordetella* species to human and other animal respiratory tissue has not been reported but this might reveal the potential contribution of the specificity of adherence to causation of PLD.

11.8 Resistance to host immunity

Persistence within the human host requires that the PLD-causing *Bordetella* species are able to resist the clearance actions of the human immune system. Again, available evidence suggests that this is achieved by different mechanisms by the three species. For example, *B. pertussis* expresses BrkA, an autotransporter protein that contributes to the resistance of *B. pertussis* to killing by human complement[31]. *Bordetella parapertussis* does not contain the *brkA* gene but is resistant to complement-mediated killing. This is conferred by the O-PS region of *B. parapertussis* LPS[32] that also contributes to the resistance of *B. parapertussis* to neutrophil-mediated killing[33], something to which PT contributes for *B. pertussis*. Thus, similar traits are achieved by different mechanisms for *B. pertussis* and *B. parapertussis*. Very probably, due to its more substantive difference from *B. pertussis* and *B. parapertussis*, this is also the case for *B. holmesii*.

A complication in defining the contribution of individual factors to *Bordetella* infection is that a number of factors have multiple different activities. For example, FHA is a dominant adhesin for colonization of the ciliated respiratory mucosa but it binds to other cell types too with distinct consequences. FHA binds to macrophages resulting in inhibition of secretion of proinflammatory cytokines[34] and has other immunomodulatory effects[35,36]. Thus, subtle differences between homologues of factors common to multiple *Bordetella* species may result in important differences in host–bacteria interactions that would not be evident from simplistic genome sequence analyses. Furthermore, tertiary products are difficult to decipher from genome sequences alone. For example, the lipid A regions of the LPSs of *B. bronchiseptica*, *B. parapertussis*, and *B. pertussis* are structurally distinct and have different biological activities (e.g. see[37]). The genes encoding lipid A biosynthesis in these bacteria appear highly conserved but allelic differences result in different activities; for example, substrate specificity of the acyl-transferase LpxA[38], a number of genes are non-functional in *B. pertussis*[37], and the distinct lipid A acceptor substrate of *B. parapertussis* alters the activity of the palmitate transferase PagP[39].

Thus, while at present there is a lack of clear traits that are specific to the PLD-causing *Bordetella* species, care must be taken in using genome sequence analyses alone to define common functions and differences. Probably, multiple factors required for causation of PLD combine in these bacteria and the lack of factors present in other *Bordetella* species, with loss occurring through genome reduction, may be as important as the presence of specific bacterial factors. Furthermore, host-specific factors are also important for manifestation of PLD. *B. pertussis* colonizes, replicates, and persists within the respiratory tract of mice with no obvious signs of pathology whereas in both naïve humans and infant baboons *B. pertussis* induces a very similar disease[40].

11.9 Going forward

The recent, well-reviewed re-emergence of pertussis in many countries has rekindled interest in this disease. Knowledge of bacterial factors involved in

Bordetella biology has increased substantially in the genomics era and much has been revealed about the complexity of *Bordetella*–host immunity interactions. However, an encapsulating narrative of the detailed events that occur during *Bordetella* infection in a human host to produce PLD is lacking. There are few detailed studies that attempt to carefully distinguish between cases of PLD that might enable associations to be made between specific manifestations and specific bacterial or host factors. The recently developed infant baboon model of pertussis offers great promise for conducting cause-and-effect studies[40], but logistical and ethical considerations will limit the size and scope of such studies. A unique human-challenge model of *B. pertussis* colonization is about to begin[41] that promises to reveal important *in vivo*-derived information about early events in the bacteria–host interaction. Importantly, a much discussed but poorly studied aspect is asymptomatic infection/carriage of PLD-causing *Bordetella* species. Numerous serological studies, in which raised levels of anti-PT antibodies in asymptomatic people is taken as evidence of recent infection, have identified surprisingly high incidences of positive samples[42]. In addition, in the baboon model *B. pertussis* colonized immunized animals without inducing any signs of disease[43] and simulations using available data suggest that asymptomatic transmission of *B. pertussis* is common[44]. Analysis of differences in host responses, host genetics, and environmental and bacterial factors between asymptomatic infections and those that result in disease will be hugely informative towards understanding the mechanisms involved in the triggering of PLD. A comparative approach, involving other *Bordetella* species, could provide a novel additional dimension to understanding classic whooping cough in infants and other cough illness in older age groups. In particular, defining the traits common to the *Bordetella* species that cause PLD, and distinguishing these from those that do not, is likely to be very revealing towards understanding the causes of this disease.

References

1. Eldering G, Kendrick P. Bacillus para-pertussis: a species resembling both Bacillus pertussis and Bacillus bronchiseptica but identical with neither. *J Bacteriol* 1938;**35**:561–72.

2. Porter JF, Connor K, Donachie W. Isolation and characterization of Bordetella parapertussis-like bacteria from ovine lungs. *Microbiology* 1994;**140**:255–61.

3. Moran NA, Plague GR. Genomic changes following host restriction in bacteria. *Curr Opin Genet Dev* 2004;**14**: 627–33.

4. Heininger U, Stehr K, Schmitt-Grohé S, et al. Clinical characteristics of illness caused by Bordetella parapertussis compared with illness caused by Bordetella pertussis. *Pediatr Infect Dis J* 1994;**13**:306–9.

5. Paddock CD, Sanden GN, Cherry JD, et al. Pathology and pathogenesis of fatal Bordetella pertussis infection in infants. *Clin Infect Dis* 2008;**47**:328–38.

6. Koepke R, Bartholomew ML, Eickhoff JC, et al. Widespread Bordetella parapertussis infections – Wisconsin, 2011–2012: clinical and epidemiologic features and antibiotic use for treatment and prevention. *Clin Infect Dis* 2015;**61**:1421–31.

7. Karalius VP, Rucinski SL, Mandrekar JN, et al. Bordetella parapertussis outbreak in Southeastern Minnesota and the United States, 2014. *Medicine (Baltimore)* 2017;**96**:e6730.

8. Parkhill J, Sebaihia M, Preston A, et al. Comparative analysis of the genome sequences of Bordetella pertussis, Bordetella parapertussis and Bordetella bronchiseptica. *Nat Genet* 2003;**35**:32–40.

9. Hausman SZ, Cherry JD, Heininger U, et al. Analysis of proteins encoded by the ptx and ptl genes of Bordetella bronchiseptica and Bordetella parapertussis. *Infect Immun* 1996;**64**:4020–6.

10. Preston A, Allen AG, Cadisch J, et al. Genetic basis for lipopolysaccharide O-antigen biosynthesis in bordetellae. *Infect Immun* 1999;**67**:3763–7.

11. Preston A, Petersen BO, Duus JØ, et al. Complete structures of Bordetella bronchiseptica and Bordetella parapertussis lipopolysaccharides. *J Biol Chem* 2006;**281**: 18135–44.

12. Weyant RS, Hollis DG, Weaver RE, et al. Bordetella holmesii sp. nov, a new gram-negative species associated with septicemia. *J Clin Microbiol* 1995;**33**:1–7.

13. Pittet LF, Emonet S, Schrenzel J, et al. Bordetella holmesii: an under-recognised Bordetella species. *Lancet Infect Dis* 2014;**14**:510–9.

14. Harvill ET, Goodfield LL, Ivanov Y, et al. Genome sequences of nine Bordetella holmesii strains isolated in the United States. *Genome Announc* 2014;**2**:e00438-14.

15. Gerlach G, Janzen S, Beier D, et al. Functional characterization of the BvgAS two-component system of Bordetella holmesii. *Microbiology* 2004;**150**:3715–29.

16. Link S, Schmitt K, Beier D, et al. Identification and regulation of expression of a gene encoding a filamentous hemagglutinin-related protein in Bordetella holmesii. *BMC Microbiol* 2007;**7**:100.

17. Rodgers L, Martin SW, Cohn A, et al. Epidemiologic and laboratory features of a large outbreak of pertussis-like illnesses associated with cocirculating Bordetella holmesii and Bordetella pertussis—Ohio, 2010–2011. *Clin Infect Dis* 2013;**56**:322–31.

18. Faulkner AE, Skoff TH, Tondella ML, et al. Trends in pertussis diagnostic testing in the United States, 1990 to 2012. *Pediatr Infect Dis J* 2016;**35**:39–44.

19. Williams MM, Taylor TH Jr, Warshauer DM, et al. Harmonization of Bordetella pertussis real-time PCR diagnostics in the United States in 2012. *J Clin Microbiol* 2015;**53**:118–23.

20. Cherry JD, Seaton BL. Patterns of Bordetella parapertussis respiratory illnesses: 2008–2010. *Clin Infect Dis* 2012;**54**:534–7.

21. Linz B, Ivanov YV, Preston A, et al. Acquisition and loss of virulence-associated factors during genome evolution and speciation in three clades of Bordetella species. *BMC Genomics* 2016;**17**:767.

22. Mira A, Pushker R, Rodríguez-Valera F. The Neolithic revolution of bacterial genomes. *Trends Microbiol* 2006;**14**:200–6.

23. Moran NA. Accelerated evolution and Muller's rachet in endosymbiotic bacteria. *Proc Natl Acad Sci U S A* 1996;**93**:2873–8.

24. Diavatopoulos DA, Cummings CA, Schouls LM, et al. Bordetella pertussis, the causative agent of whooping cough, evolved from a distinct, human-associated lineage of B. bronchiseptica. *PLoS Pathog* 2005;**1**:e45.

25. Register KB, Ivanov YV, Jacobs N, et al. Draft genome sequences of 53 genetically distinct isolates of Bordetella bronchiseptica representing 11 terrestrial and aquatic hosts. *Genome Announc* 2015;**3**:e00152-15.

26. Preston A, Parkhill J, Maskell DJ. The bordetellae: lessons from genomics. *Nat Rev Microbiol* 2004;**2**: 379–90.

27. Thalen M, van den IJssel J, Jiskoot W, et al. Rational medium design for Bordetella pertussis: basic metabolism. *J Biotechnol* 1999;**75**:147–59.

28. Gross R, Guzman CA, Sebaihia M, et al. The missing link: Bordetella petrii is endowed with both the metabolic versatility of environmental bacteria and virulence traits of pathogenic Bordetellae. *BMC Genomics* 2008;**9**:449.

29. Mattoo S, Cherry JD. Molecular pathogenesis, epidemiology, and clinical manifestations of respiratory infections due to Bordetella pertussis and other Bordetella subspecies. *Clin Microbiol Rev* 2005;**18**: 326–82.

30. Tuomanen EI, Nedelman J, Hendley JO, et al. Species specificity of Bordetella adherence to human and animal ciliated respiratory epithelial cells. *Infect Immun* 1983;**42**:692–5.

31. Fernandez RC, Weiss AA. Cloning and sequencing of a Bordetella-pertussis serum resistance locus. *Infection Immun* 1994;**62**:4727–38.

32. Goebel EM, Wolfe DN, Elder K, et al. O antigen protects Bordetella parapertussis from complement. *Infect Immun* 2008;**76**:1774–80.

33. Gorgojo J, Lamberti Y, Valdez H, et al. Bordetella parapertussis survives the innate interaction with human neutrophils by impairing bactericidal trafficking inside the cell through a lipid raft-dependent mechanism mediated by the lipopolysaccharide O antigen. *Infect Immun* 2012;**80**:4309–16.

34. McGuirk P, McCann C, Mills KH. Pathogen-specific T regulatory 1 cells induced in the respiratory tract by a bacterial molecule that stimulates interleukin 10 production by dendritic cells: a novel strategy for evasion of protective T helper type 1 responses by Bordetella pertussis. *J Exp Med* 2002;**195**:221–31.

35. Abramson T, Kedem H, Relman DA. Modulation of the NF-kappaB pathway by Bordetella pertussis filamentous hemagglutinin. *PLoS One* 2008;**3**:e3825.

36. Dieterich C, Relman DA. Modulation of the host interferon response and ISGylation pathway by B. pertussis filamentous hemagglutinin. *PLoS One* 2011;**6**: e27535.

37. MacArthur I, Mann PB, Harvill ET, et al. IEIIS Meeting minireview: Bordetella evolution: lipid A and Toll-like receptor 4. *J Endotoxin Res* 2007;**13**:243–7.

38. Sweet CR, Preston A, Toland E, et al. Relaxed acyl chain specificity of Bordetella UDP-N-acetylglucosamine acyltransferases. *J Biol Chem* 2002;**277**:18281–90.

39. Hittle LE, Jones JW, Hajjar AM, et al. Bordetella parapertussis PagP mediates the addition of two palmitates

40. to the lipopolysaccharide lipid A. *J Bacteriol* 2015;**197**: 572–80.

40. Warfel JM, Beren J, Kelly VK, et al. Nonhuman primate model of pertussis. *Infect Immun* 2012;**80**:1530–6.

41. Periscope. *Pertussis Human Challenge Colonisation Study.* 2017. http://periscope-project.eu/patients/clinical-study-1/.

42. Barkoff AM, Grondahl-Yli-Hannuksela K, He Q. Seroprevalence studies of pertussis: what have we learned from different immunized populations. *Pathog Dis* 2015;**73**:ftv050.

43. Warfel JM, Zimmerman LI, Merkel TJ. Acellular pertussis vaccines protect against disease but fail to prevent infection and transmission in a nonhuman primate model. *Proc Natl Acad Sci U S A* 2013;**111**: 787–92.

44. Althouse BM, Scarpino SV. Asymptomatic transmission and the resurgence of Bordetella pertussis. *BMC Med* 2015;**13**:146.

45. Siguier P, Perochon J, Lestrade L, et al. ISfinder: the reference centre for bacterial insertion sequences. *Nucleic Acids Res* 2006;**34**:D32–6.

46. Diavatopoulos DA, Cummings CA, van der Heide HG, et al. Characterization of a highly conserved island in the otherwise divergent Bordetella holmesii and Bordetella pertussis genomes. *J Bacteriol* 2006;**188**: 8385–94.

47. Carver TJ, Rutherford KM, Berriman M, et al. ACT: the Artemis Comparison Tool. *Bioinformatics* 2005;**21**: 3422–3.

CHAPTER 12

Surveillance and diagnostics

Shelly Bolotin, Helen Quinn, and Peter McIntyre

Abstract

This chapter discusses the characteristics and limitations of diagnostic tests for *Bordetella pertussis* infection, including bacterial culture, polymerase chain reaction, and serology. It then discusses surveillance for pertussis at the population level in high- and low-income settings, outlining the interplay of clinical, laboratory, and surveillance criteria in the identification and reporting of pertussis cases. The characteristics of pertussis surveillance systems are related to both detection and reporting of pertussis and can change, even within countries, if changes in diagnostic practice occur, making comparisons over time problematic. These considerations have significant implications for optimal surveillance of pertussis in different settings and assessment of resurgence, as also discussed in Chapter 4 on pertussis epidemiology.

12.1 Introduction

Whooping cough has been described in medical texts for at least 500 years[1]. This recognition is likely to have been based on three factors: first, distinctive characteristics of the cough (paroxysms, inspiratory whoop, vomiting); second, extended duration (the '100-day cough' in Chinese); and third, perhaps most importantly, clusters of such cases in epidemics. The earliest routinely recorded data, which allow population incidence to be estimated, is for deaths recorded as due to pertussis in statutory collections. Deaths so recorded would have been limited to those following a typical coughing illness, and therefore minimal estimates[2,3]. In both Australia and Italy, where continuous records exist, mortality rates of approximately 40 per 100,000 total population was documented during epidemic years in the late nineteenth century[4,5]. During the pre-vaccine era of the twentieth century,

recorded pertussis mortality rates decreased substantially in industrialized countries, in line with improvements in healthcare and living conditions,[3] and more dramatically in the post-vaccine era[6].

For pertussis cases reported to government health authorities, the earliest systematic records are available from the 1920s onwards at the national level from Canada[7], the United States,[8] and the United Kingdom[9]. Although *Bordetella pertussis* was first isolated and identified as the causative agent of epidemic whooping cough in 1906[1], identification and enumeration of whooping cough cases almost entirely relied on reporting by physicians for most of the twentieth century. Since their publication in the mid 1990s, randomized trials evaluating acellular pertussis vaccines have given new insights into the sensitivity and specificity of case definitions, measured against a wider range of methods for laboratory diagnosis of pertussis infection. In some high-income

Bolotin, S., Quinn, H., and McIntyre, P., *Surveillance and diagnostics*. In: *Pertussis: epidemiology, immunology, and evolution*. Edited by Pejman Rohani and Samuel V. Scarpino: Oxford University Press (2019). © Crown copyright is held by Public Health Ontario. DOI: 10.1093/oso/9780198811879.003.0012

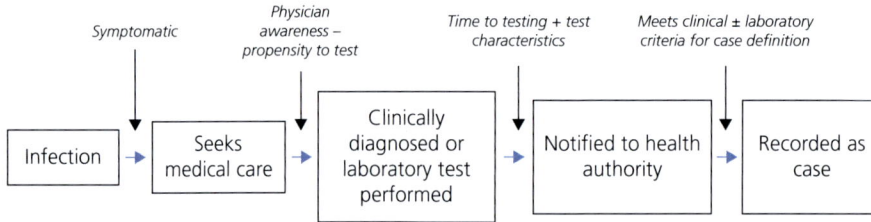

Figure 12.1 Necessary steps for capture of pertussis cases by surveillance systems.

countries, where more sensitive diagnostic tests for pertussis have become increasingly available, this has potentially changed the landscape, although still reliant on presentation to, and clinical recognition and test ordering by, physicians. Decisions at national and sub-national levels about surveillance case definitions and case reporting mechanisms, although related to considerations about clinical presentation and laboratory testing, often operate independently and influence estimates of age-specific incidence over time within and between countries. Figure 12.1 summarizes the multiple steps required for pertussis cases to be captured by surveillance systems—these apply differently according to diagnostic awareness, testing, reporting, and case definitions in the relevant setting.

In this chapter, we first discuss the characteristics and limitations of diagnostic tests for *B. pertussis* infection. We then discuss surveillance for pertussis at the population level, and outline the interplay of clinical, laboratory and surveillance criteria in the identification and capture of pertussis cases. These considerations have significant implications for assessment of the presence or absence of pertussis resurgence, as also discussed in Chapter 4 on pertussis epidemiology.

12.2 Diagnostic tests

12.2.1 Detection of *Bordetella pertussis* in the respiratory tract

The primary site of infection with *B. pertussis* is the upper respiratory tract, and the nasopharynx and nasal passages are the preferred sites from which to collect specimens[10,11], although throat swabs have also been used[12]. Specific requirements differ

for the two primary methods for direct detection of *B. pertussis*—bacterial culture and nucleic acid detection. Detection of fluorescent antibody is no longer recommended for diagnosis[11].

For bacterial culture, nasopharyngeal specimens (swabs or aspirates) are best for collecting ciliated epithelial cells to which *B. pertussis* attaches. If a nasopharyngeal specimen is taken using a swab, flocked swabs gives the best yield of cells, but aspirates yield even more cells[11,13]. Throat swabs are sometimes used, although they usually do not obtain ciliated cells, and may include normal oral flora, making isolation of *B. pertussis* in media more difficult and potentially inhibiting laboratory assays[14]. Swabs used for isolation of *B. pertussis* can be made of polyester (Dacron®), rayon, nylon, or calcium alginate[11,13,15]. Cotton swabs contain inhibitors that can limit *B. pertussis* growth in culture[11,15], while calcium alginate swabs have been shown to inhibit *B. pertussis* polymerase chain reaction (PCR)[15].

For PCR, nasopharyngeal swabs, aspirates, induced sputum, or throat swabs have all been shown to be adequate for nucleic acid detection[16,17], with nasopharyngeal aspirates again yielding the highest bacterial load[11]. Although specimens destined for bacterial culture can also be used for molecular testing, there are slight differences in the pre-analytical steps between the two methods. Swab specimens intended only for PCR need not be immersed in transport media[11]. Prior to molecular testing, the specimen is suspended in saline or molecular grade water and boiled to lyse the bacteria and extract DNA[11]. The DNA must be purified prior to testing to avoid inhibition of the PCR reaction[11,17].

12.2.2 Laboratory diagnostics

Bacterial cell culture

Bordetella pertussis has traditionally been diagnosed using bacterial culture, which remains the gold standard[11]. Despite low sensitivity and slow turnaround time, and although strain-typing directly from the specimen is sometimes possible[18], bacterial isolation by culture remains the only method reliably permitting more extensive downstream genomic analyses, and antibiotic susceptibility testing[13,19]. Isolation is more likely if specimens are plated immediately after collection, or if specific transport media (such as half strength Regan–Lowe, Casamino acid, or Amies medium with charcoal) are used[13,19]. Bordet–Gengou or Regan–Lowe (charcoal blood) media, with an added antimicrobial to inhibit growth of commensal flora[19] are optimal for *B. pertussis* growth, and also allow growth of other clinically relevant *Bordetella* congener species, including *B. holmesii*, *B. bronchiseptica*, and *B. parapertussis*[11,20]. Although colonies typically become visible by 3–4 days after incubation at 35–36°C with low levels of CO_2[11,19], incubation for 7 days is usually recommended[19], and incubation for 12 days may result in more isolate recovery[21,22].

Although highly specific, culture of *B. pertussis* is insensitive, as *B. pertussis* is a fastidious organism, with several factors contributing to the likelihood of isolation in culture[13]. Recovery of isolates is inversely correlated to bacterial load[22,23], which in turn is related to factors such as patient age and vaccination status[24]. Young children have the highest bacterial load[25,26], with culture sensitivity reported as high as 60 per cent in unvaccinated infants[27]. In contrast, sensitivity of culture in adolescents and adults may be less than 1 per cent[27]. Time from symptom onset to sample collection[13] and disease severity also affect sensitivity. Bacterial isolation is most likely during the initial catarrhal stage of disease[13,27], and is highest for hospitalized patients[25], most commonly unvaccinated infants[24], who are also more likely to present early to healthcare. Older children and adults often present after several weeks of unresolved coughing[28], and may not present typically, delaying diagnosis. Not surprisingly, treatment with antibiotics prior to sample collection lowers sensitivity[13,19].

Direct fluorescent antibody

Testing of nasopharyngeal secretions using direct fluorescent antibody (DFA) is a rapid method for pertussis diagnosis[19]; however, due to limited sensitivity and specificity its use is no longer recommended[13,29–31]. For example, in Chile, false-positive DFA results may have contributed to an overestimation of infant deaths due to pertussis[32].

Polymerase chain reaction

PCR first emerged in the late 1980s and early 1990s as a tool for detection of *B. pertussis*[33–36] and has revolutionized pertussis diagnostics. Its superior sensitivity over culture[27], improved sensitivity and specificity over DFA[37], and rapid turnaround time have resulted in PCR replacing culture for most routine testing[20].

PCR assays have evolved from end-point PCR, prevalent in the 1990s and early 2000s, to quantitative real-time (RT)-PCR, more widely used from the mid 2000s[11,38]. Although the sensitivity of the two methods is similar[39], RT-PCR has a shorter turnaround time[17,40]. For this reason, and since RT-PCR platforms have become ubiquitous in diagnostic microbiology laboratories, they have become the preferred method from a diagnostic and efficiency perspective.

Specificity of PCR: single and multiple target sequences

Different approaches can be used for designing PCR targets to detect *B. pertussis* and other clinically relevant *Bordetella* species. Many laboratories incorporate insertion sequences (ISs) into their assays. ISs are small transposable elements of approximately 1000–2000 base pairs, which are present in multiple copies in the genome[11,20,41]. This makes them excellent candidates as PCR targets, as the multiple copies increase assay sensitivity. However, many of the ISs commonly used for molecular pertussis diagnosis are present across several *Bordetella* species of clinical importance in humans,[11,42] representing a challenge to assay specificity[11,17]. While use of a single IS will detect the presence of *Bordetella*, it may not allow species determination. However, use of a combination of targets can differentiate between the *Bordetella* species that cause human disease.

Table 12.1 Expected results of PCR testing using insertion sequences IS481, IS1001, and IS1002 to differentiate between Bordetella species of clinical significance

Bordetella species	IS481	IS1001	IS1002
B. pertussis	+		+
B. parapertussis		+	+
B. bronchiseptica	+	+	+
B. holmesii	+		

This is an important point because while *B. pertussis* causes the majority of pertussis illness in humans, other related *Bordetella* species also contribute[43–45].

IS481 is the most commonly used primer and probe target for pertussis molecular diagnostics. It is present in approximately 50 to over 200 copies in *B. pertussis*[46–48], resulting in a highly sensitive assay. However, since IS481 is also found in *B. holmesii*[11,47], and in a minority of isolates of *B. bronchiseptica*[17,20], and some other *Bordetella* species (see Table 12.1),[17] specificity is reduced. IS481 can be combined in a duplex assay with IS1001, which is not present in *B. pertussis* but is present in *B. parapertussis*[49] and *B. bronchiseptica*[11] to increase specificity. If the IS1001 target is positive, this rules out *B. pertussis*. Differentiation between *B. parapertussis* and *B. bronchiseptica* can then be made by looking at the IS1002, IS1001, and/or IS481 result. *B. parapertussis* carries IS1001 and IS1002 but never IS481, whereas *B. bronchiseptica* may carry either IS1001 or IS1002 but not both and rarely may carry IS481. Since *B. pertussis* and *B. holmesii* both carry IS481, but not IS1001, a duplex assay will still not differentiate between them. Adding a third target for IS1002, which is carried by *B. pertussis* and *B. parapertussis*, but not *B. holmesii*, will therefore raise specificity considerably. In addition to the examples given above, several other ISs have been used for PCR diagnosis of pertussis, including IS1663 (present in *B. pertussis*), and hIS1001, ISBho1, and IS1001Bhii (present in *B. holmesii*)[11].

In addition to ISs, single-copy genes are also commonly used to detect *Bordetella* species. These tend to be less sensitive than multi-copy ISs, and may suffer from target gene mutation[50,51] and gene loss[11]. The most common single target used is the pertussis toxin promoter (*ptxP*). Although the pertussis toxin gene also exists in *B. parapertussis* and

B. bronchiseptica, due to mutations in the promoter region it is not expressed[20]. A *ptxP* probe that is specific to *B. pertussis* can therefore be designed. However, several studies have shown mutations in the *ptxP* region, meaning that false-negative results are possible[50,51]. Many other single-copy gene targets have been described. Some, such as open reading frames BP283, BP485, and *ptxS1* are specific to *B. pertussis*[52,53], but most are not specific to *B. pertussis* and have been detected in other *Bordetella* species. For some targets, like filamentous haemagglutinin[54], flagellin[55], porin[56], and recA[57], it is possible to design species-specific targets[20].

A common diagnostic algorithm to increase PCR test specificity combines an IS with single-copy gene targets in a multiplex PCR. For example, diagnostic assays that combine *ptxP* and IS481 have been described[58]. If both targets are positive, this is indicative of *B. pertussis*, while if only IS481 is positive, some laboratories interpret the test as positive for DNA of other *Bordetella* species[18,59], while others interpret it as indeterminate[60]. However, since *ptxP* is a single-copy target, a test that is positive for IS481 alone may also be indicative of *B. pertussis* infection, in particular if the quantity of DNA (i.e. RT-PCR cycle threshold (Ct)) is low[59]. Interpreting a PCR result that is only positive for IS481 as indeterminate may therefore potentially lower assay sensitivity[20]. Another approach uses a triplex assay to combine both IS481 and IS1001, detecting *B. pertussis* and *B. parapertussis*, respectively, with a *B. holmesii*-specific *recA* target[25,61,62].

As indicated above, the most significant factor contributing to *B. pertussis* PCR test specificity is the chosen primer-probe targets, particularly if they are based on ISs[60,63–65]. Another issue, which is common to all molecular tests, is false positives due to contamination of laboratory reagents, consumables, or surfaces, with PCR amplicon or other clinical samples[66,67]. Appropriate negative controls included in each assay should identify these, and use of closed-system platforms that require no downstream handling, such as RT-PCR platforms, should reduce their occurrence[11,17]. Outside of the laboratory, environmental contamination from clinical surfaces, contamination of the specimen during collection, and even aerosolized vaccine, which can contain *B. pertussis* DNA, have been reported[68–70].

Sensitivity of PCR

Bordetella pertussis PCR is a substantially more sensitive method of detection than culture[22,23,30,37,58,71]. In one study, PCR was positive in 95.8 per cent of infant and child cases and 61.5 per cent of adult cases that met a clinical case definition, whereas culture detected 54.1 per cent and 15.4 per cent, respectively[37]. In routine practice, a switch from culture to PCR in one diagnostic laboratory resulted in an increase in cases of 9–26 per cent[58]. In another, culture of PCR-positive nasopharyngeal samples resulted in the isolation of *B. pertussis* from only 45 of 223 (20.2 per cent) of the samples[22].

Like bacterial cell culture, PCR sensitivity is dependent on bacterial load[22], which in turn is related to the inter-related factors of age, severity, and vaccination status. Samples from children often have lower Ct values and are therefore more likely to be positive than samples from adults[25,26]. PCR Ct values are also correlated with the severity of symptoms[25] and the duration between symptom onset and specimen collection[17,20]. Antibiotic treatment can also affect the Ct value, although it has been shown that positive PCR results can still be obtained several weeks after initiation of treatment[27,72].

There are several reasons for the high sensitivity of PCR methods. In general, molecular methods are inherently more sensitive than other laboratory detection methods. For pertussis PCR assays in particular, the limits of detection are incredibly low, regardless of whether end-point or RT-PCR is used[20,71]. For IS*481* and other multi-copy IS-based assays, the limit of detection is as low as one colony forming unit[61], or even less[73]. Even assays based on single-copy gene targets can detect just a few colony forming units[61,73].

PCR positivity in asymptomatic or minimally symptomatic individuals

Unlike bacterial cell culture, PCR does not require viable bacteria to obtain a positive result[74]. For these reasons, the interpretation of high Ct positive results can be problematic, particularly in vaccinated individuals, older children, adolescents, and adults. Presuming that PCR contamination has been ruled out, in these populations there are several possible interpretations of high-Ct PCR results. While this result could represent an individual with true disease, it could also indicate an immune individual, whose immune system has already cleared viable bacteria from the upper respiratory tract. Alternatively, a high Ct PCR result could represent an individual who is transiently colonized with *B. pertussis*, but is not symptomatic[62,75]. While the carrier state in pertussis was hypothesized by Fine and Clarkson in the early 1980s[76], it was often reported for a very small proportion of those tested[77,78], likely due to the low sensitivity of bacterial culture methods. However, in recent years it has been described more frequently, most often relating to outbreak investigations that included specimens from epidemiological contacts who were asymptomatic[62,79–81]. The more frequent description of asymptomatic cases is likely related in part to the switch to highly sensitive PCR diagnosis using multi-copy ISs[62,79,82]. The proportion of asymptomatic PCR positive cases varies by study[62,79,83], but has been reported to be as high as 20 per cent[81].

12.2.3 Detection of *Bordetella pertussis* by serological testing

Serological tests for pertussis based on enzyme-linked immunosorbent assay (ELISA) methods were first developed in the 1980s. Although some early methods used whole *B. pertussis* organisms as antigen[84], purified antigens are generally recommended, with pertussis toxin (PT) the only antigen unique to *B. pertussis*[85]. A recent review summarized the history of development of serological testing, including assays for antigens other than PT, the use and interpretation of assays for immunoglobulin (Ig)-A and IgM as opposed to IgG, and the detailed international work on reference standards and determination of

Table 12.2 Comparison of laboratory methods for detection and identification of *B. pertussis*

Diagnostic method	Relative sensitivity	Relative specificity	Advantages	Disadvantages
Bacterial detection				
Culture[a]	++	++++	Isolated strain may be subtyped	Varying sensitivity, depending on vaccination status
DFA	+	+++	Rapid	Microscopist dependent
PCR	+++	+++	Rapid; increases the rate of case finding	Risk of false positives
Serology				
ELISA:				
For antibodies to PT[a]	+++	++++	Positive late in illness	Diagnosis possible only several weeks after onset; influenced by immunostatus
For antibodies to FHA	++++	+++	Positive late in illness	Non-specific response (e.g., to *Haemophilus influenzae*)

DFA, direct fluorescent antibody; FHA, filamentous haemagglutinin; PT, pertussis toxin.

[a] Varies in vaccinated versus unvaccinated populations, with age, and also with duration of symptoms before sampling.

cut-off thresholds for positivity[11]. Although a four-fold rise in antibody titre in paired sera is usually used as the diagnostic standard for serological diagnosis, careful clinical epidemiological and clinical studies from the Netherlands[86] and Massachusetts, United States,[87] have demonstrated the validity of high single-titre IgG antibody PT for diagnosis, as titres of greater than 100 international units are highly specific for recent pertussis infection, in the presence of appropriate symptoms and the absence of recent vaccination. In an effort to make serological assays for pertussis more available, use of oral swabs sent by post for detection of PT antibody has been explored in the United Kingdom. Although detection of PT antibody in oral fluid is less sensitive than detection in serum, this method was shown to have acceptable characteristics as a test for diagnostic and surveillance purposes[85]. The primary value of serological testing for pertussis is in older children and adults when symptoms have been present for 3 weeks or longer, with culture and PCR strongly preferred when symptoms have been present for less than 2 weeks[85]. Recently, the European Centres for Disease Control have issued recommendations for the use of serological tests in surveillance of pertussis[88]. Table 12.2 summarizes the characteristics of culture, DFA, PCR, and ELISA serology[89].

12.3 Surveillance for pertussis

12.3.1 Insights from clinical trials

Children

Randomized clinical trials comparing candidate acellular and/or whole-cell pertussis vaccines with diphtheria–tetanus vaccine conducted in Gothenburg, Sweden[90], and Italy[91] in the early 1990s provided important insights into the true clinical spectrum of pertussis in vaccinated and unvaccinated infants. This was because detection bias was minimized by active follow-up of participants to identify cough illness of any duration, with subsequent uniform diagnostic testing. In Gothenburg, where no pertussis vaccination had been in place for 16 years, 2037/3450 (59 per cent) of trial participants had at least one cough illness lasting longer than 7 days during the study period of whom 10.3 per cent had culture-proven pertussis. Among those with 21 days or more of cough, 18 per cent had culture or serological confirmation. In Italy, where pertussis vaccine was recommended but had very low coverage, the proportion of children with more than 7 days of cough who were culture positive was similar (474/5147; 9.2 per cent), but only 288/474 (60.8 per cent) went on to have a cough illness of greater than 21 days. In Gothenburg, PCR was performed as well as culture. Among placebo recipients, culture (154/242; 63.6 per cent) and PCR (155/205; 75.6 per cent) positivity was similar and significantly higher than among vaccine recipients (49 per cent and 51 per cent respectively). These and other vaccine trials in Sweden and Germany, all settings with little or no population use of pertussis vaccine, shed light on important considerations about the interaction between symptoms and diagnostic testing in pertussis[92]. First, cough illness was common, and with systematic testing, pertussis was found in about 20 per cent of total cases, but case definitions requiring 21 days of paroxysmal cough missed around half of these. Second, pertussis vaccines were substantially less effective against any laboratory-confirmed cough illness of greater than 7 days (approximately 70 per cent) than against cough illness meeting the World Health Organization (WHO) case definition (approximately 85 per cent), of longer than 21 days of cough and laboratory confirmation by culture or paired serological tests.

Adults

There is only one randomized clinical trial of pertussis vaccine with disease endpoints in adults[93]. In contrast to the paediatric trials, this trial was conducted in the United States, a country with relatively high coverage using a long-standing five-dose schedule of pertussis-containing vaccines in children under 5 years. Among 2781 participants, 63 per cent had a cough illness of at least 7 days during the study, but only ten laboratory-confirmed cases of pertussis were identified, 9 in the placebo group (four culture, five PCR). Cough was very common and neither total, nor prolonged, cough illness differed between vaccine and control groups. The extensive surveillance for cough episodes and laboratory testing performed means it is very unlikely that pertussis was undetected. Nevertheless, the culture or PCR-positive cases among the placebo group extrapolate to an incidence of approximately 300,000 per 100,000 person years—many-fold higher than any incidence estimates from routine reporting, and substantiating significant under-reporting among adults.

12.3.2 Surveillance at the case level

Almost all countries accept notification by a clinician of a case of pertussis without requiring information on specific clinical features, although such cases are often listed as probable, with the confirmed category restricted to cases with laboratory evidence or who have been epidemiologically linked to a laboratory-confirmed case. As referred to above, evidence from clinical trials shows that especially in adults, but also among children in settings where pertussis is endemic, clinical syndromes typical of pertussis occur in the absence of any laboratory evidence of pertussis infection and laboratory-proven pertussis may not be associated with typical symptoms. However, pre-test probability is an important consideration—during a pertussis outbreak, a simple case definition such as cough of at least 14 days' duration can have acceptable sensitivity and specificity for pertussis, even when judged by the relatively insensitive, though specific, gold standard of culture. However, such a case definition will be poorly specific outside an outbreak situation[94]. In general, as the number of criteria required to meet a clinical case definition increase, specificity increases at the expense of sensitivity[95]. Adding to the challenges of routine case-based surveillance is the consistent finding that clinicians report only a fraction of even laboratory-confirmed cases[96–98].

12.3.3 Case definitions and reporting at country level

There are important interactions between diagnostic tests for pertussis, criteria for clinical diagnosis and recorded pertussis case notifications. In high-income countries with well-established surveillance systems, cases accepted as valid for reporting can be severely constrained by requiring highly specific laboratory and clinical criteria, exemplified by the experience of the United States. At the other extreme, exemplified by the Australian experience, reported cases can be greatly expanded if first, increased diagnostic testing is coupled with compulsory laboratory notification of cases with positive tests, and second, a positive test is sufficient to satisfy the case definition. In low- and middle-income countries, and even in some high-income countries, case capture is limited to clinician diagnosis, with reports seldom reaching public health authorities.

A recent review compared reporting of measles and pertussis by member countries to the WHO using the joint reporting form[99]. This demonstrated

Table 12.3 Countries with the highest pertussis incidence reported to WHO, 2008–2012

Country	Total pertussis case count, 2008–2012 JRF	Annualized pertussis incidence, 2008–2012, per 1000 population
Australia	140,160	1.26
Norway	20,930	0.86
Papua New Guinea	18,978	0.55
Netherlands	38,053	0.46
Estonia	3036	0.45
New Zealand	8708	0.40
Switzerland	14,300	0.37
Israel	9670	0.26
Somalia	8300	0.18
Slovenia	1695	0.17

JRF, Joint Reporting Form.

Data derived from[99].

that for measles, the highest incidence during the 5-year period 2008–2012 was recorded from low-income countries and largely corresponded with low vaccine coverage, whereas for pertussis the highest incidence was in high-income countries with high coverage, with two exceptions—Papua New Guinea and Somalia (see Table 12.3). Further, reported incidence data from the 194 member states were much less complete for pertussis, with 86/194 (44 per cent) of countries reporting either zero cases[52] or no data[34] whereas for measles, only 11 countries did not provide data and many of the 62 reporting zero cases had met measles elimination targets[99].

Despite the limitations of the WHO/UNICEF joint reporting system, it is sufficiently sensitive to detect known resurgences of pertussis, such as that which occurred in the United Kingdom, related to steep falls in vaccine coverage in the 1980s[9], and in Canada, related to a poorly performing whole-cell pertussis vaccine in the 1980s and 1990s (see Figure 12.2)[100]. Of the countries with the highest reported incidence for pertussis in 2008–2012, 6 were in the European region (Norway, Netherlands, Estonia, Switzerland, Israel, and Slovenia), 3 were in the Western Pacific region (Australia, New Zealand, and Papua New Guinea), and 1 was in Africa (Somalia). Additional countries with high case

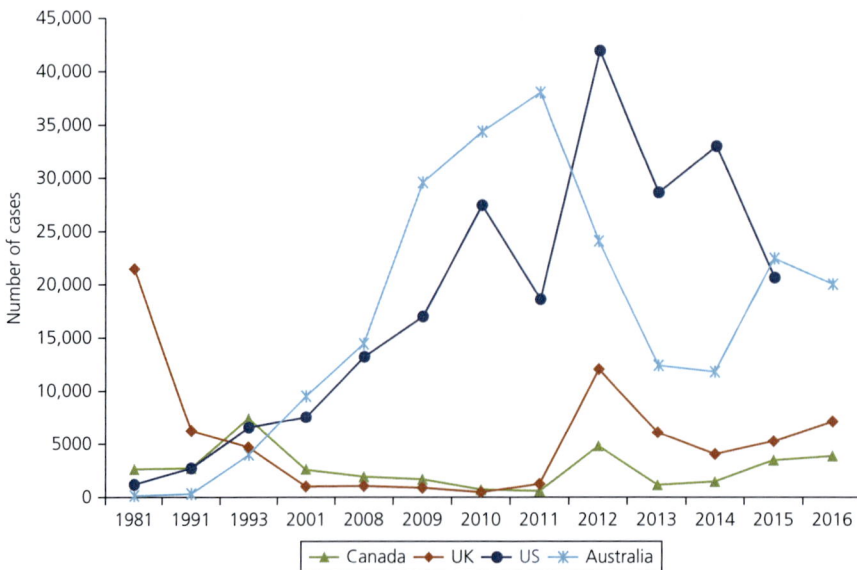

Figure 12.2 Pertussis case counts, by country, from WHO reporting, 1981–2016. UK, United Kingdom; US, United States.

counts included the United States and the United Kingdom. To provide insight into the characteristics driving high case reporting, relevant literature pertaining to surveillance in these countries is summarized in the following sections.

United States

Case definitions and notification practices

The Centers for Disease Control and Prevention (CDC) adopted the first Council of State and Territory Epidemiologists (CSTE) case definition of pertussis in 1990. For a confirmed case, this required isolation of *B. pertussis* in culture, in the presence of at least 14 days of cough and one or more of whoop, paroxysms, or vomiting. A probable case only required the clinical case definition. The case definition was expanded in 1995 to include cough of any duration in culture-positive cases, which increased acceptance of notifications in young, unimmunized infants who are most likely to be culture positive, but present much earlier than 14 days. In 1997, identification of *B. pertussis* by PCR was added to culture as a laboratory criterion, but required clinical confirmation until 2014, when PCR-positive infants were classified as probable cases with a cough of any duration[101]. In one state in the United States (Massachusetts), serological diagnosis has been accepted for case reporting since 1988, increasing incidence by fourfold in persons over 10 years of age[87]. Massachusetts cases based on serology are accepted in National Notifiable Diseases Surveillance System (NNDSS) data by the CDC, providing they meet the clinical case definition. They comprised almost 80 per cent of total serologically confirmed cases in the United States in the 1990s[101].

Diagnostic tests

There were substantial changes in diagnostic tests supporting pertussis notification in the United States between 1990 and 2012, dominated by progressive decreases in culture confirmation and increases in PCR confirmation. Among pertussis cases recorded by the NNDSS, the proportion with laboratory confirmation by culture or DFA decreased progressively from 100 per cent in the early 1990s to less than 10 per cent by 2010. Serology and PCR were first reported in 1995. Serologically confirmed

cases peaked in 2006 at 20 per cent, before declining to less than 5 per cent by 2012, whereas PCR progressively increased, accounting for 30 per cent of cases by 2005 and more than 60 per cent by 2012[101].

Australia

Case definitions and notification practices

In the early 1990s, Australia catapulted to leadership of the world league table, not just for pertussis notification rates, but also for total numbers of notifications irrespective of population (see Figure 12.2). This was driven by a key change in notifications process for its NNDSS, whereby laboratories were mandated to notify all persons who had any positive laboratory test for pertussis. This change coincided with increasing use and availability of a locally produced serological assay[84] and reimbursement for its use through national medical insurance. The case definition allowed for any combination of laboratory, clinical, or epidemiological evidence modified in 2013 to include only culture, PCR, or seroconversion as definitive laboratory evidence.

Diagnostic tests

In the decade between 1986 and 1996, acceptance of single-point serology, typically IgA antibody to a whole-cell pertussis antigen[84], had a substantial impact on total notifications, which increased in one state by tenfold, with the proportion based on serological criteria increasing from 15 to 90 per cent[5]. At national level, serology continued to predominate from 2000 to 2005, among individuals over the age of 5 years[102]. However, from 2006, PCR became the dominant diagnostic method in all age groups less than 20 years[103], driven by full remuneration of PCR testing in primary care from 2007[104]. The ready availability of pertussis diagnostic tests in primary care, and through multiple private sector laboratories, coupled with mandatory reporting by laboratories directly to public health authorities rather than via clinicians, appears to be an almost unique characteristic of pertussis surveillance in Australia, and has driven very high case numbers.

Europe: a case study in variability

A diverse array of surveillance systems are in place in the European region[105]. Historically, case surveillance in many countries was limited to reports from physicians, with all accepted irrespective of clinical symptoms[106,107]. A European case definition was first developed in 2008[95] and in 2014, all but 3 of 28 countries reporting to the European Centre for Disease Control reported using the European case definition[108].

Six European countries were included among the ten with the highest incidence reported to the WHO for the period 2008 to 2012—these were the Netherlands, Norway, Switzerland, Estonia, Slovenia, and Israel[99]. Each is notable for having a broad range of diagnostic tests available, especially serology[105,109,110]. All of these six high-incidence countries have mandatory population-based surveillance, except Switzerland which conducts sentinel surveillance through participating primary care practices[105,111]. Surveillance of pertussis which includes only infants as in Denmark[105] and France[112] or focuses on enhanced surveillance of hospitalized infants as in the United Kingdom[106] is also prevalent in Europe.

This variation in surveillance practice between countries and over time makes comparison of reported pertussis incidence in Europe problematic. In a WHO review by the Strategic Advisory Group of Experts (SAGE)[32], which assessed the presence of a resurgence in pertussis or an increase in infant mortality or hospitalizations unexplained by other factors, Portugal and England and Wales were the only European countries fulfilling this definition, but neither was among the six high incidence countries identified by WHO reports[99]. This would be explained by the WHO Joint Reporting Form not differentiating cases by age group or severity, so countries with the highest overall rates have greater reporting across the age range, as would typically arise from use of serological diagnosis, whereas the SAGE report examined only severe infant cases.

Low-income countries

Two countries with high infant mortality, low income, and low vaccine coverage—Papua New Guinea (PNG) and Somalia—were identified amongst the ten countries with the highest incidence of pertussis reported to WHO for 2008–2012[99]. It is not clear what led to high clinical reporting in these two countries, although during this period a pertussis outbreak in PNG was reported, and similar to the situation prevailing in Somalia, outbreaks were identified in the Democratic Republic of Congo, Afghanistan, and South Sudan, all countries experiencing humanitarian emergencies associated with breakdown of public health services including immunization[113]. It may be that capacity to identify outbreaks is associated with capacity for reporting.

More recently, enhanced surveillance studies using PCR for diagnosis have been conducted in a range of settings in Asia and Africa, with the aim of evaluating the presence of unidentified infant deaths from pertussis potentially preventable through maternal immunization[114]. The community-based studies amongst these, conducted in Pakistan[115] and Zambia[116], found evidence of high rates of pertussis infection in young infants but most identified cases did not meet criteria for severe disease or require hospitalization in these settings.

12.3.4 Surveillance of hospitalized cases

Hospitalized cases can be identified by a number of means. Surveillance may be limited to a hospital network responsible for all higher-level paediatric care, as is the case in France[117] and Canada[118]. Alternatively, hospitalized cases, particularly in infants, may be identified among all cases notified to public health authorities through enhanced surveillance, usually linked to higher levels of diagnostic testing as is the case in the United Kingdom[106]. Coding of hospitalizations as due to whooping cough or pertussis was shown to be reliable when judged against clinical criteria, in a study from a single tertiary care children's hospital in Australia[119], but criteria for hospitalization as opposed to outpatient treatment likely vary even among high-income countries. A higher incidence of pertussis in infants based on coded hospitalizations than from case-based surveillance in the same age group has been found in several countries[106,107,120,121], consistent with low levels of clinician reporting to public health authorities. In Australia, where mandatory laboratory reporting is in place, hospitalized infants who were culture positive were reported but not others[119], but with the emergence of PCR, infant cases reported by laboratories based on either PCR or culture resulted in incidence based on case surveillance exceeding that from coded hospitalizations[103].

An early study from the United Kingdom, using routine PCR testing of infants with respiratory symptoms admitted to intensive care demonstrated that many cases of pertussis were not clinically suspected, and would have escaped detection if only clinician-initiated ordering was in place, especially if this was limited to culture as a diagnostic test[122]. Few studies report on the highest severity hospitalized cases, those requiring intensive care admission, and only one has correlated with this with case-based surveillance[123]. This study, of all intensive care admissions among infants in Australia and New Zealand from 2002–2014, found that the incidence of intensive care admission correlated with peaks in notification and that more than 80 per cent of infant admissions occurred before 3 months of age.

Overall, surveillance of hospitalized cases has the advantage of capturing, usually more completely, the most severe cases which contribute most to total disease burden from pertussis. However, it has some important limitations. First, case capture is influenced by availability and use of diagnostic tests. A study from one hospital, where culture remained routine for all specimens submitted for PCR, demonstrated that apparent dramatic increases in pertussis hospitalization were almost eliminated when only culture-positive cases were considered over the whole time period examined[124]. Another limitation is the strong age restriction of cases captured, which are largely limited to infants and young children. Severe disease resulting in hospitalization in high-income countries with long-standing pertussis immunization programmes is being increasingly recognized in both older[125] and younger[126] adults and in turn relates to greater use of more sensitive diagnostic tests.

12.3.5 Pertussis deaths

Historically, data on deaths from pertussis have come from statutory death certificates. More recently, this has been supplemented by data from public health surveillance and capture-recapture methodology has demonstrated significant under-ascertainment of deaths in both in the United States[127] and the United Kingdom[128]. The availability of intensive care and especially more advanced intensive care such as high-frequency ventilation likely means that some cases previously resulting in death now survive, so it may be more appropriate to consider confirmed pertussis cases requiring high-level intensive care support and resulting in death together for the purposes of surveillance, especially over time. The Australia–New Zealand study found that the proportion of pertussis intensive care admissions who died decreased over a 13-year period and that of those who received high-frequency ventilation, 60 per cent survived[123].

12.4 Resurgence and outbreaks: interplay between surveillance practices and diagnostic testing

As previously mentioned, periodic localized outbreaks associated with high case fatality were historically a feature of the pre-vaccine era in both high-income and low-income countries. Currently, such outbreaks continue to be reported from regions where infant vaccination is not established or has broken down[113,129] and are based on clinical features rather than diagnostic test positivity. In contrast, reports of resurgences of pertussis are exclusively from high-income countries with long-standing high vaccine coverage and are typically more widespread at the country or large regional level. Resurgence has been reported due to whole-cell vaccines of suboptimal potency from both Canada[130] and the Netherlands[131], which in both countries appeared to abate following the introduction of acellular pertussis vaccines. However, in the United Kingdom[132], the United States[133], and Australia, resurgence has been reported following several years of acellular vaccine use and attributed to more rapidly waning vaccine-acquired immunity compared with whole-cell vaccines[134].

In Australia, the presence of widespread diagnostic testing in primary care[104] combined with discontinuation of a booster dose in the second year of life[135] contributed to the magnitude of resurgence. These factors are likely to have resulted in diagnosis of less severe cases, against which acellular vaccines are known to be less effective[136]. Thresholds for test positivity are also a factor in case identification for both serological tests and PCR. In Australia, raising the threshold for seropositivity resulted in a substantial reduction in the proportion of adults deemed to have diagnostic single-titre serology[137]. In Canada, unnecessary public health investigations were deemed to have resulted from acceptance of asymptomatic or mildly symptomatic cases with high Ct values as meeting notification criteria. Modification of the cut-off for PCR positive results[138] improved this in some instances, although the positive predictive value for meeting a clinical pertussis case definition remains largely the same if only small changes are made to the Ct cut-off[25].

At the population level, the impact of the widespread use of PCR diagnostics remains unclear. Although in some jurisdictions the resurgence of pertussis activity did not necessarily coincide with the introduction of molecular diagnostic methods[139], it seems that the availability of PCR has resulted in increased testing by clinicians[38], a trend that is particularly amplified during outbreaks[38,62]. The use of PCR has also revealed a reservoir of infection in older children and adults that was previously largely ignored[140,141], leading to greater awareness of infections in these age groups, and the addition of vaccine boosters for adolescents and adults in many Jurisdictions[20]. PCR diagnostics has also alerted us to the circulation of congeners, including *B. holmesii*, *B. bronchiseptica*, and *B. parapertussis*. It seems that these species contribute to clinical illness more than was previously thought[142,143]. However, it is unclear whether this is a temporal trend or whether more specific diagnostic methods have uncovered this[144], and as it is not reported in most settings, no information about trends over time is available.

12.5 Population-based serological surveillance

Studies in the Netherlands demonstrated that levels of antibody to PT, only produced following infection with *B. pertussis* or vaccination, remained higher for much longer after infection than after vaccination[86]. This allowed the use of serological thresholds of PT antibody to demonstrate infection with *B. pertussis* in the previous 6 or 12 months and to differentiate this from vaccine-induced antibody in immunized cohorts. A large number of seropreva-lence studies using anti-PT antibody in specimens from population-based cohorts, unrelated to testing for pertussis, have now been done worldwide[145] and can reasonably be presumed to be free from testing bias. Although the presence of a given antibody titre does not differentiate asymptomatic or mildly symptomatic infection from more clinically severe infection, age-specific estimates of clinically important pertussis episodes can be made[146]. Correlation with case-reported surveillance has suggested substantial under-ascertainment in multiple settings, consistent with those obtained from systematic studies of cough illness[147]. Population serosurveillance in Australia was able to identify increased proportions of children with undetectable PT antibody values in immunized cohorts following discontinuation of the 18-month pertussis booster[148] and in Sweden to identify an increased proportion of individuals across most age groups with undetectable PT antibody in the decade following re-introduction of pertussis vaccines[149].

12.6 Conclusion

Detection of pertussis is a function of both diagnostic awareness and diagnostic testing, with clinical trial data and serosurveillance studies demonstrating that mildly or asymptomatic pertussis infection is relatively common. The characteristics of pertussis surveillance systems are in turn related to both detection and reporting of pertussis and can change, even within countries, if changes in diagnostic practice occur, making comparisons over time problematic. Reference standards for pertussis surveillance include both measures of disease severity (for disease burden assessment) and population serosurveillance (for tracking of patterns of pertussis infection) and are probably most useful in identifying trends, assuming that sensitivity of detection has not changed substantially over time.

References

1. Versteegh FGA, Schellekens JFP, Fleer A, et al. Pertussis: a concise historical review including diagnosis, incidence, clinical manifestations and the role of treatment and vaccination in management. *Rev Med Microbiol* 2005;**16**:79–89.

2. Cherry JD. Pertussis in young infants throughout the world. *Clin Infect Dis* 2016;**63**:S119–22.

3. Chow MY, Khandaker G, McIntyre P. Global childhood deaths from pertussis: a historical review. *Clin Infect Dis* 2016;**63**:S134–41.

4. Gonfiantini MV, Villani A, Gesualdo F, et al. Attitude of Italian physicians toward pertussis diagnosis. *Hum Vaccin Immunother* 2013;**9**:1485–8.

5. Scheil W, Cameron S, Roberts C, et al. Pertussis in South Australia 1893 to 1996. *Commun Dis Intell* 1998;**22**:76–80.

6. van Wijhe M, McDonald SA, de Melker HE, et al. Effect of vaccination programmes on mortality burden among children and young adults in the Netherlands during the 20th century: a historical analysis. *Lancet Infect Dis* 2016; **16**:592–598

7. Varughese P. Incidence of pertussis in Canada. *Can Med Assoc J* 1985;**32**:1041–2.

8. Cherry JD. Pertussis in the preantibiotic and prevaccine era, with emphasis on adult pertussis. *Clin Infect Dis* 1999;**28**:S107-S11.

9. Amirthalingam G, Gupta S, Campbell H. Pertussis immunisation and control in England and Wales, 1957 to 2012: a historical review. *Eurosurveillance* 2013;**18**:1–9.

10. Mattoo S, Cherry JD. Molecular pathogenesis, epidemiology, and clinical manifestations of respiratory infections due to Bordetella pertussis and other Bordetella subspecies. *Clin Microbiol Rev* 2005;**18**:326–82.

11. van der Zee A, Schellekens JFP, Mooi FR. Laboratory diagnosis of pertussis. *Clin Microbiol Rev* 2015;**28**:1005–26.

12. Holberg-Petersen M, Jenum PA, Mannsaker T, et al. Comparison of PCR with culture applied on nasopharyngeal and throat swab specimens for the detection of Bordetella pertussis. *Scand J Infect Dis* 2011;**43**:221–4.

13. Faulkner A, Skoff T, Martin S, et al. Pertussis. In: Roush SW, Baldy LM (eds), *Manual for the Surveillance of Vaccine-Preventable Diseases*, 6th ed. Atlanta, GA: Centers for Disease Control and Prevention; 2015, pp 1–12.

14. Marcon MJ, Hamoudi AC, Cannon HJ, et al. Comparison of throat and nasopharyngeal swab specimens for culture diagnosis of Bordetella pertussis infection. *J Clin Microbiol* 1987;**25**:1109–10.

15. Cloud JL, Hymas W, Carroll KC. Impact of nasopharyngeal swab types on detection of Bordetella pertussis by PCR and culture. *J Clin Microbiol* 2002;**40**:3838–40.

16. Nunes MC, Soofie N, Downs S, et al. Comparing the yield of nasopharyngeal swabs, nasal aspirates, and induced sputum for detection of Bordetella pertussis in hospitalized infants. *Clin Infect Dis* 2016;**63**:S181–6.

17. Riffelmann M, Wirsing von Konig CH, et al. Nucleic acid amplification tests for diagnosis of Bordetella infections. *J Clin Microbiol* 2005;**43**:4925–9.

18. Litt DJ, Jauneikaite E, Tchipeva D, et al. Direct molecular typing of Bordetella pertussis from clinical specimens submitted for diagnostic quantitative (real-time) PCR. *J Med Microbiol* 2012;**61**:1662–8.

19. Murray PR, Baron EJ, Jorgensen JH, et al. *Manual of Clinical Microbiology*, 9th ed. Washington DC: American Society for Microbiology; 2007. http://www.immunise.health.gov.au/internet/immunise/publishing.nsf/Content/Handbook10-home.

20. Loeffelholz M. Towards improved accuracy of Bordetella pertussis nucleic acid amplification tests. *J Clin Microbiol* 2012;**50**:2186–90.

21. Katzko G, Hofmeister M, Church D. Extended incubation of culture plates improves recovery of Bordetella spp. *J Clin Microbiol* 1996;**34**:1563–4.

22. Vestrheim DF, Steinbakk M, Bjornstad ML, et al. Recovery of Bordetella pertussis from PCR-positive nasopharyngeal samples is dependent on bacterial load. *J Clin Microbiol* 2012;**50**:4114–5.

23. Heininger U, Schmidt-Schläpfer G, Cherry JD, et al. Clinical validation of a polymerase chain reaction assay for the diagnosis of pertussis by comparison with serology, culture, and symptoms during a large pertussis vaccine efficacy trial. *Pediatrics* 2000;**105**:E31.

24. Brotons P, de Paz HD, Toledo D, et al. Differences in Bordetella pertussis DNA load according to clinical and epidemiological characteristics of patients with whooping cough. *J Infect* 2016;**72**:460–7.

25. Bolotin S, Deeks SL, Marchand-Austin A, et al. Correlation of real time PCR cycle threshold cut-off with Bordetella pertussis clinical severity. *PLoS ONE* 2015;**10**:e0133209.

26. Nakamura Y, Kamachi K, Toyoizumi-Ajisaka H, et al. Marked difference between adults and children in Bordetella pertussis DNA load in nasopharyngeal swabs. *Cli Microbiol Infect* 2011;**17**:365–70.

27. Wirsing von Konig CH. Pertussis diagnostics: overview and impact of immunization. *Expert Rev Vaccines* 2014;**13**:1167–74.

28. Lasserre A, Laurent E, Turbelin C, et al. Pertussis incidence among adolescents and adults surveyed in general practices in the Paris area, France, May 2008 to March 2009. *Eurosurevillance* 2011;**16**::19783.

29. Halperin SA, Bortolussi R, Wort AJ. Evaluation of culture, immunofluorescence, and serology for the diagnosis of pertussis. *J Clin Microbiol* 1989;**27**:752–7.

30. Knorr L, Fox JD, Tilley PA, et al. Evaluation of real-time PCR for diagnosis of Bordetella pertussis infection. *BMC Infect Dis* 2006;**6**:62.

31. Loeffelholz MJ, Thompson CJ, Long KS, et al. Comparison of PCR, culture, and direct fluorescent-antibody testing for detection of Bordetella pertussis *J Clin Microbiol* 1999;**37**:2872–6.

32. World Health Organization. *WHO SAGE Pertussis Working Group Background Paper*. Geneva: WHO; 2014. http://www.who.int/immunization/sage/meetings/2014/april/2_SAGE_April_Pertussis_Miller_Resurgence.pdf?ua=1.

33. Ewanowich CA, Chui LW, Paranchych MG, et al. Major outbreak of pertussis in northern Alberta, Canada: analysis of discrepant direct fluorescent-antibody and culture results by using polymerase chain reaction methodology. *J Clin Microbiol* 1993;**31**:1715–25.

34. Glare EM, Paton JC, Premier RR, et al. Analysis of a repetitive DNA sequence from Bordetella pertussis and its application to the diagnosis of pertussis using the polymerase chain reaction. *J Clin Microbiol* 1990;**28**:1982–7.

35. Houard S, Hackel C, Herzog A, et al. Specific identification of Bordetella pertussis by the polymerase chain reaction. *Res Mcirobiol* 1989;**140**:477–87.

36. Olcén P, Bäckman A, Johansson B, et al. Amplification of DNA by the polymerase chain reaction for the efficient diagnosis of pertussis. *Scand J Infect Dis* 1992;**24**:339–45.

37. Grimprel E, Bégué P, Anjak I, et al. Comparison of polymerase chain reaction, culture, and western immunoblot serology for diagnosis of Bordetella pertussis infection. *J Clin Microbiol* 1993;**31**:2745–50.

38. Fisman DN, Tang P, Hauck T, et al. Pertussis resurgence in Toronto, Canada: a population-based study including test-incidence feedback modeling. *BMC Public Health* 2011;**11**:694.

39. Anderson TP, Beynon and KA, Murdoch DR. Comparison of real-time PCR and conventional hemi-nested PCR for the detection of Bordetella pertussis in nasopharyngeal samples. *Clin Microbiol Infect* 2003;**9**:746–9.

40. Muyldermans G, Soetens O, Antoine M, et al. External quality assessment for molecular detection of Bordetella pertussis in European laboratories. *J Clin Microbiol* 2005;**43**:30–5.

41. Bennett PM. Genome plasticity: insertion sequence elements, transposons and integrons, and DNA rearrangement. *Methods in Molecular Biology* 2004;**266**:71–113.

42. Diavatopoulos DA, Cummings CA, van der Heide HG, et al. Characterization of a highly conserved island in the otherwise divergent Bordetella holmesii and Bordetella pertussis genomes. *J Bacteriol* 2006;**188**:8385–94.

43. Centers for Disease Control and Prevention. Pertussis epidemic—Washington, 2012. *MMWR Morb Mortal Wkly Rep* 2012;**61**:517–22.

44. Njamkepo E, Bonacorsi S, Debruyne M, et al. Significant finding of Bordetella holmesii DNA in nasopharyngeal samples from French patients with suspected pertussis. *J Clin Microbiol* 2011;**49**:4347–8.

45. Tartof SY, Gounder P, Weiss D, et al. Bordetella holmesii bacteremia cases in the United States, April 2010-January 2011. *Clin Infect Dis* 2014;**58**:e39–43.

46. Parkhill J, Sebaihia M, Preston A, et al. Comparative analysis of the genome sequences of Bordetella pertussis, Bordetella parapertussis and Bordetella bronchiseptica. *Nat Genet* 2003;**35**:32–40.

47. Reischl U, Lehn N, Sanden GN, et al. Real-time PCR assay targeting IS481 of Bordetella pertussis and molecular basis for detecting Bordetella holmesii. *J Clin Microbiol* 2001;**39**:1963–6.

48. Tizolova A, Guiso N, Guillot S. Insertion sequences shared by Bordetella species and implications for the biological diagnosis of pertussis syndrome. *Eur J Clin Microbiol Infect Dis* 2013;**32**:89–96.

49. Templeton KE, Scheltinga SA, van der Zee A, et al. Evaluation of real-time PCR for detection of and discrimination between Bordetella pertussis, Bordetella parapertussis, and Bordetella holmesii for clinical diagnosis. *J Clin Microbiol* 2003;**41**:4121–6.

50. Mooi FR, van Loo IH, van Gent M, et al. Bordetella pertussis strains with increased toxin production associated with pertussis resurgence. *Emerg Infect Dis* 2009;**15**:1206–13.

51. Nygren M, Reizenstein E, Ronaghi M, et al. Polymorphism in the pertussis toxin promoter region affecting the DNA-based diagnosis of Bordetella infection. *J Clin Microbiol* 2000;**38**:55–60.

52. Probert WS, Ely J, Schrader K, et al. Identification and evaluation of new target sequences for specific detection of Bordetella pertussis by real-time PCR. *J Clin Microbiol* 2008;**46**:3228–31.

53. Tatti KM, Sparks KN, Boney KO, et al. Novel multi-target real-time PCR assay for rapid detection of Bordetella species in clinical specimens. *J Clin Microbiol* 2011;**49**:4059–66.

54. Koidl C, Bozic M, Burmeister A, et al. Detection and differentiation of Bordetella spp. by real-time PCR. *J Clin Microbiol* 2007;**45**:347–50.

55. Hozbor D, Fouquem F, Guiso N. Detection of Bordetella bronchiseptica by the polymerase chain reaction. *Res Microbiol* 1999;**150**:333–41.

56. Farrell DJ, McKeon M, Daggard G, et al. Rapid-cycle PCR method to detect Bordetella pertussis that fulfills all consensus recommendations for use of PCR in diagnosis of pertussis. *J Clin Microbiol* 2000;**38**:4499–502.

57. Qin X, Galanakis E, Martin ET, et al. Multitarget PCR for diagnosis of pertussis and its clinical implications. *J Clin Microbiol* 2007;**45**:506–11.

58. Fry NK, Duncan J, Wagner K, et al. Role of PCR in the diagnosis of pertussis infection in infants: 5 years' experience of provision of a same-day real-time PCR service in England and Wales from 2002 to 2007. *J Med Microbiol* 2009;**58**:1023–9.

59. Pertussis Guidelines Group Public Health England. *Guidelines for the Public Health Management of Pertussis in England*. London: Public Health England; 2018. https://www.gov.uk/government/uploads/system/uploads/attachment_data/file/576061/Guidelines_for_the_Public_Health_Management_of_Pertussis_in_England.pdf.

60. Centers for Disease Control and Prevention. Outbreaks of respiratory illness mistakenly attributed to pertussis—New Hampshire, Massachusetts, and Tennessee, 2004–2006. *MMWR Morb Mortal Wkly Rep* 2007;**56**: 837–42.

61. Guthrie JL, Robertson AV, Tang P, et al. Novel duplex real-time PCR assay detects Bordetella holmesii in specimens from patients with Pertussis-like symptoms in Ontario, Canada. *J Clin Microbiol* 2010;**48**:1435–7.

62. Waters V, Jamieson F, Richardson SE, et al. Outbreak of atypical pertussis detected by polymerase chain reaction in immunized preschool-aged children. *Pediatr Infect Dis J* 2009;**28**:582–7.

63. Register KB, Nicholson TL. Misidentification of Bordetella bronchiseptica as Bordetella pertussis using a newly described real-time PCR targeting the pertactin gene. *J Med Microbiol* 2007;**56**:1608–10.

64. Roorda L, Buitenwerf J, Ossewaarde JM, et al. A real-time PCR assay with improved specificity for detection and discrimination of all clinically relevant Bordetella species by the presence and distribution of three insertion sequence elements. *BMC Res Notes* 2011;**4**:11.

65. Weber DJ, Rutala WA, Schaffner W. Lessons learned: protection of healthcare workers from infectious disease risks. *Crit Care Med* 2010;**38**:S306–14.

66. Lievano FA, Reynolds MA, Waring AL, et al. Issues associated with and recommendations for using PCR to detect outbreaks of pertussis. *J Clin Microbiol* 2002;**40**:2801–5.

67. Taranger J, Trollfors B, Lind L, et al. Environmental contamination leading to false-positive polymerase chain reaction for pertussis. *Pediatr Infect Dis J* 1994;**13**:936–7.

68. Mandal S, Tatti KM, Woods-Stout D, et al. Pertussis pseudo-outbreak linked to specimens contaminated by Bordetella pertussis DNA from clinic surfaces. *Pediatrics* 2012;**129**:e424–30.

69. Salimnia H, Lephart PR, Asmar BI, et al. Aerosolized vaccine as an unexpected source of false-positive Bordetella pertussis PCR results. *J Clin Microbiol* 2012;**50**:472–4.

70. Tatti KM, Slade B, Patel M, et al. Real-time polymerase chain reaction detection of Bordetella pertussis DNA in acellular pertussis vaccines. *Pediatr Infect Dis J* 2008;**27**:73–4.

71. He Q, Mertsola J, Soini H, et al. Comparison of polymerase chain reaction with culture and enzyme immunoassay for diagnosis of pertussis. *J Clin Microbiol* 1993;**31**:642–5.

72. Bidet P, Liguori S, De Lauzanne A, et al. Real-time PCR measurement of persistence of Bordetella pertussis DNA in nasopharyngeal secretions during antibiotic treatment of young children with pertussis. *J Clin Microbiol* 2008;**46**:3636–8.

73. Fry NK, Tzivra O, Li YT, et al. Laboratory diagnosis of pertussis infections: the role of PCR and serology. *J Med Microbiol* 2004;**53**:519–25.

74. Centers for Disease Control and Prevention. *Pertussis (Whooping Cough); Diagnostic Testing*. 2015. https://www.cdc.gov/pertussis/clinical/diagnostic-testing/diagnosis-pcr-bestpractices.html.

75. Papenburg J. What is the significance of a high cycle threshold positive IS481 PCR for Bordetella pertussis? *Pediatr Infect Dis J* 2009;**28**:1143.

76. Fine PE, Clarkson JA. The recurrence of whooping cough: possible implications for assessment of vaccine efficacy. *Lancet* 1982;**1**:666–9.

77. Krantz I, Alestig K, Trollfors B, et al. The carrier state in pertussis. *Scand J Infect Dis* 1986;**18**:121–3.

78. Linnemann CC, Bass JW, Smith MH. The carrier state in pertussis. *Am J Epidemiol* 1968;**88**:422–7.

79. He Q, Schmidt-Schläpfer G, Just M, et al. Impact of polymerase chain reaction on clinical pertussis research: Finnish and Swiss experiences. *J Infect Dis* 1996;**174**:1288–95.

80. Horby P, MacIntyre CR, McIntyre PB, et al. A boarding school outbreak of pertussis in adolescents: value of laboratory diagnostic methods. *Epidemiol Infect* 2005;**133**:229–36.

81. Klement E, Uliel L, Engel I, et al. An outbreak of pertussis among young Israeli soldiers. *Epidemiol Infect* 2003;**131**:1049–54.

82. He Q, Arvilommi H, Viljanen MK, et al. Outcomes of Bordetella infections in vaccinated children: effects of bacterial number in the nasopharynx and patient age. *Clin Diagn Lab Immunol* 1999;**6**:534–6.

83. Srugo I, Benilevi D, Madeb R, et al. Pertussis infection in fully vaccinated children in day care centres, Israel. *Emerg Infect Dis* 2000;**6**:526–9.

84. Robertson PW, Goldberg H, Jarvie BH, et al. Bordetella pertussis infection: a cause of persistent cough in adults. *Med J Aust* 1987;**146**:522–5.

85. Guiso N, Berbers G, Fry NK, et al. What to do and what not to do in serological diagnosis of pertussis: recommendations from EU reference laboratories. *Eur J Clin Microbiol Infect Dis* 2011;**30**:307–12.

86. de Melker HE, Versteegh FG, Conyn-Van Spaendonck MA, et al. Specificity and sensitivity of high levels of immunoglobulin G antibodies against pertussis toxin in a single serum sample for diagnosis of infection with Bordetella pertussis. *J Clin Microbiol* 2000;**38**:800–6.

87. Marchant CD, Loughlin AM, Lett SM, et al. Pertussis in Massachusetts, 1981–1991: incidence, serologic diagnosis, and vaccine effectiveness. *J Infect Dis* 1994;**169**:1297–305.

88. European Centre for Disease Prevention and Control. *Guidance and Protocol for the Serological Diagnosis of Human Infection with Bordetella Pertussis.* Stockholm: ECDC; 2012.

89. Hallander HO. Microbiological and serological diagnosis of pertussis. *Clin Infect Dis* 1999;**28**:S99–106.

90. Trollfors B, Taranger J, Lagergård T, et al. A placebo-controlled trial of a pertussis-toxoid vaccine. *N Engl J Med* 1995;**333**:1045–50.

91. Greco D, Salmaso S, Mastrantonio P, et al. A controlled trial of two acellular vaccines and one whole-cell vaccine against pertussis. *N Engl J Med* 1996;**334**:341–8.

92. Cherry JD, Grimprel E, Guiso N, Heininger U, Mertsola J. Defining pertussis epidemiology: clinical, microbiologic and serologic perspectives. *Pediatr Infect Dis J* 2005;**24**:S25–34.

93. Ward JI, Cherry JD, Chang SJ, et al. Efficacy of an acellular pertussis vaccine among adolescents and adults. *N Engl J Med* 2005;**353**:1555–63.

94. Patriarca PA, Biellik RJ, Sanden G, et al. Sensitivity and specificity of clinical case definitions for pertussis. *Am J Public Health* 1988;**78**:833–6.

95. Cherry JD, Tan T, Wirsing von König CH, et al. Clinical definitions of pertussis: summary of a Global Pertussis Initiative roundtable meeting, February 2011. *Clin Infect Dis* 2012;**54**:1756–64.

96. Allen CJ, Ferson MJ. Notification of infectious diseases by general practitioners: a quantitative and qualitative study. *Med J Aust* 2000;**172**:325–8.

97. Blogg S, Trent M. Doctors' notifications of pertussis. *N S W Public Health Bull* 1998;**9**:53–4.

98. Deeks S, De Serres G, Boulianne N, et al. Failure of physicians to consider the diagnosis of pertussis in children. *Clin Infect Dis* 1999;**28**:840–6.

99. MacNeil A, Dietz V, Cherian T. Vaccine preventable diseases: time to re-examine global surveillance data? *Vaccine* 2014;**32**:2315–20.

100. Ntezayabo B, De Serres G, Duval B. Pertussis resurgence in Canada largely caused by a cohort effect. *Pediatr Infect Dis J* 2003;**22**:22–7.

101. Faulkner AE, Skoff TH, Tondella ML, et al. Trends in pertussis diagnostic testing in the United States, 1990 to 2012. *Pediatr Infect Dis J* 2016;**35**:39–44.

102. Quinn HE, McIntyre PB. Pertussis epidemiology in Australia over the decade 1995–2005—trends by region and age group. *Commun Dis Intell* 2007;**31**: 205–15.

103. Pillsbury A, Quinn HE, McIntyre PB. Australian vaccine preventable diseases epidemiological review series: pertussis 2006–2012. *Commun Dis Intell Q Rep* 2014;**38**:E179–94.

104. Kaczmarek MC, Valenti L, Kelly HA, et al. Sevenfold rise in likelihood of pertussis test requests in a stable set of Australian general practice encounters, 2000–2011. *Med J Aust* 2013;**198**:624–8.

105. Tozzi AE, Pandolfi E, Celentano LP, et al. Comparison of pertussis surveillance systems in Europe. *Vaccine* 2007;**25**:291–7.

106. Campbell H, Amirthalingam G, Andrews N, et al. Accelerating control of pertussis in England and Wales. *Emerg Infect Dis* 2012;**18**:38–47.

107. de Melker HE, Schellekens JF, Neppelenbroek SE, et al. Reemergence of pertussis in the highly vaccinated population of the Netherlands: observations on surveillance data. *Emerg Infect Dis* 2000;**6**:348–57.

108. European Centre for Disease Prevention and Control. *Annual Epidemiological Report 2016—Pertussis.* Stockholm: ECDC; 2016. https://ecdc.europa.eu/sites/portal/files/documents/Pertussis%20AER.pdf.

109. Heininger U, André P, Chlibek R, et al. Comparative epidemiologic characteristics of pertussis in 10 Central and Eastern European countries, 2000–2013. *PLoS One* 2016;**11**:e0155949.

110. Moerman L, Leventhal A, Slater PE, et al. The re-emergence of pertussis in Israel. *Israel Med Assoc J* 2006;**8**:308–11.

111. Wymann MN, Richard JL, Vidondo B, et al. Prospective pertussis surveillance in Switzerland, 1991–2006. *Vaccine* 2011;**29**:2058–65.

112. Bonmarin I, Poujol I, Levy-Bruhl D. Nosocomial infections and community clusters of pertussis in France, 2000–2005. *Eurosurveillance* 2007;**12**:E11–2.

113. Datta SS, Toikilik S, Ropa B, et al. Pertussis outbreak in Papua New Guinea: the challenges of response in a remote geo-topographical setting. *Western Pac Surveill Response* 2012;**3**:3–6.

114. Sobanjo-Ter Meulen A, Duclos P, McIntyre P, et al. Assessing the evidence for maternal pertussis immunization: a report from the Bill & Melinda Gates Foundation Symposium on Pertussis Infant Disease Burden in Low- and Lower-Middle-Income Countries. *Clin Infect Dis* 2016;**63**:S123–33.

115. Omer SB, Kazi AM, Bednarczyk RA, et al. Epidemiology of pertussis among young Pakistani infants: a community-based prospective surveillance study. *Clin Infect Dis* 2016;**63**:S148–53.

116. Gill CJ, Mwananyanda L, MacLeod W, et al. Incidence of severe and nonsevere pertussis among HIV-exposed and -unexposed Zambian infants through 14 weeks of age: results from the Southern Africa Mother Infant Pertussis Study (SAMIPS), a longitudinal birth cohort study. *Clin Infect Dis* 2016;**63**:S154–64.

117. Bonmarin I, Levy-Bruhl D, Baron S, et al. Pertussis surveillance in French hospitals: results from a 10 year period. *Eurosurveillance* 2007;**12**:678.

118. Bettinger JA, Halperin SA, De Serres G, et al. The effect of changing from whole-cell to acellular pertussis vaccine on the epidemiology of hospitalized children with pertussis in Canada. *Pediatr Infect Dis J* 2007;**26**:31–5.

119. Bonacruz-Kazzi G, McIntyre P, Hanlon M, et al. Diagnostic testing and discharge coding for whooping cough in a children's hospital. *J Paediatr Child Health* 2003;**39**:586–90.

120. Cortese MM, Baughman AL, Zhang R, et al. Pertussis hospitalizations among infants in the United States, 1993 to 2004. *Pediatrics* 2008;**121**:484–92.

121. Torvaldsen S, McIntyre P. Do variations in pertussis notifications reflect incidence or surveillance practices? A comparison of infant notification rates and hospitalisation data in NSW. *N S W Public Health Bull* 2003;**14**:81–4.

122. Crowcroft NS, Booy R, Harrison T, et al. Severe and unrecognised: pertussis in UK infants. [erratum appears in Arch Dis Child. 2006 May;91(5):453]. *Arch Dis Child* 2003;**88**:802–6.

123. Straney L, Schibler A, Ganeshalingham A, et al. Burden and outcomes of severe pertussis infection in critically ill infants. *Pediatr Crit Care Med* 2016;**17**: 735–42.

124. Hale S, Quinn HE, Kesson A, et al. Changing patterns of pertussis in a children's hospital in the post PCR era. *J Pediatr* 2016;**170**:161–5.

125. Liu BC, McIntyre P, Kaldor JM, et al. Pertussis in older adults: prospective study of risk factors and morbidity. *Clin Infect Dis* 2012;**55**:1450–6.

126. Skoff TH, Baumbach J, Cieslak PR. Tracking pertussis and evaluating control measures through Enhanced Pertussis Surveillance, Emerging Infections Program, United States. *Emerg Infect Dis* 2015;**21**:1568–73.

127. Sutter RW, Cochi SL. Pertussis hospitalizations and mortality in the United States, 1985–1988. Evaluation of the completeness of national reporting. *JAMA* 1992;**267**:386–91.

128. van Hoek AJ, Campbell H, Amirthalingam G, et al. The number of deaths among infants under one year of age in England with pertussis: results of a capture/recapture analysis for the period 2001 to 2011. *Eurosurveillance* 2013;**18**:20414.

129. Takum T, Gara D, Tagyung H, Murhekar MV. An outbreak of pertussis in Sarli Circle of Kurung-kumey district, Arunachal Pradesh, India. *Indian Pediatr* 2009;**46**:1017–20.

130. De Serres G, Boulianne N, Duval B, et al. Effectiveness of a whole cell pertussis vaccine in child-care centers and schools. *Pediatr Infect Dis J* 1996;**15**:519–24.

131. de Melker HE, Conyn-Van Spaendonck MA, Rumke HC, et al. Pertussis in the Netherlands: an outbreak despite high levels of immunization with whole-cell vaccine. *Emerg Infect Dis* 1997;**3**:175–8.

132. Choi YH, Campbell H, Amirthalingam G, et al. Investigating the pertussis resurgence in England and Wales, and options for future control. *BMC Med* 2016;**14**:121.

133. Libster R, Edwards KM. Re-emergence of pertussis: what are the solutions. *Expert Rev Vaccines* 2012;**11**:1331–46.

134. Sheridan SL, Ware RS, Grimwood K, et al. Number and order of whole cell pertussis vaccines in infancy and disease protection. *JAMA* 2012;**308**:454–6.

135. Quinn HE, Snelling TL, Macartney KK, et al. Duration of protection after first dose of acellular pertussis vaccine in infants. *Pediatrics* 2014;**133**:e513–9.

136. Zhang L, Prietsch SOM, Axelsson I, et al. Acellular vaccines for preventing whooping cough in children. *Cochrane Database Syst Rev* 2012;**3**:CD001478.

137. Wylks CE, Ewald B, Guest C. The epidemiology of pertussis in the Australian Capital Territory, 1999 to 2005—epidemics of testing, disease or false positives? *Commun Dis Intell* 2007;**31**:383–91.

138. Public Health Ontario. *Bordetella Molecular Testing – Changes to Result Reporting.* 2008. https://www.publichealthontario.ca/en/eRepository/LAB_SD_047_Bordetella_molecular_testing_reporting_changes.pdf.

139. Rohani P, Drake JM. The decline and resurgence of pertussis in the US. *Epidemics* 2011;**3**:183–8.

140. Fine PE. Adult pertussis: a salesman's dream—and an epidemiologist's nightmare. *Biologicals* 1997;**25**:195–8.

141. Schmidt-Schlapfer G, Matter HC, Zimmermann HP, et al. Bordetella pertussis infections in Switzerland: comparison of PCR with diagnosis of whooping cough based on WHO-defined clinical symptoms. *J Microbiol Methods* 1996;**27**:97–107.

142. Rodgers L, Martin SW, Cohn A, et al. Epidemiologic and laboratory features of a large outbreak of pertussis-like illnesses associated with cocirculating Bordetella holmesii and Bordetella pertussis—Ohio, 2010–2011. *Clin Infect Dis* 2013;**56**:322–31.

143. Spicer KB, Salamon D, Cummins C, et al. Occurrence of 3 Bordetella species during an outbreak of cough illness in Ohio: epidemiology, clinical features,

laboratory findings and antimicrobial susceptibility. *Pediatr Infect Dis J* 2014;**33**:e162–7.

144. Domenech de Celles M, Magpantay FM, King AA, et al. The pertussis enigma: reconciling epidemiology, immunology and evolution. *Proc Biol Sci* 2016;**283**: 20152309.

145. Barkoff AM, Grondahl-Yli-Hannuksela K, He Q. Seroprevalence studies of pertussis: what have we learned from different immunized populations. *Pathog Dis* 2015;**73**.

146. de Greeff SC, de Melker HE, van Gageldonk PG, et al. Seroprevalence of pertussis in the Netherlands: evidence for increased circulation of Bordetella pertussis. *PLoS One* 2012;**5**:e14183.

147. Strebel P, Nordin J, Edwards K, et al. Population based incidence of pertussis among adolescents and adults, Minnesota, 1995–1996. *J Infect Dis* 2001;**183**: 1353–9.

148. Campbell P, McIntyre P, Quinn H, et al. Increased population prevalence of low pertussis toxin antibody levels in young children preceding a record pertussis epidemic in Australia. *PLoS One* 2012;**7**: e35874.

149. Hallander HO, Andersson M, Gustafsson L, et al. Seroprevalence of pertussis antitoxin (anti-PT) in Sweden before and 10 years after the introduction of a universal childhood pertussis vaccination program. *APMIS* 2009;**117**:912–22.

CHAPTER 13

Contrasting ecological and evolutionary signatures of whooping cough epidemiological dynamics

Benjamin M. Althouse and Samuel V. Scarpino

Abstract

The enigmatic global pattern of whooping cough incidence presents a unique set of challenges for controlling the disease and uncovering the mechanisms underlying its epidemiological dynamics. In countries experiencing an increase in cases, five hypotheses have been proposed to explain the resurgence: (1) there has been an increase in *Bordetella pertussis* reporting rates, (2) waning of protective immunity from vaccination or natural infection over time, (3) evolution of *B. pertussis* to escape protective immunity, (4) vaccines that fail to induce sterilizing (mucosal) immunity to *B. pertussis*, and (5) asymptomatic transmission from individuals vaccinated with the currently used acellular *B. pertussis* vaccines. Each of these five hypotheses can leave contrasting signatures in both epidemiological and genomic data; however, these hypotheses must also be evaluated against data from locations that are either not experiencing a resurgence or are witnessing declining incidence. This chapter discusses how to—and whether it is possible to—disentangle the various mechanisms proposed for whooping cough's resurgence. It identifies a pathological lack of data sufficient for testing hypotheses and demonstrates how detailed, high-resolution data (in geography, time, and age) are required to distinguish even the most basic models. The chapter further discusses how approaches linking genomic and epidemiological data (i.e. phylodynamic models) may prove beneficial. The results suggest that while evidence exists for each of the five proposed hypotheses, it is unlikely that any single mechanism can account for the global pattern of whooping cough incidence and that determining the relative importance of each mechanism remains uniquely challenging.

13.1 Drivers of pertussis transmission dynamics and models of transmission

Despite steadily declining global prevalence and persistently high vaccination coverage[1], whooping cough—primarily caused by *Bordetella pertussis*—remains among the deadliest vaccine-preventable diseases and is resurging in many countries[1-3]. In the United States, there have been more *B. pertussis* cases in the past 10 years than in the preceding 40 years combined (Figure 13.1)[4] and the 2014 outbreak of *B. pertussis* in California saw nearly 10,000 cases—more than in any year since vaccination began in the mid 1940s[5]. The United Kingdom, Australia,

Althouse, B. M. and Scarpino, S. V., *Contrasting ecological and evolutionary signatures of whooping cough epidemiological dynamics*. In: *Pertussis: epidemiology, immunology, and evolution*. Edited by Pejman Rohani and Samuel V. Scarpino: Oxford University Press (2019). © Oxford University Press. DOI: 10.1093/oso/9780198811879.003.0013

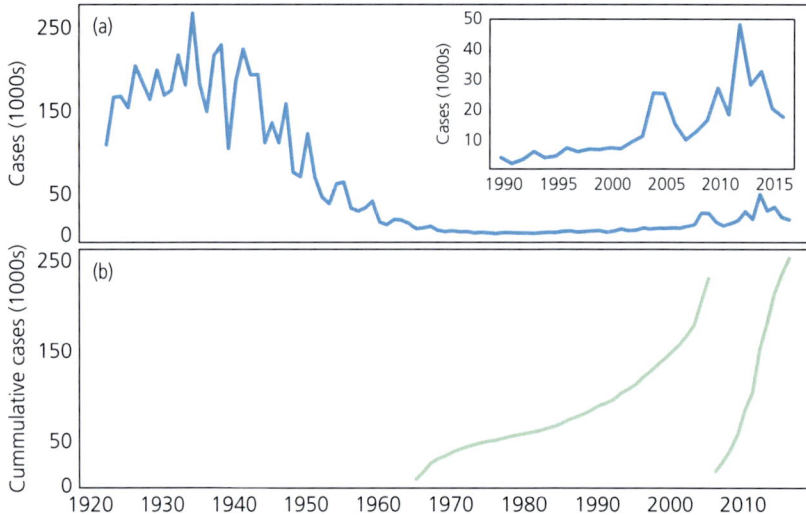

Figure 13.1 Reported *B. pertussis* cases. Figure shows the numbers of reported cases of *B. pertussis* in the United States since 1922. Panel (a) shows the reported cases by year with an inset of data since 1990, and panel (b) shows the cumulative cases from 1965 to 2005 as compared to 2006–2016. Data from the CDC[4].

Portugal, Israel, and South Korea have all seen similar rises in *B. pertussis* prevalence in recent years[2]. However, other countries including Italy, Sweden, and Japan have seen declines in *B. pertussis* over the same time period—although southern Italy has seen dramatic surges in cases over the past 2 years[6]. Due to the severe nature of symptomatic *B. pertussis* infections—especially in infants too young to be fully vaccinated—it is critically important to understand the drivers of *B. pertussis* transmission and why this pathogen exhibits seemingly incompatible epidemiological characteristics in different locations. This chapter will present several competing hypotheses to explain the resurgence, including waning vaccine-derived immunity, imperfect immunological responses to acellular (aP) versus whole-cell pertussis (wP) vaccination, evolution of the *B. pertussis* bacteria to escape vaccine-derived immunity, changes in reporting, and asymptomatic transmission. We further discuss the role of phylodynamic analyses using an example of *B. pertussis* in the United States. Finally, we end with recommendations for prospective data collection efforts, which could aid in untangling the question of *B. pertussis* resurgence.

As summarized in Chapter 1, *B. pertussis* transmission models have generally included age structure and vaccination, which are both key drivers

of *B. pertussis* transmission dynamics. Grenfell and Anderson[7] and Hethcote[8] formulated age-stratified models of *B. pertussis* transmission including various efficacies of wP vaccination and waning immunity and found repeated cohort vaccination with high coverage levels would be necessary to bring *B. pertussis* transmission to low levels. Various models have since been formulated to answer questions about *B. pertussis* transmission or biology. These include models focused on non-linear dynamics and chaotic behavior[9,10], changing contact patterns[11], the consequences of asymptomatic transmission[3], economic evaluations of *B. pertussis* burden[12,13], and potential novel vaccine strategies for *B. pertussis* mitigation[14,15]. A key area of exploration in these models is to understand the natural and vaccine-mediated transmission dynamics of *B. pertussis*, the roles of seasonality and weather, and the recruitment of susceptibles on the observed 2–5-year cycles of *B. pertussis* incidence.

Seasonality is a strong component of many childhood infections including measles[16], influenza[17,18] and respiratory syncytial virus[19,20]. Several studies have examined the seasonality of *B. pertussis* transmission, finding peaks in transmission in the autumn in the northern hemisphere[21]. Climatological factors may play a role in the persistence of *B. pertussis* in the environment, but the role of climate in *B. pertussis* transmission remains understudied[21]. *Bordetella. pertussis* transmission does not seem to be influenced by school holidays—a known driver of measles seasonality[22]. However, De Greeff et al. cite the effects of idiosyncratic waning natural and vaccine-derived immunity and the variability in infectious period, along with the presence of asymptomatic transmission, as reasons for the lack of an observable school-based signal[21]. Aside from the recruitment of susceptible individuals, there is still little consistent evidence for drivers of seasonal or multi-annual variability in *B. pertussis* outbreaks[21,23].

Beyond temporal patterns driven by climate or school calendars, numerous studies report a 2–5-year periodic cycle for pertussis outbreaks (see[23–25] and ref-

erences therein) observable in the pre-vaccine eras. However, these cycles can be transient[23,24], may be affected by vaccination coverage[23,25], and may differ based on whether wP or aP vaccines are in use[3,26] (see Figure 13.2). The likely mechanism for these cycles is the recruitment of susceptible individuals through births, where the inter-epidemic period and magnitude of epidemic peak change as the birth rate declines or increases due to an effective change in the force of infection (i.e. as a birth rate changes from μ to μ', the force of infection, β, changes to $\beta \cdot (\mu'/\mu)$, see[9]). Control of *B. pertussis* transmission through birth rate is clearly unfeasible and importantly, cannot explain the ongoing *B. pertussis* resurgence: birth rates in the United States and United Kingdom

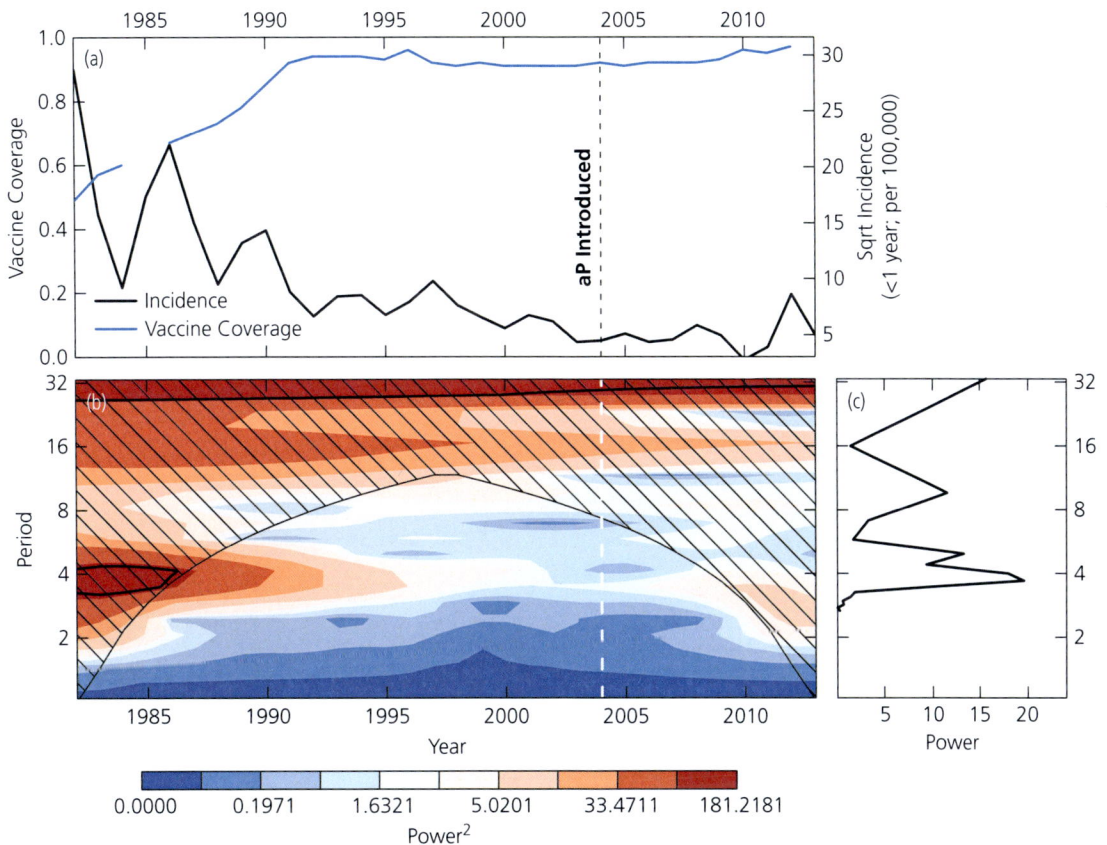

Figure 13.2 Return of pre-vaccine epidemiological periodicity after the switch from wP to aP vaccines. Using data on infants under 1 year of age, Althouse and Scarpino[3] found that after switching from the wP to the aP vaccine, the United Kingdom witnessed a return to pre-vaccination era cycling in pertussis cases. Similarly, using population-wide data from Israel, Langsam et al.[26] again found a return to pre-vaccine era cycling coinciding with the switch from wP to aP vaccination, data not shown see[24]. The periodicity of cycling, especially in un- or undervaccinated infants is strong evidence that transmission is ongoing. In both countries, vaccination coverage was above 95 per cent.

Table 13.1 Studies reporting undiagnosed pertussis infection

Percentage of cases reported	Age	Vaccine	Location	Reference
32–46	Infants	aP	Germany	47
38–44	Schoolchildren (grades 1–6)	aP	Taiwan	34
12–32	Adolescents	wP	United States	35
23–28	Adolescents and adults	aP	United Kingdom	40
13–20	Adolescents and adults	wP	United States	36
17 (children), 7 (adults)	5–49 years	aP	New Zealand	43
15	<12 years	wP	Uganda	42

aP, acellular pertussis; wP, whole-cell pertussis.

(as well as elsewhere) have been steady or declining in the past 20 years[27].

13.2 Pertussis resurgence and potential explanations

To date, there are five broad mechanisms proposed to explain the ongoing *B. pertussis* resurgence:

1. A marked increase in *B. pertussis* reporting rates, primarily due to enhanced molecular detection methods (e.g. PCR-based diagnostics)[2,3,23,28,29].
2. Vaccines elicit a sterilizing immune response, which wanes over time[30,31].
3. *B. pertussis* evolved to escape sterilizing immunity induced by the vaccine[32].
4. Vaccines fail to induce sterilizing immunity to the pathogen[33].
5. Asymptomatically infectious individuals are contributing significantly to *B. pertussis* transmission[3].

Here, we briefly review evidence for each of these hypotheses, along with potential challenges associated with evaluating them with available data. As many of these hypotheses are addressed in earlier chapters of this book, we refer the reader to those chapters for more detailed information. Lastly, an added difficulty in testing and distinguishing between these hypotheses is that they are not mutually exclusive, and are not easily assessed with extant observed data (more on this later). That said, the hypotheses do seem to vary in their potential effects and relative plausibility.

13.2.1 Under-reporting

Perhaps the most challenging aspect of *B. pertussis* epidemiology is accounting for under-reporting. Clearly, systematic biases in case report data—primarily resulting from differences in reporting probabilities—can easily bias the results of epidemiological studies. Although pronounced differences exist in age-specific rates of infection, hospitalization, and mortality—with the highest rates occurring among infants and children[2,23,28,29]—there is well-documented evidence of high levels of under-reporting, especially in adolescents and adults[2,3,23,28,29]. For example, a serosurvey of school children in Taiwan found low rates (<1 per cent) of reported pertussis, but high rates (approaching 40 per cent) of seropositivity[34]. Taiwan switched to aP vaccine in 1997–1998 and has sustained very high (> 95 per cent) vaccination coverage[34]. Strikingly, a study by Langsam et al.—using data from Israel—found no correlation between the number of laboratory-confirmed pertussis cases and the number reported to the Israeli Ministry of Health[26].

Studies in the United States on pertussis reporting rates suggest that between 12 and 32 per cent of all persistent coughs may be associated with undiagnosed *B. pertussis* infection[35–37]. Nearly identical percentages have been found in numerous locations, spanning all continents (aside from Antarctica), including Canada[38,39], the United Kingdom[40,41], Uganda[42], New Zealand[43], France[44], Australia[45], Bangladesh[46], Taiwan[34], Germany[47], and Peru[48] (see Table 13.1). In addition, *B. pertussis* is a common cause of adult community-acquired pneumonia, associated with 2–9 per cent of such episodes[49,50]. Lastly, studies using data from both the wP and aP vaccine eras in the United States indicate that up to 91 per cent of *B. pertussis*-related complications may not be reported for people 10 years of age and older[51,52]. This high, but variable, degree of under-reporting leads to the conclusion that while *B. pertussis* testing has increased over the last 25 years, a substantial burden remains unrecognized. When determining whether a location is experiencing a resurgence, or the severity of ongoing resurgence, clearly studies must account for under-reporting.

13.2.2 Waning immunity

It is clear that waning immunity plays a role in the epidemiology of *B. pertussis*, although estimates of the duration of protection from *B. pertussis* are highly varied[30,31,53,54]. A recent study by Domenech de Cellès et al., using epidemiological data from Massachusetts, United States, concluded that aP vaccination wanes slowly, with around 90 per cent of individuals retaining immunity after 10 years and 45 per cent of individuals being protected for life[31]. Interestingly, they found a higher level of whooping cough under-reporting, with an estimate of greater than 70 per cent under-reporting in adolescents and adults[31], than reported in empirical studies (see section 13.2.1 on under-reporting). Despite finding a relatively slow rate of waning and low level of vaccine failure, they concluded that the current aP vaccine is insufficient to eradicate *B. pertussis*.

An earlier meta-analysis of published studies on the duration of immunity to *B. pertussis* estimated a range of 7–20 years of protection for natural infection, 4–12 years for wP vaccine-induced immunity, and 6 years for aP vaccine-induced immunity[55]. Using a mechanistic transmission model, Wearing and Rohani estimated the duration of immunity to be 30–100 years for naturally acquired immunity[30], which is in line with recent estimates from Domenech de Cellès et al.[31]. Wearing and Rohani explain the discrepancy in their estimate, as compared to earlier studies, by the fact that their models assumed an exponential duration of immunity, which will be strongly influenced by long-duration immune indi-

viduals. Instead, if Wearing and Rohani had assumed a gamma distribution of immune duration, their estimates would be much shorter, and largely in agreement with previous studies. Taken together, these results highlight the critical importance of understanding heterogeneities in immune duration following *B. pertussis* vaccination.

Despite the fact that vaccine-derived *B. pertussis* immunity appears to be less than lifelong, and that this observation could explain the resurgence of *B. pertussis* in general[31], waning immunity alone cannot account for the age distributions of cases skewed towards younger individuals[2,23,28,29], the failure of 'cocooning' newborn infants by vaccinating all close contacts[56–58], the observed synchrony in age-specific attack rates after the switch to the aP vaccine in the United Kingdom and Israel[3,26] (Figure 13.2), nor the observed *B. pertussis* genomic patterns[3]. Lastly, findings of long-lived, vaccine-derived immunity appear at odds with serological studies in the United States. For example, a study performed on 100 adults and children during the 2016 Minnesota State Fair found that greater than 50 per cent of children aged between 1 and 6 years, greater than 70 per cent of adolescents, and around 25 per cent of adults had no detectable antibodies (< 5 IU/mL on an immunoglobulin G enzyme-linked immunosorbent assay for pertussis toxin) to *B. pertussis*[59].

Perhaps most at odds with the findings of Domenech de Cellès et al.[31] that aP vaccination confers long-lived immunity, are results from a study co-authored by researchers from the United States Centers for Disease Control and Prevention and Kaiser Permanente (a managed healthcare consortium representing more than 10 million individuals in the United States), which estimated very low vaccine efficacy for aP vaccine[60]. Using medical and vaccination records from more than 160,000 individuals between 11 and 18 years of age during the period 2005–2012, they estimated aP vaccine to be 52 per cent effective at preventing symptomatic *B. pertussis* infection[60]. Nevertheless, and as with much of the work on pertussis, the vaccine effectiveness calculations in this analysis and the waning immunity parameters estimated by Domenech de Cellès et al. are challenging to compare directly because of differences in methodological approaches, modelling assumptions, and underlying data sets.

13.2.3 *Bordetella pertussis* evolution

There exists substantial genetic variation in *B. pertussis* isolates between countries[61], and a potential explanation for the current resurgence is that some *B. pertussis* populations have evolved resistance to vaccine-derived antibodies[32]. Evolution of *B. pertussis* seems at least a partial explanation for the whooping cough resurgence in Australia where pertactin-deficient strains have increasingly circulated since 2008[61,62]. These genetic variations become important for understanding the epidemiology of *B. pertussis* transmission above and beyond protection from vaccines.

Clinicians and public health agencies increasingly rely on genetic-based tests to rapidly and accurately determine whether an individual is infected with *B. pertussis* and monitor both outbreaks and the effectiveness of interventions[63]. However, despite advances in next-generation sequencing technology and expanded specimen collection, the majority of *B. pertussis* genetic studies focus on detecting various genetic markers associated with only a handful of known virulence factors and vaccine components[64,65]. This approach is unable to identify other genomic regions involved in *B. pertussis* pathogenicity and/or vaccine resistance potentially responsible for clonal expansion and persistence in a vaccinated population.

Acellular *B. pertussis* vaccines contain one to five purified *B. pertussis* protein antigens: pertussis toxin, filamentous haemagglutinin, pertactin, and fimbrial type 2 (Fim2) and Fim3. Recently, Sealey et al. analysed a set of *B. pertussis* isolates from the United Kingdom and found genes encoding aP vaccine antigens are evolving more rapidly than other cell surface protein-coding genes, with a significantly higher frequency of single nucleotide polymorphisms in the pre- versus post-vaccination period[66,67]. They additionally found that the United Kingdom isolates were polyclonal, indicating multiple introductions of genetic variation and not a single mutation leading to a potentially hypervirulent strain. They point out that while the mutation rate of aP vaccine antigen-encoding genes are evolving at higher rates than other surface protein-coding genes, this rate increased markedly after the switch to aP vaccine. These results are further corroborated by a study by van Gent et al., where analyses of *B. pertussis* isolates from 13 European countries by multilocus

antigen sequence typing, fimbrial serotyping, multilocus variable-number tandem repeat analysis, and pulsed-field gel electrophoresis revealed high similarity in isolates from the 12 countries using aP vaccines and dissimilarity between these isolates and *B. pertussis* from Poland, the only country in their study still using wP vaccine[68].

13.2.4 Vaccine-induced immunology

Animal models of natural infection and wP and aP vaccination reveal substantially different immune responses between natural infection or wP vaccination and aP vaccination[69,70]. Whereas natural infection or wP vaccination triggers T-cell responses that are primarily Th17 and Th1 responses, aP vaccination triggers mostly Th2 responses. Th17 T-cell-mediated pathways have been shown to confer mucosal immunity for bacterial pathogens[71,72]. It is possible that aP-primed individuals are less likely to mount a sterilizing immune response after natural infection when compared to a naïve or wP-vaccinated child[73]. The potential for this type of vaccine failure has been observed in humans where reanalyses of aP vaccine studies revealed that individuals vaccinated with components of the aP vaccine were protected against disease, but not bacterial colonization[74,75]. This is in addition to the extant, but limited, evidence for natural asymptomatic infection[76–78].

The epidemiological and biological link between these immune pathways and decreased protection against *B. pertussis* colonization and/or disease has not been rigorously established and remains an open area for future research[73]. If differences exist between aP-primed and naturally infected or wP-primed individuals in the ability to mount sustained sterilizing immunity, questions arise about the effectiveness of maternal vaccination in conferring protective immunity to infants in the first months of life. There is evidence that maternal vaccination of women originally vaccinated with wP may be both safe and partially effective in protecting infants from *B. pertussis* infection[79,80]. However, in adolescents and adults, the adjusted estimate of aP vaccine effectiveness against pertussis was only 53–64 per cent[81], with poorer rates among adolescents who had been primed and boosted solely with aP vaccine. Because wP vaccine is no longer administered in the United

States, there is an emerging cohort of mothers who will be primed and boosted only with aP vaccine, which may reduce the effectiveness of maternal vaccination during pregnancy.

13.2.5 Asymptomatic transmission

The often high prevalence of individuals subclinically infected by *B. pertussis* has been demonstrated both in large studies in the United States[82] and in highly vaccinated populations around the globe, including China[83], Brazil[84], and the United Kingdom[85]. A second line of evidence in support of a large, undiagnosed population of individuals infectious with *B. pertussis* comes from intensive case investigations associated with infant- or child-diagnosed *B. pertussis* infection. Again, both in the United States[86–88] and across a similarly broad range of countries outside of the United States[89–92], the source of infection could only be identified in 24–66 per cent of cases (mean 42 per cent). The similarity of these estimates—both the prevalence of detectable *B. pertussis* in individuals with persistent cough and the high fraction of primary cases where a source could not be identified across more than ten countries—implies a mechanism that transcends all of the differences between *B. pertussis* epidemiology in these countries and points to a persistent, and now growing, public health burden of undiagnosed—asymptomatic—*B. pertussis* infection.

Using a primate model of *B. pertussis* infection and transmission, Warfel et al. determined that asymptomatic infection in aP vaccinated individuals, and subsequent transmission, could account for the increase in observed *B. pertussis* incidence. However, from a public health perspective, the presence of vaccine-induced or naturally infected asymptomatic individuals who transmit disease could have consequences beyond facilitating an increase in incidence. In response to Warfel et al., Domenech de Cellès et al.[53] argue that a reduction in incidence among unvaccinated individuals in a population with high aP vaccine coverage shows that aP vaccine must reduce *B. pertussis* transmission to some extent. Although epidemiological evidence does exist to suggest aP vaccine blocks some *B. pertussis* transmission, nevertheless, as discussed in greater detail by Althouse and Scarpino[3]—and

recreated in Figure 13.3—relying on a reduction in symptomatic cases in whooping cough to provide evidence for herd immunity is tricky if there are high rates of subclinical, asymptomatic, unreported infections.

The inability of current aP vaccines to block transmission is supported by five additional lines of evidence. First, immunizing close infant contacts (termed 'cocooning') with the aP vaccine provides no detectable increase in protection against *B. pertussis* infection in cocooned infants[57,58]. Regarding the 'cocooning' strategy and quoting directly from Carione et al.: 'In our setting, vaccinating parents with dTpa during the four weeks following delivery did not reduce pertussis diagnoses in infants [1.9 vs 2.2 infections per 1000 infants]'[58]. Second, large populations of individuals are subclinically infected with *B. pertussis* in highly vaccinated populations globally, including the United States[82], China[83], Brazil[84], and the United Kingdom[85]. Third, in the United States[86–88] and a growing number of other countries[89–92], the source of infection in detailed household epidemiological studies of childhood *B. pertussis* infection could only be identified in 42 per cent of cases. Fourth, studies in both the United States and New Zealand on the prevalence of *B. pertussis* infection in infants under 12 months, concluded that prevalence was much higher than expected given vaccination coverage[93,94]. Fifth, and as discussed above, pre-vaccine cycling of pertussis prevalence in under-vaccinated infants has returned, despite high vaccine coverage, in the UK (see Figure 13.2) and Israel[3,24]. Lastly, a study in Brazil suggests that the majority of *B. pertussis* infections in children may originate from subclinically infected adult contacts[84] and a case investigation of a neonatal whooping cough case in Taiwan concluded that asymptomatic transmission was the most likely explanation[95].

However, a clear challenge exists for determining whether aP vaccine effectively blocks transmission through reduced colonization or through mere reduction in symptoms. There is epidemiological evidence of indirect protection from aP vaccination: upon introduction of aP vaccine in Sweden, a drop in *B. pertussis* incidence was seen in the infants who received aP vaccine (direct vaccine protection) as well as in age groups over 20 years of age (cited as

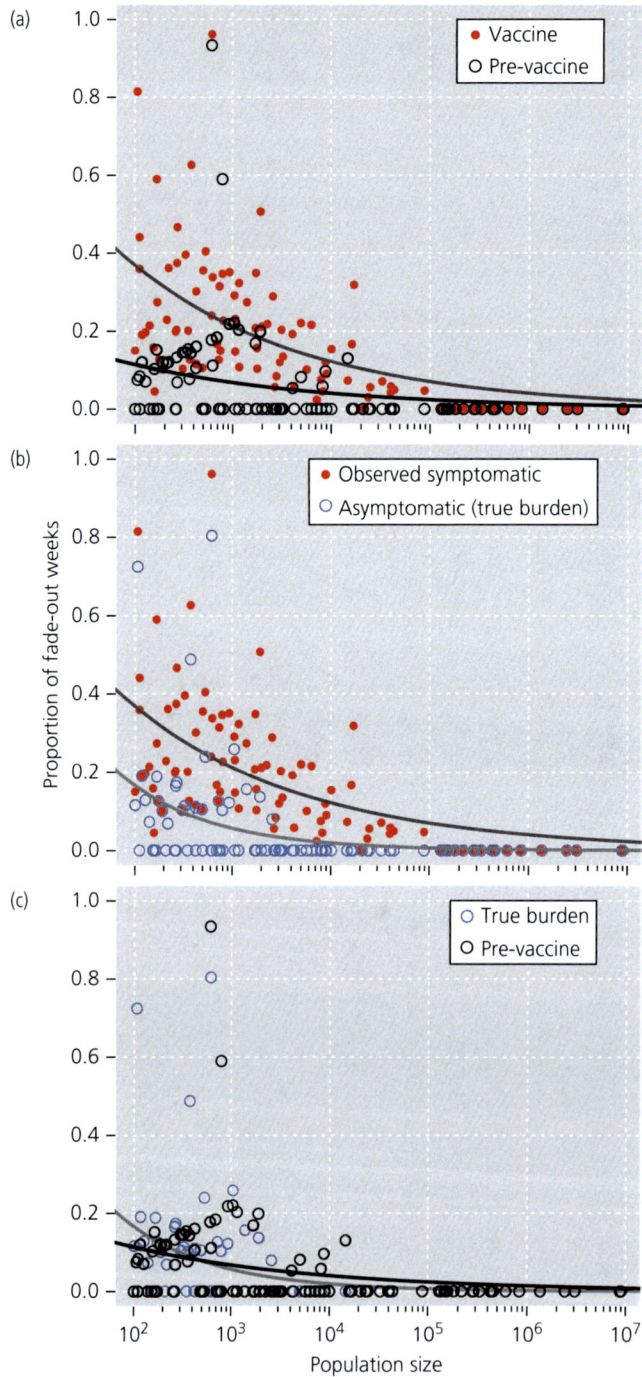

Figure 13.3 Changes in prevalence in pre- and post-vaccination simulations. Figure shows the proportion of disease-free weeks (i.e. stochastic fade-outs) for various population sizes using the model described by Althouse and Scarpino[3]. Panel (a) compares the symptomatic cases in the aP vaccination era with those in the pre-vaccine era; panel (b) compares the symptomatic to asymptomatic cases in the vaccine era; panel (c) compares the asymptomatic cases in the post-vaccine era with those in the pre-vaccine era. These results demonstrate no changes in transmission due to vaccination, despite seeming evidence from symptomatic, that is, reported, case data that transmission has been interrupted. Figure recreated from[3].

evidence for indirect protection)[96]—note, however, an *increase* in *B. pertussis* incidence was seen in ages 10–19 years. The biological evidence points away from reductions in colonization, with aP vaccine reducing contagiousness (mediated by Th2 responses generating antibodies against pertussis toxin and other factors) rather than shifts in mucosal immunity (via Th17 responses) conferring resistance to exposure[69]. Again, distinguishing these two competing hypotheses from routinely collected epidemiological surveillance data introduces an identifiability problem, as both mechanisms could give rise to identical reductions in *B. pertussis* incidence. Detailed studies of the immune responses in humans are necessary to identify the role of immunity response in shaping epidemiological patterns.

13.3 Phylodynamics of hidden *Bordetella pertussis* transmission

Genetic data from pathogens—now routinely collected during ongoing outbreaks[97]—can provide an additional, and highly valuable, source of information regarding unreported cases (i.e. hidden transmission events). As a pathogen moves through a population, its DNA or RNA accumulates mutations mostly as a function of the number of infected hosts and, thus, the population genomic variation

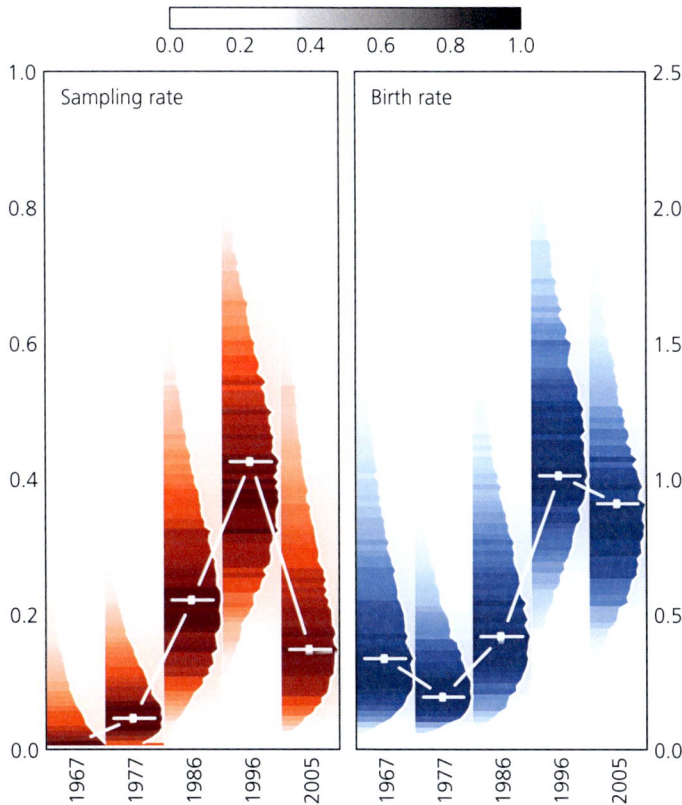

Figure 13.4 Phylodynamic analysis of *B. pertussis* in the United States. Using *B. pertussis* genomes collected from infected individuals in the United States between 1978 and 2012, Althouse and Scarpino[3] performed a phylodynamic analysis of *B. pertussis* transmission before and after the aP vaccine switch. The figure plots the sampling rate and birth rate derived from the phylodynamic analysis. Solid white lines with square boxes indicate the posterior median, with the shaded region indicating the 95% highest posterior density. Darker colours are associated with regions of higher posterior density, with the shape representing the full posterior density. Despite the birth rate remaining higher after the switch to aP, the sampling rate declines. This pattern would be expected with an increasing rate of asymptomatic transmission and is statistically significant at the *p* < 0.001 level. The switch from wP to aP is indicated with the dashed line and occurred in 1997. Figure recreated from[3].

in the pathogen reflects the total number of infected individuals (irrespective of whether those individuals are reported and/or symptomatic). Put simply, under-reporting can cause a mismatch between the prevalence rate estimated from case counts versus genetic data[3]. For example, if there is a constant level of under-reporting, case count data will reflect lower transmission rates and lead to underestimation proportional to the under-reporting rate. However, the extent of genetic variation among pathogen sequences taken from the same set of cases will reflect the true, larger population size of circulating pathogens, and lead to estimates closer to the true incidence.

This discrepancy between reported cases and infection prevalence reconstructed from population genomic variation exists even if the clinical samples are only collected from individuals reported to an epidemiological monitoring system (e.g. a public health agency), assuming reported and unreported individuals are social contacts with each other and thus can transmit infection. More specifically, selective, demographic, and neutral processes will alter the shape of a pathogen's genealogy. As a result, coalescent theory can be used to extract information about phenotypic evolution from the genetic variability of pathogen populations.

Figure 13.4 shows an increase in prevalence estimated using genomic data after the switch to aP vaccine ('birth rate'), as expected given the rise in cases; however, we see a decrease in the estimated fraction of cases sampled for genome sequencing ('sampling rate'). This estimated decrease in genome sampling coincides with the historical period containing the largest number of sampled *B. pertussis* genomes[98]. In addition, natural selection driving a vaccine escape mutant to fixation would result in an increased estimated sampling rate. Lastly, without systematic under-reporting, waning immunity could not account for the observed discordance between the sampling and birth rates.

13.4 Recommendations and conclusions

Vaccination for *B. pertussis* has led to a dramatic reduction in global incidence and saved millions of lives. However, whooping cough remains among the leading causes of mortality from vaccine-preventable diseases and its resurgence in many countries is rapidly becoming a public health crisis. Regardless of the mechanism, the emerging scientific consensus points towards the inability of the aP vaccine to prevent large-scale whooping outbreaks. There are several promising new *B. pertussis* vaccines in development[99,100], including those that induce mucosal immunity[101,102]. In the years before a new vaccine is ready for clinical use, other options are necessary for reducing new *B. pertussis* infections, including vaccination of pregnant women[103,104], or potentially a switch back to wP vaccination as a priming dose[14,105–107], although in many locations, switching back to the wP vaccine is likely to be politically unpalatable. Despite this emerging consensus surrounding the need for a new *B. pertussis* vaccine, substantial disagreement exists regarding the mechanisms behind resurgence and persistence. In order to better evaluate the various hypotheses outlined in this chapter, more detailed *B. pertussis* epidemiological and genomic studies are necessary. The identified potential surveillance bias associated with *B. pertussis* incidence due to under-reporting highlights a critical need for population-wide serological surveys to detect recent infection, studies to examine the genetic diversity of the *B. pertussis* bacterium, more detailed studies of the incidence rate in unvaccinated individuals, and increased active surveillance of attenuated, symptomatic *B. pertussis* infections. Our results suggest that while evidence exists for each of the proposed hypotheses, it is unlikely any single mechanism can account for the global pattern of whooping cough incidence and that determining the relative importance of each mechanism remains uniquely challenging.

References

1. World Health Organization. Pertussis vaccines: WHO position paper, August 2015. *Wkly Epidemiol Rec* 2015;**90**:433–58.
2. Jackson DW, Rohani P. Perplexities of pertussis: recent global epidemiological trends and their potential causes. *Epidemiol Infect* 2013;**142**:672–84.
3. Althouse BM, Scarpino SV. Asymptomatic transmission and the resurgence of Bordetella pertussis. *BMC Med* 2015;**13**:146.
4. Centers for Disease Control and Prevention. *Pertussis (Whooping Cough): Surveillance and Reporting*. 2018. http://www.cdc.gov/pertussis/surv-reporting.html.

5. California Department of Public Health. *Pertussis (Whooping Cough)*. 2018. https://www.cdph.ca.gov/Programs/CID/DCDC/Pages/Immunization/pertussis.aspx.

6. Loconsole D, De Robertis AL, Morea A, et al. Resurgence of pertussis and emergence of the ptxp3 toxin promoter allele in south Italy. *Paediatr Infect Dis J* 2018;**37**:e126–31.

7. Grenfell B. Pertussis in England and Wales: an investigation of transmission dynamics and control by mass vaccination. *Proc R Soc Lond B* 1989;**236**:213–52.

8. Hethcote HW. An age-structured model for pertussis transmission. *Math Biosci* 1997;**145**:89–136.

9. Earn DJ, Rohani P, Bolker BM, et al. A simple model for complex dynamical transitions in epidemics. *Science* 2000;**287**:667–70.

10. Olsen LF, Truty GL, Schaffer WM. Oscillations and chaos in epidemics: a nonlinear dynamic study of six childhood diseases in Copenhagen, Denmark. *Theor Pop Biol* 1988;**33**:344–70.

11. Rohani P, Zhong X, King AA. Contact network structure explains the changing epidemiology of pertussis. *Science* 2010;**330**:982–5.

12. Oliveira SM, Gonçalves-Pinho M, Freitas A, et al. Trends and costs of pertussis hospitalizations in Portugal, 2000 to 2015: from 0 to 95 years old. *Infect Dis* 2018;**50**:625–33.

13. Lee GM, Lett S, Schauer S, et al. Societal costs and morbidity of pertussis in adolescents and adults. *Clin Infect Dis* 2004;**39**:1572–80.

14. DeAngelis H, Scarpino SV, Fitzpatrick MC, et al. Epidemiological and economic effects of priming with the whole-cell Bordetella pertussis vaccine. *JAMA Pediatr* 2016;**170**:459–65.

15. Fitzpatrick MC, Wenzel NS, Scarpino SV, et al. Cost-effectiveness of next-generation vaccines: the case of pertussis. *Vaccine* 2016;**34**:3405–11.

16. Grenfell B, Bjørnstad O, Kappey J. Travelling waves and spatial hierarchies in measles epidemics. *Nature* 2001;**414**:716–23.

17. Lofgren E, Fefferman NH, Naumov YN, et al. Influenza seasonality: underlying causes and modeling theories. *J Virol* 2007;**81**:5429–36.

18. Tamerius J, Nelson MI, Zhou SZ, et al. Global influenza seasonality: reconciling patterns across temperate and tropical regions. *Environ Health Perspect* 2011;**119**:439–45.

19. Weinberger DM, Klugman KP, Steiner CA, et al. Association between respiratory syncytial virus activity and pneumococcal disease in infants: a time series analysis of us hospitalization data. *PLoS Med* 2015;**12**:e1001776.

20. Pitzer VE, Viboud C, Alonso WJ, et al. Environmental drivers of the spatiotemporal dynamics of respiratory syncytial virus in the United States. *PLoS Pathog* 2015;**11**:e1004591.

21. de Greeff SC, Dekkers AL, Teunis P, et al. Seasonal patterns in time series of pertussis. *Epidemiol Infect* 2009;**137**:1388–95.

22. Bjørnstad ON, Finkenstädt BF, Grenfell BT. Dynamics of measles epidemics: estimating scaling of transmission rates using a time series sir model. *Ecol Monogr* 2002;**72**:169–84.

23. Domenech de Cellès M, Magpantay F, King A, et al. The pertussis enigma: reconciling epidemiology, immunology and evolution. *Proc Royal Soc B* 2016;**283**:20152309.

24. Broutin H, Guégan J-F, Elguero E, et al. Large-scale comparative analysis of pertussis population dynamics: periodicity, synchrony, and impact of vaccination. *Am J Epidemiol* 2005;**161**:1159–67.

25. Broutin H, Viboud C, Grenfell B, et al. Impact of vaccination and birth rate on the epidemiology of pertussis: a comparative study in 64 countries. *Proc Biol Sci* 2010;**277**:3239–45.

26. Langsam D, Anis E, Yamin D, et al. Re-emergence of pertussis in Israel retains periodicity of pre-vaccine era. *SSRN* 2017;**17 June**. https://ssrn.com/abstract=2988298.

27. Hamilton BE, Martin JA, Osterman MJ, et al. Births: provisional data for 2017. *Vital Stat Rapid Release* 2018;**May**:4. https://www.cdc.gov/nchs/data/vsrr/report004.pdf

28. Rothstein E, Edwards K. Health burden of pertussis in adolescents and adults. *Paediatr Infect Dis J* 2005;**24**:S44–7.

29. Tan T, Trindade E, Skowronski D. Epidemiology of pertussis. *Paediatr Infect Dis J* 2005;**24**:S10–8.

30. Wearing HJ, Rohani P. Estimating the duration of pertussis immunity using epidemiological signatures. *PLoS Pathog* 2009;**5**:e1000647.

31. de Cellès MD, Magpantay FM, King AA, et al. The impact of past vaccination coverage and immunity on pertussis resurgence. *Sci Transl Med* 2018;**10**:eaaj1748.

32. Mooi FR, Van Loo I, King AJ. Adaptation of Bordetella pertussis to vaccination: a cause for its reemergence? *Emerg Infect Dis* 2001;**7**:526–8.

33. Warfel JM, Zimmerman LI, Merkel TJ. Acellular pertussis vaccines protect against disease but fail to prevent infection and transmission in a nonhuman primate model. *Proc Natl Acad Sci U S A* 2014;**111**:787–92.

34. Lu C-Y, Tsai HC, Huang YC, et al. A national seroepidemiologic survey of pertussis among school children in Taiwan. *Paediatr Infect Dis J* 2017;**36**:e307–12.

35. Cherry JD, Grimprel E, Guiso N, et al. Defining pertussis epidemiology: clinical, microbiologic and serologic perspectives. *Paediatr Infect Dis J* 2005;**24**:S25–34.

36. Cherry JD. The epidemiology of pertussis: a comparison of the epidemiology of the disease pertussis with

the epidemiology of Bordetella pertussis infection. *Pediatrics* 2005;**115**:1422–7.

37. Strebel P, Nordin J, Edwards K, et al. Population-based incidence of pertussis among adolescents and adults, Minnesota, 1995–1996. *J Infect Dis* 2001;**183**:1353–9.

38. Senzilet LD, Halperin SA, Spika JS, et al. Pertussis is a frequent cause of prolonged cough illness in adults and adolescents. *Clin Infect Dis* 2001;**32**:1691–7.

39. Crowcroft NS, Johnson C, Chen C, et al. Under-reporting of pertussis in Ontario: a Canadian immunization research network (CIRN) study using capture-recapture. *PLoS One* 2018;**13**:e0195984.

40. Wang K, Birring SS, Taylor K, et al. Montelukast for postinfectious cough in adults: a double-blind randomised placebo-controlled trial. *Lancet Respir Med* 2014;**2**:35–43.

41. Harnden A, Grant C, Harrison T, et al. Whooping cough in school age children with persistent cough: prospective cohort study in primary care. *BMJ* 2006;**333**:174–7.

42. Kayina V, Kyobe S, Katabazi FA, et al. Pertussis prevalence and its determinants among children with persistent cough in urban Uganda. *PLoS One* 2015;**10**:e0123240.

43. Philipson K, Goodyear-Smith F, Grant CC, et al. When is acute persistent cough in school-age children and adults whooping cough? *Br J Gen Pract* 2013;**63**:e573–9.

44. Gilberg S, Njamkepo E, Du Châtelet IP, et al. Evidence of Bordetella pertussis infection in adults presenting with persistent cough in a French area with very high whole-cell vaccine coverage. *J Infect Dis* 2002;**186**:415–8.

45. Cagney M, MacIntyre CR, McIntyre P, et al. Cough symptoms in children aged 5–14 years in Sydney, Australia: non-specific cough or unrecognized pertussis? *Respirology* 2005;**10**:359–64.

46. Yesmin K, Shamsuzzaman S, Chowdhury A, et al. Isolation of potential pathogenic bacteria from nasopharynx from patients having cough for more than two weeks. *Bangladesh J Med Microbiol* 2012;**4**:13–8.

47. Schielke A, Takla A, von Kries R, et al. Marked underreporting of pertussis requiring hospitalization in infants as estimated by capture–recapture methodology, Germany, 2013–2015. *Paediatr Infect Dis J* 2018;**37**:119–25.

48. Castillo ME, Bada C, Del Aguila O, et al. Detection of Bordetella pertussis using a PCR test in infants younger than one year old hospitalized with whooping cough in five Peruvian hospitals. *Int J Infect Dis* 2015;**41**:36–41.

49. Beynon KA, Young SA, Laing RT, et al. Bordetella pertussis in adult pneumonia patients. *Emerg Infect Dis* 2005;**11**:639–41.

50. Holter JC, Müller F, Bjørang O, et al. Etiology of community-acquired pneumonia and diagnostic yields of microbiological methods: a 3-year prospective study in Norway. *BMC Infect Dis* 2015;**15**:64.

51. Purdy KW, Hay JW, Botteman MF, et al. Evaluation of strategies for use of acellular pertussis vaccine in adolescents and adults: a cost-benefit analysis. *Clin Infect Dis* 2004;**39**:20–8.

52. Sutter RW, Cochi SL. Pertussis hospitalizations and mortality in the United States, 1985–1988: evaluation of the completeness of national reporting. *JAMA* 1992;**267**:386–91.

53. Domenech de Cellès M, Riolo MA, Magpantay FMG, et al. Epidemiological evidence for herd immunity induced by acellular pertussis vaccines. *Proc Natl Acad Sci U S A* 2014;**111**:E716–7.

54. van Boven M, de Melker HE, Schellekens JF, et al. Waning immunity and sub-clinical infection in an epidemic model: implications for pertussis in the Netherlands. *Math Biosci* 2000;**164**:161–82.

55. Wendelboe AM, Van Rie A, Salmaso S, et al. Duration of immunity against pertussis after natural infection or vaccination. *Paediatr Infect Dis J* 2005;**24**:S58–61.

56. Castagnini LA, Healy CM, Rench MA, et al. Impact of maternal postpartum tetanus and diphtheria toxoids and acellular pertussis immunization on infant pertussis infection. *Clin Infect Dis* 2012;**54**:78–84.

57. Healy CM, Rench MA, Wootton SH, et al. Evaluation of the impact of a pertussis cocooning program on infant pertussis infection. *Paediatr Infect Dis J* 2015;**34**:22–6.

58. Carcione D, Regan AK, Tracey L, et al. The impact of parental postpartum pertussis vaccination on infection in infants: a population-based study of cocooning in western Australia. *Vaccine* 2015;**33**:5654–61.

59. Sanstead E, Basta NE, Martin K, et al. Pertussis and the Minnesota state fair: demonstrating a novel setting for efficiently conducting seroepidemiologic studies. *J Community Health* 2018. 7 April. doi: 10.1007/s10900-018-0508-y. [Epub ahead of print]

60. Briere EC, Pondo T, Schmidt M, et al. Assessment of Tdap vaccination effectiveness in adolescents in integrated health-care systems. *J Adolesc Heal* 2018;**62**:661–6.

61. Bart MJ, Harris SR, Advani A, et al. Global population structure and evolution of Bordetella pertussis and their relationship with vaccination. *MBio* 2014;**5**:e01074–14.

62. Lam C, Octavia S, Ricafort L, et al. Rapid increase in pertactin-deficient Bordetella pertussis isolates, Australia. *Emerg Infect Dis* 2014;**20**:626–33.

63. Van der Zee A, Agterberg C, Peeters M, et al. Polymerase chain reaction assay for pertussis: simultaneous detection and discrimination of Bordetella pertussis and Bordetella parapertussis. *J Clin Microbiol* 1993;**31**:2134–40.

64. van Loo IHM, Heuvelman KJ, King AJ, et al. Multilocus sequence typing of Bordetella pertussis based on surface protein genes. *J Clin Microbiol* 2002;**40**:1994–2001.

65. Schouls LM, van der Heide HGJ, Vauterin, L, et al. Multiple-locus variable-number tandem repeat analysis

of Dutch Bordetella pertussis strains reveals rapid genetic changes with clonal expansion during the late 1990s. *J Bacteriol* 2004;**186**:5496–505.

66. Sealey KL, Belcher T, Preston A. Bordetella pertussis epidemiology and evolution in the light of pertussis resurgence. *Infect Genet Evol* 2016;**40**:136–43.

67. Sealey KL, Harris SR, Fry NK, et al. Genomic analysis of isolates from the United Kingdom 2012 pertussis outbreak reveals that vaccine antigen genes are unusually fast evolving. *J Infect Dis* 2014;**212**:294–301.

68. Van Gent M, Heuvelman CJ, van der Heide HG, et al. Analysis of Bordetella pertussis clinical isolates circulating in European countries during the period 1998–2012. *Eur J Clin Microbiol Infect Dis* 2015;**34**:821–30.

69. Gill C, Rohani P, Thea DM. The relationship between mucosal immunity, nasopharyngeal carriage, asymptomatic transmission and the resurgence of Bordetella pertussis. *F1000Research* 2017;**6**:1568.

70. Warfel J, Merkel T. Bordetella pertussis infection induces a mucosal il-17 response and long-lived th17 and th1 immune memory cells in nonhuman primates. *Mucosal Immunol* 2013;**6**:787–96.

71. Higgs R, Higgins S, Ross P, et al. Immunity to the respiratory pathogen Bordetella pertussis. *Mucosal Immunol* 2012;**5**:485–500.

72. Kolls JK, Khader SA. The role of th17 cytokines in primary mucosal immunity. *Cytokine Growth Factor Rev* 2010;**21**:443–8.

73. Platt L, Thun M, Harriman K, et al. A population-based study of recurrent symptomatic Bordetella pertussis infections in children in California, 2010–2015. *Clin Infect Dis* 2017;**65**:2099–104.

74. Storsaeter J, Hallander H, Farrington CP, et al. Secondary analyses of the efficacy of two acellular pertussis vaccines evaluated in a Swedish phase III trial. *Vaccine* 1990;**8**:457–61.

75. Von Linstow M-L, Pontoppidan PL, von König C-HW, et al. Evidence of Bordetella pertussis infection in vaccinated 1-year-old Danish children. *Eur J Pediatr* 2010;**169**:1119–22.

76. Zhang Q, Yin Z, Li Y, et al. Prevalence of asymptomatic Bordetella pertussis and Bordetella parapertussis infections among school children in china as determined by pooled real-time PCR: a cross-sectional study. *Scand J Infect Dis* 2014;**46**:280–7.

77. de Melker HE, Versteegh FG, Schellekens JF, et al. The incidence of Bordetella pertussis infections estimated in the population from a combination of serological surveys. *J Infect* 2006;**53**:106–13.

78. Cortese MM, Baughman AL, Brown K, et al. A 'new age' in pertussis prevention: new opportunities through adult vaccination. *Am J Prev Med* 2007;**32**:177–85.

79. Centers for Disease Control and Prevention (CDC). Updated recommendations for use of tetanus toxoid, reduced diphtheria toxoid, and acellular pertussis vaccine (Tdap) in pregnant women–Advisory Committee on Immunization Practices (ACIP), 2012. *MMWR Morb Mortal Wkly Rep* 2013;**62**:131–5.

80. Amirthalingam G, Andrews N, Campbell H, et al. Effectiveness of maternal pertussis vaccination in England: an observational study. *Lancet* 2014;**384**:1521–8.

81. Baxter R, Bartlett J, Rowhani-Rahbar A, et al. Effectiveness of pertussis vaccines for adolescents and adults: case-control study. *BMJ* 2013;**347**:f4249.

82. Ward JI, Cherry JD, Chang SJ, et al. Efficacy of an acellular pertussis vaccine among adolescents and adults. *N Engl J Medicine* 2005;**353**:1555–63.

83. Zhang Q, Yin Z, Li Y, et al. Prevalence of asymptomatic Bordetella pertussis and Bordetella parapertussis infections among school children in china as determined by pooled real-time PCR: a cross-sectional study. *Scand J Infect Dis* 2014;**46**:280–7.

84. Berezin EN, de Moraes JC, Leite D, et al. Sources of pertussis infection in young babies from São Paulo state, Brazil. *Paediatr Infect Dis J* 2014;**33**:1289–91.

85. Wendelboe AM, Van Rie A, Salmaso S, et al. Duration of immunity against pertussis after natural infection or vaccination. *Paediatr Infect Dis J* 2000;5;**24**, S58–61.

86. Skoff TH, Kenyon C, Cocoros N, et al. Sources of infant pertussis infection in the United States. *Pediatr* 2015;**136**:635–41.

87. Bisgard KM, Pascual FB, Ehresmann KR, et al. Infant pertussis: who was the source? *Paediatr Infect Dis J* 2004;**23**:985–9.

88. Staudt A, Mangla AT, Alamgir H. Investigation of pertussis cases in a Texas county, 2008–2012. *South Med J* 2015;**108**:452–7.

89. Wendelboe AM, Njamkepo E, Bourillon A, et al. Transmission of Bordetella pertussis to young infants. *Paediatr Infect Dis J* 2007;**26**:293–9.

90. Crowcroft N, Booy R, Harrison T, et al. Severe and unrecognised: pertussis in UK infants. *Arch Dis Child* 2003;**88**:802–6.

91. Baron S, Njamkepo E, Grimprel E, et al. Epidemiology of pertussis in French hospitals in 1993 and 1994: thirty years after a routine use of vaccination. *Paediatr Infect Dis J* 1998;**17**:412–18.

92. Kowalzik F, Barbosa AP, Fernandes VR, et al. Prospective multinational study of pertussis infection in hospitalized infants and their household contacts. *Paediatr Infect Dis J* 2007;**26**:238–42.

93. Masseria C, Martin CK, Krishnarajah G, et al. Incidence and burden of pertussis among infants less than 1 year of age. *Paediatr Infect Dis J* 2017;**36**:e54–61.

94. Macdonald-Laurs E, Ganeshalingham A, Lillie J, et al. Increasing incidence of life-threatening pertussis: a retrospective cohort study in New Zealand. *Paediatr Infect Dis J* 2017;**36**:282–9.

95. Ma H-Y, Pan SC, Wang JT, et al. Lack of pertussis protective antibodies in healthcare providers taking care of neonates and infants in a children's hospital. *Paediatr Infect Dis J* 2017;**36**:433–5.

96. Carlsson R-M, Trollfors B. Control of pertussis—lessons learnt from a 10-year surveillance programme in Sweden. *Vaccine* 2009;**27**:5709–18.

97. Ip CL, Pybus OG, Gardy JL. Virus genomics and evolution: the transformative effect of new technologies and multidisciplinary collaboration on virus research and outbreak management. *Genome Biol* 2016;**17**:159.

98. Cherry JD. Epidemic pertussis in 2012—the resurgence of a vaccine-preventable disease. *N Engl J Med* 2012;**367**:785–7.

99. Locht C, Mielcarek N. Live attenuated vaccines against pertussis. *Expert Rev Vaccines* 2014;**13**:1147–58.

100. Meade BD, Plotkin SA, Locht C. Possible options for new pertussis vaccines. *J Infect Dis* 2014;**209**:S24–7.

101. Feunou PF, Kammoun H, Debrie, A-S, et al. Long-term immunity against pertussis induced by a single nasal administration of live attenuated B. pertussis BPZE1. *Vaccine* 2010;**28**:7047–53.

102. Fedele G, Bianco M, Debrie A-S, et al. Attenuated Bordetella pertussis vaccine candidate BPZE1 promotes human dendritic cell CCL21-induced migration and drives a Th1/Th17 response. *J Immunol* 2011;**186**: 5388–96.

103. Abu Raya B, Srugo I, Kessel A, et al. The effect of timing of maternal tetanus, diphtheria, and acellular pertussis (Tdap) immunization during pregnancy on newborn pertussis antibody levels – a prospective study. *Vaccine* 2014;**32**:5787–93.

104. Munoz FM, Bond NH, Maccato M, et al. Safety and immunogenicity of tetanus diphtheria and acellular pertussis (Tdap) immunization during pregnancy in mothers and infants: a randomized clinical trial. *JAMA* 2014;**311**:1760–9.

105. Sheridan SL, Ware RS, Grimwood K, et al. Number and order of whole cell pertussis vaccines in infancy and disease protection. *JAMA* 2012;**308**:454–6.

106. Liko J, Robison SG, Cieslak PR. Priming with whole-cell versus acellular pertussis vaccine. *N Engl J Medicine* 2013;**368**:581–2.

107. Witt MA, Arias L, Katz PH, et al. Reduced risk of pertussis among persons ever vaccinated with whole cell pertussis vaccine compared to recipients of acellular pertussis vaccines in a large US cohort. *Clin Infect Dis* 2013;**56**:1248–54.

CHAPTER 14

Pertussis immunity and the epidemiological impact of adult transmission: statistical evidence from Sweden and Massachusetts

Aaron A. King, Matthieu Domenech de Cellès, Felicia M.G. Magpantay, and Pejman Rohani

Abstract

An understanding of the consequences of infection and vaccination on host immunity sets the stage for interpreting pertussis epidemiology. Yet, with no known serological marker of protection, such an understanding is currently not possible. This chapter interrogates longitudinal age-stratified pertussis incidence reports from Sweden and Massachusetts, United States, with the aim of quantifying the impact of infection and immunization on protective immunity. The analysis of data from Sweden during the vaccination hiatus period (1986–1996) indicates that adults contribute little to transmission. This may either be because infection-derived immunity is very long-lasting, or that individuals whose immunity has waned are subsequently less susceptible. The analysis of data from Massachusetts (1990–2005) identifies the primary mechanism of vaccine failure—for both whole-cell and acellular pertussis vaccines—to be waning. However, the average duration of immunity is identified as many decades, though the model predicts substantial individual variability in this trait. Finally, the chapter demonstrates the estimates to be consistent with those obtained from popular measures of vaccine effectiveness, though the interpretation of these findings is quite different.

14.1 Introduction

The ongoing resurgence of pertussis in some locales, despite high vaccination coverage, underscores the incompleteness of our understanding of this important disease. In particular, key unknowns in pertussis epidemiology and vaccinology include:

1. the nature, efficacy, and durability of infection- and vaccine-derived immunity
2. the identity of the groups disproportionately responsible for transmission
3. the extent to which the observed resurgence is an artefact of increased attention to the disease and

King, A. A., Domenech de Cellès, M., Magpantay, F. M. G., and Rohani, P., *Pertussis immunity and the epidemiological impact of adult transmission: Statistical evidence from Sweden and Massachusetts*. In: *Pertussis: epidemiology, immunology, and evolution*. Edited by Pejman Rohani and Samuel V. Scarpino: Oxford University Press (2019). © Oxford University Press.
DOI: 10.1093/oso/9780198811879.003.0014

more sensitive, but perhaps less specific, ascertainment methods.

Efforts to resolve these uncertainties have been conducted at various scales, from the individual infection at the clinical level, to case–control and cohort studies at the level of small populations, to epidemiological studies at the level of communities, regions, and nations, and to comparative studies at the international scale. The lessons gleaned at different scales can be seemingly at odds with one another. Such discrepancies highlight the fact that interpretation of the data is far from straightforward.

Thus, for example, there is considerable uncertainty as to the significance of serological data: while these data do contain information about the occurrence of recent infection[1], the intensity of the implied infection, as well as its duration, severity, and transmissibility, remain highly uncertain. Moreover, the absence of reliable serological correlates of protection means that serological data are silent, or nearly so, with respect to the implications of such infections for subsequent immunity. With respect to epidemiological data, the commonness of mild or asymptomatic infections, combined with differential symptomatology in different age groups, means that such data give us an incomplete and biased perspective on the underlying mechanisms of transmission. The common practice of using different ascertainment methods on different age groups compounds this uncertainty. From genetic data, again, we obtain only an incomplete and potentially skewed view due to limitations in the abundance of sequences and sampling biased towards severe infections and particular locations.

These uncertainties and differences in perspective fuel controversy. Much stands to be gained, therefore, from efforts to reconcile and synthesize information obtained from data gathered at disparate scales. In this chapter, we explore a thread in the literature embodying such an effort. Specifically, we first show how age-stratified incidence data, combined with behavioural data, can be used to shed light on otherwise obscure heterogeneities in susceptibility and transmission. We then work towards a more formal and rigorous synthesis, using data

stratified by both age and time, showing that such data on the dynamics of the pertussis system can be particularly informative with respect to the questions enumerated above.

14.2 Heterogeneities in susceptibility and transmission

Age-specific pertussis incidence data display characteristic patterns. A case in point is afforded by data from Sweden, which halted mandatory vaccination with whole-cell pertussis (wP) vaccine in 1979 and reinstated vaccination with an acellular pertussis (aP) vaccine in 1997, achieving more than 98 per cent coverage almost immediately. When vaccine coverage was low, incidence was concentrated in children aged 0–5 years. With the resumption of mass vaccination, the age distribution shifted to older age categories, and a small 'bump' among teenagers became visible (Figure 14.1).

What does this pattern of age-specific incidence tell us about the risk of infection faced by people of different ages? Although the data tell us something about the incidence of infections, they suffer both from under-reporting (since not all pertussis infections are symptomatic, and not all symptomatic infections are reported). Moreover, to translate these incidence numbers into risk, we must estimate the number of people *at risk* of infection or, equivalently, the number protected from infection by immunity. Since the degree of immunity depends on the history of prior infection and vaccination, in ways that are not fully understood, the problem is not a straightforward one.

However, we can exploit the fact that, in Sweden, both the demographic structure of the population, and the age-distribution of pertussis cases remained relatively constant across the entire vaccine hiatus period (Figure 14.2). Therefore, these data give us an excellent picture of the age profile of pertussis incidence. To quantify the risk of pertussis in each age group, we can compare the number of cases in that group with the number at risk. The latter, of course, depends in turn upon the fraction of each age group that remains protected due to infection earlier in life. This protection, in turn, depends on the duration of infection-induced immune protection,

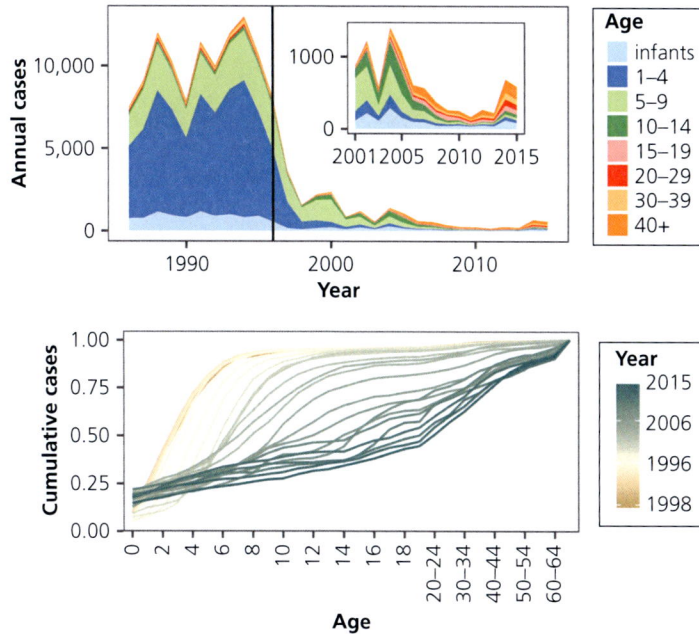

Figure 14.1 Swedish incidence data from 1986 to 2015, stratified by age. The most recent DTaP immunization programme was initiated in January of 1996 (vertical line). Bottom panel depicts cumulative cases through time, highlighting the shift from youngest age groups during the vaccine-free period (1986–1996) to older ages in the vaccine era. Data from reference[2].

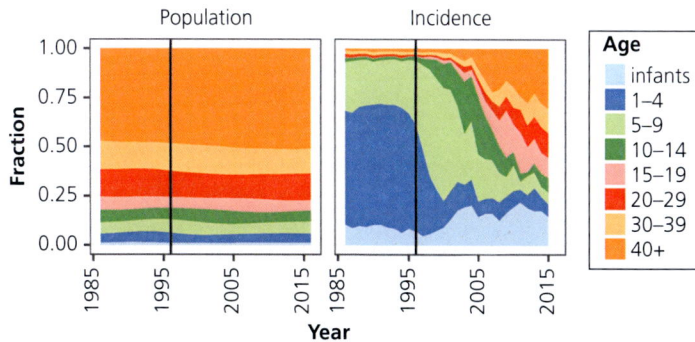

Figure 14.2 Age structure in Sweden. During the vaccine hiatus (1986–1996), the age structures of the population and of pertussis incidence within the population were remarkably stable.

which is not well known. However, we can quantify the age profile of infection risk under a variety of assumptions regarding the durability of protection, seeking conclusions robust to variation in this parameter. In the next section, we go into some mathematical detail on two ways of accomplishing this. The section can be skipped by those interested only in the results of the analysis.

14.2.1 Two approaches to inferring age-specific infection risk based on the catalytic model

The first approach, a modification of that of Van Boven et al.[3], tracks the fractions of each cohort susceptible to, or immune from, infection, respectively. Its basis is the standard compartmental (SIR) modelling framework, within which individuals are categorized in terms of their infection status (Figure 14.3a). At birth, hosts enter the naïve susceptible compartment, S_1, where they face a risk, λ of infection. Because the duration of infection is short relative to the span of life, there is no need to explicitly track the infections. Rather, these individuals pass after infection into the first recovered class, R_1, within which they are temporarily

protected from infection. This immunity wanes at rate σ, and having lost their immunity (S_2), hosts are susceptible to infection at the same rate as naïve hosts. Infections of naïve hosts are assumed to be reported with probability ρ_1, while subsequent infections are visible with (possibly lower) probability, ρ_2. The differential equations corresponding to Figure 14.3a are:

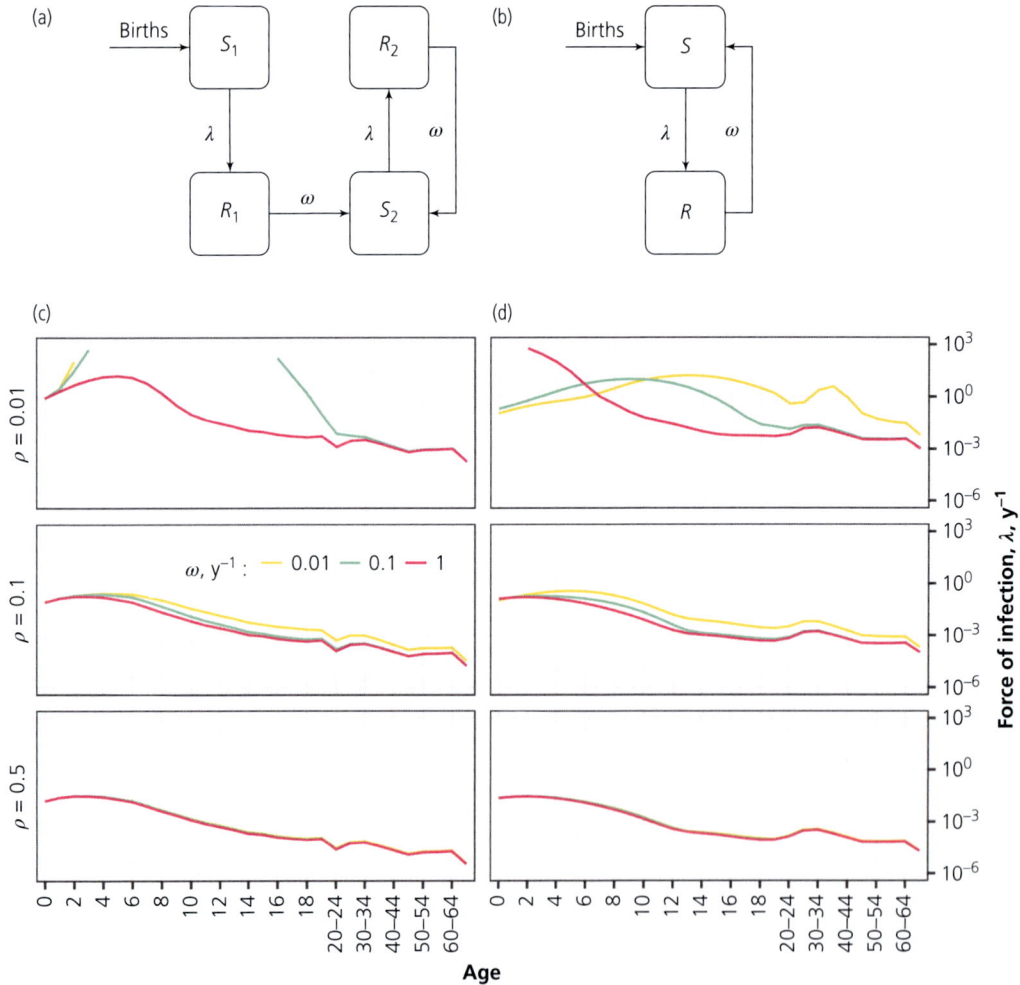

Figure 14.3 Two methods for estimating the age-specific force of infection. (a) Model representation of the life cycle of a single person with respect to pertussis infection and immunity under the modified method of Van Boven et al.[3]. Births occur at rate B, feeding the naïve class S_1. The force of infection, λ, is the per capita risk of infection. Infection induces protection that wanes at rate ω. Primary (i.e. $S_1 \rightarrow R_1$ infections are observed with probability ρ_1; repeat infections ($S_2 \rightarrow R_2$ are observed with probability ρ_2. The short duration of infection, relative to that of time spent in the S and R classes, allows us to neglect it in this model. (b) Model representation for the modified method of Rohani et al.[4]. (c) Estimated age-specific force of infection, $\lambda(a)$, under model (a) and various assumptions about the reporting rate, $\rho_1 = \rho_2 = \rho$, and waning rate, ω. (d) Estimated age-specific force of infection under model (b) and various assumptions about the reporting rate, ρ, and waning rate, ω.

$$\frac{dS_1(a)}{da} = -\lambda(a)S_1(a)$$

$$\frac{dR_1(a)}{da} = \lambda(a)S_1(a) - \omega R_1(a)$$

$$\frac{dS_2(a)}{da} = \omega R_1(a) + \omega R_2(a) - \lambda(a)S_2(a)$$

$$\frac{dR_2(a)}{da} = \lambda(a)S_2(a) - \omega R_2(a) \qquad (1)$$

Equations (1) determine the populations of each of the four classes of Figure 14.3a as a function of age, a. If we have data on the incidence, H_i, of infection within age categories (a_i, a_{i+1}), for $i = 1,\ldots, n$, and we are willing to assume that $\lambda(a)$ is constant over each interval $a_i \leq a < a_{i+1}$, then we can integrate equations (1) from a_i to a_{i+1} to obtain:

$$S_1(a_{i+1}) = e^{-\lambda_i \delta_i} S_1(a_i)$$

$$S_2(a_{i+1}) = \frac{\omega}{\lambda_i + \omega}\left(1 - e^{-(\lambda_i + \omega)\delta_i}\right)$$

$$- e^{-\lambda_i \delta_i}\left(1 - e^{-\omega\delta_i}\right)S_1(a_i) + e^{-(\lambda_i + \omega)\delta_i} S_2(a_i)$$

$$R_1(a_{i+1}) = e^{-\omega\delta_i} R_1(a_i) - \lambda_i \frac{e^{-\lambda_i \delta_i} - e^{-\omega\delta_i}}{\lambda_i - \omega} S_1(a_i), \qquad (2)$$

where $a_1 = 0$ and we write $\delta_i = a_{i+1} - a_i$ and $\lambda_i = \lambda(a_i)$. Note that R_2 is redundant, so there is no need to compute it. Similarly, we integrate the new infections occurring between age a_i and age a_{i+1} to obtain the incidence, H_i, in age group i:

$$H_i = \int_{a_i}^{a_{i+1}} \lambda(a)\big(\rho_1 S_1(a) + \rho_2 S_2(a)\big)da$$

$$= \lambda_i\big(\rho_1 S_1(a_i) + \rho_2 S_2(a_i)\big)\delta_i + O(\delta_i^2). \qquad (3)$$

Now, truncating equation (3) at first order, and solving for λ_i, we obtain:

$$\lambda_i \approx \frac{H_i}{\big(\rho_1 S_1(a_i) + \rho_2 S_2(a_i)\big)\delta_i}. \qquad (4)$$

Taken together, equations (2) and (4) form a recursion: setting $S_1(a_1) = 1$, $S_2(a_1) = R_1(a_1) = R_2(a_1) = 0$, and solving equations (2) and (4) iteratively for $i = 1,\ldots, n$, we obtain an estimate of the age-specific force of infection $\lambda_i = \lambda(a_i)$ for each i. Note that this estimate depends upon our assumptions about the visibility of infections (via ρ_1 and ρ_2) and the duration of infection-induced immunity (via ω). Figure 14.3c

shows the results of this exercise applied to the data from the vaccine hiatus period in Sweden.

Interpretation of the waning rate, ω, is not without its subtleties. In effect, equations (1) assume that the duration of immunity—the residence time in R_1 or R_2—is exponentially distributed. Because the variance of the exponential distribution is so large, the *average* duration of immunity is not particularly *typical*. Thus, for example, with a mean duration of immunity equal to 10 years (corresponding to $\omega = 0.1$), fully 18 per cent have lost their immunity within 2 years and nearly 14 per cent have immunity that lasts more than 20 years. Similarly, when $\omega = 0.01$ (so that the mean duration of immunity is 100 years), roughly half of the population will have lost its immunity within 70 years and almost 10 per cent within 10 years.

Beneath the mathematics, this approach amounts to a straightforward exercise in bookkeeping: it simply accounts for the at-risk population, $S_1 + S_2$, by subtracting the number of infections and re-inserting them once their immunity has waned. Thus, although it cannot assist us in estimating waning rates (ω) or reporting probabilities (ρ_1, ρ_2), it can reveal those features of the risk profile that are robust to assumptions about the these unknowns. In the present case, Figure 14.3c shows that for all but the longest-lived immunity and the most extreme under-reporting, the infection risk profile shows a modest increase among the youngest age groups (< 8 years old) and then decays exponentially with age. The latter is a simple consequence of the paucity of infections observed in the older age groups. When $\rho = \rho_1 = \rho_2$ is small and immunity is long-lived, there are so many infections in the younger age groups that the susceptible pool becomes depleted early. In this case, the risk of infection must become very large among schoolchildren to account for the infections we observe there. To give some idea of how large, note that if $\lambda > 10^3$, as it must be for some age groups if $\rho = 0.01$, for example, then the odds of avoiding infection over the course of any week are about 225×10^6 to 1 against. As we will see, this feature, an unavoidable consequence of the bookkeeping, will crop up elsewhere.

This first approach, while straightforward and transparent, has two unfortunate features. First, errors in the estimation of λ made at one age category cascade to errors at larger age categories. This means that uncertainties in the risk profile tend to grow larger as one moves from left to right in Figure 14.3c. Second, because the method in effect simply translates incidence data, H_i, to infection-risk rates, λ_i, at the same resolution, it affords no means of estimating the parameters ρ or ω. A second approach, originally developed by Rohani et al.[4] and improved upon here, fits the age profile of infection risk, $\lambda(a)$, directly to the data using a smoothing approach. It yields a risk profile with more uniform uncertainty and allows estimation of the two unknown parameters.

The compartment diagram underlying this method is simpler than that of the previous one (Figure 14.3b). No distinction is made between infections in naïve hosts and those in previously infected ones. The age-specific force of infection, $\lambda(a)$, is now assumed to be a smooth function of age, represented using B-splines:

$$\lambda(a) = \sum_{i=1}^{K} b_k s_k(a), \tag{5}$$

where $s_k(a)$, $k = 1, \ldots, K$ are cubic B-splines[5] and b_k are the corresponding coefficients. The function λ appears in the equations corresponding to Figure 14.3b:

$$\frac{dS(a)}{da} = B - \lambda(a)S(a) + \omega R(a)$$
$$\frac{dR(a)}{da} = \lambda(a)S(a) - \omega R(a) \tag{6}$$

As before, the model-predicted incidence, $\bar{H}_i$, in age category (a_i, a_{i+1}) is an integral:

$$\bar{H}_i = \int_{a_i}^{a_{i+1}} \lambda(a) S(a) da. \tag{7}$$

In this approach, we make no attempt solve equation (7). Rather, we relate $\bar{H}_i$ to the observed incidence, H_i, via:

$$H_i \sim \text{Binomial}\left(\bar{H}_i, \rho\right). \tag{8}$$

One might then attempt to estimate ρ, ω, and the coefficients b_k using maximum likelihood. However, if λ is allowed to be flexible, i.e., K in equation (5) is large, the model will over-fit the data. To protect

against this, we adopt a penalized likelihood approach[6]. Specifically, we estimate the coefficients b_k by maximizing:

$$\mathcal{U} = \sum_{i=1}^{n-1} \left(\log \binom{\bar{H}_i}{H_i} + H_i \log \rho + (\bar{H}_i - H_i) \log(1 - \rho) \right)$$
$$- \eta \int_0^{a_n} \left| \sum_{k=1}^{K} b_k \frac{d^2 s_k}{da^2}(a) \right|^2 da. \tag{9}$$

In the first term of $\mathcal{U}$ we recognize the logarithm of the binomial likelihood. The second term is a roughness penalty: it is proportional to the average squared second derivative of $\lambda(a)$. The constant of proportionality, η, scales the penalty relative to the likelihood.

One cannot consistently estimate η by maximum likelihood: one must use another criterion to choose its value. Although one can employ cross-validation, for example, as a criterion for selecting η, in the present case, the λ estimate is stable over a broad range of η and we can simply choose one value of η from within this range.

The results of fitting this model to the Swedish data are shown in Figure 14.3d. Again, we observe that infection risk is high among the youngest age groups and declines exponentially with age among adults, though a small bump in early adulthood is evident. Comparing these risk profiles with those in Figure 14.3c, we see that the extreme estimates of λ seen in the latter are mollified by the regularizations of the second method. However, we still see $\lambda \approx 10^3$ for infants when the reporting rate is low ($\rho = 0.01$) and immunity short lived ($\omega = 1$). In this regimen, the number of infections is so high (100 infections per observed case) that the model has had to resort to extreme measures. In particular, by making λ so large, individuals can be infected on average, up to twice per year. When immunity wanes more slowly, this operating mode is no longer available to the model. Nevertheless, even in this regimen, Figure 14.3c,d shows that the exponential decline in infection risk with age among adults follows from the observed pattern of age-specific incidence in Sweden under a wide range of assumptions about reporting and immunity.

We can translate these infection risk curves into more intuitive terms if we make the very reasonable assumption that infection risk comes from exposure

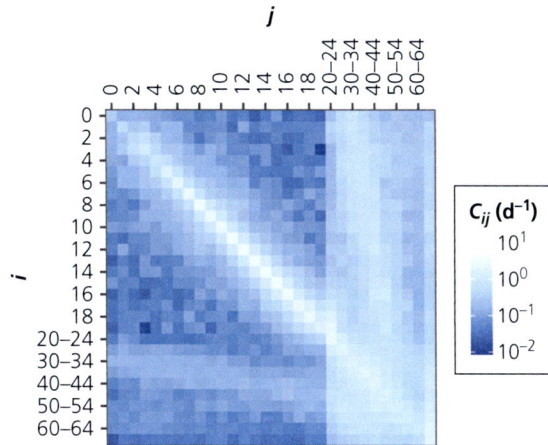

Figure 14.4 The per capita contact rate matrix derived from the POLYMOD study[7]. C_{ij} is the average number of unique contacts with members of age group j experienced by a member of group i per day. All contacts from all eight European countries included in the study were incorporated into these estimates.

to infected persons. Then, if we have an estimate of the rates at which people of age a_i make contact with people of age a_j, we can attribute which portion of the aggregated risk experienced by the former is due to contacts with the latter. Formally, we write:

$$\lambda(a_i) = \int q(a_i) C(a_i, a_j) \frac{I(a_j)}{N(a_j)} da_j, \qquad (10)$$

where $C(a_i, a_j)$ is the rate at which individuals of age a_j contact those of age a_i, I/N is age-specific infection prevalence, and q is the age-dependent probability that a person is infected, given contact. We can discretize equation (10) using a fixed set of age categories, which gives

$$\lambda_i = \sum_j q_i C_{ij} \frac{I_j}{N_j} \approx \frac{q_i}{\rho \tau} \sum_j C_{ij} \frac{H_j}{N_j}, \qquad (11)$$

where in the last expression we have approximated the prevalence in terms of the incidence rate, H_j/N_j, the reporting rate, ρ, and the infectious period τ.

We can estimate the age-specific contact rates, C_{ij}, using the results of the POLYMOD study[7]. This diary-based study reveals, among other things, the rates of contact between different age groups in western European societies. In particular, it provides an estimate of the per capita rate, m_{ij}, at

which a member of age class i has contacts with members of age class j. Because the total number of contacts between i and j must equal that between j and i, the per capita rate is better estimated by:

$$C_{ij} = \frac{1}{2}\left(m_{ij} + m_{ji} \frac{N_j}{N_i} \right), \qquad (12)$$

where N_i and N_j are the sizes of the populations in the respective age classes. Mossong et al.[7] demonstrated a more sophisticated approach to estimating C_{ij}, but the results we describe below are quite robust, so that the simplicity of equation (12) is sufficient for our purposes. Figure 14.4 depicts the contact rate matrix C_{ij}. Armed with estimates of λ_i (e.g. Figure 14.3d), the age-specific contact rate matrix C_{ij}, and incidence rate H_j/N_j (from the data), we use equation (11) to estimate q_i and plot the results in Figure 14.5.

14.2.2 Results from the catalytic model approaches

The results of the foregoing analysis, as shown in Figure 14.3, make it clear that the force of infection declines exponentially with age in adults. This conclusion is robust to the method used and to variation in the assumptions about the duration of protection

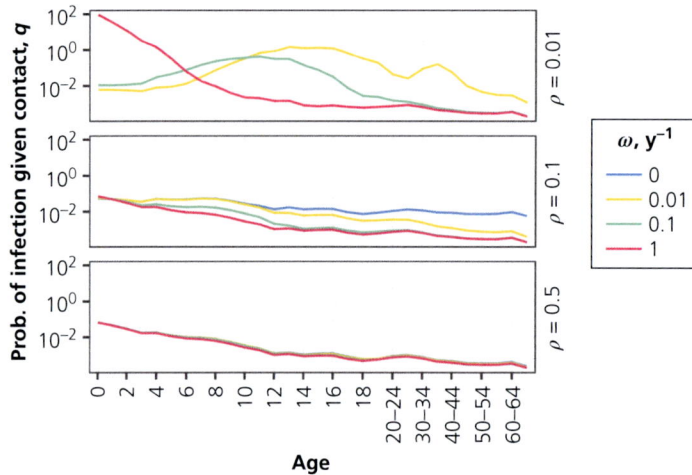

Figure 14.5 The age-specific risk of infection given exposure, q, as estimated using the smoothing approach. When $\rho = 0.01$ and immunity is long-lived $\omega \leq 0.01$), one must have $q > 1$, which is implausible. This is a consequence of the fact that, with these assumptions, there are too many infections and too few susceptibles. For immunity that wanes more quickly or higher reporting rates, q declines steadily from a maximum of around 0.1. Note that for all plausible values of ρ and ω shown here, adult infections are of minimal importance. This is because either q takes very small values in the adult age categories—so that adults are very unlikely to be infected—or because immunity lasts so long that few adults are susceptible.

and the degree of underreporting. Examining Figure 14.5, we observe, when the reporting rate, ρ, is very small (top panel), that the age-specific susceptibility, q, nonsensically exceeds 1 in the younger age groups, that is, it is necessary to assume many infections result from each contact. Inspection of Figure 14.5 makes it clear that adult transmission must be minor in comparison to that from younger age groups. This is because, for all plausible values of the immunity and reporting parameters ω and ρ, either immunity lasts so long that few adults are susceptible or the probability, q, of infection given exposure reaches such low values by adulthood, that susceptible adults are unlikely to become infected. In either case, adults can play little role in transmission.

Though the previous equations may appear to be quite technical, in effect, we are performing a bookkeeping exercise in which we attempt to estimate the probability that a contact results in an infection. Although more sophisticated analyses would doubtless refine these estimates, the qualitative conclusions of this crude analysis are very robust. These findings are consistent with epidemiological studies

of repeat pertussis infections. For instance, the study of a cohort of children in California by Platt et al.[8] identified 0.1 per cent repeat infections over a 4-year period. If the duration of immunity is assumed to be exponentially distributed, this implies a mean duration of protection against pertussis disease of c. 4000 years. Similarly, examining clinical pertussis reports from Niakhar, Senegal, Broutin et al.[9] noted that 2.5 per cent of those who experienced one episode of pertussis had a second episode before the age of 18, the average time between episodes being c. 7 years. Again under the exponential distribution assumption, this frequency of repeat infections translates to an average duration of immunity of more than 700 years. We note that the findings of Broutin et al. were misinterpreted by Wendelboe and colleagues in their highly cited study of the duration of pertussis immunity[10]. Ignoring the fact that the vast majority of individuals experienced no second clinical infection, Wendelboe et al. cited the mean time elapsed between each episode in the Broutin et al. study as evidence for duration of immunity of c. 7 years.

14.3 Inferring vaccine traits by fitting dynamic models

The foregoing results shed light on the traits of infection-derived immunity, making a strong case for long-lasting protection (Figure 14.5). These findings reduce the uncertainty with respect to one key aspect of pertussis epidemiology, but shed no light on perhaps the most hotly debated issue: the effectiveness and duration of vaccine-derived protection, especially that elicited by aP vaccines[11,12]. While one might imagine extending the approaches described above to accommodate the reintroduction of vaccination, the evident disruption of the age profile of incidence (Figure 14.2) indicates the need for a more dynamic approach. The fact that the Swedish data are aggregated annually limits their usefulness as a window on the dynamics triggered by the resumption of mass vaccination.

Recently, we have focused attention on vaccine-derived immunity by fitting transmission models to temporally well-resolved incidence data from Italy. In particular, the paper by Magpantay et al.[13] applied a likelihood-based inference approach to monthly pertussis notifications aggregated by region, to explore whether the dynamics of pertussis are better described by a model in which aP vaccine-derived immunity is perfect in degree, but transient (the *waning model*) or one in which protection is permanent but imperfect in degree (the *leaky model*). It was found that both models were able to explain the data equally well despite strongly contrasting conclusions regarding the overall impact of aP vaccines. Because the two models make very different predictions with respect to the distribution of cases among age groups, Magpantay et al. argued that age-specific data would allow better discrimination between these very distinct modes of vaccine failure.

The project of fitting mechanistic models to data highly resolved in both time and age represents a substantial technical challenge, primarily due to the high dimension of the data and the complexity of the underlying dynamical system. Additionally, because of the explosion in the number of parameters in such models, efficient extraction of information necessitates data regarding age-specific patterns of mixing. However, much is to be gained from the increased information content of age-resolved data. We next elaborate on a study by Domenech de Cellès et al.[14] using age-specific pertussis incidence data from Massachusetts, United States over the years 1990–2005.

14.3.1 The Massachusetts data

The age-specific incidence records from Massachusetts are of unusually high quality because of a long-standing active surveillance programme (Figure 14.6 and references[15,16]). These data display the trends typical of pertussis re-emergence, namely increases in both size and frequency of epidemics between 1990 and 2005, with a fourfold rise in overall incidence (relative increase 9.7 per cent/year). Adolescents aged 10–20 years accounted for more than 50 per cent of these cases (Figure 14.6b,c). The trends in adults were even more pronounced: cases in those aged 20 years or older increased by more than 16 per cent/year and more than tenfold over the 15-year interval. By contrast, trends were less obvious in children less than 10 years old and no systematic increase was observed in infants (<1 year old), despite the consistently high incidence rate in that age group (average 57×10^{-5}/year). As previously noted by Rohani and Drake[17], these trends continue the pattern of resurgence that began in the mid 1970s: no direct effects of the 1996 switch from wP to aP vaccine are immediately evident. Conclusions regarding age-specific disease burden, however, must take into account the differential sensitivities of ascertainment methods used in different age groups. Specifically, the serological enzyme-linked immunosorbent assay used to identify infections in adolescents and adults is considerably more sensitive, and less specific, than the culture-based ascertainment methods used in children less than 11 years old[15]. We limited our attention to pre-2006 data to avoid having to deal with (1) the introduction of the Tdap (tetanus, diphtheria, and acellular pertussis) booster for teenagers in 2006 and (2) the switch to the polymerase chain reaction technique as the method of infection ascertainment among adolescents and adults, with concomitant increased sensitivity and diminished specificity[18].

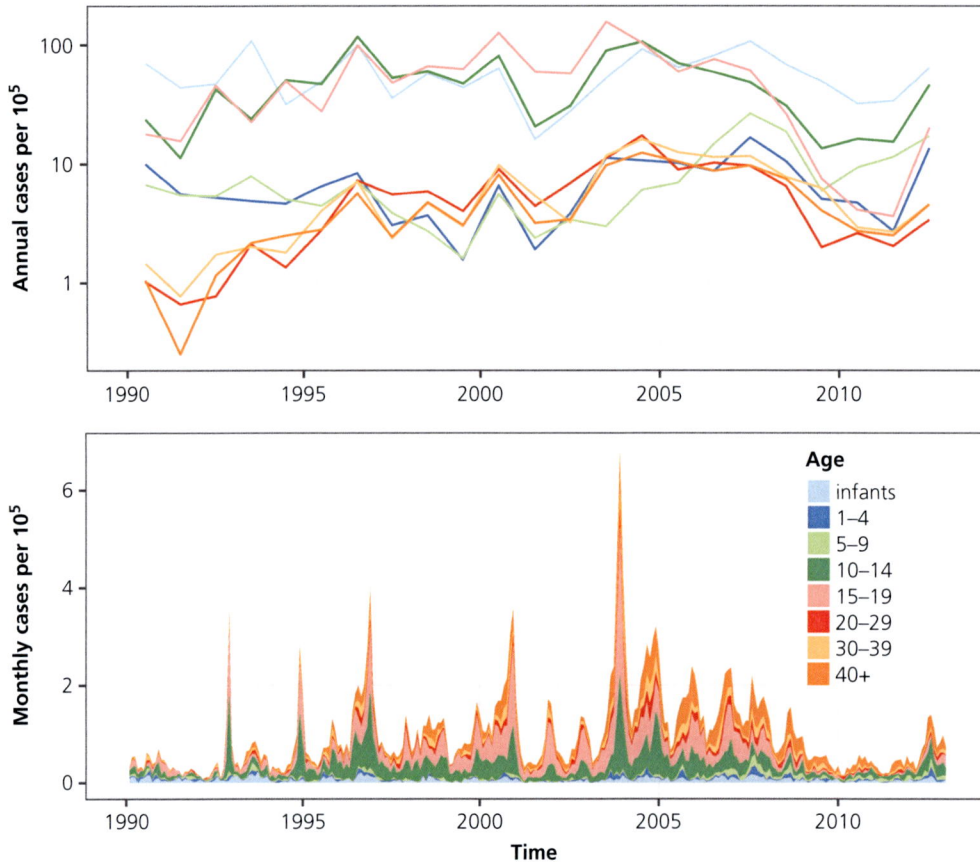

Figure 14.6 Age-specific annual (top panel) and monthly (bottom panel) incidence in Massachusetts from 1990 to 2006. Adapted from Domenech de Cellès et al.[14].

To set the model-based inference in context, we examined the relative patterns of seasonality by age (Figure 14.7)[19]. The data display a peak in pertussis incidence in late summer among infants and their young siblings (1–4 years old), while two peaks are observed in the 5–9 years age group, one in July and the other in November. There is a single seasonal peak in late fall among 10–19-year-olds. Among adults (≥20 years old), the second half of the year is marked by higher incidence, but the pattern is less distinct. These patterns of seasonality are suggestive of core transmission groups among young children (<5 years old) and among teenagers.

14.3.2 Age-structured transmission model

We implemented an age-structured, compartmental model of pertussis transmission (Figure 14.8)[14], built

on previously described models[3,20,21]. The model is an extension of the classic SEIR model that allows for post-vaccination infections in previously vaccinated individuals. The population of susceptibles is divided into those naïve to exposure (S) and those whose immune system has been previously primed by vaccination (S_V). Exposed and infected individuals are similarly divided into those who experience a naïve infection (E and I) or a post-vaccination infection (E_V and I_V). Upon recovery from either type of infection, individuals move to the recovered class, R.

Individuals are categorized by 5-year age groups from 5 years old to 75 years old, with the 0–5-year-olds further divided into 1–5-year-olds and infants aged 0–4 months and 4–12 months. The 0–4 months age group is included to represent the fact that infants are fully vulnerable to infection before

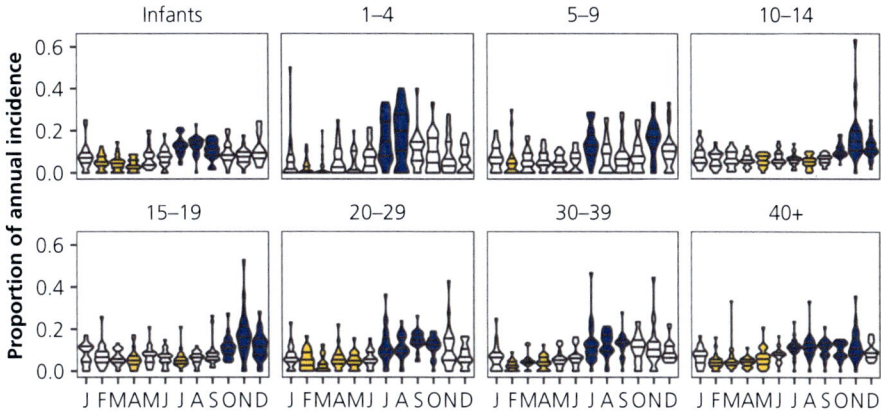

Figure 14.7 Age-stratified seasonality in data from Massachusetts, United States. The violin plots show the distribution, across years, of the fraction of incidence in each month. Blue indicates that the given month has a significantly higher proportion of the total incidence than average; yellow indicates a significantly lower proportion. Statistical significance was assessed using a non-parametric, blocked permutation test. Note that the seasonality of infants and preschool children is strikingly different from that of teenagers. The seasonality in the 5–9-year-old and adult groups resembles a blend of the two. After Domenech de Cellès et al.[14]

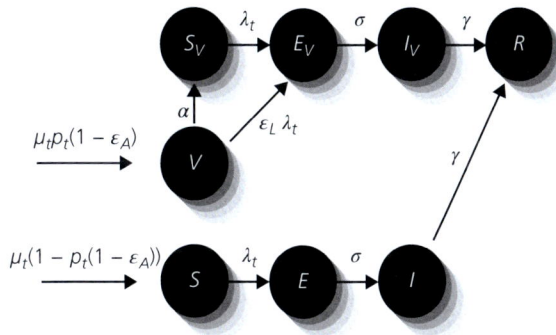

Figure 14.8 Age structured transmission model used in likelihood-based inference. Here, μ_t represents the time varying per capita birth rate, p_t is the proportion of infants that were immunized, $1/\gamma$ and $1/\sigma$ determine the mean latent and infectious periods, respectively. λ_t is the time varying hazard of infection: $\lambda_t = \beta_t (I + \theta I_V + \iota)$, where β_t is the seasonally driven transmission rate, θ quantifies the relative transmissibility of infections in individuals who have previously been vaccinated, and ι represents transmission due to imports. The vaccine failure parameters are ε_W, the probability that vaccine-derived immunity wanes within an individual's lifetime ($\varepsilon_W = \alpha / (\alpha + \mu)$); ε_A, the primary vaccine failure; and ε_L, the vaccine leakiness, defined as the probability of transmission to a vaccinated individual upon exposure.

receiving the second dose of DTP (diphtheria, tetanus, and pertussis vaccine) at age 4 months[4]. Overall, the model tracks 17 age groups, labelled $i = 1, \ldots, 17$. Ageing occurs continuously, at rates $\delta_i = 1 / \Delta a_i$, where Δa_i is the age span in age group i. To model the effect of the primary vaccination course, a fraction v_1 of susceptible individuals (S and S_V) is moved to the vaccinated class on ageing from 0–4 months to 4–12 months. Similarly, the effect of the preschool booster dose is modelled by moving a

fraction v_2 of susceptible individuals ageing from 1–5 years to 5–10 years to the vaccinated class. Because the paediatric booster dose (at age 18–24 months) is administered shortly after the primary course, we ignored the effect of this dose.

14.3.3 Vaccine failure hypotheses

To account for possible differences between infection- and vaccine-derived immunity, vaccinated individuals are explicitly modelled (V). We estimated the probability of primary vaccine failure, also referred to as failure in 'take'[22]. As previously described[13,22–26], we examined additional mechanisms of vaccine failure, including imperfect immunity (failure in 'degree' or 'leakiness') and temporary protection (failure in duration). Specifically, we weighed the evidence for four alternative hypotheses of vaccinal immunity, following vaccine take:

1. Vaccine protection is perfect in both duration and degree ('no loss model').
2. Vaccine-derived immunity is perfect in degree, but transient ('waning model').
3. Protection is permanent but imperfect in degree ('leaky model').
4. Vaccinal immunity is imperfect both in degree and duration ('waning plus leaky model').

A schematic of the model structure is presented in Figure 14.8.

Parallel data: immunization plus demography

Immunization levels of children entering kindergarten were provided by the Massachusetts Department of Public Health from school year 1975/1976 for children having received more than four doses of vaccine, and from school year 1995/1996 onwards for children having received five doses. In the absence of immunization data before 1970, we assumed that routine vaccination had started in 1940 and had ramped up until 1955. Indeed, although mass production of the wP vaccine began in 1950 in Massachusetts[15], it was already distributed across the United States from 1940[27]. This assumption is also consistent with historical incidence data in Massachusetts, which demonstrate a decline in pertussis incidence in the 1940s, with further steep decreases after 1950[17,28,29]. We also assumed that the preschool booster dose started being administered in 1967[30].

The model incorporated detailed demographic data (age-specific mid-year population estimates[31] and yearly number of births[32]) during 1990–2005. These data were used as covariates, so that the simulated age-specific population sizes approximately matched the observed demographic trends during that period.

The final ingredient was accounting for age-specific patterns of contacts. Here, we report results obtained by using the data obtained from the POLYMOD study in Great Britain[7], but in Domenech de Cellès et al.[14] we demonstrated the consistency of our conclusions to using a contact matrix derived from detailed household census surveys from Massachusetts[33].

14.3.4 Statistical inference

We assumed an observation model—relating model trajectories to the observed incidence data—that was age specific to account for known variations of reporting fidelity with age[34]. In particular, it accounted for the specific use of serological testing to detect cases in adolescents and adults[15]. Models were started 70 years before the start of vaccination, assuming that the system was stationary in the pre-vaccine era[4,26]. Complete model details and equations are provided in Domenech de Cellès et al.[14].

The model was implemented as a continuous-time Markov process, approximated via a multinomial tau-leap algorithm with a fixed time step of 0.01 years[35]. The model was fitted using the iterated filtering algorithm[36] as implemented in the 'pomp' package[37,38]. Approximate 95 per cent confidence intervals (CIs) were calculated using a parametric bootstrap.

14.3.5 Results of dynamic model fitting

The results concerning vaccine traits were clear cut (Table 14.1): the waning model received substantially higher support than other models (ΔAIC > 140). The best-fitting model predicts (1) that the probability of primary vaccine failure is 4 per cent (95 per cent CI 1–8 per cent), and (2) that vaccine protection wanes slowly on average, though there is substantial variability among individuals. Specifically, the model predicts a 10 per cent risk (95 per cent CI 3–19 per cent) of protection waning to zero within 10 years of completing routine vaccination and a 55 per cent chance that protection remains perfect for life. Although the waning plus leaky model allows for a mixture of all three vaccine failure modes, its leakiness parameter is estimated at 0 (95 per cent CI 0–3 per cent) and its waning rate is identical to that

Table 14.1 Model comparison. The best Akaike information criterion (AIC) value is indicated in boldface

Quantity	No-loss model	Leaky model	Waning model	Waning plus Leaky model
logL	−3726.9 (se: 0.4)	−3664.9 (se: 0.5)	−3594.5 (se: 0.6)	−3598.5[a] (se: 1.8)
AIC	7474	7356	**7215**	7217[a]
Δ AIC	259	141	0	2
R_p	2.4 (1.8, 2.7)	1.6 (1.3, 2.2)	1.8 (1.5, 2.0)	1.8 (1.6, 2.0)
R_0	13.6 (7.5, 23.0)	12.6 (9.0, 19.4)	10.1 (6.5, 17.2)	9.1 (5.3, 16.2)
Vaccine impact (ϕ)	0.85 (0.70, 0.95)	0.90 (0.81, 0.95)	0.85 (0.75, 0.93)	0.83 (0.70, 0.92)

[a] Because the two models are nested, the likelihood of the full model should be higher or equal to that of the waning model. The small difference indicates that the maximum likelihood estimates (95 per cent confidence interval) of the leakiness for the full model is 0. Consequently, the AIC was calculated with the likelihood of the waning model.

of the waning model. Thus, the additional complexity of this model is not supported by the data (ΔAIC = −2).

Our fitted model also informed the ongoing debate[39–41] regarding post-vaccine infections, defined as infections in individuals in whom the vaccine took, but whose immunity subsequently waned. These infections are as transmissible, but less visible, than naïve infections (relative transmissibility 0.99 (95 per cent CI 0.40–1.00), relative observability 0.39 (95 per cent CI 0.19–1.00)). This finding is consistent with evidence from animal challenge studies[42,43], though wide confidence intervals for our estimates preclude definitive conclusions on this question. Also consistent with previous evidence[15], we estimated high detection rates of pertussis in adolescents and in adults, with 24 per cent (95 per cent CI 10–66 per cent) of post-vaccination infections reported.

To quantify vaccine effectiveness in reducing transmission, we computed ϕ, the recently proposed quantity to summarize *vaccine impact*[13,25]. This population-wide measure comprises all modes of vaccine failure:

$$\phi = \frac{1}{p}(1 - R_p/R_0). \qquad (13)$$

Here, p is the vaccine uptake, R_0 is the basic reproduction ratio in the absence of vaccination[44,45] and R_p is the reproduction ratio in the presence of vaccination[13,25]. The estimated vaccine impact was 0.85 (95 per cent CI 0.75–0.93) for the waning model, and similarly high for the leaky model (0.90 (0.81–0.95), leakiness 0.06 (0.02–0.14), primary vaccine failure rate 0.06 (0.02–0.14)) and the no-loss model (0.85 (0.70–0.95)). The vaccine impact modifies the theoretical vaccination effort required for eradication according to $(1 - 1/R_0)/\phi$. One of our key results, therefore, is that despite the effectiveness of the vaccine, eradication via routine immunization alone is not possible given the relatively large estimated value of R_0 (10.1 (95 per cent CI 6.5, 17.2)). The robustness of estimates to variation in model structure strengthens the evidence for the effectiveness of vaccination in reducing pertussis transmission.

To place our results within the context of recent pertussis epidemiology, we note that Klein et al., in a highly cited publication[11], reported a 42 per cent annual increase in the odds of acquiring pertussis after the fifth booster dose, which has been interpreted as evidence for rapid loss of DTaP immunity, in apparent contradiction of our results. To investigate this, we simulated our best-fitting model from 2006 to 2015 and estimated the annual change in the odds of acquiring pertussis with age. As shown in Figure 14.9, our model, with its assumption of a slow-waning, high-impact DTaP vaccine, predicts an odds increase of 32 per cent per year, comparable to the Klein et al. study and a variety of others[11,12,46,47]. Thus, not only are our results, perhaps surprisingly, quite consistent with these case–control and cohort studies, but they also show that time series data aggregated at the population scale can be more informative about the quantities of interest than the data from these smaller scale studies. Our finding calls into question the standard but naïve

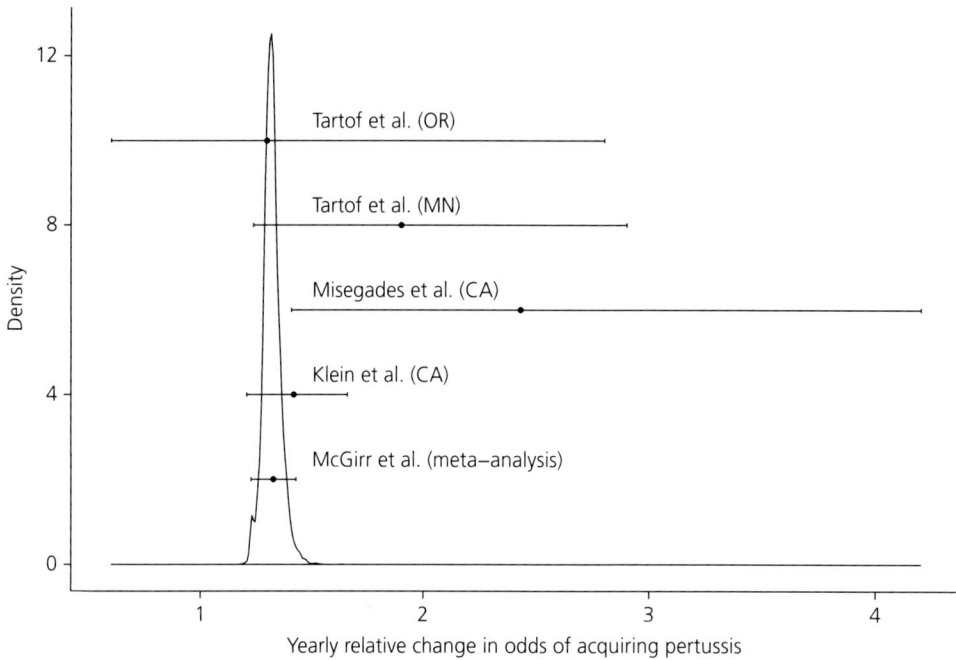

Figure 14.9 Comparison of model predictions with empirical quantification of DTaP failure by estimating relative changes (over age) in the odds of acquiring pertussis. We simulated the waning model from 2006 to 2015 and used log-linear regression to calculate the yearly relative change (over age) in the odds of acquiring pertussis in children aged 5–10 years, that is, within 5 years of receiving the fifth vaccine dose. The distribution is based on 10^4 simulations, accounting for parametric uncertainty by sampling from the bootstrap distribution. Also presented are estimates from three empirical studies in the United States (Klein et al.[11], Misegades et al.[46], and Tartof et al.[47]) and from a meta-analysis (McGirr et al.[12]).

interpretation of odds ratio/age slopes as quantification of the speed of loss of vaccine-derived immunity: it demonstrates that the slopes observed in case–control and cohort studies are consistent with immunity as long-lived as we estimate from the time series. Further, our fitted model represents an alternative explanation for empirical observations: increased infection risk in schoolchildren need not be due to rapid decreases in vaccine protection and can arise from increased contact rates following school entry.

14.4 Conclusion

There remains much discussion and uncertainty regarding pertussis immunity, both as elicited by natural infection and by vaccines. In this chapter, we have attempted to bring to bear information gleaned from longitudinal age-stratified incidence reports. Overall, we submit that the following robust conclusions are supported by the foregoing:

1. Our analysis of data from populations during a vaccine-free period (Sweden) reveals that susceptibility to infection in adults must be low, whether this is due to long-lasting immunity following infection or for some other reason.
2. There is strong evidence supporting the proposition that the current generation of vaccines provide effective—though imperfect—protection against transmissible infection.
3. However, given the large estimated basic reproductive number, R_0, routine infant immunization with aP vaccines, by itself, is insufficient for elimination or eradication of pertussis.
4. At least in the locales considered, the core transmission group, responsible for sustaining chains of transmission, comprises school-age children and teenagers. Adults appear to occupy a more peripheral position with respect to transmission. This conclusion clearly affects the design of age-specific booster schedules.
5. Data on the frequency and nature of human–human contacts, especially data stratified by age,

are of particular utility inasmuch as they modify the perspective on transmission and allow greater resolution of issues of transmission and immunity.

6. Epidemiological approaches can play a pivotal role in quantifying the propensity of phenomena observed in smaller-scale clinical- or cohort-based studies.

7. Model-based approaches to inference in pertussis have arrived at conflicting conclusions, which are due in part to real differences among focal regions and time periods, but also to differences in models and methodology. Therefore, we propose that much is to be gained from a systematic comparison of alternative model formulations using rigorous and objective criteria such as those advocated here.

References

1. de Melker HE, Versteegh FGA, Schellekens JFP, et al. The incidence of *Bordetella pertussis* infections estimated in the population from a combination of serological surveys. *J Infect* 2006;**53**:106–13.

2. Carlsson RM, Trollfors B. Control of pertussis–lessons learnt from a 10-year surveillance programme in Sweden. *Vaccine* 2009;**27**:5709–18.

3. van Boven M, De Melker HE, Schellekens J, et al. Waning immunity and subclinical infection in an epidemic model: implications for pertussis in The Netherlands. *Math Biosci* 2000;**164**:161–82.

4. Rohani P, Zhong X, King AA Contact network structure explains the changing epidemiology of pertussis. *Science* 2010;**330**:982–5.

5. de Boor C. *A Practical Guide to Splines*. New York: Springer; 2001.

6. Ramsay JO, Silverman BW. *Functional Data Analysis*, 2nd ed. New York: Springer-Verlag; 2005.

7. Mossong J, Hens N, Jit M, et al. Social contacts and mixing patterns relevant to the spread of infectious diseases. *PLoS Med* 2008;**5**:e74.

8. Platt L, Thun M, Harriman K, et al. A population-based study of recurrent symptomatic *Bordetella pertussis* infections in children in California, 2010–2015. *Clin Infect Dis* 2017;**65**:2099–104.

9. Broutin H, Elguero E, Simondon F, et al. Spatial dynamics of pertussis in a small region of Senegal. *Proc R Soc Lond B* 2004;**271**:2091–8.

10. Wendelboe A, Van Rie A, Salmaso S, et al. Duration of immunity against pertussis after natural infection or vaccination. *Pediatr Infect Dis J* 2005;**24**:S58.

11. Klein NP, Bartlett J, Rowhani-Rahbar A, et al. Waning protection after fifth dose of acellular pertussis vaccine in children. *N Engl J Med* 2012;**367**:1012–9.

12. McGirr A, Fisman DN. Duration of pertussis immunity after DTaP immunization: a metaanalysis. *Pediatrics* 2015;**135**:331–43.

13. Magpantay FMG, Domenech de Cellès M, Rohani P, et al. Pertussis immunity and epidemiology: mode and duration of vaccine-induced immunity. *Parasitol* 2016;**143**:835–49.

14. Domenech de Cellès M, Magpantay FMG, King AA, et al. The impact of past vaccination coverage and immunity on pertussis resurgence. *Sci Transl Med* 2018;**10**:eaaj1748.

15. Marchant CD, Loughlin AM, Lett SM, et al. Pertussis in Massachusetts, 1981–1991: incidence, serologic diagnosis, and vaccine effectiveness. *J Infect Dis* 1994;**169**:1297–305.

16. Yih WK, Lett SM, des Vignes FN, et al. The increasing incidence of pertussis in Massachusetts adolescents and adults, 1989–1998. *J Infect Dis* 2000;**182**:1409–16.

17. Rohani P, Drake JM The decline and resurgence of pertussis in the US. *Epidemics* 2011;**3**:183–8.

18. Njamkepo E, Bonacorsi S, Debruyne M, et al. Significant finding of *Bordetella holmesii* DNA in nasopharyngeal samples from French patients with suspected pertussis. *J Clin Microbiol* 2011;**49**:4347–8.

19. Tanaka M, Vitek CR, Pascual FB, et al. Trends in pertussis among infants in the United States, 1980–1999. *JAMA* 2003;**290**:2968–75.

20. Wearing HJ, Rohani P. Estimating the duration of pertussis immunity using epidemiological signatures. *PLoS Path* 2009;**5**:e1000647.

21. Blackwood JC, Streicker DG, Altizer S, et al. Resolving the roles of immunity, pathogenesis, and immigration for rabies persistence in vampire bats. *Proc Natl Acad Sci U S A* 2013;**110**:20837–42.

22. Halloran ME, Longini I, Struchiner CJ. *Design and Analysis of Vaccine Studies*. New York: Springer; 2010.

23. McLean AR, Anderson RM. Measles in developing countries. Part I. Epidemiological parameters and patterns. *Epidemiol Infect* 1988;**100**:111–33.

24. Águas R, Gonçalves G, Gomes M. Pertussis: increasing disease as a consequence of reducing transmission. *Lancet Infect Dis* 2005;**6**:112–7.

25. Magpantay FMG, Riolo MA, de Cellès MD, et al. Epidemiological consequences of imperfect vaccines for immunizing infections. *SIAM J Appl Math* 2014;**74**:1810–30.

26. Riolo MA, King AA, Rohani P. Can vaccine legacy explain the British pertussis resurgence? *Vaccine* 2013;**31**:5903–8.

27. Shapiro-Shapin CG. Pearl Kendrick, Grace Eldering, and the pertussis vaccine. *Emerg Infect Dis* 2010;**16**:1273–8.

28. Lavine JS, King AA, Bjornstad ON. Natural immune boosting in pertussis dynamics and the potential for long-term vaccine failure. *Proc Natl Acad Sci U S A* 2011;**108**:7259–64.

29. Magpantay FMG, Rohani P. Dynamics of pertussis transmission in the United States. *Am J Epidemiol* 2015;**181**:921–31.

30. Karzon DT. Immunization practice in the United States and Great Britain: a comparative study. *Postgrad Med J* 1969;**45**:147–60.

31. US Census Bureau. Population Estimates. 2014. https://www.census.gov/topics/population.html.

32. Martinez-Bakker M, Bakker KM, King AA, et al. Human birth seasonality: latitudinal gradient and interplay with childhood disease dynamics. *Proc R Soc Lond B* 2014;**281**:1573–8.

33. Fumanelli L, Ajelli M, Manfredi P, et al. Inferring the structure of social contacts from demographic data in the analysis of infectious diseases spread. *PLoS Comput Biol* 2012;**8**:e1002673.

34. Sutter R, Cochi S. Pertussis hospitalizations and mortality in the United States, 1985–1988. *JAMA* 1992;**267**:386–91.

35. He D, Ionides EL, King AA. Plug-and-play inference for disease dynamics: measles in large and small populations as a case study. *J R Soc Interface* 2010;**14**:20170115.

36. Ionides EL, Nguyen D, Atchadé Y, et al. Inference for dynamic and latent variable models via iterated, perturbed Bayes maps. *Proc Natl Acad Sci U S A* 2015;**112**:719–24.

37. King AA, Domenech de Cellès M, Magpantay FMG, et al. Avoidable errors in the modelling of outbreaks of emerging pathogens, with special reference to Ebola. *Proc R Soc Lond B* 2015;**282**:20150347.

38. King AA, Nguyen D, Ionides EL. Statistical inference for partially observed Markov processes via the R package pomp. *J Stat Softw* 2016;**69**:1–43.

39. Tan T, Trindade E, Skowronski D. Epidemiology of pertussis. *Pediatr Infect Dis J* 2005;24:S10–8.

40. Broutin H, Viboud C, Grenfell BT, et al. Impact of vaccination and birth rate on the epidemiology of pertussis: a comparative study in 64 countries. *Proc R Soc Lond B* 2010;**277**:3239–45.

41. Gill C, Rohani P, Thea DM. The relationship between mucosal immunity, nasopharyngeal carriage, asymptomatic transmission and the resurgence of *Bordetella pertussis*. *F1000Research* 2017;**6**:1568.

42. Smallridge WE, Rolin OY, Jacobs NT, et al. Different effects of whole-cell and acellular vaccines on *Bordetella* transmission. *J Infect Dis* 2014;**209**:1981–8.

43. Jason M Warfel, Lindsey I Zimmerman, and Tod J Merkel. Acellular pertussis vaccines protect against disease but fail to prevent infection and transmission in a nonhuman primate model. Proc Natl Acad Sci USA, 2014 vol. 111 pp. 787–792.

44. Anderson RM, May RM. *Infectious Diseases of Humans*. Oxford: Oxford University Press; 1991.

45. Keeling MJ, Rohani P. *Modelling Infectious Diseases in Humans and Animals*. Princeton, NJ: Princeton University Press; 2008.

46. Misegades LK, Winter K, Harriman K, et al. Association of childhood pertussis with receipt of 5 doses of pertussis vaccine by time since last vaccine dose, California, 2010. *JAMA* 2012;**308**:2126–32.

47. Tartof SY, Lewis M, Kenyon C, et al. Waning immunity to pertussis following 5 doses of DTaP. *Pediatrics* 2013;**131**:e1047–52.

CHAPTER 15

Public health consequences

Tami H. Skoff, Colin S. Brown, and Gayatri Amirthalingam

Abstract

Although pertussis can cause morbidity and mortality across all age groups, infants are at greatest risk for severe disease, especially during the first months of life. Despite the implementation of effective vaccination programmes, pertussis remains a significant global health problem and protecting those at highest risk is a priority. Numerous public health strategies, including cocooning, maternal vaccination during pregnancy, targeted vaccination of healthcare personnel, and post-exposure antibiotic prophylaxis, have been used to control the burden of pertussis with varying degrees of success. While maternal immunization during pregnancy has been demonstrated to be highly effective at preventing disease among infants during the first months of life, no single strategy alone is sufficient to control pertussis across age groups. In the setting of a resurgence in disease, a combination of approaches is needed to minimize the burden of disease, especially among those at highest risk for severe morbidity and mortality.

15.1 Introduction

Historically, a collection of public health measures have been used to control the spread of pertussis. While these control measures initially focused largely on removing infectious people from the susceptible population through isolation of cases, the widespread use of this approach eventually waned with the introduction of antimicrobials and mass vaccination. Antibiotics with anti-pertussis action were discovered during the 1930s and 1940s as an effective tool for eradicating *Bordetella pertussis* from patients with the disease, though the recommended timing of administration has limited the utility of this approach as an effective public health intervention[1,2]. In the 1930s, the first pertussis vaccine was developed and by the mid 1940s, a vaccine was routinely administered in combination with diphtheria and tetanus antigens, changing the landscape of pertussis prevention and control[3,4].

Despite the successful implementation of pertussis vaccination and a significant reduction in the global burden of pertussis, protection from natural infection or vaccination is not lifelong and cyclical peaks in disease continue to be reported from countries with well-established vaccination programmes. Although pertussis can cause disease across all age groups, infants at are highest risk for severe pertussis-related morbidity and mortality, especially during the first months of life. Pertussis childhood immunizations generally commence between 6 and 8 weeks of age, but may be administered as late as 3 months in some parts of the world, leaving a significant period of infant vulnerability. Decreasing the burden of pertussis and interrupting disease

Skoff, T. H., Brown, C. S., and Amirthalingam, G., *Public health consequences*. In: *Pertussis: epidemiology, immunology, and evolution*. Edited by Pejman Rohani and Samuel V. Scarpino: Oxford University Press (2019). DOI: 10.1093/oso/9780198811879.003.0015 transmission is important across all age groups; however, protecting infants in the period before active immunization remains a critical component of pertussis prevention and control, especially in the setting of ongoing disease transmission.

In this chapter, the authors will provide a comprehensive overview of the key public health strategies that have been used to control the burden of pertussis. Emphasis will be placed on strategies specifically designed to protect young infants, including cocooning and maternal vaccination during pregnancy, as well as strategies that have been implemented to control and limit disease transmission in healthcare and community settings, including targeted vaccination of healthcare personnel (HCP) and post-exposure antibiotic prophylaxis. The authors will discuss the rationale behind each strategy, the strategy's intended target group(s), and highlight key strengths and limitations that impact the ability of each strategy to effectively control disease.

15.2 Cocooning

Because pertussis childhood vaccines do not provide lifelong protection, the development of acellular pertussis (aP) booster vaccines targeting adolescents and adults marked an important advancement in the prevention of pertussis. Less reactogenic, reduced-dose formulations offered a way to control disease outside of childhood, with the potential added benefit of indirectly protecting young infants. Numerous studies have investigated the dynamics of pertussis transmission, highlighting the critical role that mothers, fathers, and other family members play in infecting infants with disease[5]. Targeted vaccination of close contacts of infants, a strategy known as cocooning, was devised as a method to protect young infants by interrupting the transmission of pertussis from infant caregivers and household contacts. The strategy offered the added benefit of directly reducing pertussis morbidity in vaccine recipients, an important advantage for new parents.

The Global Pertussis Initiative first discussed cocooning as a potential strategy for decreasing infant pertussis morbidity and mortality in 2001, and by 2005, recommendations were made for introducing cocooning in countries where the strategy was economically and logistically feasible[6]. Studies modelled

the impact of cocooning, demonstrating its potential to significantly reduce the burden of infant pertussis[7–10]. Since the mid 2000s, variations of the cocooning strategy have been adopted by a number of high-income countries, including Canada, the United States, Australia, and throughout Europe (e.g. France, Germany, Austria, Belgium, and Switzerland)[11,12]. Several Latin American countries (e.g. Chile, Panama, Costa Rica, and Argentina) have also implemented the strategy, many in response to notable increases in infant pertussis cases[12,13]. While cocooning has also been recommended in parts of Asia, data describing implementation of the strategy in this region are currently limited.

Although simple in concept, the cocooning strategy has encountered substantial logistical hurdles making widespread implementation difficult. Achieving high vaccine uptake in adults can be challenging, especially in the absence of routine adult immunization platforms. Tdap (tetanus–diphtheria–acellular pertussis combination vaccine) has been recommended by the United States Advisory Committee on Immunization Practices since 2006 for routine and targeted vaccination of adults; however, overall coverage barely exceeded 20 per cent of the adult population in 2015[14]. Even in countries where pertussis vaccines are publicly funded for adults, high vaccine coverage remains difficult to achieve outside of childhood; in Canada, only 6.7 per cent of adults reported pertussis booster vaccination in 2012[15].

While some cocooning programmes have been successful at achieving high vaccine uptake among new mothers[16–20], suboptimal coverage among other cocoon members is not uncommon, impeding the ability to achieve complete cocooning around an infant and potentially jeopardizing the effectiveness of the strategy[16–19,21,22]. Two years following the implementation of a national cocooning programme in Switzerland, a cross-sectional survey revealed that cocooning was complete in only 7 per cent of surveyed families, and in most families, less than 50 per cent of infant contacts had been vaccinated[21]. Even in hospital-based programmes where targeted vaccination of defined populations might be expected to increase vaccine coverage, reaching family members can be a significant challenge, especially in large families or when individuals are referred to other locations for vaccination. In a study by Healy et al., less than 30 per cent of families were fully cocooned when vaccine was provided free of charge during a hospital-based cocooning programme in the United States[16]. Many studies have explored determinants of pertussis vaccine acceptance among parents and other key infant contacts, highlighting the importance of an individual's perception of severe disease risk and benefit of vaccination, low perceived barriers to vaccination, recommendations to vaccinate from a healthcare provider, and educational efforts that promote adult vaccination[17,20,21,23–26]. Interestingly, but perhaps not surprisingly, some cocooning programmes have demonstrated higher success when implemented in the setting of increased pertussis activity. A hospital-based clinic offering free vaccination to cocoon members during the 2010 California, United States, epidemic achieved an 84 per cent uptake rate of vaccine among family members, with 76 per cent of households reporting a complete cocoon[27]. Similarly, a government-funded vaccination programme targeting new parents in Victoria, Australia, achieved greater than 70 per cent vaccine coverage among both mothers and fathers following a 2009 epidemic[28].

Published data on the effectiveness of cocooning remain limited, and the ability of the strategy to prevent pertussis transmission to young infants has not been well established. A case–control study conducted in Australia found a 50 per cent reduction in infant disease when both mother and father were vaccinated[29]. And in Chile, cocooning has been estimated to prevent 84 per cent of pertussis deaths among infants less than 6 months of age[12]. However, there are also data that suggest no benefit of cocooning, especially when vaccine is administered only to the mother during the postpartum period[30–32]. This reported lack of effectiveness may not be entirely unexpected given key limitations of the strategy. Because antibody production does not peak until 2 weeks post vaccination, delaying vaccination of mothers and other key caregivers until the postpartum period leaves newborns susceptible to infection for a short, but critical, period of time. Additionally, studies consistently report an unknown source of pertussis infection for more than half of infant pertussis cases, suggesting that targeted vaccination of known contacts may not be sufficient[5]. Lastly, recent

data from a non-human primate model have revealed that vaccination with aP vaccine, unlike whole-cell pertussis (wP) vaccination, may not prevent *B. pertussis* colonization and further transmission of the organism[33]. Therefore, aP vaccine-boosted individuals could remain a significant source of pertussis transmission to young infants.

Although some evaluations have suggested an economic value to cocooning, others have failed to support the cost-effectiveness of the strategy[34–37]. Numerous attempts have been made to calculate the number of adults needed to vaccinate to prevent morbidity and pertussis-related complications in infants. In the setting of low pertussis incidence where infant hospitalizations and deaths are rare, it is estimated that extremely large numbers are needed to avert a single infant pertussis case, hospitalization, or death[7,8,35,38]. Researchers have speculated that the strategy could be more cost-effective in the context of a pertussis epidemic; however, many of the key parameters in the existing economic models may be outdated given rapidly changing pertussis epidemiology and new information regarding aP vaccines. Questions about the cost-effectiveness of the strategy currently remain unanswered, generating further uncertainty about the usefulness of cocooning.

In a 2010 report, the Strategic Advisory Group of Experts on Immunization (SAGE) of the World Health Organization (WHO) concluded that there was insufficient evidence for countries to include cocooning as part of a national immunization programme[39]. Cocooning alone is unlikely to provide adequate protection to infants in the first months of life and should therefore not be relied on as the primary prevention strategy to control pertussis in this vulnerable age group. However, despite key limitations, cocooning may be used to augment newer strategies, such as maternal immunization during pregnancy, and may still remain important in settings where vaccination during pregnancy is not recommended or not feasible[13,32].

15.3 Maternal immunization during pregnancy

Maternal immunization during pregnancy is not a new concept, and immunization programmes to protect infants beginning at birth from diseases such as neonatal tetanus and to protect mothers and their infants from influenza are well established in many parts of the world. The first studies investigating the safety and immunogenicity of wP vaccination in pregnant women were undertaken in the 1940s. Although studies using aP vaccines in the 1990s demonstrated the efficient transplacental transfer of antibodies following vaccination in pregnancy, the lack of an accepted correlate of protection for pertussis and limited safety data in pregnancy meant that maternal immunization was not actively pursued by policymakers until a much later date[40].

Although routine immunization programmes have greatly reduced the childhood burden of pertussis, infants remain vulnerable from birth until they can be fully protected through primary immunization. Protection in the first months of life is derived from maternal antibodies which are actively transferred transplacentally during pregnancy, most efficiently during the last trimester[41]. In the pre-vaccine era, maternal antibodies from natural pertussis infection were thought to provide at least short-term protection for newborn infants, as a lower proportion of pertussis deaths were observed in children less than 1 month of age compared with those aged 1–3 months[42]. However, this pattern changed following the introduction of mass pertussis vaccination programmes, with no substantial differences in pertussis mortality between the age groups, which likely reflected a reduction in maternal antibodies[43]. Low maternal pertussis antibody levels and their rapid waning in infants leaves neonates with little humoral protection and at increased risk of severe disease and death[44]. Closing this early susceptibility gap through strategies such as maternal immunization has been advocated for in countries where the infant burden of pertussis remains high, despite sustained high coverage of routine childhood vaccination programmes. The aim of maternal immunization is to sufficiently boost immunity at the right stage in pregnancy to maximize the transplacental transfer of maternal antibodies and passively protect infants during the first months of life. Several immunogenicity studies have demonstrated higher concentrations of pertussis antibodies in cord blood of neonates of mothers vaccinated during pregnancy when compared with non-vaccinated mothers and

these antibodies persist until the primary series begins[45–50]. An added benefit of maternal immunization is the potential for indirect protection conferred through reduced maternal exposure by providing a cocooning effect.

In response to resurgences of pertussis and concurrent rises in infant disease observed in several countries, a range of control strategies were reviewed by the WHO SAGE workgroup in a 2015 position paper in order to optimize control and prevent infant disease[12]. Strategies included adult booster doses, neonatal vaccination, cocooning, and maternal immunization, some of which had already been implemented with varying degrees of success. Based on the available evidence at the time, vaccination of pregnant women was considered to be the most cost-effective strategy to protect infants in the first months of life and for a number of reasons, was more favourable than the strategy of cocooning[12].

A number of countries currently recommend maternal immunization during pregnancy using Tdap vaccine, with most recommending vaccination during every pregnancy. In 2011, the United States became the first country to advise that Tdap could be safely administered to pregnant women who had not previously received their recommended adult dose[51]. Since 2012, numerous other countries have made similar recommendations including Australia, New Zealand, the United Kingdom, Spain, Belgium, and Portugal, and several Latin American countries including Argentina, Brazil, Colombia, Mexico, and Panama.

The pathway to maternal immunization has varied across countries and has been influenced by country-specific vaccination schedules and local pertussis epidemiology, particularly the burden and rate of increase in infant disease. In the United States and Australia, routine maternal vaccination was introduced following the adoption of other strategies, such as cocooning, to optimize infant control. In contrast, in the United Kingdom and several Latin American countries, implementation of maternal immunization occurred in response to rapid increases in infant disease. The United Kingdom introduced maternal immunization in 2012 as a temporary emergency measure in response to a significant rise in pertussis morbidity and mortality in infants. This strategy was considered by the United

Kingdom Joint Committee on Vaccination and Immunization to be the only option to protect infants from birth as other approaches, such as cocooning, were less favourable and had not been previously recommended[52].

Historical data on vaccination during pregnancy with wP vaccines demonstrated the ability of the strategy to effectively protect young infants from disease[53]. However, it was not until the availability of data from two observational studies in the United Kingdom that the first evidence of the effectiveness of maternal aP immunization in preventing infant disease was demonstrated[54,55]. Evaluation of the first year of the programme in England using a dT5aP-inactivated polio vaccine (IPV) estimated that the protection conferred to infants less than 2 months of age was 90 per cent[54,55]. Furthermore, there was a 78 per cent decline in laboratory-confirmed cases and a 68 per cent decline in hospital admissions among infants less than 3 months of age[54]. Vaccine effectiveness was sustained at greater than 90 per cent 3 years following programme implementation, despite the transition to another aP vaccine product with a different antigen composition (dT3aP-IPV)[56]. The United Kingdom was also able to demonstrate the first estimates of effectiveness against infant deaths, estimated at 95 per cent (95 per cent confidence interval (CI) 79–100 per cent). Further evidence from the United States and Argentina has also supported the impact and effectiveness of maternal immunization during pregnancy. Following the introduction of maternal Tdap immunization in Argentina, infant pertussis cases declined by 51 per cent in high-coverage states (>50 per cent) compared with low-coverage states (≤50 per cent), and an 87 per cent overall decline in pertussis deaths was observed in 2013 compared with 2011[57]. In the United States, a retrospective cohort study found that Tdap vaccination at 27–36 weeks' gestation was 85 per cent (95 per cent CI 33–98 per cent) more effective at preventing pertussis in infants less than 8 weeks of age compared with postpartum vaccination[58]. Infants whose mothers received vaccine during pregnancy were also found to have a lower risk of hospitalization, with vaccine effectiveness against hospitalization estimated at 58 per cent (95 per cent CI 15–80 per cent)[59]. In another observational study in the United States,

the effectiveness of Tdap administered during pregnancy was 91.4 per cent (95 per cent CI 19.5–99.1 per cent) during the first 2 months of life[60].

Most countries with maternal programmes in place currently recommend vaccination during the third trimester of pregnancy to maximize antibody transfer to the infant which is minimal before 30 weeks. However, recent immunogenicity studies assessing levels of transplacentally derived antibodies in infants born to mothers vaccinated at different stages of pregnancy suggest that vaccination during the second trimester provides higher levels of antibodies to infants compared with third-trimester vaccination, and early third-trimester vaccination is more effective than later in pregnancy[61–63]. More recent observational data have shown that second-rather than third-trimester vaccination also results in higher anti-pertussis toxin titres among preterm neonates at birth, suggesting that earlier vaccination may be of particular benefit to this vulnerable group[64]. In light of these findings and in an attempt to provide more opportunities for women to be vaccinated and improve maternal vaccination coverage, some countries such as the United Kingdom have updated their guidance to recommend vaccination from as early as 16 weeks' gestation[65]. However, the comparative clinical protection conferred from second- versus third-trimester vaccination is yet to be established.

A 2017 systematic review assessing the safety of pertussis-containing vaccines administered during pregnancy on maternal, fetal, and infant outcomes identified 13 observational studies from Europe and North America[66]. Although these largely retrospective observational studies showed methodological heterogeneity and were prone to bias, the findings were consistent across studies and were reassuring, with no significant increased risk by point estimation of preterm birth (0.47–1.50), small for gestational age (0.65–1.00), stillbirth (0.36–0.85), neonatal death (0.76–1.20), low birth weight (0.20–0.91), and congenital anomalies[66]. While one large study of over 25,000 vaccinated women[67] demonstrated a small but significant increased risk of chorioamnionitis, this finding was not replicated in two other studies, and there was no evidence of increased risk of preterm birth which would be expected in association with chorioamnionitis.

Furthermore, when accounting for the predictive value of coding by the tenth revision of the International Classification of Diseases for chorioamnionitis, this association was no longer significant. Evidence of the safety of repeat vaccination during pregnancy provides further reassurance given that most countries have implemented programmes recommending vaccination in every pregnancy[68]. However, despite reassuring evidence of the safety of maternal pertussis vaccination, accurate estimates of the background incidence of potential adverse outcomes and ongoing post-marketing surveillance is essential for countries considering implementation of such programmes.

The presence of high maternal antibodies from natural infection has previously been shown to 'blunt' the infant immune response to wP vaccine[40,42]. Concerns about the potential impact from maternal vaccination during pregnancy on the infant immune response prompted a number of studies to assess levels of immune responses in infants of vaccinated mothers who have received three doses of aP vaccines according to various infant pertussis vaccination schedules[47–49,69,70]. While these studies have demonstrated that high levels of pertussis antibodies are transferred to the infant and persist until primary immunization commences, most studies have found lower pertussis responses in infants born to vaccinated mothers after completion of the primary schedule when compared to infants of unvaccinated mothers. Evidence on the persistence of this blunting after a booster dose remains inconclusive, and the clinical significance of these immunological findings remains unclear. Some studies have suggested that this blunting disappears following a booster dose, though one Belgian study demonstrated that responses were still impaired after the booster dose[49]. Therefore, for countries which do not offer a pertussis booster in the second year of life, understanding the clinical relevance and longer-term impact of lower responses post primary schedule is a priority. Additionally, the impact of maternal immunization on other routine non-pertussis antigens has also been assessed and further investigation is needed to confirm the observation of enhanced tetanus responses and declines in diphtheria and CRM_{197}-conjugated vaccine responses[70].

More recently, two studies have evaluated the protection conferred from maternal vaccination during pregnancy among infants who have commenced their primary schedule[56–60]. Evaluation of the United Kingdom programme indicated that maternal vaccine continues to offer protection to children who have received a first primary dose (vaccine effectiveness 82 per cent (95 per cent CI 65–91 per cent)), but that protection declines with subsequent doses[56]; similar findings were published from a retrospective cohort study in the United States which estimated maternal vaccine effectiveness at 81.4 per cent (95 per cent CI 42.5–94 per cent) between doses one and two of the infant schedule but declining to 65.9 per cent (95 per cent CI 4.5–87.8 per cent) after the infants completed three DTaP (diphtheria and tetanus toxoids and acellular pertussis vaccine) doses[60]. While these findings provide reassurance of no increased risk of pertussis among infants born to vaccinated mothers compared to children of unvaccinated mothers, continued monitoring of the epidemiology of infant disease will be important to detect any shifts in the age distribution of infant cases and relative increase in toddlers and children, which has been predicted from a recent modelling study[71]. However, any potential increase in these age groups where disease is less severe needs to be weighed against the strong evidence of preventing severe and fatal cases in young infants.

Although a number of countries now have maternal immunization programmes in place, these have been implemented with varying degrees of success. In the United States, results from a recent Internet panel survey suggest that 48.8 per cent of pregnant women received a dose of Tdap during the 2015–2016 influenza season, an increase of 21.8 per cent from the previous year[72]. In Argentina, the first country in Latin America to recommend a pertussis vaccination programme for pregnant women, national Tdap coverage was over 67 per cent by 2013[57]. In Australia, estimates of coverage in Queensland, the first jurisdiction to implement a maternal programme, was reported at 40–50 per cent as of May 2015[73]. And in England, monthly uptake at the national level increased during the first 5 months of the programme from 43.7 per cent in October 2012 to 59.4 per cent in February 2013; coverage has increased even further during 2016 with the latest estimates reaching 76.2 per cent[74,75].

A number of factors are important in the successful implementation of a maternal vaccination programme, although their relative importance may vary in different settings. In a review of factors influencing acceptance of vaccine during pregnancy, the key factors for women included evidence of the safety of the vaccine, the need and effectiveness of the vaccine, recommendations by healthcare professionals, and access to services[76]. Availability and access to appropriate training were identified as important factors for healthcare professionals. While this was also an important factor in the United Kingdom where maternal immunization was introduced as an emergency measure with limited opportunity for training[77], the success of the programme likely reflects the effective communication to mothers of the benefits and importance of the programme to protect their babies in light of the growing outbreak. And finally, the service configuration to deliver the programme and the involvement of health professionals are also important factors. In the United Kingdom, the maternal programme has largely been delivered through general practice staff who are responsible for delivery of other routine vaccines and are therefore confident to discuss vaccination issues with women. This contrasts with the approach adopted in countries such as the United States, where specialist obstetricians are primarily responsible for discussing and administering Tdap to pregnant women.

In recent years, there has been growing evidence supporting the effectiveness and safety of maternal immunization as a strategy to protect infants in the first months of life. As a result, there has been increasing international interest in such an approach, particularly in low-income settings where the burden of pertussis remains high. Despite this, however, important questions remain such as the optimal timing of vaccination during pregnancy as demonstrated by clinical protection; the impact and effectiveness of maternal vaccination among infants receiving wP vaccination, a question that is particularly relevant in low-income countries; and the effectiveness of maternal immunization among aP-primed women, an important question in countries that have used aP vaccines in their routine programmes

for several years. In the meantime, countries with maternal immunization programmes in place should continue to promote maternal immunization to increase uptake among pregnant women in order to optimize the benefits of the strategy in protecting young infants.

15.4 Vaccination of healthcare personnel

Since the introduction of pertussis booster vaccines for adults, another strategy that has been used to control the burden of pertussis, particularly in high-risk settings, is targeted vaccination of HCP. Nosocomial transmission of pertussis is well described and outbreaks have been reported in a variety of settings[78,79]. Because of the risk of widespread transmission to highly susceptible patient populations, nosocomial pertussis outbreaks are of great concern, especially when they occur in neonatal or paediatric settings where the probability of severe pertussis-related outcomes is high[78,80]. While HCP, patients, and family members have all been identified as sources of pertussis transmission in healthcare settings[78,79,81–83], HCP are thought to play an important role in nosocomial outbreaks since they are at increased risk for acquiring and further spreading pertussis to at-risk patients[80]. Given the clinical course of infection in adults, HCP may lack classic pertussis symptoms, resulting in delayed diagnosis, treatment, and exclusion from work[84]. Recent data from the Enhanced Pertussis Surveillance system in the United States shows that HCP account for 5.8 per cent of adult pertussis cases reported between 2014 and 2016, with 60 per cent providing direct patient care (Enhanced Pertussis Surveillance (EPS)/Emerging Infections Program (EIP) Network, unpublished data). Serosurveys among HCP have found seroprevalence rates ranging from 14 to 73 per cent, suggesting not only that infections occur in HCP, but that a significant population is susceptible to disease[85]; a recent survey of HCP in three Spanish hospitals found that 31.2 per cent of staff were seropositive for pertussis, 3.3 per cent of whom had enzyme-linked immunosorbent assay values suggestive of current or recent infection[86].

Nosocomial pertussis outbreaks are disruptive and generate substantial cost to an institution since control measures, such as contact tracing, diagnostic testing, post-exposure prophylaxis (PEP), and furlough of employees, are expensive and labour intensive[78,80,87–92]. In the setting of an adult pertussis vaccination programme, numerous countries have recommended vaccination of HCP to help minimize the transmission of pertussis and prevent outbreaks of illness in the healthcare setting. While some countries have adopted recommendations for vaccinating all HCP, others have emphasized targeted vaccination of staff working in neonatal, paediatric, or obstetric settings, or who have direct patient contact[12,85,93]. A recent review by Haviari et al. reported that pertussis booster vaccination coverage among HCP ranges from 14 to 72 per cent globally[85]. High vaccine coverage is necessary for programmes to be effective, and numerous strategies to increase pertussis vaccine uptake in the healthcare setting, including collaboration with influenza vaccination programmes, providing vaccines to employees at reduced cost or free of charge, and implementation of vaccination mandates, have been discussed[80,94,95]. While mandatory vaccination programmes may result in higher vaccine coverage[96,97], there is much controversy surrounding this approach, and institutions may face challenges such as defining acceptable criteria for vaccination exemptions[96]. However, even without mandatory vaccination, institutions have been able to successfully achieve high pertussis vaccine uptake among HCP. A recent evaluation from the United Kingdom found that the implementation of a non-mandatory programme in two hospitals following the 2012 epidemic resulted in greater than 85 per cent coverage, suggesting that substantial gains in HCP vaccination coverage can be achieved in the setting of a pertussis epidemic[98].

To date, there is no direct evidence to support vaccination of HCP as an effective measure for controlling nosocomial transmission of pertussis[12,85,99]. An agent-based simulation model from Canada in 2009 suggested that booster vaccination of neonatal HCP could decrease the likelihood of secondary pertussis transmission in the hospital setting by 47 per cent[84]. Other models have shown that implementation of a HCP vaccination programme in paediatric settings can be cost saving[87,100]. Given the substantial costs associated with nosocomial outbreaks[88,90,101], implementation of routine HCP vaccination may be more sensible from an economic perspective than a post-exposure approach for the control of nosocomial pertussis[102]. However, similar to concerns raised in the interpretation of cocooning model data, model outputs are dependent on the underlying parameters, and incorporating updated information on pertussis epidemiology and aP vaccine performance could produce different findings. Additionally, further evidence is needed to determine whether vaccination of HCP is better at preventing nosocomial transmission of pertussis than other strategies, such as masking procedures and furloughing ill employees.

While pertussis vaccination provides direct protection to the individual healthcare worker in the short term, immunity from aP vaccines wanes over time, highlighting key limitations of current vaccine formulations. With new pertussis vaccines still on the distant horizon, considerations for revaccination of HCP may be needed to optimize the prevention and control of pertussis in the healthcare setting. In countries with an adult pertussis vaccination programme in place, targeted vaccination of HCP could be used as one strategy to help control nosocomial transmission of pertussis in settings where high coverage can be attained; however, vaccination of HCP alone cannot be relied on to have a wider population impact.

15.5 Post-exposure prophylaxis

Chemoprophylaxis, or PEP, has long been used as a public health intervention to control pertussis, the rationale being to prevent secondary transmission of *B. pertussis* by eradicating the organism from exposed contacts before the development of clinical infection. Public health action is recommended for contacts within 21 days of a pertussis exposure and macrolide antibiotics are generally the antimicrobial of choice for both the treatment of and prophylaxis against *B. pertussis*[103–105]. Newer macrolides, particularly azithromycin and clarithromycin, are preferred to erythromycin given the benefits of greater absorption, longer half-life, shorter duration of treatment, good *in vitro* activity, an improved side effect profile[1,106–108] and modelling studies suggest that azithromycin is the most cost-effective PEP agent against *B. pertussis*[109]. While co-trimoxazole has also been shown to be effective in eradicating the organism from the nasopharynx and can be used for those unable to tolerate macrolides or when a macrolide is contraindicated[1,110,111], penicillins and early-generation cephalosporins remain ineffective against *B. pertussis*. The choice of antibiotic used for treatment and chemoprophylaxis generally reflects antimicrobial resistance patterns which vary globally. Although no macrolide resistance was reported prior to 1994, erythromycin resistance has more recently been reported in a number of countries including the United States[112], France[113], Iran[114], China[115,116], and possibly Brazil[117]. Review of representative historical samples from Japan (2008–2012)[118], Australia (1971–2006)[119], the United Kingdom (2001–2009)[120], and Romania (2012)[121], however, have shown no resistance.

To date, evidence supporting the efficacy of chemoprophylaxis is limited. A 1998 systematic review of 11 studies looking at the use of erythromycin in persons exposed to pertussis reported little effect of the antibiotic in preventing secondary transmission of *B. pertussis*; effects were modest, short term, and associated with gastrointestinal side effects[122]. Further evidence was summarized in a 2007 Cochrane review of antibiotics used for the management of pertussis which also concluded that there was insufficient evidence to determine the benefit of prophylaxis among contacts of a pertussis case[1]; in the one additional prophylaxis study included in the Cochrane review, the number of household contacts of a pertussis case who became culture positive was less in the erythromycin prophylaxis group (3/142, 2.1 per cent) compared to the placebo group (8/158, 5.1 per cent). However, there was no evidence of a statistical association (risk ratio 0.42; 95 per cent CI 0.11–1.54 per cent)[123]. A more recent Spanish study from 2012 to 2013, which looked at the effectiveness of the newer macrolide azithromycin, demonstrated that the effectiveness of azithromycin chemoprophylaxis at preventing disease in contacts of a pertussis case was 62.1 per cent (95 per cent CI 40.3–75.9 per cent), although it was not clear the timing within which chemoprophylaxis was given following exposure to a pertussis case[124].

Evidence supporting the benefit of chemoprophylaxis outside of the household setting is also limited.

In Spain in 2015, chemoprophylaxis in addition to vaccination were considered important factors in ending a school outbreak of pertussis[125], and similar findings have been reported from the United States[126,127], and Estonia[128]. Chemoprophylaxis has also been evaluated within the healthcare setting, but the evidence of effectiveness remains mixed. Of 23 Swiss healthcare workers who underwent post-exposure prophylaxis following a pertussis exposure in a hospital setting, none developed symptoms[129], and mass prophylaxis of 120 workers and exclusion from work of symptomatic staff in a British neonatal unit was believed to curtail the spread of disease beyond two index patients[81]. However, in three large hospital outbreaks in the United States, the benefits of chemoprophylaxis could not be directly separated from other interventions, such as exclusion from work of symptomatic staff and increased awareness following outbreak recognition, even though none of the combined 666 contacts who accepted prophylaxis had a confirmed pertussis diagnosis[130].

Given the available evidence, national public health authorities have adopted various approaches with regard to the prophylaxis of contacts. In addition, the global drive towards antimicrobial stewardship, particularly in high-income settings, has heavily influenced national PEP guidelines in recent years. In many countries, prophylaxis is generally recommended for all close contacts of a suspected or laboratory-confirmed case of pertussis in the household as well as for 'high-risk' or 'vulnerable' contacts or those who come into contact with high-risk or vulnerable groups outside the household, regardless of age and immunization status. This includes public health guidelines from New Zealand[131], Canada, Ireland[132], Australia[104], the Netherlands[133], Brazil[134], Argentina[135], Chile*, and the United States[136]. In contrast, the United Kingdom has taken a much more conservative approach with regard to the use of antibiotic prophylaxis around pertussis cases, recommending that prophylaxis only be offered in households where there is an identified 'vulnerable contact'. While all contacts, including high-risk and vulnerable contacts, are

---

* Personal communication, pertussis surveillance point, Chile.

offered chemoprophylaxis with the aim of preventing transmission to those at highest risk of severe disease, definitions of 'high-risk' or 'vulnerable' populations tend to vary across countries. For example, in the United States and Latin America, immunosuppressed individuals are classified under the 'high-risk' category[134-136]; however, these individuals as well as those with chronic diseases are not considered 'high risk' in the United Kingdom guidelines. More recently, the United Kingdom has limited the indications for chemoprophylaxis even further, taking into consideration the emerging data on the effectiveness of maternal vaccination during pregnancy[54,55]. Identification of 'vulnerable' infants in the United Kingdom guidelines now take into account infant age, maternal vaccine status, and gestational age at birth, and prophylaxis is only recommended for infants not optimally protected at birth through maternal vaccination during pregnancy. In addition to differences in the definition of high-risk or vulnerable contacts between countries, there are also considerable differences in how significant exposures outside of the household setting are defined, which may take into account proximity to and duration of contact with a confirmed case[104,131,137].

Key factors in the success of chemoprophylaxis as a public health tool include both the acceptability of and compliance with recommended antibiotic regimens. One survey of American parents highlighted that mass antimicrobial prophylaxis, if offered in school outbreaks, would not be universally accepted[138]. Compliance can also be challenging, especially in the setting of erythromycin use. During large-scale prophylaxis of staff in a Japanese hospital outbreak in 2007, 26 per cent of 942 questionnaire respondents indicated that they had stopped taking erythromycin due to side effects[139]. Similar findings were reported following widespread erythromycin prophylaxis in a French hospital outbreak[108]. The availability of newer macrolides with better side effect profiles and with shorter duration of therapy regimens is likely to improve patient compliance.

Several other methods have been suggested as alternatives to chemoprophylaxis, often discussed in the Pan American Health Organization region countries within the context of non-medical countermeasures to pandemic influenza preparedness[140].

These include community mitigation strategies, such as cough exclusion for 2–5 days after starting appropriate antimicrobial therapy or 21 days without treatment[103-105], but also cough surveillance and watchful waiting. A 2012 United States trial of PEP versus watchful waiting found that among Tdap-vaccinated HCP, non-inferiority was not demonstrated for watchful waiting when compared to PEP; however, as only one of 42 (2.4 per cent) workers given azithromycin developed pertussis compared to six of 44 (13.6 per cent) who were not, more research into this approach is warranted[141]. Early implementation of cough exclusion within schools has also been suggested for areas with low vaccination coverage[142].

Overall, there is limited evidence of efficacy in the use of chemoprophylaxis for preventing disease in contacts of pertussis cases. In addition, the global drive towards antimicrobial stewardship and ensuring judicious use of antibiotics is increasingly relevant[143]. Therefore, while PEP can play a supplementary role in controlling the burden of pertussis, the primary control measure remains achieving and sustaining high pertussis vaccine coverage, including maternal vaccination coverage, as this is a sounder investment for disease control than the intensive use of PEP, especially in the setting of resurgences in disease.

15.6 Conclusion

Pertussis presents many unique challenges from the perspective of prevention and control. Despite the success of routine pertussis childhood vaccination programmes, pertussis remains a global public health concern as increases in reported disease have recently been observed in a number of countries, particularly those that have transitioned to aP vaccine formulations. With global vaccine coverage now reaching more than 85 per cent for three doses[144], the resurgence of pertussis in some countries suggests that current strategies are suboptimal for controlling disease, and that effective strategies are urgently needed to protect young infants.

Until new pertussis vaccines with improved efficacy and duration of protection are made available, countries must work to optimize the use of current public health tools. No single strategy alone is sufficient to effectively control pertussis across age groups. In the post-vaccination era, a combination of approaches must be used to minimize the burden and severity of *B. pertussis* infections, and combat the re-emergence of disease, especially among those at highest risk for severe morbidity and mortality. As highlighted by this chapter, novel approaches to prevention and control have been implemented over the years with varying degrees of success. The recent introduction of maternal vaccination during pregnancy offers tremendous promise as it has been demonstrated in high-income settings to be highly effective in protecting young infants. Although challenges persist in achieving high Tdap coverage in pregnant women and questions remain as to the longer-term impact of this approach, it is highly likely even with the advent of new vaccines that maternal immunization will remain a key component of pertussis control. Differences across countries with respect to vaccine products, schedules, and available resources mean that a 'one-size-fits-all' approach to prevention and control is currently not realistic. What is clear, however, is the need for continued monitoring of pertussis trends through high-quality surveillance to reliably determine the burden of disease across age groups, evaluate the impact of new or existing vaccination and PEP strategies, and target prevention and control strategies appropriately, especially among infants who are at highest risk for severe disease.

Disclaimer

The findings and conclusions in this chapter are those of the authors and do not necessarily represent the official position of the Centers for Disease Control and Prevention.

References

1. Altunaiji SM, Kukuruzovic RH, Curtis NC, et al. Antibiotics for whooping cough (pertussis). *Cochrane Database Syst Rev* 2007;**3**:CD004404.
2. Lewis K. Platforms for antibiotic discovery. *Drug Discov* 2013;**12**:371–87.
3. Cherry JD. Historical review of pertussis and the classical vaccine. *J Infect Dis* 1996;**174**:S259–63.
4. Immunization Action Coalition. *Pertussis (Whooping Cough): Questions and Answers. Information about the Disease and Vaccines*. 2017. St Paul, MN: IAC. http://www.immunize.org/catg.d/p4212.pdf.
5. Wiley KE, Zuo Y, Macartney KK, et al. Sources of pertussis infection in young infants: a review of key evidence informing targeting of the cocoon strategy. *Vaccine* 2013;**31**:618–25.
6. Forsyth KD, Wirsing von Konig CH, Tan T, et al. Prevention of pertussis: recommendations derived from the second Global Pertussis Initiative roundtable meeting. *Vaccine* 2007;**25**:2634–42.
7. Lim GH, Deeks SL, Crowcroft NS. A cocoon immunisation strategy against pertussis for infants: does it make sense for Ontario? *Euro Surveill* 2014;**19**:20688.
8. Meregaglia M, Ferrara L, Melegaro A, et al. Parent "cocoon" immunization to prevent pertussis-related hospitalization in infants: the case of Piemonte in Italy. *Vaccine* 2013;**31**:1135–7.
9. Van Rie A, Hethcote HW. Adolescent and adult pertussis vaccination: computer simulations of five new strategies. *Vaccine* 2004;**22**:3154–65.
10. de Greeff SC, de Melker HE, Westerhof A, et al. Estimation of household transmission rates of pertussis and the effect of cocooning vaccination strategies on infant pertussis. *Epidemiology* 2012;**23**:852–60.
11. Chiappini E, Stival A, Galli L, et al. Pertussis re-emergence in the post-vaccination era. *BMC Infect Dis* 2013;**13**:151.
12. World Health Organization. Pertussis vaccines: WHO position paper—September 2015. *Wkly Epidemiol Rec* 2015;**90**:433–59.
13. Ulloa-Gutierrez R, Gentile A, Avila-Aguero ML. Pertussis cocoon strategy: would it be useful for Latin America and other developing countries? *Expert Rev Vaccines* 2012;**11**:1393–6.
14. Williams WW, Lu PJ, O'Halloran A, et al. Surveillance of vaccination coverage among adult populations—United States, 2015. MMWR. *Surveill Summ* 2017;**66**:1–28.
15. Public Health Agency of Canada. Vaccine Coverage Amongst Adult Canadians: Results from the 2012 Adult National Immunization Coverage (aNIC) Survey. 2017. http://www.phac-aspc.gc.ca/im/nics-enva/vcac-cvac-eng.php.
16. Healy CM, Rench MA, Baker CJ. Implementation of cocooning against pertussis in a high-risk population. *Clin Infect Dis* 2011;**52**:157–62.
17. Donnan EJ, Fielding JE, Rowe SL, et al. A cross sectional survey of attitudes, awareness and uptake of the parental pertussis booster vaccine as part of a cocooning strategy, Victoria, Australia. *BMC Public Health* 2013;**13**:676–6.

18. Hayles EH, Cooper SC, Wood N, et al. What predicts postpartum pertussis booster vaccination? A controlled intervention trial. *Vaccine* 2015;**33**:228–36.

19. Cheng PJ, Huang SY, Su SY, et al. Increasing postpartum rate of vaccination with tetanus, diphtheria, and acellular pertussis vaccine by incorporating pertussis cocooning information into prenatal education for group B streptococcus prevention. *Vaccine* 2015;**33**: 7225–31.

20. O'Leary ST, Pyrzanowski J, Brewer SE, et al. Influenza and pertussis vaccination among pregnant women and their infants' close contacts: reported practices and attitudes. *Pediatr Infect Dis J* 2015;**34**:1244–9.

21. Urwyler P, Heininger U. Protecting newborns from pertussis—the challenge of complete cocooning. *BMC Infect Dis* 2014;**14**:397.

22. Blain AE, Lewis M, Banerjee E, et al. An assessment of the cocooning strategy for preventing infant pertussis-United States, 2011. *Clin Infect Dis* 2016;**63**:S221–6.

23. Wong CY, Thomas NJ, Clarke M, et al. Maternal uptake of pertussis cocooning strategy and other pregnancy related recommended immunizations. *Hum Vaccines Immunother* 2015;**11**:1165–72.

24. Visser O, Kraan J, Akkermans R, et al. Assessing determinants of the intention to accept a pertussis cocooning vaccination: a survey among Dutch parents. *Vaccine* 2016;**34**:4744–51.

25. Hayles EH, Cooper SC, Sinn J, et al. Pertussis vaccination coverage among Australian women prior to childbirth in the cocooning era: a two-hospital, cross-sectional survey, 2010–2013. *Aust N Z J Obstet Gynaecol* 2016;**56**:185–91.

26. Ko HS, Jo YS, Kim YH, et al. Knowledge and acceptability about adult pertussis immunization in Korean women of childbearing age. *Yonsei Med J* 2015;**56**: 1071–8.

27. Rosenblum E, McBane S, Wang W, et al. Protecting newborns by immunizing family members in a hospital-based vaccine clinic: a successful Tdap cocooning program during the 2010 California pertussis epidemic. *Public Health Rep* 2014;**129**:245–51.

28. Rowe SL, Cunningham HM, Franklin LJ, et al. Uptake of a government-funded pertussis-containing booster vaccination program for parents of new babies in Victoria, Australia. *Vaccine* 2015;**33**:1791–6.

29. Quinn HE, Snelling TL, Habig A, et al. Parental Tdap boosters and infant pertussis: a case-control study. *Pediatrics* 2014;**134**:713–20.

30. Carcione D, Regan AK, Tracey L, et al. The impact of parental postpartum pertussis vaccination on infection in infants: a population-based study of cocooning in Western Australia. *Vaccine* 2015;**33**:5654–61.

31. Castagnini LA, Healy CM, Rench MA, et al. Impact of maternal postpartum tetanus and diphtheria toxoids

and acellular pertussis immunization on infant pertussis infection. *Clin Infect Dis* 2012;**54**:78–84.

32. Healy CM, Rench MA, Wootton SH, et al. Evaluation of the impact of a pertussis cocooning program on infant pertussis infection. *Pediatr Infect Dis J* 2015;**34**: 22–6.

33. Warfel JM, Zimmerman LI, Merkel TJ. Acellular pertussis vaccines protect against disease but fail to prevent infection and transmission in a nonhuman primate model. *Proc Natl Acad Sci U S A* 2014;**111**: 787–92.

34. Westra TA, de Vries R, Tamminga JJ, et al. Cost-effectiveness analysis of various pertussis vaccination strategies primarily aimed at protecting infants in the Netherlands. *Clin Ther* 2010;**32**:1479–95.

35. Skowronski DM, Janjua NZ, Tsafack EP, et al. The number needed to vaccinate to prevent infant pertussis hospitalization and death through parent cocoon immunization. *Clin Infect Dis* 2012;**54**:318–27.

36. Lugnér AK, van der Maas N, van Boven M, et al. Cost-effectiveness of targeted vaccination to protect newborns against pertussis: comparing neonatal, maternal, and cocooning vaccination strategies. *Vaccine* 2013;**31**: 5392–7.

37. Coudeville L, Van Rie A, Getsios D, et al. Adult vaccination strategies for the control of pertussis in the United States: an economic evaluation including the dynamic population effects. *PLoS One* 2009;**4**:e6284.

38. Fernández-Cano MI, Armadans Gil L, Campins Martí M. Cost-benefit of the introduction of new strategies for vaccination against pertussis in Spain: cocooning and pregnant vaccination strategies. *Vaccine* 2015;**33**: 2213–20.

39. World Health Organization. Meeting of the Strategic Advisory Group of Experts on Immunization, April 2010 – Conclusions and recommendations. *Wkly Epidemiol Rec* 2010;**85**:197–212.

40. Englund JA, Anderson EL, Reed GF, et al. The effect of maternal antibody on the serologic response and the incidence of adverse reactions after primary immunization with acellular and whole-cell pertussis vaccines combined with diphtheria and tetanus toxoids. *Pediatrics* 1995;**96**:580–4.

41. Faucette AN, Unger BL, Gonik B, et al. Maternal vaccination: moving the science forward. *Hum Reprod Update* 2015;**21**:119–35.

42. Sako W. Early immunization against pertussis with alum precipitated vaccine. *JAMA* 1945;**127**:379–84.

43. Van Rie A, Wendelboe AM, Englund JA. Role of maternal pertussis antibodies in infants. *Pediatr Infect Dis J* 2005;**24**:S62–5.

44. Healy CM, Munoz FM, Rench MA, et al. Prevalence of pertussis antibodies in maternal delivery, cord, and infant serum. *J Infect Dis* 2004;**190**:335–40.

45. Gall SA, Myers J, Pichichero M. Maternal immunization with tetanus-diphtheria-pertussis vaccine: effect on maternal and neonatal serum antibody levels. *Am J Obstet Gynecol* 2011;**204**:334.e1–e5.

46. Leuridan E, Hens N, Peeters N, et al. Effect of a prepregnancy pertussis booster dose on maternal antibody titers in young infants. *Pediatr Infect Dis J* 2011;**30**:608–10.

47. Munoz FM, Bond NH, Maccato M, et al. Safety and immunogenicity of tetanus diphtheria and acellular pertussis (Tdap) immunization during pregnancy in mothers and infants: a randomized clinical trial. *JAMA* 2014;**311**:1760–9.

48. Hoang HTT, Leuridan E, Maertens K, et al. Pertussis vaccination during pregnancy in Vietnam: results of a randomized controlled trial pertussis vaccination during pregnancy. *Vaccine* 2016;**34**:151–9.

49. Maertens K, Caboré RN, Huygen K, et al. Pertussis vaccination during pregnancy in Belgium: results of a prospective controlled cohort study. *Vaccine* 2016;**34**:142–50.

50. Vilajeliu A, Goncé A, López M, et al. Combined tetanus-diphtheria and pertussis vaccine during pregnancy: transfer of maternal pertussis antibodies to the newborn. *Vaccine* 2015;**33**:1056–62.

51. Centers for Disease Control and Prevention (CDC). Updated recommendations for use of tetanus toxoid, reduced diphtheria toxoid and acellular pertussis vaccine (Tdap) in pregnant women and persons who have or anticipate having close contact with an infant aged <12 months – Advisory Committee on Immunization Practices (ACIP), 2011. *MMWR Morb Mortal Wkly Rep* 2011;**60**:1424–6.

52. Amirthalingam G. Strategies to control pertussis in infants. *Arch Dis Child* 2013;**98**:552–5.

53. Cohen P, Scadron SJ. The effects of active immunization of the mother upon the offspring. *J Pediatr* 1946;**29**:609–19.

54. Amirthalingam G, Andrews N, Campbell H, et al. Effectiveness of maternal pertussis vaccination in England: an observational study. *Lancet* 2014;**384**:1521–8.

55. Dabrera G, Amirthalingam G, Andrews N, et al. A case-control study to estimate the effectiveness of maternal pertussis vaccination in protecting newborn infants in England and Wales, 2012–2013. *Clin Infect Dis* 2015;**60**:333–7.

56. Amirthalingam G, Campbell H, Ribeiro S, et al. Sustained effectiveness of the maternal pertussis immunization program in England 3 years following introduction. *Clin Infect Dis* 2016;**63**:S236–43.

57. Vizzotti C, Neyro S, Katz N, et al. Maternal immunization in Argentina: a storyline from the prospective of a middle income country. *Vaccine* 2015;**33**:6413–9.

58. Winter K, Nickell S, Powell M, et al. Effectiveness of prenatal versus postpartum tetanus, diphtheria, and acellular pertussis vaccination in preventing infant pertussis. *Clin Infect Dis* 2017;**64**:3–8.

59. Winter K, Cherry JD, Harriman K. Effectiveness of prenatal tetanus, diphtheria, and acellular pertussis vaccination on pertussis severity in infants. *Clin Infect Dis* 2017;**64**:9–14.

60. Baxter R, Bartlett J, Fireman B, et al. Effectiveness of vaccination during pregnancy to prevent infant pertussis. *Pediatrics* 2017;**139**:e20164091.

61. Naidu MA, Muljadi R, Davies-Tuck ML, et al. The optimal gestation for pertussis vaccination during pregnancy: a prospective cohort study. *Am J Obstet Gynecol* 2016;**215**:237.e1–237.

62. Abu Raya B, Srugo I, Bamberger E. Optimal timing of immunization against pertussis during pregnancy. *Clin Infect Dis* 2016;**63**:143–4.

63. Eberhardt CS, Blanchard-Rohner G, Lemaître B, et al. Maternal immunization earlier in pregnancy maximizes antibody transfer and expected infant seropositivity against pertussis. *Clin Infect Dis* 2016;**62**:829–36.

64. Eberhardt CS, Blanchard-Rohner G, Lemaître B, et al. Pertussis antibody transfer to preterm neonates after second- versus third-trimester maternal immunization. *Clin Infect Dis* 2017;**64**:1129–32.

65. Joint Committee on Vaccination and Immunization. *Minutes of the Joint Committee on Vaccination and Immunization; February 2016*. 2016. https://app.box.com/s/iddfb4ppwkmtjusir2tc/1/2199012147/66698939189/1.

66. McMillan M, Clarke M, Parrella A, et al. Safety of tetanus, diphtheria, and pertussis vaccination during pregnancy: a systematic review. *Obstet Gynecol* 2017;**129**:560–73.

67. Kharbanda EO, Vazquez-Benitez G, Lipkind HS, et al. Evaluation of the association of maternal pertussis vaccination with obstetric events and birth outcomes. *JAMA* 2014;**312**:1897–904.

68. Sukumaran L, McCarthy NL, Kharbanda EO, et al. Association of Tdap vaccination with acute events and adverse birth outcomes among pregnant women with prior tetanus-containing immunizations. *JAMA* 2015;**314**:1581–7.

69. Hardy-Fairbanks AJ, Pan SJ, Decker MD, et al. Immune responses in infants whose mothers received Tdap vaccine during pregnancy. *Pediatr Infect Dis J* 2013;**32**:1257–60.

70. Ladhani SN, Andrews NJ, Southern J, et al. Antibody responses after primary immunization in infants born to women receiving a pertussis-containing vaccine during pregnancy: single arm observational study with a historical comparator. *Clin Infect Dis* 2015;**61**:1637–44.

71. Bento AI, King AA, Rohani P. Maternal pertussis immunisation: clinical gains and epidemiological legacy. *Euro Surveill* 2017;**22**:30510.

72. Kahn KE, Black CL, Ding H, et al. *Pregnant Women and Tdap Vaccination, Internet Panel Survey, United States*. April 2016. https://www.cdc.gov/vaccines/imz-managers/coverage/adultvaxview/tdap-report-2016.html.

73. Beard FH. Pertussis immunization in pregnancy: a summary of funded Australian state and territory programs. *Comm Dis Intell* 2015;39:E329–E336.

74. Public Health England. *Pertussis Vaccination Programme for Pregnant Women. Vaccine coverage estimates in England. October 2012 to September 2013*. http://webarchive.nationalarchives.gov.uk/20140714001000/http://www.hpa.org.uk/hpr/archives/2013/hpr50-5113.pdf.

75. Public Health England. *Pertussis Vaccination Programme for Pregnant Women Update. Vaccine coverage in England, October to December 2016*. 2016. https://www.gov.uk/government/uploads/system/uploads/attachment_data/file/594799/hpr0817_prntl-prtsss-VC.pdf.

76. Wilson RJ, Paterson P, Jarrett C, et al. Understanding factors influencing vaccination acceptance during pregnancy globally: a literature review. *Vaccine* 2015;**33**:6420–9.

77. Amirthalingam G, Letley L, Campbell H, et al. Lessons learnt from the implementation of maternal immunization programs in England. *Hum Vaccines Immunother* 2016;**12**:2934–9.

78. Maltezou HC, Ftika L, Theodoridou M. Nosocomial pertussis in neonatal units. *J Hosp Infect* 2013;**85**:243–8.

79. Sandora TJ, Gidengil CA, Lee GM. Pertussis vaccination for health care workers. *Clin Microbiol Rev* 2008;**21**:426–34.

80. Calderón TA, Coffin SE, Sammons JS. Preventing the spread of pertussis in pediatric healthcare settings. *J Pediatr Infect Dis Soc* 2015;**4**:252–9.

81. Alexander EM, Travis S, Booms C, et al. Pertussis outbreak on a neonatal unit: identification of a healthcare worker as the likely source. *J Hosp Infect* 2008;**69**:131–4.

82. Elumogo TN, Booth D, Enoch DA, et al. Bordetella pertussis in a neonatal intensive care unit: identification of the mother as the likely source. *J Hosp Infect* 2012;**82**:133–5.

83. Mahieu L, De Schrijver K, Van den Branden D, et al. Epidemiology of pertussis in children of Flanders Belgium: can healthcare professionals be involved in the infection? *Acta Clin Belg* 2014;**69**:104–10.

84. Greer AL, Fisman DN. Keeping vulnerable children safe from pertussis: preventing nosocomial pertussis transmission in the neonatal intensive care unit. *Infect Control Hosp Epidemiol* 2009;**30**:1084–9.

85. Haviari S, Bénet T, Saadatian-Elahi M, et al. Vaccination of healthcare workers: a review. *Hum Vaccines Immunother* 2015;**11**:2522–37.

86. Rodríguez de la Pinta ML, Castro Lareo MI, Ramon Torrell JM, et al. Seroprevalence of pertussis amongst healthcare professionals in Spain. *Vaccine* 2016;**34**:1109–14.

87. Greer AL, Fisman DN. Use of models to identify cost-effective interventions: pertussis vaccination for pediatric health care workers. *Pediatrics* 2011;**128**:e591–9.

88. Baggett HC, Duchin JS, Shelton W, et al. Two nosocomial pertussis outbreaks and their associated costs—King County. Washington, 2004. *Infect Control Hosp Epidemiol* 2007;**28**:537–43.

89. Daskalaki I, Hennessey P, Hubler R, et al. Resource consumption in the infection control management of pertussis exposure among healthcare workers in paediatrics. *Infect Control Hosp Epidemiol* 2007;**28**:412–7.

90. Calugar A, Ortega-Sánchez IR, Tiwari T, et al. Nosocomial pertussis: costs of an outbreak and benefits of vaccinating health care workers. *Clin Infect Dis* 2006;**42**:981–8.

91. Ward A, Caro J, Bassinet L, et al. Health and economic consequences of an outbreak of pertussis among healthcare workers in a hospital in France. *Infect Control Hosp Epidemiol* 2005;**26**:288–92.

92. Heininger U. Vaccination of health care workers against pertussis: meeting the need for safety within hospitals. *Vaccine* 2014;**32**:4840–3.

93. Libster R, Edwards KM. Re-emergence of pertussis: what are the solutions? *Expert Rev Vaccines* 2012;**11**:1331–46.

94. Russi M, Behrman A, Buchta WG, et al. Pertussis vaccination of health care workers: ACOEM Medical Center occupational health section task force on pertussis vaccination of health care workers. *J Occup Environ Med* 2013;**55**:1113–5.

95. Walther K, Burckhardt MA, Erb T, et al. Implementation of pertussis immunization in health-care personnel. *Vaccine* 2015;**33**:2009–14.

96. Leibu R, Maslow J. Effectiveness and acceptance of a health care-based mandatory vaccination program. *J Occup Environ Med* 2015;**57**:58–61.

97. Weber DJ, Consoli SA, Sickbert-Bennett E, et al. Assessment of a mandatory tetanus, diphtheria, and pertussis vaccination requirement on vaccine uptake over time. *Infect Control Hosp Epidemiol* 2012;**33**:81–3.

98. Paranthaman K, McCarthy N, Rew V, et al. Pertussis vaccination for healthcare workers: staff attitudes and perceptions associated with high coverage vaccination programmes in England. *Public Health* 2016;**137**:196–9.

99. Rivero-Santana A, Cuéllar-Pompa L, Sánchez-Gómez LM, et al. Effectiveness and cost-effectiveness of different immunization strategies against whooping cough to reduce child morbidity and mortality. *Health Policy* 2014;**115**:82–91.

100. Tariq L, Mangen MJ, Hövels A, et al. Modelling the return on investment of preventively vaccinating healthcare workers against pertussis. *BMC Infect Dis* 2015;**15**:75.

101. Yasmin S, Sunenshine R, Bisgard KM, et al. Healthcare-associated pertussis outbreak in Arizona: challenges and economic impact, 2011. *J Pediatr Infect Dis Soc* 2014;**3**:81–4.

102. Urbiztondo L, Broner S, Costa J, et al. Seroprevalence study of B pertussis infection in health care workers in Catalonia, Spain. *Hum Vaccines Immunother* 2015;**11**:293–7.

103. Amirthalingam G, The Pertussis Guidelines Group. *Guidelines for the Public Health Management of Pertussis in England*. London: Public Health England; 2016.

104. Australia Government Department of Health. *Pertussis—CDNA National Guidelines for Public Health Units*. 2013. http://www.health.gov.au/internet/main/publishing.nsf/content/cdna-song-pertussis.htm.

105. American Academy of Pediatrics Committee on Infectious Disease. Pertussis. In: *Red Book®: 2015 Report of the Committee on Infectious Diseases*, 30th ed. Elk Grove Village, IL: American Academy of Pediatrics; 2015, pp 608–20.

106. Lebel MH, Mehra S. Efficacy and safety of clarithromycin versus erythromycin for the treatment of pertussis: a prospective, randomized, single blind trial. *Pediatr Infect Dis J* 2001;**20**:1149–54.

107. Langley JM, Halperin SA, Boucher FD, et al. Azithromycin is as effective as and better tolerated than erythromycin estolate for the treatment of pertussis. *Pediatrics* 2004;**114**:e96–101.

108. Giugliani C, Vidal-Trecan G, Traore S, et al. Feasibility of azithromycin prophylaxis during a pertussis outbreak among healthcare workers in a university hospital in Paris. *Infect Control Hosp Epidemiol* 2006;**27**:626–9.

109. Thampi N, Gurol-Urganci I, Crowcroft NS, et al. Pertussis post-exposure prophylaxis among household contacts: a cost-utility analysis. *PLoS One* 2015;**10**:e0119271.

110. Hoppe JE, Halm U, Hagedorn HJ, et al. Comparison of erythromycin ethylsuccinate and co-trimoxazole for treatment of pertussis. *Infection* 1989;**17**:227–31.

111. Henry RL, Dorman DC, Skinner JA, et al. Antimicrobial therapy in whooping cough. *Med J Aust* 1981;**2**:27–8.

112. Wilson KE, Cassiday PK, Popovic T, et al. Bordetella pertussis isolates with a heterogeneous phenotype for erythromycin resistance. *J Clin Microbiol* 2002;**40**:2942–4.

113. Guillot S, Descours G, Gillet Y, et al. Macrolide-resistant Bordetella pertussis infection in newborn girl, France. *Emerg Infect Dis* 2012;**18**:966–8.

114. Shahcheraghi F, Nakhost Lotfi M, Nikbin VS, et al. The first macrolide-resistant Bordetella pertussis strains isolated from Iranian patients. *Jundishapur J Microbiol* 2014;**7**:e10880.

115. Yang Y, Yao K, Ma X, et al. Variation in Bordetella pertussis susceptibility to erythromycin and virulence-related genotype changes in China (1970–2014). *PLoS One* 2015;**10**:e0138941.

116. Wang Z, Cui Z, Li Y, et al. High prevalence of erythromycin-resistant Bordetella pertussis in Xi'an, China. *Clin Microbiol Infect* 2014;**20**:O825–30.

117. Andrade BGN, Marin MF, Cambuy DD, et al. Complete genome sequence of a clinical Bordetella pertussis isolate from Brazil. *Mem Inst Oswaldo Cruz* 2014;**109**:972–4.

118. Horiba K, Nishimura N, Gotoh K, et al. Clinical manifestations of children with microbiologically confirmed pertussis infection and antimicrobial susceptibility of isolated strains in a Regional Hospital in Japan, 2008–2012. *Jpn J Infect Dis* 2014;**67**:345–8.

119. Sintchenko V, Brown M, Gilbert GL. Is Bordetella pertussis susceptibility to erythromycin changing? MIC trends among Australian isolates 1971–2006. *J Antimicrob Chemother* 2007;**60**:1178–9.

120. Fry NK, Duncan J, Vaghji L, et al. Antimicrobial susceptibility testing of historical and recent clinical isolates of Bordetella pertussis in the United Kingdom using the Etest method. *Eur J Clin Microbiol Infect Dis* 2010;**29**:1183–5.

121. Dinu S, Guillot S, Dragomirescu CC, et al. Whooping cough in South-East Romania: a 1–year study. *Diagn Microbiol Infect Dis* 2014;**78**:302–6.

122. Dodhia H, Miller E. Review of the evidence for the use of erythromycin in the management of persons exposed to pertussis. *Epidemiol Infect* 1998;**120**:143–9.

123. Halperin SA, Bortolussi R, Langley JM, et al. A randomized, placebo-controlled trial of erythromycin estolate chemoprophylaxis for household contacts of children with culture-positive Bordetella pertussis infection. *Pediatrics* 1999;**104**:e42.

124. Godoy P, García-Cenoz M, Toledo D, et al. Factors influencing the spread of pertussis in households: a prospective study, Catalonia and Navarre, Spain, 2012–2013. *Euro Surveill* 2016;**21**:30393.

125. Miguez Santiyan A, Ferrer Estrems R, Chover Lara JL, et al. Early intervention in pertussis outbreak

with high attack rate in cohort of adolescents with complete acellular pertussis vaccination in Valencia, Spain, April to May 2015. *Euro Surveill* 2015;**20**:21183.

126. Christie CD, Marx ML, Daniels JA, et al. Pertussis containment in schools and day care centers during the Cincinnati epidemic of 1993. *Am J Public Health* 1997;**87**:460–2.

127. Centers for Disease Control and Prevention (CDC). School-associated pertussis outbreak-Yavapai County, Arizona, September 2002–February 2003. *MMWR Morb Mortal Wkly Rep* 2004;**53**:216–9.

128. Torm S, Meriste S, Tamm E, et al. Pertussis outbreak in a basic school in Estonia: description, contributing factors and vaccine effectiveness. *Scand J Infect Dis* 2005;**37**:664–8.

129. Crameri S, Heininger U. Successful control of a pertussis outbreak in a university children's hospital. *Int J Infect Dis* 2008;**12**:e85–7.

130. Centers for Disease Control and Prevention (CDC). Outbreaks of pertussis associated with hospitals—Kentucky, Pennsylvania, and Oregon, 2003. *MMWR Morb Mortal Wkly Rep* 2005;**54**:67–71.

131. Ministry of Health of New Zealand. *Pertussis – Communicable Disease Control Manual.* 2012. https://www.health.govt.nz/system/files/documents/publications/cd-manual-pertussis-may2012.pdf.

132. Public Health Medicine Communicable Disease Group. *Guidelines for the Public Health Management of Pertussis.* 2013. https://www.hpsc.ie/a-z/vaccine-preventable/pertussiswhoopingcough/informationforhealthcareworkers/File,13577,en.pdf.

133. Netherlands National Institute for Public Health and the Environment. *Pertussis.* 2016. http://www.rivm.nl/Documenten_en_publicaties/Professioneel_Praktisch/Richtlijnen/Infectieziekten/LCI_richtlijnen/LCI_richtlijn_Pertussis_kinkhoest.

134. Ministério Da Saúde. *Volume.* UNICO; 2016. Guia de Vigilância Em Saúde.

135. Ministério Da Saúde. *Recomendaciones Nacionales de Vacunación Argetina 2012.* 2012. http://www.msal.gob.ar/images/stories/bes/graficos/0000000451cnt-2013–06_recomendaciones-vacunacion-argentina-2012.pdf.

136. Centers for Disease Control and Prevention. *Pertussis—Postexposure Antimicrobial Prophylaxis.* 2015. https://www.cdc.gov/pertussis/outbreaks/pep.html.

137. Public Health Agency of Canada. *Canadian Immunization Guide. Part 4 Active Vaccines: Pertussis Vaccines.* 2014. http://www.phac-aspc.gc.ca/publicat/cig-gci/p04–pert-coqu-eng.php.

138. Borchardt SM, Polyak G, Dworkin MS. Parental attitude towards mass antimicrobial prophylaxis during a school-associated pertussis outbreak. *Epidemiol Infect* 2007;**135**:11–16.

139. Tanaka H, Kaji M, Higuchi K, et al. Problems associated with prophylactic use of erythromycin in 1566 staff to prevent hospital infection during the outbreak of pertussis. *J Clin Pharm Ther* 2009;**34**:719–22.

140. Qualls N, Levitt A, Kanade N, et al. Community mitigation guidelines to prevent pandemic influenza—United States, 2017. Morbidity and mortality weekly report. *MMWR Recomm Rep* 2017;**66**:1–34.

141. Goins WP, Edwards KM, Vnencak-Jones CL, et al. A comparison of 2 strategies to prevent infection following pertussis exposure in vaccinated healthcare personnel. *Clin Infect Dis* 2012;**54**:938–45.

142. Centers for Disease Control and Prevention (CDC). Use of mass Tdap vaccination to control an outbreak of pertussis in a high school—Cook County, Illinois, September 2006–January 2007. *MMWR Morb Mortal Wkly Rep* 2008;**57**:796–9.

143. Tiong JJL, Loo JSE, Mai CW. Global antimicrobial stewardship: a closer look at the formidable implementation challenges. *Front Microbiol* 2016;**7**:1860–60.

144. Locht C. Live pertussis vaccines: will they protect against carriage and spread of pertussis? *Clin Microbiol Infect* 2016;**22** S:S96–102.

Epilogue: the road ahead

Pejman Rohani and Samuel V. Scarpino

"Today it is possible to predict one thing for this book with confidence. It will reveal controversy over the causative organism(s) of the disease, the diagnosis of infection, the relative importance of possible virulence factors, the suitability of animal models, therapeutic procedures, and the composition and administration of vaccines."

These words were written by Noel Preston in his introduction[1] to the opening chapter of the book *Pathogenesis and Immunity in Pertussis*, published in 1988[2]. While we have learned a great deal in the intervening three decades, the controversies predicted by Preston linger in this present volume. In this epilogue, we present an opinionated and idiosyncratic perspective on potential reasons underlying differing perspectives on pertussis, highlight a number of inherent complexities that render pertussis a challenging study system, and, finally, suggest potential avenues for greater synthesis.

Contradictions and counter-narratives

There remain contradictory narratives at different scales of biological organization, such as organismal, within-host, and population levels. One specific example of such a contradiction would be the immune impacts of acellular pertussis (aP) vaccines. Animal studies have documented the absence of infection-blocking immunity when subjects have been challenged shortly following aP vaccination (see Chapters 3 and 6)[3,4]. In contrast, in countries such as Sweden and Italy, the widespread use of aP vaccines has been associated with clear declines

in pertussis incidence among infants too young to be immunized, indicating vaccine-induced herd immunity (Chapter 14)[5,6]. A partial list of other inconsistencies would include the following:

- *Interpretation of sequence data*: as discussed in Chapters 1 and 13, the field of phylodynamics is a rapidly expanding area focusing on the simultaneous exploration of phylogenetic and epidemiological dynamics[7]. With pertussis, however, such an approach would first need to overcome a number of technical hurdles, including: (1) *patchy temporal distribution*—available pertussis sequences are typically very few in number, especially in comparison with viral infections such as influenza and HIV. For instance, the most comprehensive analysis of global pertussis sequences carried out by Bart et al.[8] included only 36 pertussis sequences from the United States, which spanned seven decades from 1935 to 2005. Furthermore, there are often only a few sequences from the whole-cell pertussis (wP) vaccine era or from those countries currently administering wP vaccines; hence, there is a substantial over-representation of aP vaccine-era sequences[8]. Such temporal sampling imbalances

Rohani, P. and Scarpino, S. V., *Epilogue: the road ahead. In: Pertussis: epidemiology, immunology, and evolution.*
Edited by Pejman Rohani and Samuel V. Scarpino: Oxford University Press (2019). © Oxford University Press.
DOI: 10.1093/oso/9780198811879.003.0016

can bias phylogenetic reconstruction and the interpretation of evolutionary trends. (2) *Clustered geographic distribution of isolates*—a related problem pertains to the geographic bias in available sequences. Again, to illustrate, the sequence data analysed by Bart et al.[8], which remains the most comprehensive study to date, comprised strains from only 19 countries, with 60 out of 343 sequences from the Netherlands. (3) *Correlation versus causation*—while it is tempting to attribute the appearance of novel 'allelotypes'[9] to changes in vaccine composition[8], establishing the underlying causative factors will need the development and application of robust inferential methodology, as has recently been pioneered in the phylodynamics of dengue virus[10,11] or HIV[12].

- *Molecular typing*: another source of confusion stems from conclusions drawn using different methods for molecular typing, as reviewed in Chapter 9. Some studies conclude that pertussis is monomorphic (e.g. Chapter 9)[13], while others have identified geographic bottlenecks[14]. A better understanding of the patterns of gene flow would help in dissecting the driving forces underlying geographic heterogeneity in pertussis epidemiology and evolution. Specifically, how much of the variation in observed pertussis can be attributed to the localized impacts of national or regional factors (e.g. vaccine composition, vaccine schedule, or population demography) compared to exogenous sources of genetic variability due to human mobility?

- *Serotype evolution*: historical shifts in pertussis serotype in response to wP vaccination have been documented. For instance, Preston demonstrated a shift in dominance from Fim2 to Fim3 shortly after the inception of routine immunization with wP vaccines in the United Kingdom[15]. He noted a similar pattern in a number of other countries[15]. Similarly, the transition from vaccine-hiatus to acellular vaccines in Sweden in the mid to late 1990s was accompanied by a rapid shift from Fim2 to Fim3[16]. In this instance, the authors noted a correlation between the fimbriae serotype and the *ptxP* allele, such that bacteria featuring the combination of *fim3–ptxP3* largely replaced those with *fim2–ptxP1*. In other settings such as the United States, however, shifts in allele dominance have occurred very slowly, over many decades, and appear to be uncorrelated with changes in vaccine composition or schedule[14].

- *Epidemiological data*: (1) *under-reporting*—as summarized in Chapter 1 (Table 1.1), and discussed extensively in Chapter 4, pertussis case reports represent a fraction of true infections. What makes this problem even more challenging is the variation of symptom severity with age of case[17] and stage of infection[18], which can affect the likelihood of case ascertainment. Additionally, in a number of countries, different methods are in use for case detection in adults versus children (Chapter 12). For these reasons, some question the validity of inferences drawn from incidence data. We emphasize that mathematical models typically explicitly take into account errors in reporting (e.g.[19, 20]). (2) *False positives*—an issue that is less frequently discussed is the prevalence of incorrectly diagnosed pertussis cases, resulting from serology and polymerase chain reaction (PCR) testing[21], including misdiagnosis of *Bordetella* congeners (Chapter 11),[22] and surprising sources of contamination[23]. (3) *Data quality standards*—to us, there appears to be a somewhat ad hoc basis for accepting the reliability of some data and not others (e.g. Chapter 4)[24]. A principled and consistent approach to assessing the fidelity of incidence reports is clearly timely.

Pertussis is hard to study!

The observation that pertussis is hard to study may be both obvious and perhaps unhelpful but we would like to list some of the basic hurdles that distinguish pertussis from other well-studied infectious disease systems:

- *Extensive literature*: as discussed in a number of chapters in this book, the human association with pertussis dates back many centuries. Consequently, the relevant literature on pertussis is large and can be overwhelming (see Figure 1). To date, nearly 14,000 papers have been published on pertussis and the annual publication of 400–500 papers makes staying up to date with the literature difficult.

- *Epidemiological data*: as described above, variable age-specific clinical manifestations of pertussis inevitably lead to variation in reporting and impede the unambiguous interpretation of epidemiological data. Additionally, there are differences between public health agencies in PCR testing and detection thresholds (as outlined in Chapter 12). The absence of standardization hampers the comparison of incidence reports from place to place. Also, as discussed in Chapter 4, surveillance data are intricately linked to access to care which introduces potentially significant socioeconomic biases in data from different populations. Finally, the interpretation of incidence reports requires a better understanding of the consequences of pertussis evolution. In particular, does the production of greater pertussis toxin (Chapter 10)[13], which is expected to lead to more severe cough, imply higher rates of transmission?

- *Genomic evolution*: much of the methodology developed for the analysis of sequences focuses on single nucleotide polymorphisms. Studies of pertussis genomes have revealed, however, that a major source of variation is genome rearrangement (Chapter 9)[25,26]. We need a better understanding of the phenotypic consequences of this phenomenon as well as novel methods for appropriately taking it into account when reconstructing the evolutionary trajectory of pertussis. Further, such analyses will need to allow for variable

mutation models given evidence for more rapid mutation rates at antigenic sites[27].

- *Immunity*: at present, there is no known serological correlate of protection (Chapters 3 and 7)[28]. A number of the important uncertainties surrounding pertussis are related to immunity, both in terms of disease severity as well as infection.

- *Vaccines and immunization*: historically, there have been important variations in pertussis vaccine composition, manufacturing, and efficacy, partly arising from the number of vaccine manufacturers during the wP vaccine era. For example, according to one study, in 1946 there were 14 companies in the United States with their own pertussis vaccine[29]. Though there are currently many fewer vaccine manufacturers (two companies for aP vaccines and one for wP vaccines), there is variation between them in composition. Another potentially important difference among countries is the immunization schedule (Chapter 5). Thus, side-by-side comparison of pertussis epidemiology across populations is hard when there are differences in vaccines (including the number and concentration of antigens), the vaccination schedule, and historical uptake levels.

- *Mathematical and computational models*: in principle, the confrontation of mechanistic models with data can shed light on the underlying transmission of a pathogen. Models have reported surprisingly consistent findings concerning the impact

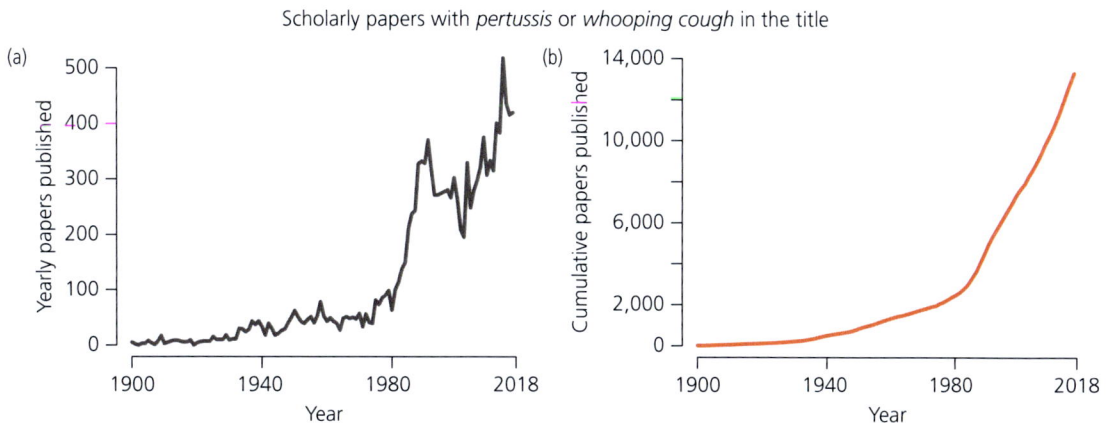

Scholarly papers with *pertussis* or *whooping cough* in the title

Figure 1 Scholarly publications through time. Data were obtained by searching Google Scholar for publications with 'pertussis' or 'whooping cough' the title (a) Raw number of annual publications. (b) Cumulative publications through time.

of infection- and wP vaccine-derived immunity, but have differed in their conclusions regarding the efficacy of aP vaccines. Contrast, for example, the studies by Gambhir et al.[30], Choi et al.[31], Campbell et al.[32], and Domenech de Celles et al.[20]. We submit that this has arisen largely because authors have formulated different models, fitted to different data, using a variety of (different) estimation methods.

Looking forward

Having acknowledged some of the intrinsic complexities of pertussis, we believe the following practical steps may make our task of understanding pertussis easier:

- *Sequence data*: given the exciting developments in the world of phylogenetic inference and phylodynamics (see Chapter 13), we believe the routine collection of sequence data is essential to painting a more complete picture of pertussis epidemiology and evolution.
- *Household studies*: we agree with the sentiments expressed in Chapter 4 and advocate performing more household transmission studies, together with whole-genome sequencing.
- *Diagnostics*: a commonly adopted standard for PCR-based diagnosis (Chapter 12), or at the very least an explicit discussion of the reporting methodology, would be very helpful.
- *Immunization data*: there is an urgent need for better population-specific data on vaccine coverage, the vaccines administered, and the associated immunization schedule. Such information is crucial to the rational interpretation of epidemiological patterns and model-based inference (e.g. Chapter 14).
- *Animal models*: the development of the baboon model is generally regarded as a game changer in the study of pertussis (Chapter 6)[33]. The next step may be the identification of a human transmission model that would be equivalent to the ferret model in influenza[34] or the mouse model in *Plasmodium chabaudi*[35].
- *New generation of computational models*: we propose that an exciting frontier in pertussis modelling

is the harnessing of information available in multiple data streams simultaneously. In this context, we have emphasized the potential benefits of phylodynamic models that assimilate both epidemiological and sequence data (e.g.[12]). We suggest that another frontier may be models that simultaneously account for serology and epidemiology (e.g.[36]).

This 'wish list' represents a combination of research studies, data collection, and methodological developments that would be transformative. Our final proposition is substantially more ambitious. We recommend the formation of a Pertussis Consortium, as has been achieved for malaria[37] and HIV[38], among others. Such an effort will require leadership from international entities, such as the World Health Organization or the Bill & Melinda Gates Foundation. We imagine that a Pertussis Consortium would spearhead greater coordination among research groups and foster a more collective approach to problem-solving. Two urgent tasks that such a consortium should consider would be the following:

- *Data and data sharing*: public health agencies and researchers routinely collect valuable information in the form of pertussis genome sequences, serosurveys, vaccine uptake, household transmission outcomes, and incidence records. Improved protocols for sharing of data and biological specimens, perhaps via a centralized repository, have the potential to be invaluable.
- *Unifying pertussis models*: how do we produce models that explain or are at least qualitatively consistent with all the data? We propose that pertussis research would benefit from greater dialogue between modelling groups. This may take the form of inviting different groups to explain the same set of pertussis data from a specific population followed by formal out-of-fit forecasts regarding relevant epidemiological features (e.g. age incidence, seasonality, or outbreak size). Such an activity is similar in spirit to recent 'prediction challenges', where multiple modelling groups have independently forecast the epidemiology of influenza[39] or Ebola[40], among others. The critical dimension of these efforts is the focus on the same geographic location for predictions. Alternatively,

attempts at finding a unified explanation may require the simultaneous fitting of multiple (competing) models to the same set of incidence data using the same statistical inference algorithm, as was achieved for models of rotavirus vaccines[41].

References

1. Preston N. Pertussis today. In: Wardlaw AC, Parton R (eds), *Pathogenesis and Immunity in Pertussis.* Chichester: John Wiley & Sons; 1988, pp 1–18.

2. Wardlaw AC, Parton R (eds). *Pathogenesis and Immunity in Pertussis.* Chichester: John Wiley & Sons; 1988.

3. Warfel JM, Zimmerman LI, Merkel TJ. Acellular pertussis vaccines protect against disease but fail to prevent infection and transmission in a nonhuman primate model. *Proc Natl Acad Sci U S A* 2014;**111**:787–92.

4. Smallridge WE, Rolin OY, Jacobs NT, et al. Different effects of whole-cell and acellular vaccines on Bordetella transmission. *J Infect Dis* 2014;**209**:1981–8.

5. Domenech de Cellès M, Riolo MA, Magpantay FMG, et al. Epidemiological evidence for herd immunity induced by acellular pertussis vaccines. *Proc Natl Acad Sci U S A* 2014;**111**:E716–7.

6. Domenech de Cellès M, Magpantay FMG, King AA, et al. The pertussis enigma: reconciling epidemiology, immunology and evolution. *Proc Biol Sci* 2016;**283**:20152309.

7. Grenfell BT, Pybus OG, Gog JR, et al. Unifying the epidemiological and evolutionary dynamics of pathogens. *Science* 2004;**303**:327–32.

8. Bart MJ, Harris SR, Advani A, et al. Global population structure and evolution of Bordetella pertussis and their relationship with vaccination. *mBio* 2014;**5**:e01074.

9. Mooi FR, van der Maas NAT, De Melker HE. Pertussis resurgence: waning immunity and pathogen adaptation—two sides of the same coin. *Epidemiol Infect* 2014;**142**:685–94.

10. Rasmussen DA, Ratmann O, Koelle K. Inference for nonlinear epidemiological models using genealogies and time series. *PLoS Comput Biol* 2011;**7**:e1002136.

11. Gill MS, Lemey P, Bennett SN, et al. Understanding past population dynamics: bayesian coalescent-based modeling with covariates. *Syst Biol* 2016;**65**:1041–56.

12. Volz EM, Ionides E, Romero-Severson EO, et al. HIV-1 transmission during early infection in men who have sex with men: a phylodynamic analysis. *PLOS Med* 2013;**10**:e1001568.

13. Mooi FR. Bordetella pertussis and vaccination: the persistence of a genetically monomorphic pathogen. *Infect Genet Evol* 2010;**10**:36–49.

14. Schmidtke AJ, Boney KO, Martin SW, et al. Population diversity among Bordetella pertussis isolates, United States, 1935–2009. *Emerg Infect Dis* 2012;**18**:1248–55.

15. Preston NW. Influence of challenge strain on potency of pertussis vaccines in mice. *Nature* 1967;**213**:830–1.

16. Advani A, Gustafsson L, Åhrén C, et al. Appearance of Fim3 and ptxP3 – Bordetella pertussis strains, in two regions of Sweden with different vaccination programs. *Vaccine* 2011;**29**:3438–42.

17. Sutter RW, Cochi SL. Pertussis hospitalizations and mortality in the United States, 1985–1988. Evaluation of the completeness of national reporting. *JAMA* 1992;**267**:386–91.

18. Kristensen B. Occurrence of the bordet Gengou Bacillus. *JAMA* 1933;**101**:204–6.

19. Lavine JS, King AA, Andreasen V, et al. Immune boosting explains regime-shifts in Prevaccine-era pertussis dynamics. *PLOS ONE* 2013;**8**:e72086.

20. Domenech de Cellès M, Magpantay FMG, King AA, et al. The impact of past vaccination coverage and immunity on pertussis resurgence. *Sci Transl Med* 2018;**10**:eaaj1748.

21. Taranger J, Trollfors B, Lind L, et al. Environmental contamination leading to false-positive polymerase chain reaction for pertussis. *Pediatr Infect Dis J* 1994;**13**:936–7.

22. Miranda C, Porte L, García P. Bordetella holmesii in nasopharyngeal samples from Chilean patients with suspected Bordetella pertussis infection. *J Clin Microbiol* 2012;**50**:1505–5.

23. Salimnia H, Lephart PR, Asmar BI, et al. Aerosolized vaccine as an unexpected source of false-positive Bordetella pertussis PCR results. *J Clin Microbiol* 2012;**50**:472–4.

24. World Health Organization. Pertussis vaccines: WHO position paper. *Wkly Epidemiol Rec* 2015;**90**:433–60.

25. Parkhill J, Sebaihia M, Preston A, et al. Comparative analysis of the genome sequences of Bordetella pertussis, Bordetella parapertussis and Bordetella bronchiseptica. *Nat Genet* 2003;**35**:32–40.

26. Weigand MR, Peng Y, Loparev V, et al. The history of Bordetella pertussis genome evolution includes structural rearrangement. *J Bacteriol* 2017;**199**:e00806.

27. Sealey KL, Harris SR, Fry NK, et al. Genomic analysis of isolates from the United Kingdom 2012 pertussis outbreak reveals that vaccine antigen genes are unusually fast evolving. *J Infect Dis* 2015;**212**:294–301.

28. Mills KHG, Ross PJ, Allen AC, et al. Do we need a new vaccine to control the reemergence of pertussis? *Trends Microbiol* 2014;**22**:49–52.

29. Geier D, Geier M. The true story of pertussis vaccination: a sordid legacy? *J Hist Med Allied Sci* 2002;**57**:249–84.

30. Gambhir M, Clark TA, Cauchemez S, et al. A change in vaccine efficacy and duration of protection explains recent rises in pertussis incidence in the United States. *PLoS Comput Biol* 2015;**11**:e1004138.

31. Choi YH, Campbell H, Amirthalingam G, et al. Investigating the pertussis resurgence in England and Wales, and options for future control. *BMC Med* 2016;**14**:1–11.

32. Campbell PT, McCaw JM, McIntyre P, et al. Defining long-term drivers of pertussis resurgence, and optimal vaccine control strategies. *Vaccine* 2015;**33**: 5794–800.

33. Warfel JM, Merkel TJ. *Bordetella pertussis* infection induces a mucosal IL-17 response and long-lived th17 and th1 immune memory cells in nonhuman primates. *Mucosal Immun* 2013;**6**:787–96.

34. Imai M, Watanabe T, Hatta M, et al. Experimental adaptation of an influenza H5 HA confers respiratory droplet transmission to a reassortant H5 HA/H1N1 virus in ferrets. *Nature* 2012;**486**:420–8.

35. Barclay VC, Sim D, Chan BH, et al. The evolutionary consequences of blood-stage vaccination on the rodent malaria Plasmodium chabaudi. *PLOS Biol* 2012;**10**:e1001368.

36. Kretzschmar M, Teunis PFM, Pebody RG. Incidence and reproduction numbers of pertussis: estimates from serological and social contact data in five European countries. *PLOS Med* 2010;**7**:e1000291.

37. Malaria Consortium. Homepage. https://www. malariaconsortium.org/.

38. Pillay D, Herbeck J, Cohen MS, et al. Pangea-HIV: phylogenetics for generalised epidemics in Africa. *Lancet Infect Dis* 2015;**15**:259–61.

39. Biggerstaff M, Alper D, Dredze M, et al. Results from the Centers for Disease Control and Prevention predict the 2013–2014 nfluenza season challenge. *BMC Infect Dis* 2016;**16**:357.

40. Viboud C, Sun K, Gaffey R, et al. The RAPIDD Ebola forecasting challenge—synthesis and lessons learnt. *Epidemics* 2018;**22**:13–21.

41. Pitzer VE, Atkins KE, de Blasio BF, et al. Direct and indirect effects of rotavirus vaccination: comparing predictions from transmission dynamic models. *PLoS One* 2012;**7**:e42320.

Index

Tables and figures are indicated by an italic *t* and *f* following the page number.